Photonics and Electronics: Emerging Tools, Techniques and Applications

Photonics and Electronics: Emerging Tools, Techniques and Applications

Edited by **Kate Brown**

Willford Press

New York

Published by Willford Press,
118-35 Queens Blvd., Suite 400,
Forest Hills, NY 11375, USA
www.willfordpress.com

Photonics and Electronics: Emerging Tools, Techniques and Applications
Edited by Kate Brown

International Standard Book Number: 978-1-68285-286-6 (Hardback)

Printed in China.

Contents

	Preface	IX
Chapter 1	**Configurable multiplier modules for an adaptive computing system** O. A. Pfänder, H.-J. Pfleiderer and S.W. Lachowicz	1
Chapter 2	**Monotonic transition based forward body bias for dual threshold voltage low power embedded processors** S. Jayapal and Y. Manoli	7
Chapter 3	**Design of UWB pulse radio transceiver using statistical correlation technique in frequency domain** M. Anis and R. Tielert	12
Chapter 4	**Towards a generic operational amplifier with dynamic reconfiguration capability** S. K. Lakshmanan and A. Koenig	20
Chapter 5	**Hardware accelerators for elliptic curve cryptography** C. Puttmann, J. Shokrollahi, M. Porrmann and U. Rückert	24
Chapter 6	**Realtime multiprocessor for mobile ad hoc networks** T. Jungeblut, M. Grünewald, M. Porrmann and U. Rückert	30
Chapter 7	**Impact of process variations and long term degradation on 6T-SRAM cells** Th. Fischer, A. Olbrich, G. Georgakos, B. Lemaitre and D. Schmitt-Landsiedel	35
Chapter 8	**A low-power and high-quality implementation of the discrete cosine transformation** B. Heyne and J. Götze	40
Chapter 9	**Physical IC debug – backside approach and nanoscale challenge** C. Boit, R. Schlangen, A. Glowacki, U. Kindereit, T. Kiyan, U. Kerst, T. Lundquist, S. Kasapi and H. Suzuki	47
Chapter 10	**ESD full chip simulation: HBM and CDM requirements and simulation approach** E. Franell, S. Drueen, H. Gossner and D. Schmitt-Landsiedel	55
Chapter 11	**Design issues of arithmetic structures in adiabatic logic** Ph. Teichmann, J. Fischer, F. Chouard and D. Schmitt-Landsiedel	62
Chapter 12	**Area- and energy-efficient CORDIC accelerators in deep sub-micron CMOS technologies** U. Vishnoi and T. G. Noll	67

Chapter 13 **A CMOS integrated impedance-to-frequency converter for sensing cellular adhesion** **74**
A. Mucha, M. Schienle and D. Schmitt-Landsiedel

Chapter 14 **Fully-integrated LDO voltage regulator for digital circuits** **81**
M. Lüders, B. Eversmann, D. Schmitt-Landsiedel and R. Brederlow

Chapter 15 **An 8 bit current steering DAC for offset compensation purposes in sensor arrays** **86**
G. Bertotti, A. Laifi, E. Di Gioia, M. Masoumi, N. Dodel, E. F. Scarselli and R. Thewes

Chapter 16 **Design and quantitative analysis of parametrisable eFPGA-architectures for arithmetic** **92**
B. Neumann, T. von Sydow, H. Blume and T. G. Noll

Chapter 17 **Modelling of the parametric yield in decananometer SRAM-Arrays** **99**
Th. Fischer, T. Nirschl, B. Lemaitre and D. Schmitt-Landsiedel

Chapter 18 **From algorithm to implementation: a case study on blind carrier synchronization** **104**
D. Schmidt, T. Brack, U. Wasenmüller and N.Wehn

Chapter 19 **Simultaneous large-scale reliability analysis of ultra-thin MOS gate dielectrics using an automated test system** **110**
A. Domdey, K. M. Hafkemeyer, W. H. Krautschneider and D. Schroeder

Chapter 20 **On the synthesis and optimization of cascaded continuous-time Sigma-Delta modulators** **113**
M. Keller, A. Buhmann, M. Kuderer and Y. Manoli

Chapter 21 **Influence of gate tunneling currents on switched capacitor integrators** **118**
W. Kraus and D. Schmitt-Landsiedel

Chapter 22 **Impact of negative and positive bias temperature stress on 6T-SRAM cells** **123**
S. Drapatz, G. Georgakos and D. Schmitt-Landsiedel

Chapter 23 **Efficiency modeling for MHz DCDC converters at 40V input voltage range** **129**
J.Wittmann, A. Seidel and B. Wicht

Chapter 24 **Investigation of carbon nanotube antennas using thin wire integral Equations** **134**
N. Fichtner, X. Zhou and P. Russer

Chapter 25 **Circuit design with independent double gate transistors** **137**
M.Weis, R. Emling and D. Schmitt-Landsiedel

Chapter 26 **A novel technique for CAD-optimization of analog circuits with bipolar transistors** **143**
B. Dimov, V. Boos, T. Reich, Ch. Lang, E. Hennig and R. Sommer

Chapter 27 **Energy modelling in sensor networks** **148**
D. Schmidt, M. Krämer, T. Kuhn and N. Wehn

Chapter 28 **Modeling of temperature scenarios in a multicore processor system** **153**
E. Glocker and D. Schmitt-Landsiedel

Chapter 29	**Simulation of RF MEMS based matching networks and a single pole double throw switch for multiband T/R Modules** S. A. Figur, F. van Raay, R. Quay, L. Vietzorreck and V. Ziegler	**160**
Chapter 30	**Comparison of in-situ delay monitors for use in adaptive voltage scaling** N. Pour Aryan, L. Heiß, D. Schmitt-Landsiedel, G. Georgakos and M. Wirnshofer	**170**
Chapter 31	**Envelope tracking of a radio frequency amplifier for long term evolution using a three-level class-G modulator** A. Bräckle, L. Rathgeber, F. Siegert, S. Heck and M. Berroth	**176**
Chapter 32	**2.5 Gbps clock data recovery using 1/4th-rate quadricorrelator frequency detector and skew-calibrated multi-phase clock generator** S. Tontisirin and R. Tielert	**182**
Chapter 33	**Power supply network design: a case study driven approach** M. Eireiner, S. Henzler, T. Missal, J. Berthold and D. Schmitt-Landsiedel	**187**
Chapter 34	**50 Gbit/s real-time test environment for integrated photonic DQPSK receivers** T. Föhn, C. Fischer and M. Berroth	**193**
Chapter 35	**Highly efficient integrated rectifier and voltage boosting circuits for energy harvesting applications** D. Maurath, C. Peters, T. Hehn, M. Ortmanns and Y. Manoli	**198**
Chapter 36	**Analog circuits using FinFETs: benefits in speed-accuracy-power trade-off and simulation of parasitic effects** M. Fulde, J. P. Engelstädter, G. Knoblinger and D. Schmitt-Landsiedel	**205**
Chapter 37	**A sensing circuit for single-ended read-ports of SRAM cells with bit-line power reduction and access-time enhancement** T. Heselhaus and T. G. Noll	**211**
Chapter 38	**An X-Band low-power and low-phase-noise VCO using bondwire inductor** K. Hu, F. Herzel and J. C. Scheytt	**218**
Chapter 39	**Fast evaluation of nonlinear functions using FPGAs** S. Lachowicz and H.-J. Pfleiderer	**223**
Chapter 40	**Impact of on-chip inductance on power supply integrity** M. Eireiner, S. Henzler, X. Zhang, J. Berthold and D. Schmitt-Landsiedel	**228**
Chapter 41	**Countermeasures against NBTI degradation on 6T-SRAM cells** E. Glocker, D. Schmitt-Landsiedel and S. Drapatz	**234**
Chapter 42	**A low power clock generator with adaptive inter-phase charge balancing for variability compensation in 40-nm CMOS** U. Schulze, M. Broich, O. Weiss and T. G. Noll	**241**
Chapter 43	**A low-noise current preamplifier in 120 nm CMOS technology** H. Uhrmann, W. Gaberl and H. Zimmermann	**246**

Chapter 44 **Temperature modeling and emulation of an ASIC temperature monitor system for Tightly-Coupled Processor Arrays (TCPAs)** **251**
E. Glocker, S. Boppu, Q. Chen, U. Schlichtmann, J. Teich and D. Schmitt-Landsiedel

Chapter 45 **Implementation of a digital trim scheme for SAR ADCs** **258**
J. Bialek, A. Wickmann, F. Ohnhaeuser, G. Fischer, R. Weigel and T. Ussmueller

Permissions

List of Contributors

Preface

The purpose of the book is to provide a glimpse into the dynamics and to present opinions and studies of some of the scientists engaged in the development of new ideas in the field from very different standpoints. This book will prove useful to students and researchers owing to its high content quality.

This book aims to offer an integrated perspective on the multidisciplinary fields of photonics and electronics. There has been rapid advancement in the fields of photonics and electronics in the last few decades. Some of the important concepts covered in this text include power supply networks, voltage converters, signal transmission and sensor networks, etc. The case studies and researches incorporated in the book are bound to provide innovative insights for novel developments and applications. It would prove to be an excellent reference for students and researchers looking to explore these fields in detail.

At the end, I would like to appreciate all the efforts made by the authors in completing their chapters professionally. I express my deepest gratitude to all of them for contributing to this book by sharing their valuable works. A special thanks to my family and friends for their constant support in this journey.

Editor

1

Configurable multiplier modules for an adaptive computing system

O. A. Pfänder[1], H.-J. Pfleiderer[1], and S. W. Lachowicz[2]

[1]Microelectronics Department, University of Ulm, Germany

[2]School of Engineering and Mathematics, Edith Cowan University, Perth, Western Australia

Abstract. The importance of reconfigurable hardware is increasing steadily. For example, the primary approach of using adaptive systems based on programmable gate arrays and configurable routing resources has gone mainstream and high-performance programmable logic devices are rivaling traditional application-specific hardwired integrated circuits. Also, the idea of moving from the 2-D domain into a 3-D design which stacks several active layers above each other is gaining momentum in research and industry, to cope with the demand for smaller devices with a higher scale of integration. However, optimized arithmetic blocks in course-grain reconfigurable arrays as well as field-programmable architectures still play an important role. In countless digital systems and signal processing applications, the multiplication is one of the critical challenges, where in many cases a trade-off between area usage and data throughput has to be made. But the a priori choice of word-length and number representation can also be replaced by a dynamic choice at run-time, in order to improve flexibility, area efficiency and the level of parallelism in computation. In this contribution, we look at an adaptive computing system called 3-D-SoftChip to point out what parameters are crucial to implement flexible multiplier blocks into optimized elements for accelerated processing. The 3-D-SoftChip architecture uses a novel approach to 3-dimensional integration based on flip-chip bonding with indium bumps. The modular construction, the introduction of interfaces to realize the exchange of intermediate data, and the reconfigurable sign handling approach will be explained, as well as a beneficial way to handle and distribute the numerous required control signals.

1 Introduction

Today's ever-increasing demand for computing power in virtually all application sectors, including the emerging photonics arena, is placing heavy demands on current integrated systems. Crucial scenarios are e.g. modern wired and wireless communication over highly complex networks, processing of content-rich multimedia, and mobile applications where area and power are primary considerations. New ways to overcome today's technological barriers of silicon integrated circuits (ICs) are emerging, aiming to reduce both non-recurring engineering costs and time-to-market periods, and also considering production factors like re-usage of lithography masks. Reconfigurable computing has gone mainstream, with the many forms of reconfigurable logic gradually rivaling traditional application-specific ICs. Thanks to the progress in IC manufacturing techniques and the shrinking feature sizes, it has become possible to integrate a multitude of different functions into one single chip, while still meeting restrictions like package size or footprint. Thus, a whole system incorporating multiple special functions that had to be realized with multiple specialized separate ICs can now be implemented in a single system-on-chip (SoC). Going one step further from SoC, the idea of systems-in-a-package (SiP) emerges, when high-end concepts such as flip-chip or chip stacks are used to reach an even higher level of design integration and minimization. While the IC density continues to grow, another key feature is becoming more and more essential for many of today's widely spread application scenarios; that is, flexibility and/or configurability. There is a paradigm change in the world of microelectronics, because the term hardware does no longer stand for the total opposite of software. The notion of ICs offering options to configure them, even by the end customer, is nothing but intriguing. As feature sizes are constantly shrinking and the design complexity is continually increasing, design methodologies are forced to develop at an increasing rate and the designer faces many new obstacles. Novel 3-D integration systems such as 3-D SoC (Joyner et al., 2004) or 3-D-SoftChip (Eshraghian et al., 2003), targeted to satisfy the enormous demand for more computation throughput by effectively manipulating the functionality of hardware primitives through

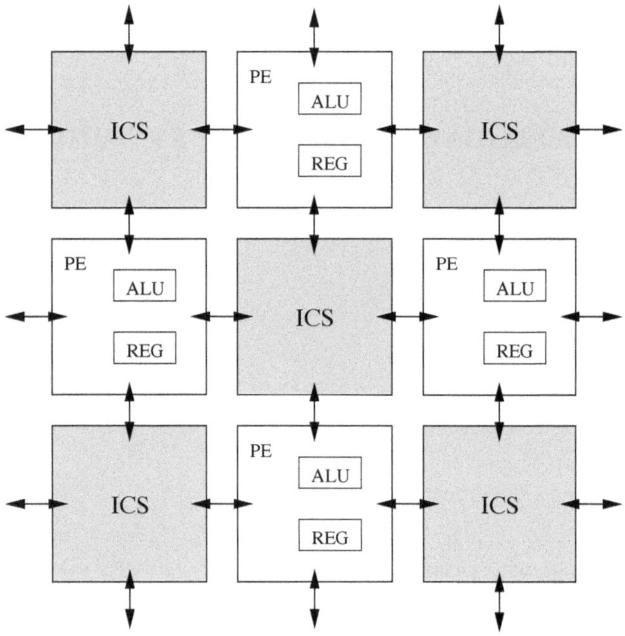

Fig. 1. The starting point is a 2-dimensional course-grain soft reconfigurable array of processing elements (PEs) and intelligent configurable switches (ICS).

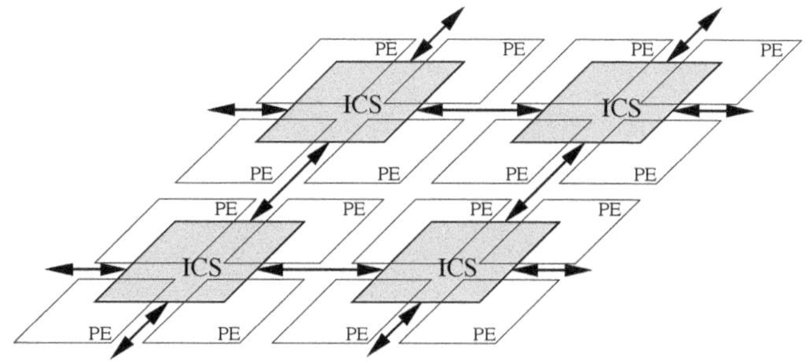

Fig. 2. Functional separation into a sea of PEs and an array of ICS blocks. Each ICS connects a group of 4 PEs using vertical wiring resources.

vertical integration of two 2-D chips, are becoming an attractive solution to combat the rising requirements for interconnect wires (Davis et al., 2005).

This paper is organized as follows: Sect. 2 reviews the 3-D-SoftChip adaptive computing system and its approach to architectural mapping and vertical integration, Sect. 3 concentrates on the design of embedded multiplier blocks into the 3-D-SoftChip's processing elements, and Sect. 4 concludes the paper and gives suggestions for further investigation.

2 3-D-SoftChip adaptive computing system

The concept of the 3-D-SoftChip adaptive computing system is located in the intersection of three major fields of research: Modern very large scale integration (VLSI) in deep sub-micron IC design, reconfigurable hardware, and 3-D integration. This section will at first highlight the basic method of transforming a 2-dimensional course-grain reconfigurable array into a sophisticated 3-dimensional integrated system, then mention the architectural mapping approach and also present the vertical integration method using indium bumps.

2.1 2-D to 3-D transformation

The starting point for the transformation into the 3-D-SoftChip system is a 2-dimensional course-grain soft reconfigurable array of processing elements as depicted in Fig. 1. Each processing element (PE) contains arithmetic and logical functional blocks combined in an arithmetic logic unit (ALU) and also buffer elements (registers, REG). The ALUs are designed for increased flexibility in order to support variable data word-lengths for various types of computation. The overall layout is very much like a checkerboard, with intelligent configuration switches (ICS) accessing the surrounding PEs through a next-neighbor interconnect scheme. Thus, each ALU is able to communicate with all adjacent ALUs through the ICS which serves as a cross point switch with embedded memory.

Looking at the checkerboard layout of Fig. 1, two functional layers are noticeable: First, the array of PEs, and second, a superordinate array of ICS blocks that covers multiple management, interface and I/O functions. However, the area usage of a complete 2-dimensional array is considerable, since there are as many ICS blocks as there are PEs and the area of the respective blocks can differ to a great extent, depending on the provided features. The area usage can be optimized if these two functional layers are separated in space. The proximate solution would be to have an arrangement in groups – one group of PEs in one part of the chip and a group of ICS blocks in another, like separating the black squares from the white squares on a checkerboard. But this solution is impracticable, since it would require an oversized amount of interconnect circuitry and wire space to connect the two groups. A better way to cope with this situation is to think 3-dimensional: By using a novel method for flip-chip bonding, as will be explained in Subsect. 2.2, the two functional layers can be realized as two separate chips that are integrated vertically. To share e.g. memory or outbound routing resources, four ICSs are merged, and the resulting blocks are placed on a second chip. On the first chip, the PEs can now move together and the area can be compacted significantly. The data exchange now happens not only internally on each of the two chips, but also between them, as will be pointed out later. Figure 2 illustrates the basic principle of transforming the planar 2-D architecture of Fig. 1 into two separate functional layers.

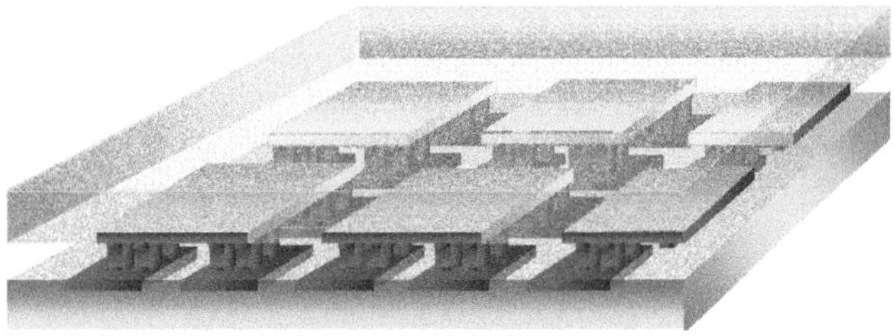

Fig. 3. To achieve a vertical 3-dimensional integration of the upper ICS chip and the lower CAP chip, indium bumps are used for flip-chip bonding.

2.2 Vertical integration

As described in the previous subsection, the sea of PEs is now separated from the array of ICS blocks. The lower chip that now contains the PEs only is called the configurable array processor (CAP). The transformation of the planar 2-D architecture through the approach depicted in Fig. 2 into two vertically integrated chips, namely the upper ICS chip and the lower CAP chip, is illustrated in Fig. 3.

Each of the two chips has an array of aluminum pads on the top metallization layer, each with a size of $10\mu m \times 10\mu m$. The pad pattern on the lower chip is a mirror image of the upper chip's pattern. When the chips are placed face-to-face on top of each other, this creates an interface for a vertical connection between them, and the actual connection is realized by depositing indium bumps onto the aluminum pads on the upper and lower chip, respectively. Creating the bumps comprises the following processing steps: After the silicon substrate has been oxidized and the aluminum pads have been patterned the normal way, there is an additional photoresist coating and patterning step involved to cover and protect the pads' surrounding die surface. Then, titanium (as a diffusion inhibitor), gold (as a contact layer) and indium get evaporized and thus brought onto the aluminum pads. After the lift-off, there is an extra reflow process to make use of the surface tension of the low-melting indium material, in order to transform the block-shaped deposits into bumps with an increased height-to-width ratio. This ensures a certain self-adjusting capability once the two chips get connected (Pfänder et al., 2005). The bumps have a diameter of about 7.5 μm after reflow. The upper ICS chip is flipped and bonded face-down to the lower CAP chip, then the space between is filled with a curing material to ensure a mechanical stability.

2.3 Architectural mapping

The upper chip's function (ICS) is to act as a massively parallel cross point switch as well as a parallel interconnected buffer memory, to allow for a very high-speed data manipulation within the plane. The lower chip (CAP) is a highly parallel array of soft-programmable processing elements, which is capable of carrying out complex calculation tasks directly on data stored in the CAP plane or – using the 3-D interconnect – stored in the top plane. Each of the PEs includes its own embedded register file, along with functional ALU blocks, glue logic and instruction decoding circuitry. Software-programmed instructions are forwarded globally to all processors from on-chip RAM. Even transforms and other processing tasks may be carried out according to embedded software instructions on the highly parallel sea-of-PE array. Two levels of hierarchy within the CAP architecture facilitate the configuration of the ALU's word-length: While at the first level, four processors and one ICS are utilized, this basic group can communicate with adjacent groups at the second level. The interconnection between the parallel array processors is provided by a bus architecture for rapid extraction or insertion of data. Due to the programmable nature of the CAP, the system is highly flexible, and as a result of the vertical interconnects and highly parallel configurable architecture, the efficiency is improved compared to the 2-D planar architecture that was the starting point. On the array level in the system, addressing will follow a switched bus architecture. The multi-tasking requirement of the system introduces a particular need for formulating the architecture that is driven by a variable and configurable number of bits. Therefore, the identification of generic primitives such as adders and multipliers suitable for word-length expansion becomes an important task to realize a higher system flexibility without compromising performance.

3 Configurable multiplier modules

3.1 Heterogeneous processing elements in the CAP

For generic ALU functions, the 3-D-SoftChip's standard PE is optimized for bit-level computation. However, when it comes to frequently used arithmetic functions – e.g. in signal processing applications where the data path contains many multiplication steps involving constantly changing filter coefficients – it becomes more efficient to implement fixed-wired and dedicated functional blocks instead of combining the configurable logic (Haynes et al., 1999). This approach is now common practice in modern high-performance FPGA designs, like the Xilinx Virtex-II Pro that uses 18×18 bit multiplier hard IP blocks strategically placed on the chip (Xilinx, 2005). Introducing a processing accelerator PE with embedded special blocks (Pfänder et al., 2005), such as a barrel shifter or an accumulator/subtractor unit, helps to reduce the data transfer and increases the computation efficiency. There is only a medium flexibility trade-off compared with a complete homogeneous type PE array.

The processing accelerator PE (PA-PE) thus provides increased performance for specific tasks thanks to its dedicated special-purpose functions. In the following, we will concentrate on the multiplication, since it is one of the most impor-

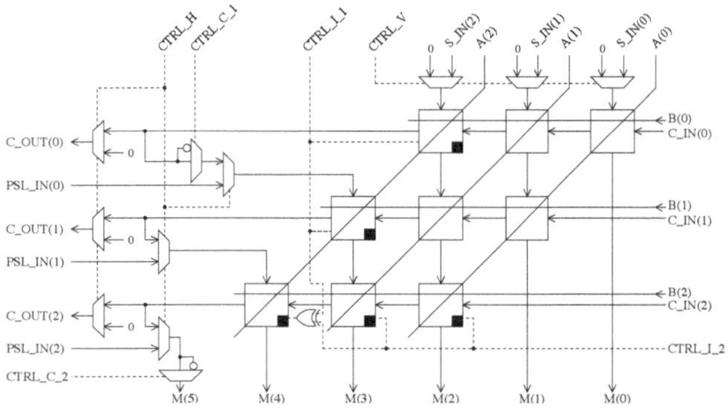

Fig. 4. A parallel-parallel modified Baugh-Wooley array multiplier with multiplexer-based interfaces

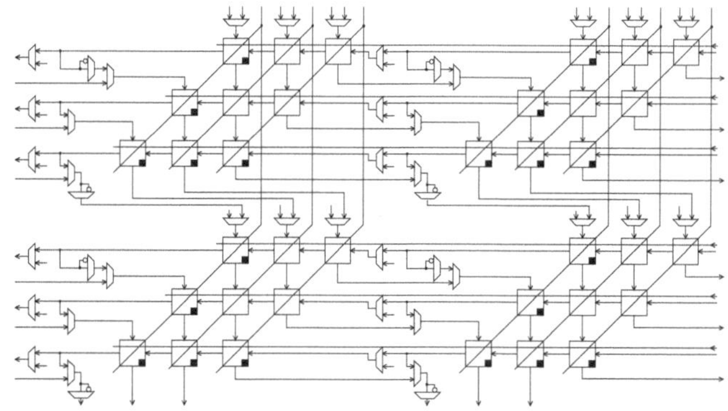

Fig. 5. A concatenation of four 3×3 bit multiplier modules creates a 6×6 bit superior multiplier. (Control signals not shown for better overview)

tant and widely used mathematical operations and requires a large amount of hardware resources when implemented in a straight-forward way. Instead of that, we have come up with a modular multiplier design that enables the system to compute at a configurable word-length, which will be explained in the following.

3.2 Basic scheme and module characteristics

With the concept of arithmetic operations accommodating word-lengths that are configurable at run-time, a digital hardware circuit can adapt to changing accuracy requirements easily and fast. With little overhead, new extensions can help to improve the embedding of multiplier architectures in the surroundings of the 3-D-SoftChip system. As was presented in Pfänder and Pfleiderer (2004), we have designed different variants of multiplier modules that rely on a multiplexer-based connectivity extension for an intermediate data exchange. The general idea is to tap the flow of carry and output bits in specific positions inside an array of basic cells. These cells compute the partial product and provide sum and carry, and with the help of multiplexers and corresponding control signals, the module's behavior can be directly influenced at run-time. The flexible word-length architecture and its dynamic reconfigurability provides either a higher throughput at low levels of precision or a higher precision by grouping multiple elements together, thus greatly increasing the design's efficiency. Also, since every module is a fully functional multiplier itself, a high parallelism can be achieved when all modules compute separately.

The basic modular building block is an $n \times n$ bit multiplier. By concatenating $m \times m$ of these uniform blocks through the use of interconnect resources, in our case also involving the vertical inter-chip interface, a superior $(m \times n) \times (m \times n)$ bit multiplier is formed that enables the computation at flexible precision in steps of n. The input operand word-lengths of the multiplier elements do not necessarily have to be equal, a scheme of $n_1 \times n_2$ bit with $n_1 \neq n_2$ is also possible.

3.3 Realization options

Starting from an unchanged parallel-parallel array multiplier core (Hwang, 1979), there are different expansion steps required in order to achieve the desired connectivity options. Here, since we are dealing with signed numbers, this demands extra sign handling circuitry. An architecture comparison in terms of hardware usage and complexity has shown that the modified Baugh-Wooley design (Baugh and Wooley, 1973) offers the least overhead of the proposed two's complement multipliers, since there is no need for a sign extension (Pfänder and Pfleiderer, 2004). The modified design incorporates the data exchange interfaces mentioned in the previous subsection and the possibility to handle different number representations, due to the modified and configurable basic cells in specific array positions. It can handle the following number systems: Unsigned (when the special cells are configured to act without negating the partial product), Signed-Magnitude (when the sign bit is calculated externally, namely utilizing an XOR gate) and Two's Complement (Pfänder et al., 2005).

Figure 4 shows an $n = 3 \Rightarrow 3 \times 3$ bit parallel array multiplier as an example, and Fig. 5 represents a concatenation of four modules to form an $(m = 2;\ n = 3) \Rightarrow 6 \times 6$ bit multiplier. The control signals are used to control the interfaces and also the configurable partial product inversion step in the special basic cells, in order to obey the mathematical requirements given in Baugh and Wooley (1973). The signals are mapped according to the position in the concatenation and the desired number system using a control decoder in order to save I/O and interconnect resources. Thus, a part of the ICS control logic is transferred onto the CAP plane and integrated into the multiplier modules.

The identical operand scheme that builds the foundation of the parallel-parallel array multiplier can also be implemented in a serial-parallel fashion as depicted in Fig. 6. However, the control signals and also the correcting term must be provided as serial-input data words (Bermak et al., 1997).

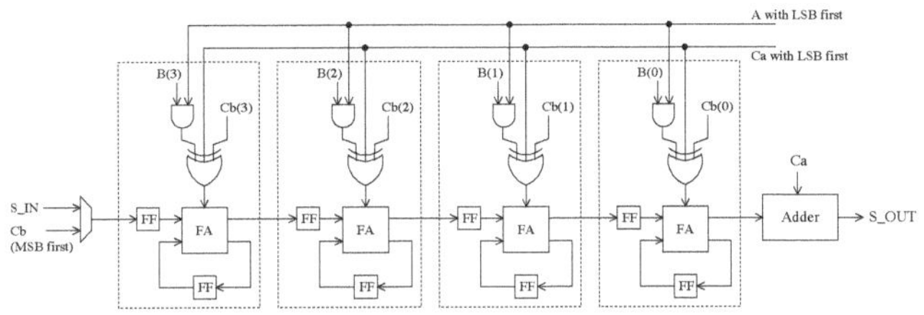

Fig. 6. The modified Baugh-Wooley multiplier can also be realized in a serial-parallel way. The first multiplicand and the control signal vector are fed in serially.

3.4 Comparison of multiplication schemes

Looking at the schemes shown in Figs. 4 and 6, it becomes evident that the number of basic cells N is a function of the word-length. For the parallel array $N \propto n^2$ and for the serial-parallel multiplier $N \propto n$. Note that the serial-parallel multiplier's alleged area advantage is compensated by the expenses for the shift registers needed to store one input operand and the control vector. However, both options have in common that considerable overhead to enable the connectivity is entailed, for example 40% at $n = 4$ for the parallel array (Pfänder and Pfleiderer, 2004). It is not sufficient to judge from the schematic hardware overhead alone; for a fair comparison, the following aspects have to be considered additionally:

- Area usage – depending on the available technology and cell library etc.
- Computation time – since the serial-parallel approach uses $2n$ clock periods
- Data throughput

In order to increase the data throughput, a fully pipelined parallel array is possible, but more than $5n^2$ extra registers would be necessary, resulting in even higher area penalty and increasing the complexity of a processing element. Thus, the decision which approach to use depends on a multitude of parameters. The range of applications plays the dominant role by asserting the claims for throughput and speed, and also other restrictions in terms of area usage and power consumption may apply. In general, the connectivity option requires a considerable extra amount of circuitry in each processing element.

4 Conclusion

In this paper, we have reviewed the 3-D-SoftChip adaptive computing system and highlighted the 2-D to 3-D transformation and its architectural mapping. To achieve a high level of integration, a novel flip-chip bonding technique based on indium bumps on aluminum pads is used to build a system-in-a-package. The indium bumps enable a 3-dimensional routing between the lower configurable array processor chip and the upper intelligent configurable switch chip. Looking at the heterogeneous processing elements located on the lower CAP chip, we have presented configurable embedded multiplier modules for accelerated processing. These modules are based on a modified Baugh-Wooley multiplier and expanded by multiplexer-based data exchange interfaces to provide a connectivity option. Both parallel-parallel and serial-parallel array multiplication schemes are possible, and three different number systems – namely unsigned, signed-magnitude and two's complement – can be handled. By connecting multiple multiplier elements together via 2-D or 3-D interconnect resources, a superior multiplier computing at an increased word-length is formed. Thus, the multiplication word-length can be chosen at run-time by dynamically realizing a concatenation of separate modules. There is a significant amount of extra hardware needed to make the dynamic array arrangement possible. But when the word-length of an optimized arithmetic unit is not fixed but can be chosen dynamically at run-time, this opens the door for multi-precision algorithms as well as a massively parallel and efficient usage of resources in the 3-D-SoftChip system.

As an outlook on future work, the impact of a certain choice of processing element architecture on the actual area usage as a function of the specific technology parameters will become necessary. Then, more detailed conclusions about hardware overhead, area efficiency and required interconnect resources can be made. A refinement of the architecture concepts as well as a design space exploration in consideration of the high-potential 3-D integration approach using indium bumps are currently in the process.

References

Baugh, C. R. and Wooley, B. A.: A Two's Complement Parallel Array Multiplication Algorithm, IEEE Trans. Computers, C-22, 1045–1047, 1973.

Bermak, A., Martinez, D., and Noullet, J.-L.: High-Density 16/8/4-bit Configurable Multiplier, IEE Proc. Circuits Devices Systems, 144, 272–276, 1997.

Davis, W. R., Wilson, J., Mick, S., Xu, J., Hua, H., Mineo, C., Sule, A. M., Steer, M., and Franzon, P. D.: Demystifying 3D ICs: The Pros and Cons of Going Vertical, IEEE Design & Test of Computers, 22, 498–509, 2005.

Eshraghian, S., Lachowicz, S., and Eshraghian, K.: Ultra High Bandwidth Image and Data Processing using 3-D Vertically Integrated Architectures, Proceedings of the SCI 2003, Orlando, FL, X, 189–195, 2003.

Haynes, S. D., Ferrari, A. B., and Cheung, P. Y. K.: Flexible Reconfigurable Multiplier Blocks suitable for enhancing the Architecture of FPGAs, Proceedings of the IEEE 1999 Custom Integrated Circuits Conference, San Diego, CA, 191–194, 1999.

Hwang, K.: Computer Arithmetic – Principles, Architecture, and Design, John Wiley and Sons, New York, 1979.

Joyner, J. W., Zarkesh-Ha, P. J., and Meindl, J. D.: Global Interconnect Design in a Three-Dimensional System-on-chip, IEEE

Transactions on VLSI systems, 2004.

Pfänder, O. A. and Pfleiderer, H.-J.: Dynamische Rekonfiguration von arithmetischen Einheiten auf Bitebene, Advances in Radio Science 2004, Miltenberg, Germany, 319–323, 2004.

Pfänder, O. A., Lachowicz, S. W., and Pfleiderer, H.-J.: Flexible Multiplier Blocks for Accelerated Processing in a 3D-SoftChip Adaptive Computing System, Proceedings of the IFIP VLSI-SoC 2005, Perth, Western Australia, 485–491, 2005.

Xilinx: Virtex™ -II Platform FPGAs and Product Specification, Xilinx Document DS031 (v3.4), 2005.

2

Monotonic transition based forward body bias for dual threshold voltage low power embedded processors

S. Jayapal and Y. Manoli

Chair of Microelectronics, Department of Microsystems Engineering (IMTEK), University of Freiburg, Georges-Koehler-Allee 102, 79110 Freiburg, Germany

Abstract. Dual threshold voltage and forward body bias techniques are effective ways to optimally balance the standby leakage power and performance. In this paper, we propose a novel fine-grained forward body biasing scheme for monotonic static logic circuits. In the proposed scheme, the forward body bias is applied to high threshold voltage of either the pull-up or the pull-down network based on the evaluation transition and the state of operation. This technique improves the low skew NAND and NOR circuit performance by 7% and 11%, high skew NAND and NOR by 8% and 13% respectively. It reduces both active and standby leakage power as compared to monotonic static CMOS with dual-V_T technique. The simulations are carried out using 130 nm mixed mode process technology to validate our proposed technique.

1 Introduction

In nanometer era, leakage power becomes gradually a major portion of the total power consumption in battery powered embedded systems. This increase in leakage power reduces the battery life time in such applications as wireless sensors with high standby to active time ratio (Jayapal et al., 2005a). To best utilize the battery source, both the active and standby leakage power has to be minimized while meeting the performance demands.

Modern CMOS process offers different types of threshold voltages and oxide thickness devices to meet the low power and high performance demands. The low and high threshold device provides high performance at the cost of high leakage and low leakage with low performance respectively. Depending on the application requirements, the designers can make optimal power performance trade-offs. For low standby power applications, dual threshold voltage with power gating reduces the standby leakage power at the cost of performance through traditional static CMOS logic circuits. On the other hand, in case of pass logic circuits an additional input gating is required at both ends of the customized pass logic blocks to reduce the standby leakage power (Jayapal et al., 2005b). To overcome this delay penalty and to reduce the transistor count, the high threshold voltage with forward body bias for improving the performance level has been introduced. The change in threshold voltage due to the bulk source potential is given by

$$\triangle V_t = \gamma\sqrt{2\Phi_F + V_{sb}} - \sqrt{2\Phi_F} \tag{1}$$

where $\gamma =$ body effect parameter, $\Phi_F =$ Fermi potential and $V_{sb} =$ SourceBulkpotential. Recent research reports indicate that the optimal forward body bias (FBB) value of 400–500 mV provides 13% and 37% performance improvement at 130 nm and 90 nm triple well process (Narenda etal., 2003; v. Arnim et al., 2004). This clearly suggests that the efficiency of body bias depends on process technology.

Monotonic Static CMOS circuits provide noise immunity low power and high performance (Solomatnikov et al., 2000), and differ from traditional static CMOS in the way transition are made. In order to have monotonic transitions at the gate outputs, the concept of pre-charge and evaluation mode resembles domino circuits. During the pre-charge mode, each gate output is either charged to V_{DD} or discharged to V_{SS}. In the evaluation mode, the output either stays at its present value or makes a transition depending on the inputs.

The rest of the paper will discuss the limitation of global body bias, the essentials of transition based fine grained FBB and simulation results using monotonic static low and high skewed gates.

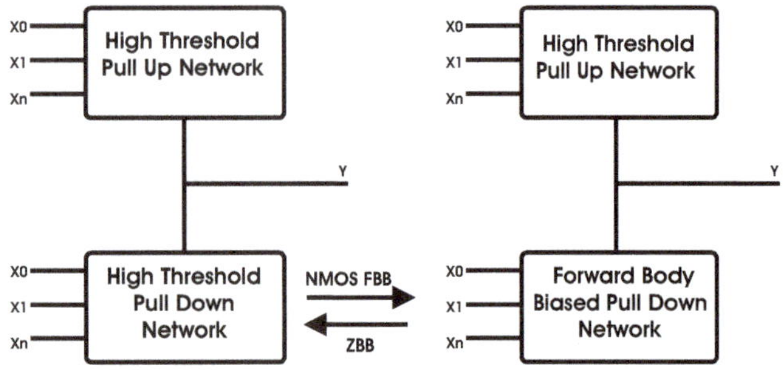

Fig. 1. Transition based fine grained forward body biasing with low skew NMOS body bias.

2 Fine-grained forward body bias

The application of global forward body bias using traditional static CMOS improves the performance at the expense of very high active leakage power. Applying forward body bias to high threshold device increases active leakage power from the non-evaluation transistors. On the other hand, it becomes challenging to place low and high threshold transistors in traditional static CMOS circuits depending on evaluation transitions. To overcome this, we chose monotonic static logic circuits in which placement of dual threshold devices based on evaluation transition becomes easier. However, this technique increases both active and standby leakage power because of static low threshold evaluation transistors. Similarly, applying FBB to the high threshold monotonic static CMOS logic causes more active leakage power similar to the monotonic static CMOS with dual-V_T technique (Thorp et al., 1999). As a result, the increase in active leakage power makes global body bias inadequate for low standby power applications.

We propose a transition based fine-grained forward body bias using monotonic static CMOS to avoid unwanted active leakage with performance improvement as shown in Fig. 1. In the proposed scheme, body bias is applied to high threshold transistors of either the pull-up or the pull-down network depending on the evaluation transitions. In monotonic static circuits, the applied forward body bias increases the transition rate without resizing the evaluation transistors. During active mode, a chosen body bias of 450 mV is applied to the evaluation transistors, which decreases the CV/I delay by increasing the I_{on}. The non-evaluation pre-charge high threshold transistor offers lower active leakage power and the body biased low threshold evaluation transistor contributes performance enhancement. During standby mode, the applied Zero Body Bias (ZBB) brings the body biased evaluation transistor into high threshold state to reduce standby leakage power. Thus, the selective forward and zero body biased evaluation transistor reduces both active and standby leakage power using monotonic property.

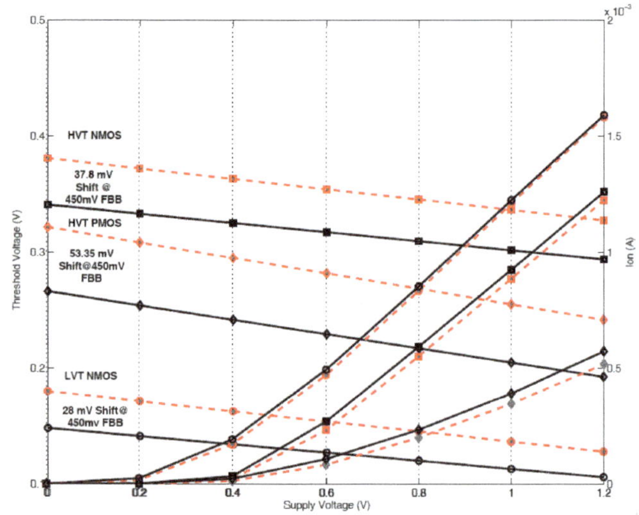

Fig. 2. The variation of threshold voltage and on-current of low and high threshold NMOS and PMOS devices.

Table 1. Threshold voltage variations and leakage currents for different supply voltages

Device under test	$\triangle V_T$ @1.2V VDD	$\triangle V_T$ @400 mV VDD	$\triangle V I_{\mathrm{off}}$ @1.2V VDD	$\triangle V I_{\mathrm{off}}$ @400 mV VDD
HVT NMOS	33.7 mV	37.8 mV	12.62 nA	41.4 nA
LVT NMOS	22 mV	28 mV	0.54 uA	0.268 uA

3 Transition based forward body bias - comparison of power and performance

Dual threshold 130 nm device measurement shows that applying 450 mV body bias to the high threshold NMOS and PMOS increases I_{on} by 3% and 11% at high drain source voltage. At the same time, the I_{on} of low threshold NMOS are virtually identical due to severe Drain Induced Barrier Lowering (DIBL) and V_T rolloff effects as shown in Fig. 2. Thus, the efficiency of body bias decreases with thinner oxide, shorter gate length and smaller V_T. Table 1 shows that the rate of change of threshold voltage with body bias at lower drain source voltage is greater due to reduced DIBL and V_T rolloff. Dual V_T high threshold PMOS measurement shows that the threshold voltage shift with 450 mV FBB is about 1.4x more than those of NMOS devices. As a result, the larger I_{on} from body bias increases robustness of leakage

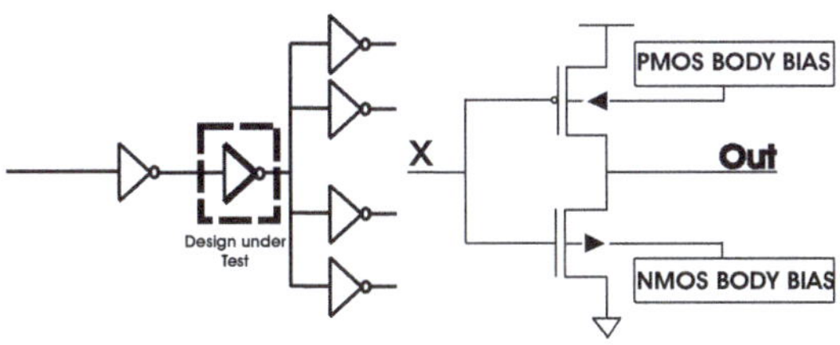

Fig. 3. Fan-out of 4 inverter delay with transition based forward body bias.

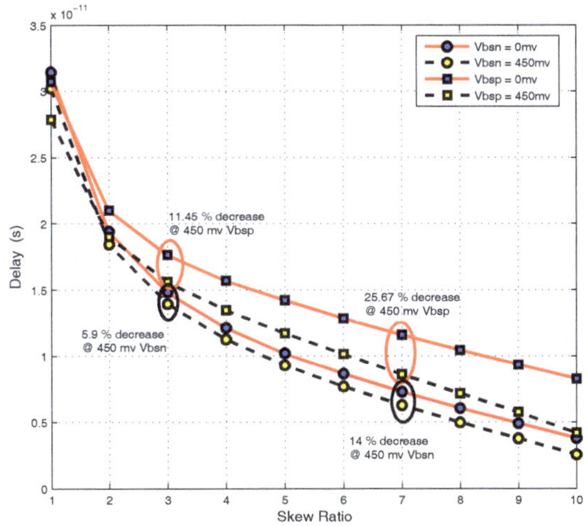

Fig. 4. FO4 Inverter delay variation with 450 mV FBB for various skew ratios.

sensitive circuits such as domino logic and highly skewed static logic gates.

Lower operating temperature improves performance, however the leakage current increase due to 450 mV FBB is about 3.2 x and 4 x at 1.2 V and 0.4 V V_{DS}. Higher temperature offers 2.6 x and 3.1 x increase in leakage current at 1.2 V and 0.4 V V_{DS}. Thus, the leakage current increase due to FBB at lower drain source voltage is slightly more than those of higher V_{DS}, irrespective of operating temperature.

To validate the proposed technique, the fan-out of 4 (FO4) inverter delay as shown in Fig. 3, is designed to compute the performance improvement for both evaluation transitions. Fig. 4 shows that the transition rate due to PMOS body bias has larger CV/I delay reduction compared to those of NMOS body bias. PMOS body bias provides the same CV/I delay with (1-2) x less skew ratio (the ratio between PMOS and NMOS transistor widths) compared to without body bias condition. On the other hand, the NMOS body bias offers the same transition rate with almost 1 x less skew ratio.

4 Monotonic static gates - design considerations

Figure 5 shows the monotonic static CMOS logic gates, each gate is biased to make faster Low to High (L-H) or High to Low (H-L) transitions. In general, there are two types of gates such as low skew (LS) gates which have faster H-L transitions and high skew (HS) gates with faster L-H transitions. Since monotonic logic is inherently non-inverting, it must be mapped to a network that does not contain intermediate inversions. This is in contrast to traditional static CMOS which can allow intermediate inverters. To utilize the monotonic property for fine-grained FBB, four different types of basic gates like LS-NAND, LS-NOR, HS-NAND and HS-NOR were chosen. Each skewed gate is analyzed for three different schemes like high threshold with transition based forward body bias, high threshold without body bias and MS-CMOS with dual V_T (low V_T in critical paths and high V_T in non-critical paths) (Thorp et al., 1999) techniques.

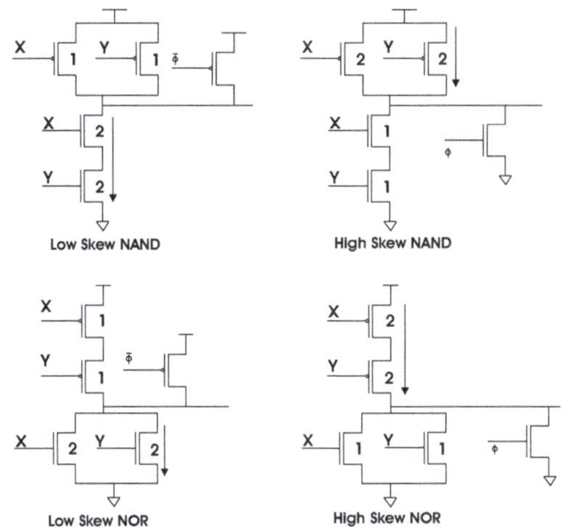

Fig. 5. Monotonic Static CMOS - Low and high skewed gates.

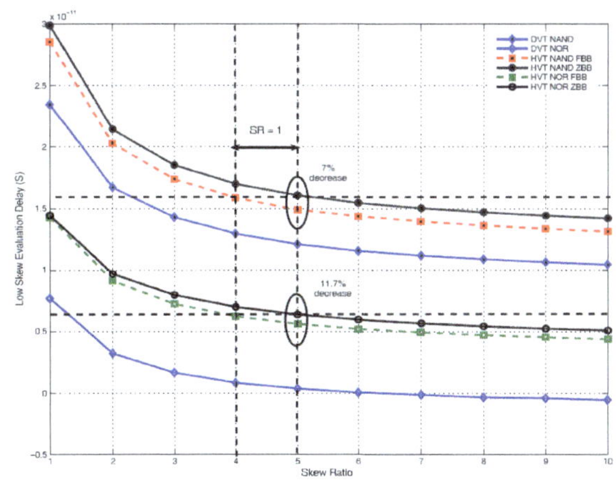

Fig. 6. Evaluation delay of low skewed gates for three different schemes.

Figures 6 and 7 show the evaluation delay and average switching power for LS-NAND and LS-NOR in three different techniques. Like FO4 inverter, the LS-NAND and LS-NOR achieves the same CV/I delay with reduced skew ratio (SR = 1). The parallel connected high fan-in LS-NOR gains

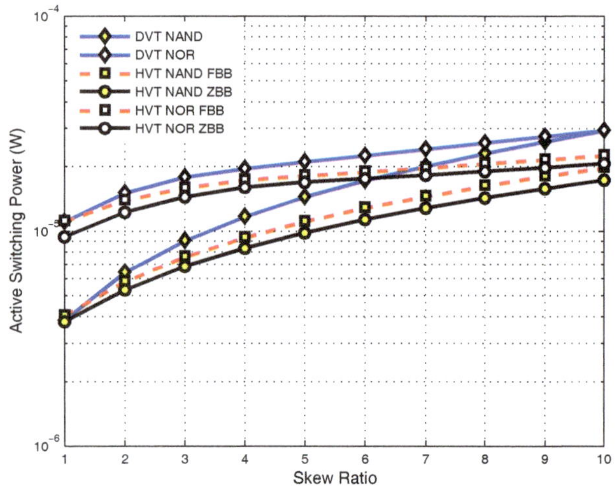

Fig. 7. Average switching power of low skewed gates for three different schemes.

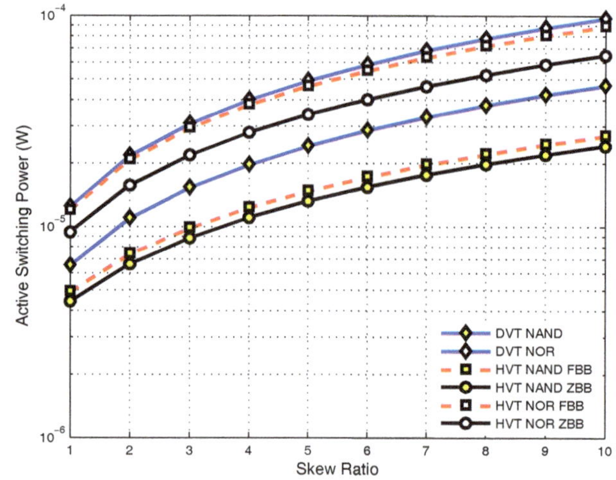

Fig. 9. Average switching power of high skewed gates for three different schemes.

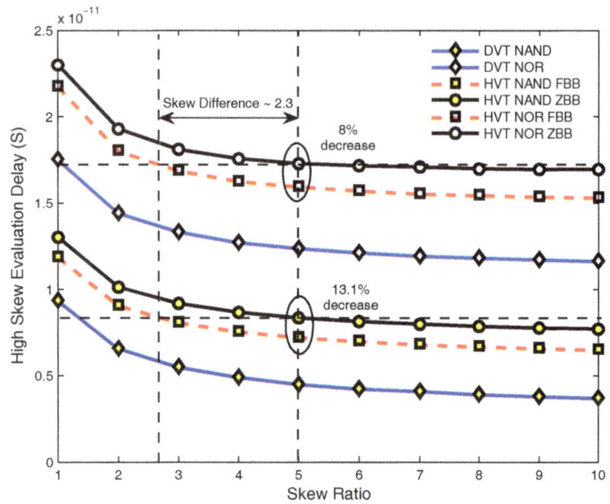

Fig. 8. Evaluation delay of high skewed gates for three different schemes.

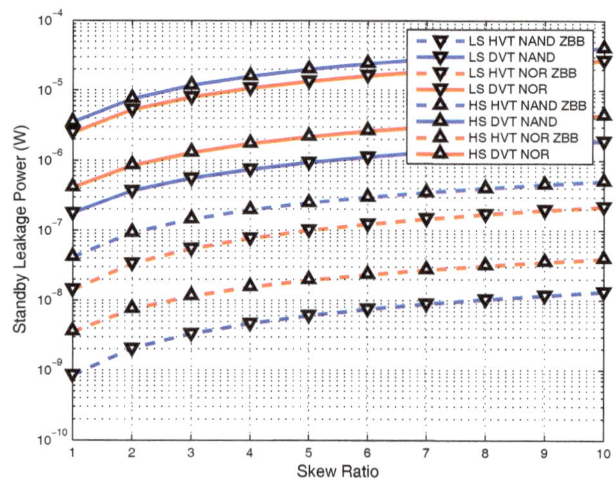

Fig. 10. The worst case standby leakage power of low and high skew gates.

0.67 times more delay reduction than the serial connected LS-NAND gate. The transition based body bias increases the switching power by 13.6% and 7.6% of LS-NAND and NOR gates, which is lower than that of MS-CMOS with dual V_T techniques. Thus, the LS-NOR attains substantial performance gain with marginal increase in switching power using transition based body bias. On the other hand, the switching power increase due to body bias in LS-NAND is greater than that of LS-NOR gate.

Like low skewed gates, the large fan-in HS-NAND PMOS body bias reduces the evaluation delay by 0.6 times than those of HS-NOR gate as shown in Fig. 8. High skewed gates with PMOS body bias achieve the same CV/I delay at reduced skew ratio (SR = 2.3). Thus, PMOS body bias gates achieve additional performance improvement with lower skew ratio than low skewed gates. From Fig. 9, the average switching power of HS-NOR is increased by a factor of 2 when compared to HS-NAND. Transition based forward body biased high skew gate consumes lower switching power than the MS-CMOS with dual V_T skewed gates.

The standby leakage power for both low and high skew gates are computed with worst case input patterns in Fig. 10. Applying Zero Body Bias (ZBB) during standby mode, brings the body biased skew gates to high V_T state. In MS-CMOS with dual V_T technique, the high dynamic and standby leakage power consumptions are mainly due to parallel connected low V_T evaluation transistors. Series connected LS-NAND and HS-NOR consumes lower standby leakage power than that of parallel connected LS-NOR and HS-NAND gates. Thus, the worst case standby leakage power for the proposed scheme is far below the MS-CMOS with dual V_T technique.

5 Conclusion

Transition based fine grained forward body bias using monotonic property has been proposed. The proposed fine-grained body bias scheme provides the same circuit performance with reduced skew ratio for both low and high skew gates. This method consumes lower dynamic and standby leakage power than MS-CMOS with dual V_T technique.

References

v. Arnim, K., Borinski, E., Seegebrecht, P., Fiedler, H., Brederlow, R., Thewes, R., Berthold, J., and Pacha, C.: Efficiency of Body Biasing in 90 nm CMOS for Low Power Digital Circuits, ESSCIRC, 175–178, 2004.

Jayapal, S., Huang, R., Ramachandran, S., Bhutada, R., and Manoli, Y.: Optimization of Electr. Power Consumption in Wireless Sensor Nodes, 8th Euromicro conf. on Digital Syst. Design (DSD), 165–169, 2005a.

Jayapal, S., Sudalaiyandi, S., and Manoli, Y.: Efficiency of Leakage Reduction on Different Static Logic Styles for Embedded Applications with High Standby to Active Ratio, Int. Symp. on System-on-Chip (SoC), 151–154, 2005b.

Narendra, S., Keshavarzi, A., Bloechel, B., Borkar, S., and De, V.: Forward body bias for microprocessors in 130 nm technology generation and beyond, IEEE J. Solid-State Circuits, 38, 696–701, 2003.

Solomatnikov, A., Somasekhar, D., Roy, K., and Koh, C.-K.: Skewed CMOS: Noise-Immune High-Performance Low-Power Static Circuit Family, Int. Conf. Comp. Design, 241–246, 2000.

Thorp, T., Yee, G., and Sechen, C.: Monotonic static CMOS and dual Vt technology, IEEE/ACM Int. Symp. on Low Power Electronics and Design, 151–155, 1999.

3

Design of UWB pulse radio transceiver using statistical correlation technique in frequency domain

M. Anis and R. Tielert

Institute for Microelectronics, Technical University of Kaiserslautern, Germany

Abstract. In this paper, we propose a new technique to extract low power UWB pulse radio signals, near to noise level, using statistical correlation technique in frequency domain. The receiver consists of many narrow bandpass filters which extract energy either from transmitted UWB signal, interfering channels or noise. Transmitted UWB data can be eliminated by statistical correlation of multiple bandpass filter outputs. Super-regenerative oscillators, tuned within UWB spectrum, are designed as bandpass filters. Summers and comparators perform statistical correlation.

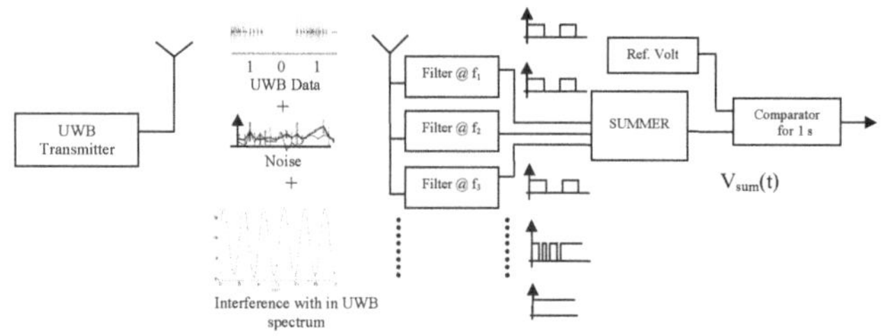

Fig. 1. OOK Low Data Rate UWB RF Front End Block Diagram.

1 Introduction

In the last few years, UWB pulse radio has become attractive for low power and low cost short range applications, especially in wireless body area networks. WBAN requires innovative solutions to meet the energy consumption budget of the autonomous sensor nodes. Today's low power radio systems like Bluetooth and Zigbee do not meet these requirements, but the pulse based UWB reaches the target. It is an effective technique to deliver short range communication, using very short pulses and wide spectrum, enables coexistence with other narrow band radio systems. To mitigate the effects of interference to UWB from noise and other radio systems, we propose statistical correlation technique in frequency domain.

This paper describes the design of the first UWB CMOS transceiver, having ability to extract the UWB signal from the noise level. Our proposed architecture is based on simple on-off keying for low and medium data rate UWB system. The transmitter generates a pseudo random pattern of UWB impulses in a sub nanosecond time scale. The receiver consists of many narrow band pass super regenerative oscillators (SRO), tuned within UWB spectrum, acting as narrow bandpass filters. These SROs can be triggered either by the noise, interference or either by transmitted UWB data. Noise and interferences due to other radio channels are uncorrelated in all filter (SRO) outputs. UWB data are extracted after the statistical correlation of multiple bandpass filter outputs. Statistical correlation is performed by summer and comparator.

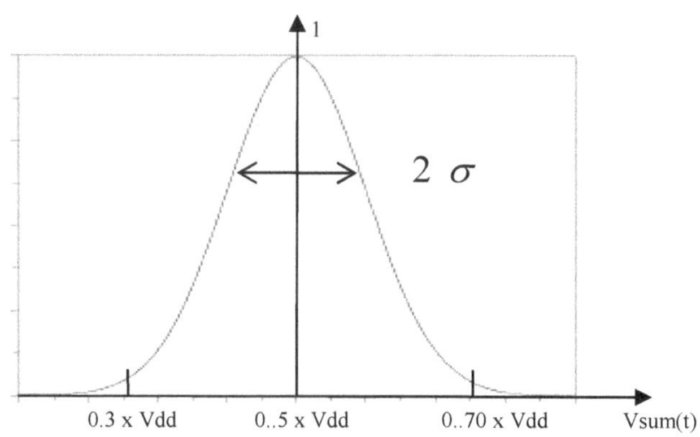

Fig. 2. The Probability distribution function of Vsum(t), in the absence of UWB data.

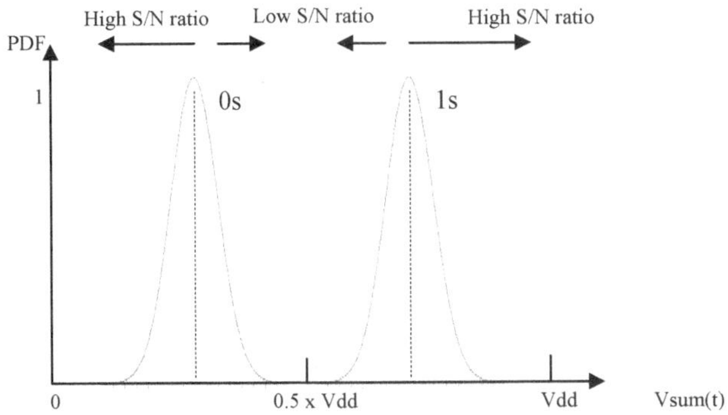

Fig. 3. Probability distribution function of Vsum(t), in the presence of UWB data.

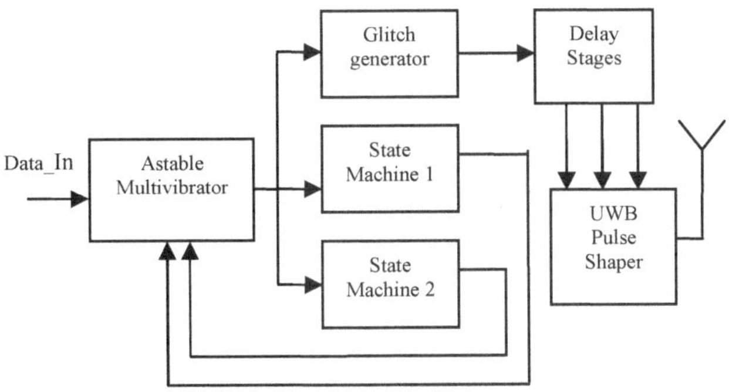

Fig. 4. Block diagram of UWB transmitter.

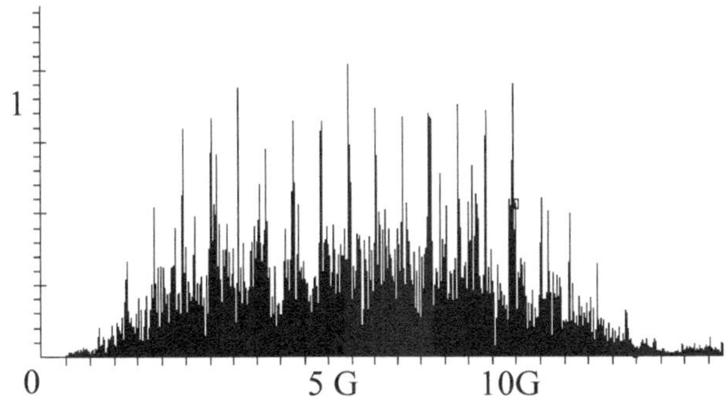

Fig. 5a. Normalize power density spectrum of periodic UWB impulses.

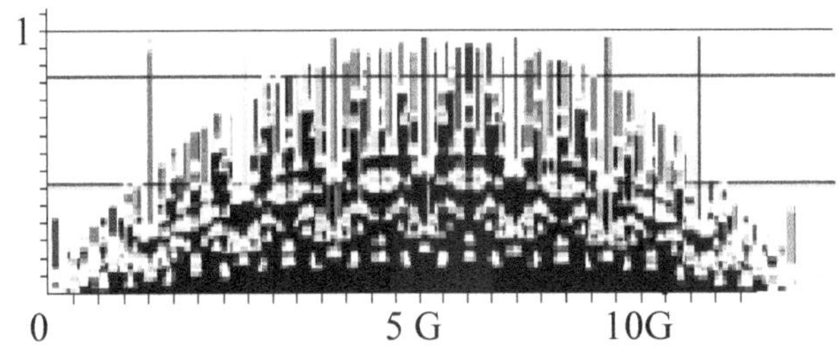

Fig. 5b. Normalize power density spectrum of PN sequence of UWB impulses

2 System overveiw

The proposed architecture based on statistical correlation in frequency domain is designed to extract the UWB signals which are embedded into noise and interfering channels. The UWB transmitter generates the pseudo random pattern of UWB impulses during on time of on-off modulation keying. The maximum energy level of UWB signal is near to noise level.

Figure 1 shows the UWB receiver which consists of multiple narrow band pass filters, tuned within the UWB spectrum. These filter banks will extract energy either from UWB transmitted data, noise or interfering channels. Outputs of filter banks due to noise are uncorrelated in each channel. Narrow band interfering channels like WLAN can disturb one or two filter channels. When the transmitted UWB data with embedded noise and interfering channels are present at antenna, a majority of the filter bank outputs will be correlated due to UWB data. Filter banks of frequency range 6 GHz–8.5 GHz are designed by super-regenerative oscillators. The SRO was invented in 1922 by Armstong, well suited for short range low power applications. The SRO is turned on and off by a quench signal. The start up time of the SRO depends on the matching of tuned LC tank frequency and signal frequency of received input. Twelve SROs, each using an LC filter tank have been integrated in our test structure. The output of every SRO is connected with envelop detectors, in order to get the output of each channel in digital form.

The most critical point is the biasing of SROs. The bias point of all SROs is adjusted so that average output voltage of each channel should be half of the supply voltage. In the absence of RF UWB signals, regulator increases the bias current of all SROs to make them sensitive to noise level, which means that every SRO has 50% probability to be triggered due to noise and the summation of all filter (SROs + envelop det.) outputs Vsum(t) has a Gaussian distribution, as shown in Fig. 2. When the modulated UWB data having energy level near to noise is radiated by transmitter, the probability to detect 1s increases in all channels. The regulator starts to decrease the bias current of all SROs, to maintain the average output voltage of each channel at the half of supply voltage. The UWB data can be extracted by comparing the summer output to the half of supply voltage.

$$V_{sum}(t) > 0.5 \times Vdd \Rightarrow 1$$

$$V_{sum}(t) < 0.5 \times Vdd \Rightarrow 0$$

Figure 3 shows the probability distribution function of Vsum(t) in the presence of UWB data. The distribution functions for detecting 1 s and 0 s move towards the 50% of supply voltage for signals having small S/N ratios and move

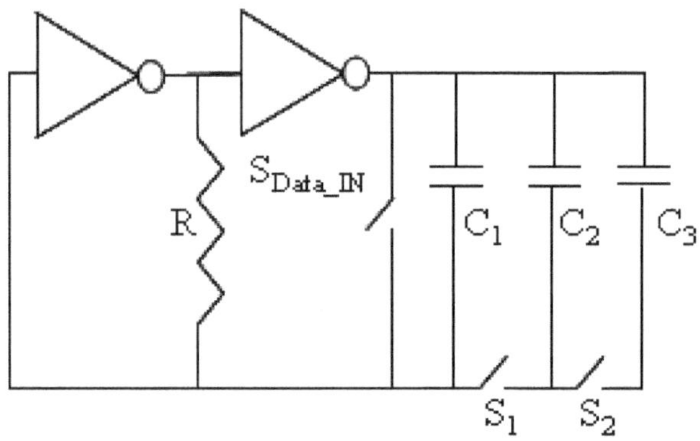

Fig. 6. Astable multivibrator with additional switches and capacitors.

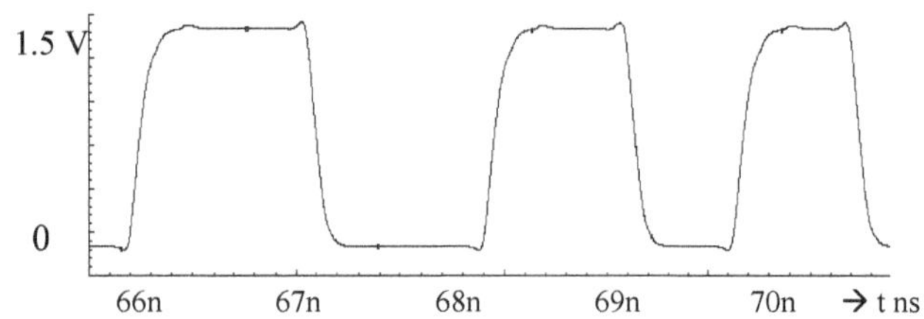

Fig. 7a. Pseudo random pattern at the output of astable multivibrator.

away from the 50% of supply voltage for signals having high S/N ratios as shown in Fig. 3.

3 UWB transmitter

A fully integrated pulse generator for impulse based UWB system, covers the spectrum of 6–8.5 GHz, and was designed on 0.18 μm CMOS technology. On/off modulation scheme is used. The basic structure consists of four parts, astable multivibrator with state machines, glitch generator, delay stages and UWB pulse shaper, shown in Fig. 4. Astable multivibrator (Baker et al., 1997) with state machines is used to generate the pseudo random (PN) pattern of UWB impulses. The periodic repetition of impulses generates the large spikes over the UWB spectrum. The PN sequence is used to flatten the UWB spectrum (Maggio et al., 2001). The simulation for power spectrum of periodic and PN sequence of impulses are shown in Figs. 5a, b.

The basic structure of astable multivibrator can generate periodic signals of pulse width t∼0.65 RC. Three switches (S_{Data_In}, S_1, S_2)and two capacitors (C_2, C_3) are added to the astable multivibrator as shown in Fig. 6. Switch S_{Data_In}, connected to input digital data, will turn the multivibrator on and off. On-off keying modulation is used to transmit UWB pattern. When the switch is closed, multivibrator will be off and UWB pulse shaper will not transmit the UWB impulse. When the switch is open, multivibrator generates the pattern

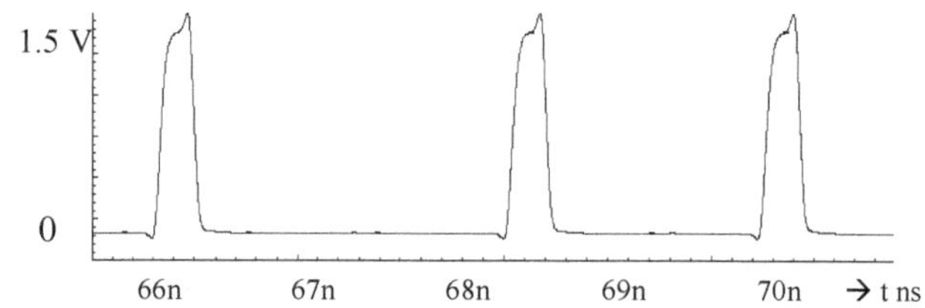

Fig. 7b. Pseudo random pattern at the output of glitch generator.

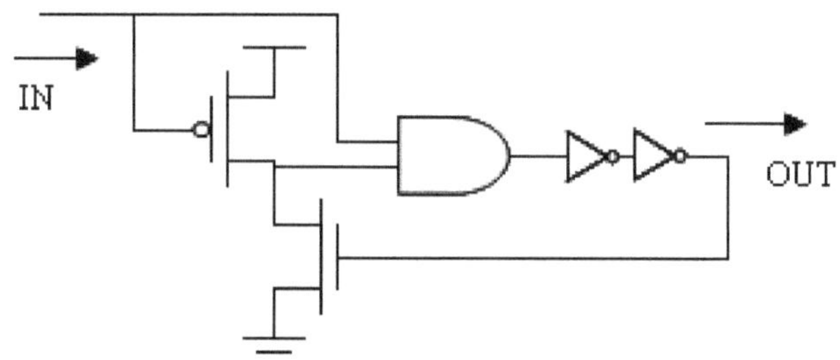

Fig. 8. Circuit diagram of basic glitch generator.

of pulse width t∼0.7 RC_{eq}, where C_{eq} is the sum of capacitances selected by state machines and parasitics of connected transistors. Initially the multivibrator generates a periodic rectangular pattern due to capacitor C_2. The output of multivibrator is connected to the clock signal of two different state machines. These state machines generate the defined pattern used to select and deselect the capacitors by switches S1 and S2. Due to the changing of C_{eq} pulse width of the multivibrator changes, resulting in different clock rates of the cycles which generates the pseudo random pattern at the output, as shown in Fig. 7a.

The UWB pulses require sharp rising and falling transitions. Due to the different rise and fall transition times of the multivibrator, two different pulse shapes and spectrum will be generated. In order to avoid this problem, very small glitches are generated on each rising edge of pattern. Figure 7b shows the simulation results of the PN sequence at the output of the glitch generator. Figure 8 shows the circuit diagram of basic glitch generator. The delay stages are designed by a chain of basic CMOS inverters (Jeong et al., 2004). Delay stages generate three delayed small glitches. The delays are adjusted by the W/L ratios of inverters. These delays in between the glitches are smaller than the widths of glitches. These three glitch stages are connected to a UWB pulse shaper, consisting of three LC resonators tuned within 6–8.5 GHz range. Chirped UWB pulses with flat spectrum of 6–8.5 GHz are achieved after mixing the outputs. Figure 9 shows the delay stages and UWB pulse shaper. Figure 10 shows the simulated pattern of UWB pulses at the load of 50Ω.

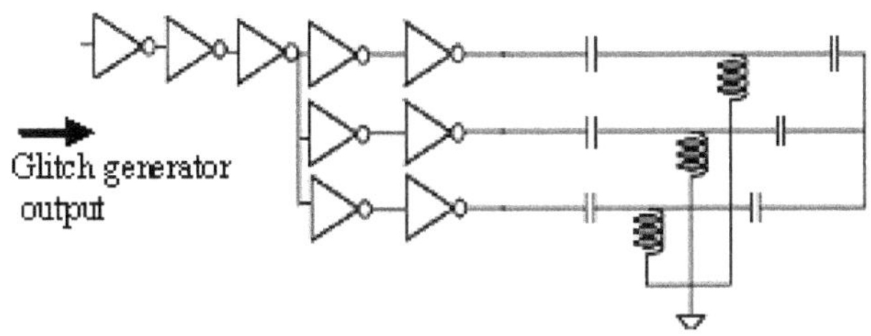

Fig. 9. Delay Stages and UWB Pulse Shaper.

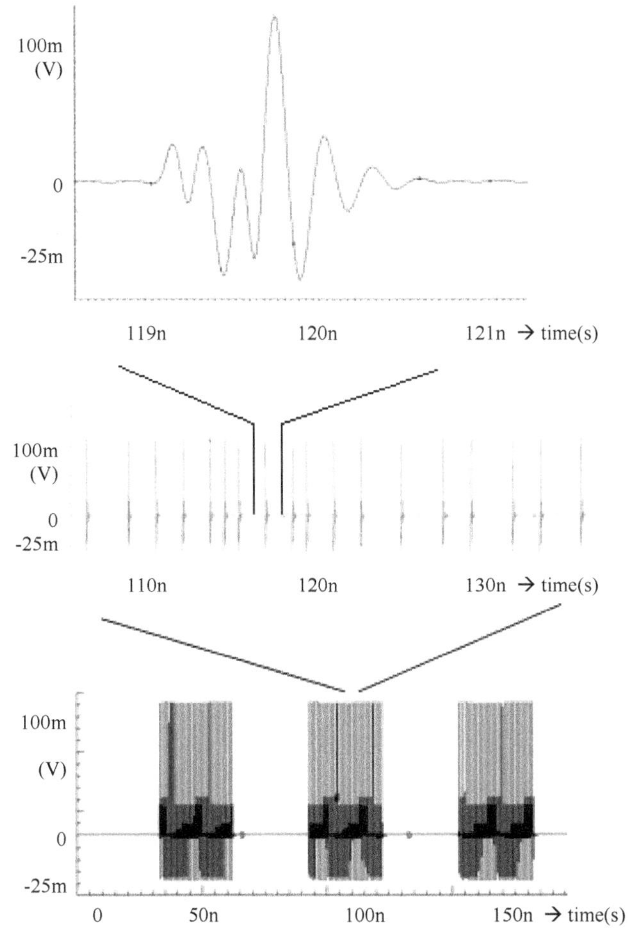

Fig. 10. The simulated pattern of UWB pulses at the load of 50Ω.

4 UWB receiver

A fully integrated UWB pulse receiver covering the spectrum from 6–8.5 GHz has been implemented in 0.18 μm CMOS technology. Figure 1 shows the block diagram of the receiver. The simple low power system consists of isolation amplifier, super regenerative oscillators, envelop detectors, data synchronizers and regulated current control units, as shown in Fig. 11. The RF UWB signal is first fed to the isolation amplifier (IA). This common gate amplifier provides isolation of

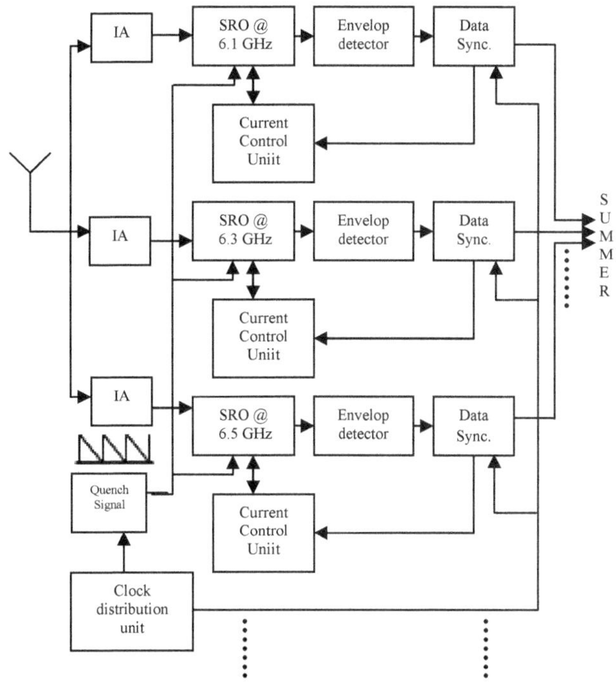

Fig. 11. Block diagram of multiple narrow band filter banks.

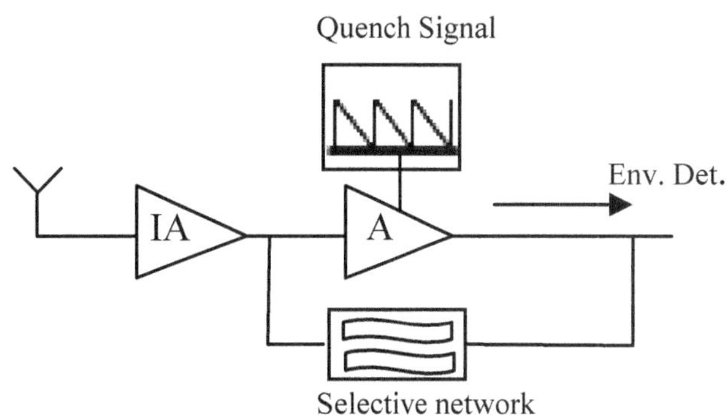

Fig. 12. Block diagram of super-regenerative oscillator.

antenna and SROs. The signal then goes to SROs controlled by a feedback network as shown in Fig. 12. These oscillators are tuned within the range of the UWB signal spectrum 6–8.5 GHz. All oscillators are turned on and off by one quench generator. The start up time of the individual oscillators depends on the received input power and its frequency. The oscillator will start early, when the received signal is near to the center frequency of the oscillator. Without any RF signal, the start up time of oscillators depends on the feedback gain and thermal noise (Moncunill et al., 2005). Figure 13 shows the simulation results of four different SROs with on –off modulation for UWB. Selectivity and sensitivity, frequency and shape of the quench signal, minimum and maximum levels of bias currents of oscillators are critical issues, require

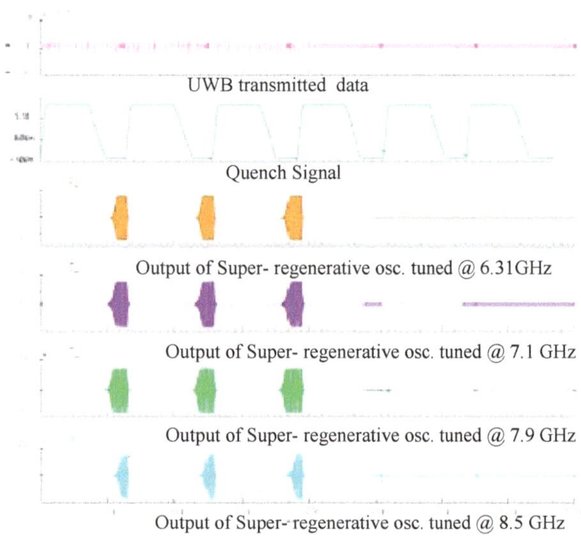

Fig. 13. The simulation results at the output of four SROs with UWB data.

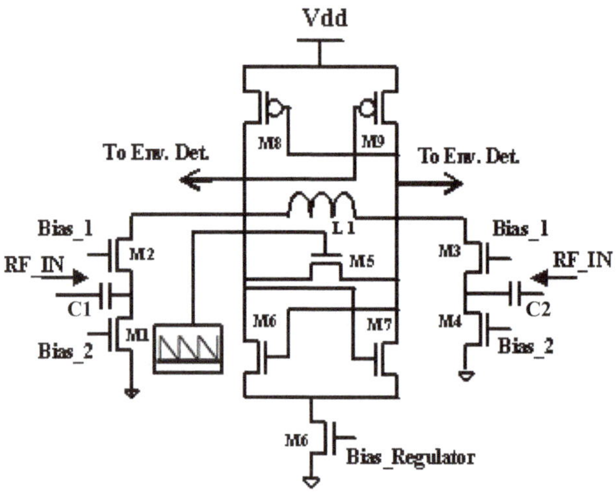

Fig. 14. The circuit diagram of isolation amplifier and super- regenerative oscillator.

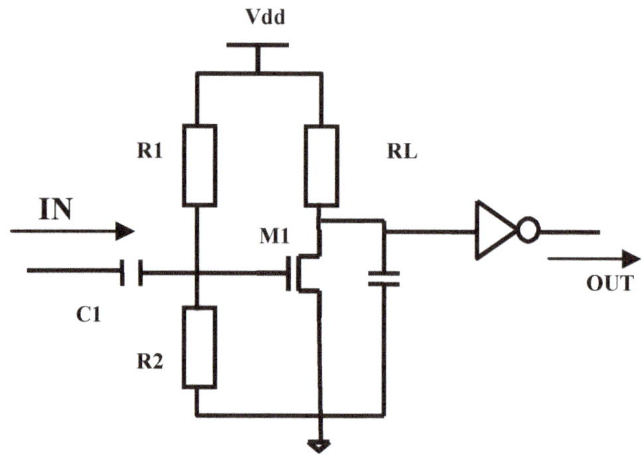

Fig. 15. The circuit diagram of envelop detector.

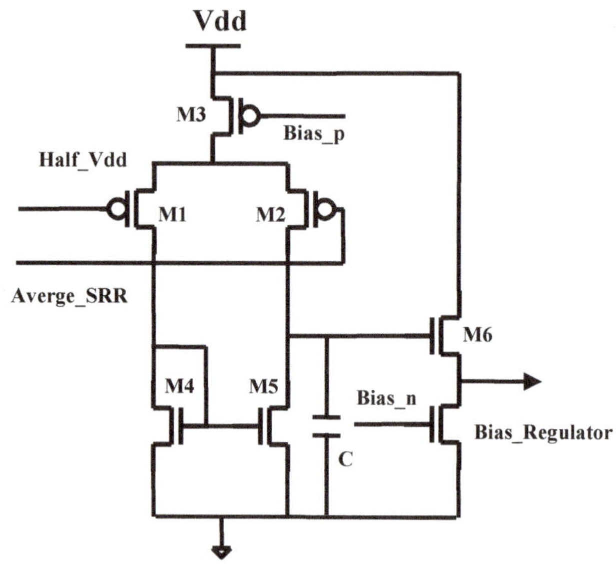

Fig. 16. The circuit diagram of current control unit.

detailed analysis as presented in (Favre et al., 1998; Vouilloz et al., 1998). The adjustment of bias points of all oscillators and the slope of the quench signal during start up of oscillators has to be optimized, in order to meet the bandwidth requirements of the filters.

The bias point of the individual channels is adjusted by the regulator so that all oscillators are sensitive to the noise level. Due to the biasing, there is 50% chance that oscillators will start to oscillate or remain off during the turn on time of the quench signal. An envelop detector senses the magnitude of the oscillations and provides the digital output. The transmitted UWB data is not synchronized with the quench signal. Data synchronizers as simple flip flops are used to synchronize the outputs of all envelop detectors with the reference clock, generated by the clock distribution unit. The average voltage at the output of data synchronizers should be half of the supply voltage. The current control unit regulates the bias currents of individual oscillators by comparing the average output voltage of data synchronizers with the half of supply voltage. In order to fulfill this condition, the input data coding of UWB modulation must contain equal number of ones and zeros, like in 8bit/10 bit coding.

The shape and frequency of quench signal has an effect on the performance of SROs. Detailed analysis of different types of quench signals is explained in (Joehl et al., 2001; Vouilloz et al., 2001). According to the nyquist criteria, the quench frequency must be two times or more than the transmitted UWB data rate. The quench signal is derived from

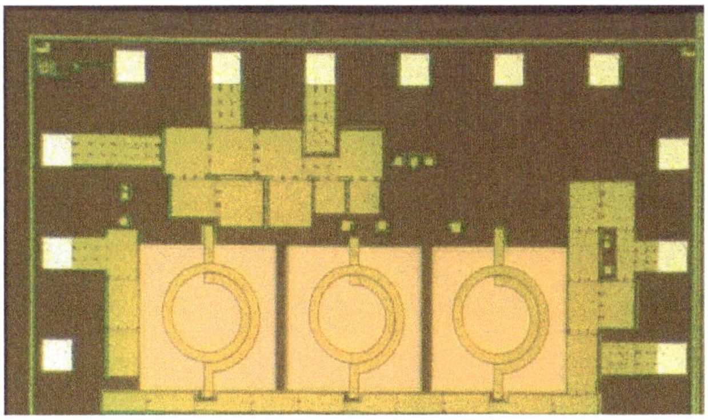

Fig. 17. Die photograph.

the reference clock, having a fast low to high transition and a slow high to low transition. When the slope of the quench signal is low, SROs are sensitive in narrow frequency bands, consuming low power and allowing low data rate. When transition is fast during starting of oscillators, SROs will be sensitive in a wide frequency band, consuming high power and allowing high data rate (Moncunill et al., 2005; Feick and Rojas, 1997). The statistical correlation between multiple channels requires a low bandwidth of the SROs in order to separate the frequency channels. The center frequencies and bandwidth of all SROs are designed in a way that cross coupling between adjacent channels is negligible. In the designed test chip, SROs of adjacent frequency bands are placed most far from each other to reduce the effect of coupling, as illustrated in Fig. 19

The schematic of SRO is shown in Fig. 14. The RF signal of the UWB antenna goes to the source of transistors M2 and M3, which are common gate isolation amplifiers. M1 and M4 provide the bias currents for this IA. The outputs of IA are directly connected to the differential LC oscillator. Twelve integrated inductors are used to cover the 6–8.5 GHz range. The quench signal is connected to the gate of transistor M5, which turns the oscillator on and off. Transistor M6 provides the regulated bias current of the oscillator. Fig. 15 shows the circuit diagram of an envelop detector. Capacitor C1 is used to remove the DC component from the oscillation. Resistors R1 and R2 are adjusted to keep the transistor M1 at its threshold voltage, hence transistor M1 acts as RF rectifier. Figure 16 shows the circuit diagram of a current control unit. The circuit is comparing the average output voltage of data synchronizers with half of Vdd. Capacitor C is used to integrate the DC-free modulation.

The output of data synchronizers from all channels are added by the summer circuit. The summer output is connected to a clocked comparator. The reference voltage of the comparator is adjustable between 0 to Vdd, used to identify ones and zeros of transmitted data. If the UWB transmitter is off, the bias current of SROs will start to increase by regulator to make them sensitive to noise level, which means

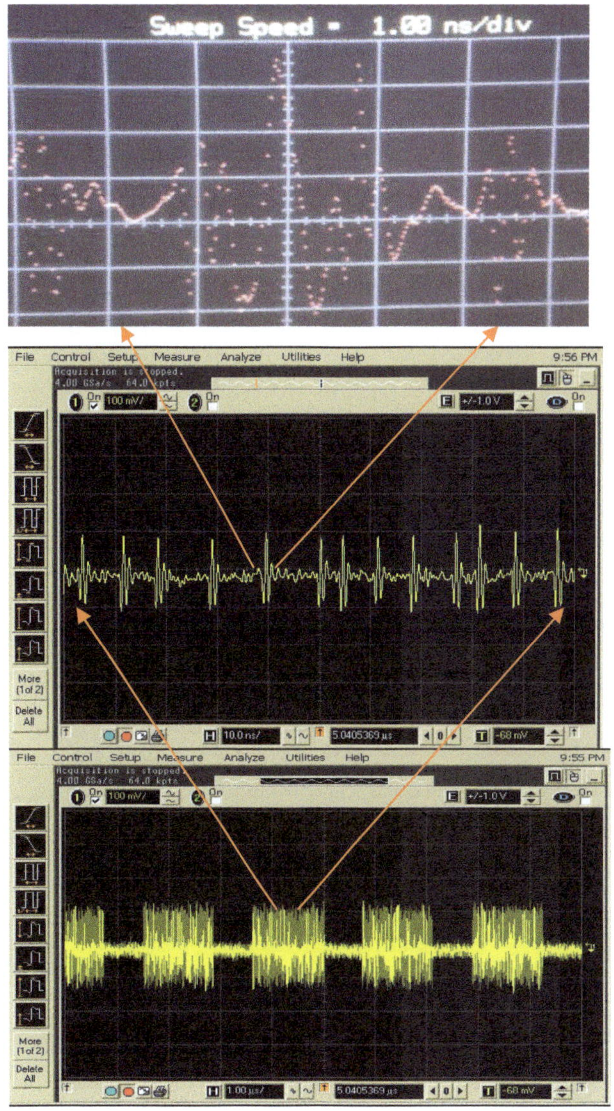

Fig. 18. The measurement results for UWB transmitter at the load of 50Ω.

that the triggering of all channels is only due to noise, and summer output has Gaussian distribution as shown in Fig. 2. When the UWB transmitter radiates pseudo random pattern of pulse during on – off keying, the bias current of all SROs starts to decrease to make the all channels less sensitive, allowing for an improved S/N ratio. The maximum number of continuous ones and zeros in modulated UWB data is limited due to the RC time constant of regulator loop.

7GHz	6GHz	6.75GHz	7.95GHz	7.1GHz	8.3GHz
6.9GHz	8.1GHz	7.37GHz	8.5GHz	6.31GHz	7.7GHz

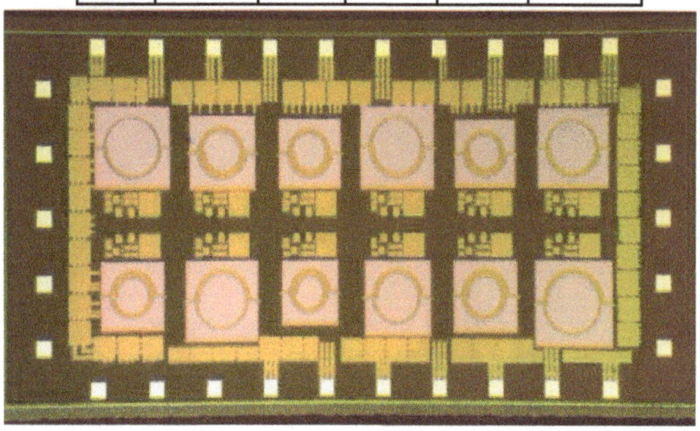

Fig. 19. Die photograph for UWB receiver.

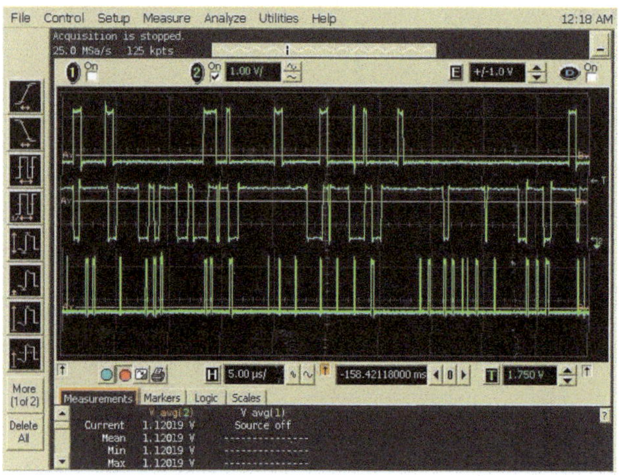

Fig. 20. The uncorrelated outputs of three adjacent channels.

5 Experimental results

5.1 UWB transmitter

The UWB (6 GHz–8.5 GHz) transmitter has been implemented on 0.18μm CMOS technology, active area of 0.65 mm^2. The power consumption is 0.6 m W for transmitting UWB pulses with on –off modulation keying at the data rate of 4 MHz. The die photo is shown in Fig. 17. The measurement of UWB transmitter, shown in Fig. 18, can be compared with the simulation results shown in Fig. 10

5.2 UWB receiver

The test structure has been implemented on 0.18 μm CMOS technology, active area of 2.5 mm^2. The UWB receiver, designed for maximum quench rate of 8 MHz, can receive maximum data rate of 4MHz with power consumption of 2 mW

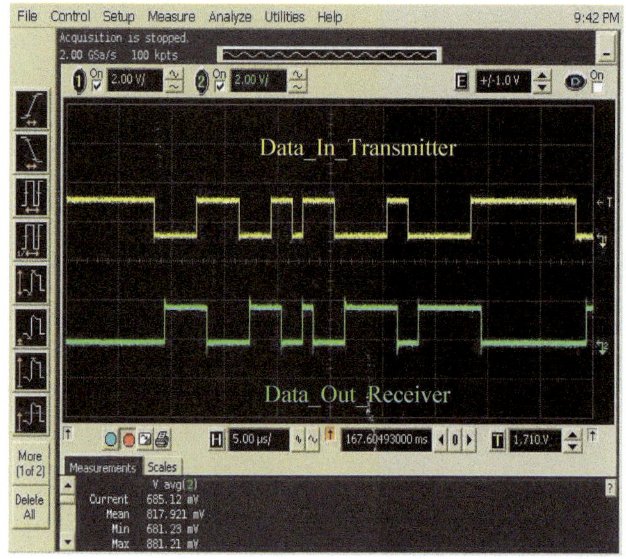

Fig. 21. The measured output at the receiver for radiated pattern from transmitter (data out is delayed and inverted).

from 1.5 V supply. The die photograph is shown in Fig. 19. The adjacent frequency channels are placed far from each other to reduce the effect of coupling. The outputs of three adjacent channels, shown in Fig. 20, are uncorrelated to each other in the absence of UWB data, triggering only due to noise. Figure 21 shows the input data pattern to the transmitter. The transmitter generates pseudo random pattern of UWB pulses during low input level and off during is high input level. The peak voltage of these pulses is around 20 mV. The receiver input has been coupled to the UWB transmitter by a 60 dB attenuator. The output of UWB receiver after statistical correlation of all channels is shown in Fig. 21.

6 Conclusions

The new architecture to receive the low power UWB pulse radio signals by statistical correlation of multiple channels has been proposed in this paper. The transmitter radiates pseudo random pattern of UWB pulse in the spectrum from 6–8.5 GHz. The receiver extracts the UWB pulses of energy level near to noise. Narrow band interfering transmitters are detected and eliminated by statistical correlation of twelve narrow bandpass SROs.

References

Baker,R. J., Li,H. W., and Boyce, D. E.: CMOS Circuit design, Layout, and Simulations, IEEE press series on Icroelectronics System, 1997.

Favre, P., Joehl, N., Vouilloz, A., Deval, P., Dehollain, C., and Declerq, M.: A 2-V 600 uA 1GHz BiCMOS super-regenerative receiver for ISM applications, IEEE J. Solid- State Circuits, 33(12), 2186–2196, 1998.

Feick, R., Rojas, O., and Mackay, A.: Modeling and simulation of the super regenerative receiver, IEEE Trans. On Consumer Electronics, 43(2), 92–102, 1997.

Jeong, Y., Jung, S., and Liu, J.: A CMOS impulse generator for UWB Wireless communication systems", in proc. ISCAS, 129–132 2004.

Joehl, N., Dehollain, C., Favre, P., Deval, P., and Declercq, M.: A low power 1-GHz super-regenerative transceiver with time-shared PLL control, IEEE J. Solid State Circuit, 36(7), 1025–1031, 2001.

Moncunill-Geniz, F. X., P-Schönwälder, P., and del Aguila- Lopez, F.: New superregenerative architectures for direct sequence spread spectrum communications, IEEE Trans. on Circuits and Systems-II Express Briefs, 52(7), 415–419, 2005.

Moncunill-Geniz, F. X.,P-Schönwälder, P., and M-Casals, O.: A generic Approach to the theory of Super regenerative reception, IEEE Tran. on Circuits and Systems-I: Regular Paper, 52(1), 54–70, 2005.

Maggio, G. M., Rulkov, N., and Reggiani, L.: Pseudo- chaotic time hopping for UWB impulse radio, IEEE Trans. On circuit and system-I, 48(12), 1025–1031, 2001.

Vouilloz, A., Declercq, M., and Dehllain, C.: A low power CMOS Super- Regenerative Receiver, IEEE J. Solid State Circuit, 36(3), 440–451, 2001.

Vouilloz, A., Declerq, M., and Dehollain, C.: Selectivity and sensitivity performance of super-regenerative receivers, in proc. ISCAS', 325–328 98.

4

Towards a generic operational amplifier with dynamic reconfiguration capability

S. K. Lakshmanan and A. Koenig

Institute of Integrated Sensor Systems, TU-Kaiserslautern, Germany

Abstract. Analog and analog-digital mixed signal electronics needed for sensor systems are indispensable components which tend to drifts from the normal phase of operation due to the impact of manufacturing conditions and environmental influences like etching, aging etc. Precise design methodology, trimming / calibration are essential to restore functionality of the system. Recent block level granular approaches using Field Programmable Analog Array and the more recent approaches from evolutionary electronics providing transistor level granularity using Field Programmable Transistor Arrays offers considerable extensions. In our work, we started on a new medium granular level approach called Field Programmable medium-granular Mixed-signal Array (FPMA) providing basic building blocks of heterogeneous array of active and passive devices to configure established circuit structures which are adaptive, biologically inspired and dynamically re-configurable. Our design objective is to create components of clear compatibility to that of the industrial standards having predictable behavior along with the incorporation of existing design knowledge. The cells can be used in as a single instance or multiple instances. Further, we will focus on a generic dynamic reconfigurable amplifier cell with flexible topology and dimension called Generic Operational Amplifier (GOPA). The incentive of our work comes from recent development in the field of measurement and instrumentation. The digital programming of analog devices is carried out using range of algorithms from simple to evolutionary. Physical realization of the basic cells is carried out in 0.35 μm CMOS technology.

1 Introduction

Primary sensor and mixed signal electronics are indispensable components, which are subjected to strong manufacturing conditions and environmental influences like etching, doping concentration, ageing etc. Subsequently the system tends to drift from its normal functionality. In order to reduce the substantial impact of the undesirable conditions or influences, usual approaches employ trimming / calibration and careful matched layouts during production time (Analog Devices, 2004). The counter measures adapted helps to restore the functionality of the system. The so adapted conservative procedures are slow, static, expensive, and have to be performed repeatedly, if dynamic. More improved approaches adapt these established procedures by compensation techniques during actual working phase in the name of self diagnosis / self-calibration. Block level granular approach entitled Field Programmable Analog Array (FPAA) uses digitally programmable resistors for compensation which are interconnected to the whole amplifier blocks (Fruehauf et al., 2002). The field programmable array structures of the commercially available Anadigm chip best suites the category. The chip allows interconnection of passive elements with the amplifier blocks to application specific in time-discrete domain (Anadigm Inc., 2003). More recent approaches comes from the field of evolutionary electronics (Zebulum et al., 2002), where the circuit synthesis are carried out through learning procedures by a suitable flexible transistor level granular hardware structure called Field Programmable Transistor Arrays (FPTA) (Langeheine et al., 2003). Substantial drawbacks here are the fact that optimisation algorithm should invent the knowledge of an experienced designer, e.g, designing a sensor signal amplifier, uses excessive switching resources resulting in large internal memory configuration. These structures also exhibit characteristics of a black box, where the underlying behaviour remains unforeseeable. Here the point of preservation is that the quality parameters essential for automatic control comes from a special flexible structure of hardware, especially dynamic reconfiguration. Our research work, focuses on developing a suitable, flexible and dynamically re-configurable hardware structure called Field Programmable medium-granular Mixed-signal

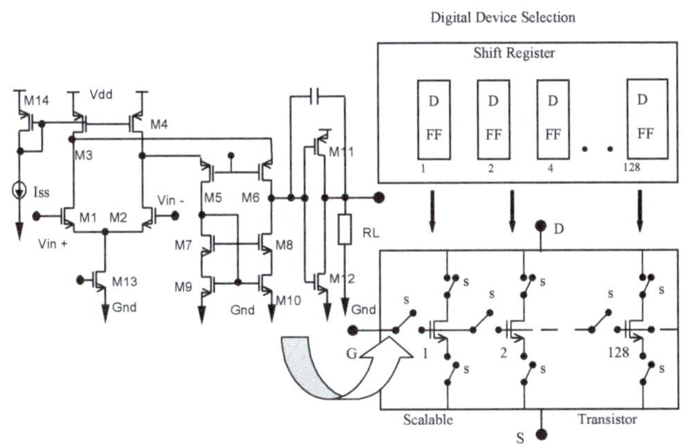

Fig. 1. Folded Cascode amplifier and the replacement scheme of ordinary by the heterogeneous array / scalable transistors (Lakshmanan and Koenig, 2004).

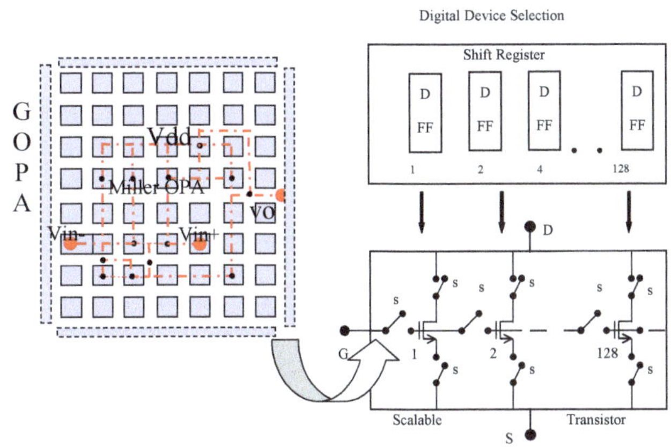

Fig. 2. Schematic representation of GOPA realizing miller operational amplifier.

Array (FPMA) providing a platform for the implementation of algorithms ranging from a simple to the more recent organic computing, paving the way in realising the favourable behaviour or characteristics of living organisms. The favourable characteristics for our interest include self-monitoring, self-testing, self-healing, self-repair, which are generally referred to as Self x characteristics. The implementation of the work is carried out in 0.35μm CMOS technology of Austriamicrosystems provided by EUROPRACTICE.

2 Re-configurable operational amplifier

The approach pursued by our work combines the concepts presented in the previous section with the objective of building flexible hardware structure suitable for sensor electronics. In our work, we focused on the design and implementation of sensor signal amplifier reducing the fine grained homogenous FPTA structure to a suitable medium grained heterogeneous structure thereby reducing the switching resources and consequently the parasitic effects of the switches. The approach also reduces the on chip storage requirement and decreases the reconfiguration time compared to that of its counterpart namely FPTA. The choice of the dimensions of the basic active and passive elements of the heterogeneous array are determined by simulations for the chosen technology (Lakshmanan and Koenig, 2004) and from the inspiration taken from A/D – D/A converters.

The dimensions of the transistors vary in the powers of two that are interconnected through switches. The selection of the appropriate dimension for the switches plays a vital role as it has subsequent impact on the overall system performance. The selection of active or passive device from the array are carried out by turning on the switches connected to their corresponding terminals. The switching patterns are fed sequentially through shift registers. Figure 1 shows the schematic of the time-continuous sensor signal amplifier with folded cascode topology using ordinary transistors and their corresponding replacement scheme by the scalable transistors / heterogeneous array.

3 Generic operational amplifier - GOPA

In the previous section, realisation of a time continuous sensor signal amplifier was carried out. In this section we focus on building a generic amplifier block which is flexible to both dimensions and topology. GOPA consists of array of scalable active and passive devices.

The next level hierarchical switches called Topology Switches (TS) interconnect the heterogeneous arrays. By switching ON and OFF these topology switches interconnecting the scalable array, various established amplifier topologies are realized ranging from a simple Miller to Telescopic operational amplifier. The flexibility is not only restricted to two stages but are also extended to three stage amplifier structures with programmable compensation capacitors and nullifying resistors. Some possible building structures of any operational amplifier topology that are realized by GOPA are simple and cascoded current mirrors, Wilson current mirror in the tail current region. Diode or current mirror loaded differential pairs and cascoded output stages. Realization of single ended and differentially ended output is also realized. The GOPA structure is capable of realizing more than 15 established simulation verified amplifier structures providing complete flexibility both in the choice of the structure and in dimensioning the devices constituting the structures. Figure 2 represents GOPA structure realizing Miller operational amplifier.

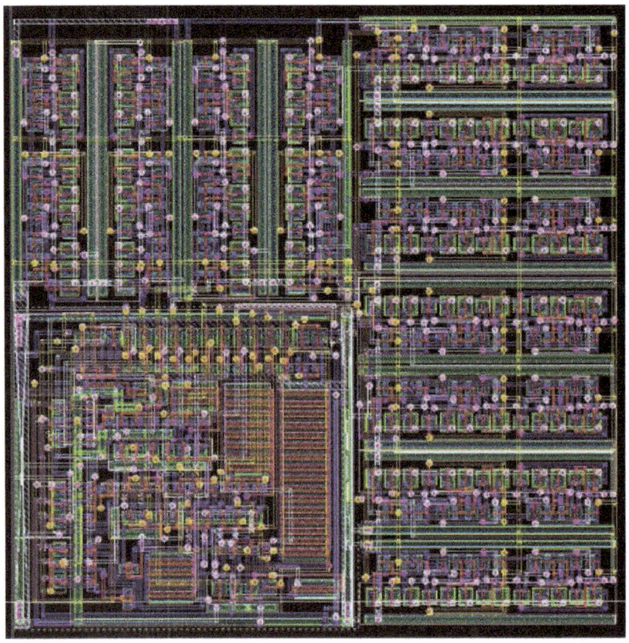

Fig. 3. Layout of NMOS scalable transistors.

4 Implementation

After promising results from the schematic level simulations of the pursued concepts in the previous sections, are thereafter implemented for a sensor signal amplifier. Fig.3 shows the layout of the NMOS scalable transistor array along with the digital interface surrounding the. heterogeneous array. The layout area of the NMOS scalable transistor is $60\,\mu\text{m} \cdot 64\,\mu\text{m}$. The flexibility does not come for free. A compromising additional layout area is consumed. The scalable transistors form the basic building blocks of FPMA. The array consists of 11 transistors interconnected by transmission switches. The minimum sized transistors are replicated 4 times in order to avoid the effects of mismatching. The remaining 7 transistors vary in the powers of two. Therefore, a total of 11 switch signal are needed for each scalable transistor array. The selection of the appropriate transistors from the array is performed by feeding in the bit patterns to the shift registers, which is a sequential process. The concept of dimensioning or programming is not refrained to active devices but also extended to passive devices like resistors and capacitors.

The passive element array plays a vital role in compensation and in feedback circuitry. Unlike the active devices, the passive array consists of 8 re-configurable bits ranging from 125 fF to 16 pF for the compensation capacitance and from 125 Ω to 16 KΩ for the scalable resistors. The heterogeneous array of passive elements, both resistors and capacitors are built in similar fashion like the transistor array through interconnection of the switches. The selection process of the devices remains identical. Combination of all scalable active

Fig. 4. Layout of complete GOPA realizing one possible OpAmp topology namely folded cascode.

and passive devices constitutes GOPA. Figure 4 shows the physical implementation of the sensor signal amplifier realized from GOPA constructed with the scalable devices.

The layout consists a total of 45 scalable devices and 59 topological switches. The performance curve of the folded cascode operational amplifier is shown in Figs. 5 and 6 for its gain and phase margin respectively. Different curves in the graph shown in Figs. 5 and 6 refers to simulation results obtained initially with ordinary transistors and later by replacing with the scalable version. In this work, simulations were carried out in the so called extrinsic fashion. A considerable degree of behavior comparability is to be expected with the manufactured chip of the dynamic re-configurable folded cascode operational amplifier performed later with intrinsic simulations. With regard to practical applications, in addition to normal pins of the OpAmp, data (D) pin, enable (EN) pin, clock (clk) pin are essential. The enable signal ensures that the digital section does not interact with operation phase of the amplifier after suitable device selections are performed. This ensures additional separation between the different domains of analog and digital is not necessary. The objective of this chip comes after studying the need and advancement in the field of measurement and instrumentation. Our design objective is to create cells of clear compatibility to that of the industrial standards with expectable behavior along with incorporation of existing design knowledge. The chip can be used in as a single or multiple instances. The prototype of the chip is under preparation.

5 Conclusion

In this paper, we have realized the implementation of the time continuous dynamically re-configurable sensor signal amplifier that is flexible in both topology and in sizing of the active and passive devices with drift compensation capabilities, compatible to that of the industrial standards. The flexibility does not come for free. A compromising additional layout area is consumed. The chip under preparation will allow implementation of algorithms ranging from a simple to Organic / Evolutionary, exhibiting the desirable self-x char-

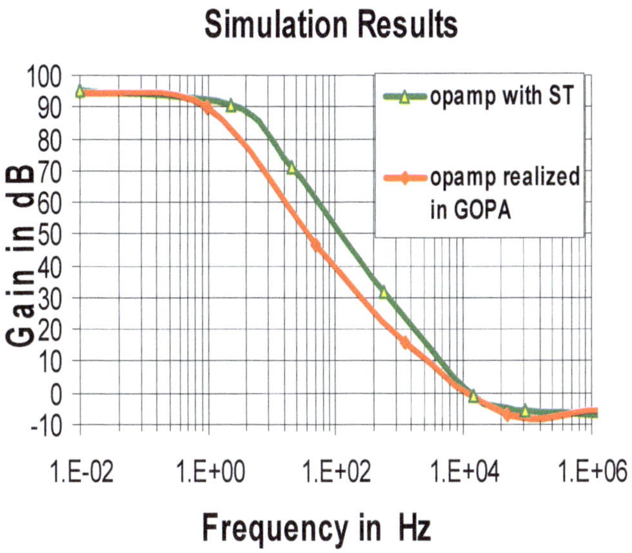

Fig. 5. Folded-cascode OPA-gain.

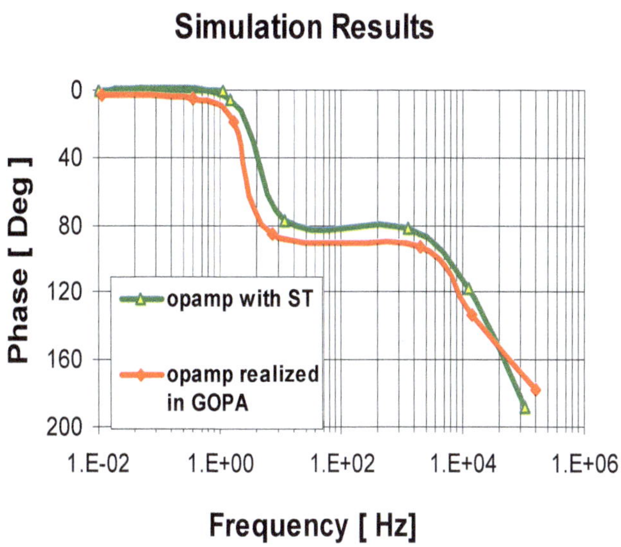

Fig. 6. Folded-cascode OPA-phase.

acteristics of biological organisms. In future works, the developed chip will be subjected to reconfigure in a mixtrinsic fashion. Result confirmation through measurements. Optimization of the area through much compact layouts. As a long-term research, adaptation could be extended to the sensor in the optimization loop.

References

Anadigm Inc,: AN221R02 – Dynamically Reconfigurable FPAA with Enhance I/O, Datasheet Entry Level, 2003.

Analog Devices: AD8555 Zero-Drift, Single Supply Sensor Signal Amplifier, with Digitally Programmable Gain and Offset, Preliminary Techn. Data, 2004.

Fruehauf, U., Kranz, E. G., and Leuterer, H.: Auto Correction and Design-for-Testability in Embedded Measurement System, Proc. Of 12th IMEKA TC4 Int. Symp., Zagreb, Croatia, 25–27 Sept., 79–82, 2002.

Lakshmanan, S. K. and Koenig, A..: Reconfigurable Amplifier Circuits for Adaptive Sensor Systems Employing Bio-Inspiration, Proc. Of the 4th Int. Conf. on Hybrid Intelligent Systems, IEEE CS, Kitakyushu, Japan, 38–43, 2004.

Langeheine, J., Meier, K., and Schemmel, J.: Evolution of Analog Electronic Circuits using a CMOS FPTA Chip, Proc. Of the 5^{th}Conf. On Evolutionary Methods for Design Optimization and Comtrol with Application to Industrial and Societal Problems, (EUROGRN 2003), 87–88, 2003.

Zebulum, R. S., Pacheco, M. A. C., and Vellasco, M. M. B. R.: Evolutionary Electronics – Automatic Design of Electronic Circuits and Systems by Generic Algorithms, CRC press, 2002.

5

Hardware Accelerators for Elliptic Curve Cryptography

C. Puttmann[1], J. Shokrollahi[2], M. Porrmann[1], and U. Rückert[1]

[1]Heinz Nixdorf Institute, University of Paderborn, Germany
[2]Chair for System Security, Ruhr-University of Bochum, Germany

Abstract. In this paper we explore different hardware accelerators for cryptography based on elliptic curves. Furthermore, we present a hierarchical multiprocessor system-on-chip (MPSoC) platform that can be used for fast integration and evaluation of novel hardware accelerators. In respect of two application scenarios the hardware accelerators are coupled at different hierarchy levels of the MPSoC platform. The whole system is implemented in a state of the art 65 nm standard cell technology. Moreover, an FPGA-based rapid prototyping system for fast system verification is presented. Finally, a metric to analyze the resource efficiency by means of chip area, execution time and energy consumption is introduced.

1 Introduction

Data security is an important requirement for many applications in our daily life. Especially internet applications such as e-commerce need to transmit secret data via insecure communication channels. Therefore, various cryptographical methods exist that allow to protect sensitive data. Asymmetric cryptography, which is also known as public-key cryptography, does not only provide algorithms for encryption and decryption of data, but also for digital signatures and authentication. Recently asymmetric cryptography based on elliptic curves is gaining interest. Compared to traditional asymmetric techniques, e.g. the RSA algorithm, the elliptic curve cryptography (ECC) achieves an equivalent level of security with smaller key sizes. Using elliptic curve cryptography therefore results in memory as well as bandwidth savings.

Nevertheless, computational intensive operations emerge during the processing of ECC protocols. The scalar multiplication on elliptic curves represents a frequently required and complex operation. The acceleration of this task by dedicated hardware units can not only speed up the execution time, but also help to save energy. In this paper, we propose different types of hardware accelerators for scalar multiplication in respect of two application scenarios. On one hand, we consider mobile devices with limited power resources such as smartcards. On the other hand, security servers are considered, where high performance is more important than power consumption and costs in terms of chip area.

Our work presents various hardware accelerators for elliptic curve cryptography, which are evaluated by integration into a multiprocessor system-on-chip (MPSoC). This proposed evaluation platform is scalable and can be adapted to suit different application scenarios. Dedicated hardware blocks can easily be integrated at different hierarchy levels of the MPSoC, which allows a comfortable evaluation of novel accelerators. Furthermore, the whole system is mapped onto an FPGA-based rapid prototyping environment in order to speed up simulation and verification.

The proposed hardware accelerators are synthesized in a modern 65 nm standard cell technology and analyzed regarding their resource efficiency in terms of chip area, execution speed and power consumption. Furthermore, a metric is introduced to compare the resource efficiency of different implementations in respect of the considered application scenario.

The paper is structured as follows. Section 2 gives a brief overview of the mathematical background of elliptic curve cryptography. In Sect. 3 the evaluation platform that is used to integrate and verify the hardware accelerators is discussed. The proposed hardware accelerators as well as the metric for analyzing their resource efficiency are explained in detail in Section 4. Finally, Sect. 5 concludes the paper.

2 Elliptic Curve Arithmetic

In this section the elliptic curve cryptography based on binary field arithmetic is introduced. The general equation for a non-supersingular elliptic curve E over the binary finite field $\mathbb{F}_{2^m}$ is given by equation:

$$E : y^2 + xy = x^3 + ax^2 + b \qquad (1)$$

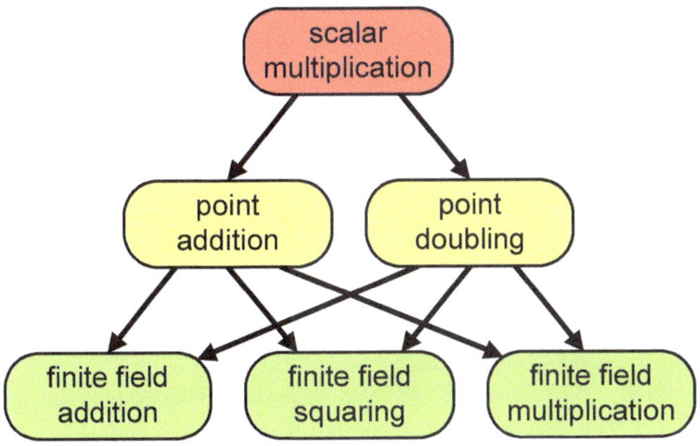

Fig. 1. Arithmetic hierarchy of elliptic curve cryptography.

for appropriate parameters $a, b \in \mathbb{F}_{2^m}$. The set of points $(x, y) \in \mathbb{F}_{2^m} \times \mathbb{F}_{2^m}$, which satisfy Eq. (1), together with the identity element $\mathcal{O}$, generate an additive abelian group. Let $\mathcal{P}=(x_p, y_p)$ and $\mathcal{Q}=(x_q, y_q)$ be two given points on the curve in Eq. (1). The point addition $\mathcal{P}+\mathcal{Q}$ as well as the point doubling $\mathcal{P}+\mathcal{P}$ are two operations defined on the elliptic curve E, which can geometrically be represented by the tangent and chord operation, respectively. By applying the point addition and point doubling operations, we are able to multiply an integer k with a point $\mathcal{P}$, which is the result of $k-1$ times adding the point $\mathcal{P}$ to itself. This operation is known as scalar or point multiplication $k \times \mathcal{P}$. Figure 1 depicts this hierarchical structure of arithmetic operations used for elliptic curve cryptography over finite fields.

2.1 Projective Coordinates

Naturally, point addition and point doubling also require a field inversion when using affine coordinates (x, y). Since inversion is a very expensive operation compared to multiplication, addition and squaring in finite fields, we use projective coordinates. In standard projective coordinates the points on the elliptic curve are represented as a triple (X, Y, Z) in such a way that $x \rightarrow X/Z$ and $y \rightarrow Y/Z$. By using projective coordinates only one finite field inversion is required at the end of a scalar multiplication in order to transform the projective coordinates back to affine coordinates.

2.2 Montgomery Ladder Algorithm

In Montgomery (1987) a very efficient method to perform the scalar multiplication is presented, which was applied to elliptic curve cryptography by López and Dahab (1999). The method is known as montgomery ladder and is shown at point level in Algorithm 1. Since in every loop iteration the same operations are performed, namely one point addition and one point doubling, the montgomery ladder algorithm is shielded against timing attacks and simple power analysis attacks.

Algorithm 1 The Montgomery ladder algorithm for scalar multiplication expressed at point level.

Input: A point $\mathcal{P}$ on the elliptic curve E, together with the binary representation of the scalar k as $(k_{i-1}k_{i-2}\dots k_1k_0)_2$.
Output: $k \cdot \mathcal{P}$

```
P1 ← P, P2 ← 2P
for j from i − 2 downto 0 do
    if kj = 1 then
        P1 ← P1 + P2, P2 ← 2P2
    else
        P2 ← P1 + P2, P1 ← 2P1
    end if
end for
```

2.3 Polynomial Basis Representation

As illustrated in Fig. 1, finite field arithmetic represents the base operations. Therefore, an efficient representation of the finite field elements in $\mathbb{F}_{2^m}$ is important. The polynomial basis representation can be described as a vector space of dimension m over the field $\mathbb{F}_2$ and is one of the most common representations in ECC. Field elements in polynomial basis representation are expressed as binary polynomials of degree $m-1$ as follows:

$$a(x) = \sum_{i=0}^{m-1} a_i \times x^i \quad \text{with} \quad a_i \in \{0, 1\} \tag{2}$$

One benefit of binary fields is that the finite field addition is calculated by a carry-free XOR operation of corresponding coefficients. Finite field squaring can be achieved by shifting each bit a_i to a_{2i} and filling the gaps with zeros. Similar to finite field multiplication, the result is a binary polynomial of degree $2m-2$, which has to be reduced modulo an irreducible, sparse polynomial of degree m. However, the finite field multiplication is equivalent to the product of the corresponding polynomials, which is, compared to addition and squaring, the most computational intensive operation.

2.4 Karatsuba Multiplication Method

In order to reduce the complexity of polynomial multiplication, the method of Karatsuba and Ofman (1963) is applied. Whereas "classically" the coefficients of the product $(a_1x+a_0)(b_1x+b_0)=a_1b_1x^2+(a_1b_0 \oplus a_0b_1)x+a_0b_0$ from the four input coefficients a_1, a_0, b_1, and b_0 are computed with 4 multiplications and 1 addition, the Karatsuba formula uses only 3 multiplications and 4 additions in binary fields:

$$\begin{aligned}(a_1x + a_0)(b_1x + b_0) &= \\ a_1b_1x^2 + ((a_1 \oplus a_0)(b_1 \oplus b_0) &\oplus a_1b_1 \oplus a_0b_0)x + a_0b_0.\end{aligned} \tag{3}$$

By applying the Karatsuba method to larger polynomials the costs of extra additions vanish compared to those of the

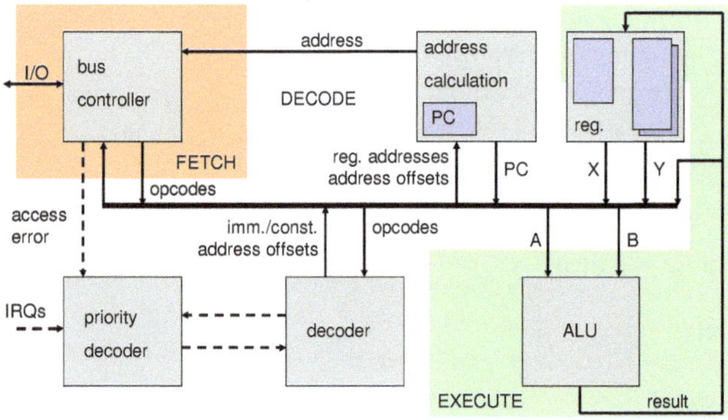

Fig. 2. N-Core processor architecture providing a three-stage pipeline.

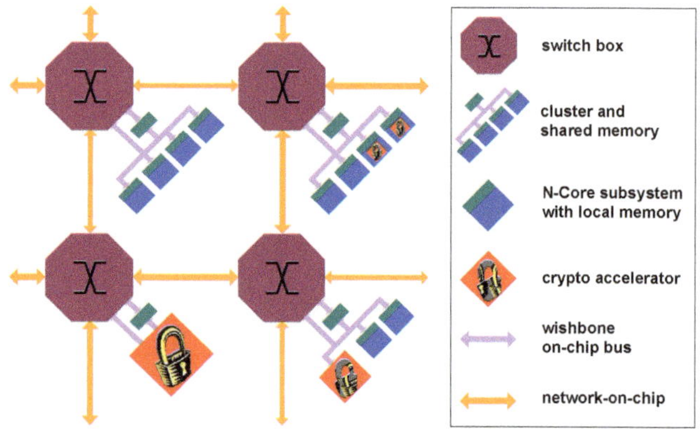

Fig. 3. Integration of hardware accelerators at different levels of our MPSoC-based evaluation platform.

saved multiplications and an asymptotical cost of $O(m^{1.59})$ compared to the classical cost of $O(m^2)$ is achieved (Hankerson et al., 2004).

3 Evaluation Platform

In Niemann et al. (2007) we presented the development of a scalable and hierarchical multiprocessor system-on-chip (MPSoC) architecture. This system architecture is used here as an evaluation platform for the integration of hardware accelerators.

3.1 N-Core Processor Element

The core processing element is our 32-bit RISC microprocessor named N-Core (Niemann et al., 2007). The N-Core comprises a common load-store architecture that is depicted in Fig. 2. The processor provides a three-stage pipeline, which performs instruction *fetch*, *decode* and *execute*. Two independent banks of sixteen 32-bit registers allow fast program context switching, e.g. for fast interrupt handling. Although the N-Core provides a 32-bit data bus, instructions have a fixed width of 16 bit, which delivers a high code density. Due to 11% free opcode space, instruction set extensions can be added to the microprocessor.

3.2 Hierarchical MPSoC Architecture

The processor core is extended with local data and instruction memory, a programmable interrupt controller (PIC) and a timer module to form a *N-Core subsystem*. The N-Core subsystem represents the lowest level of hierarchy within the MPSoC system. At the next higher level, a parametrizable amount of N-Core subsystems are connected via an on-chip Wishbone bus. Together with a shared memory this multiprocessor arrangement forms a *cluster*. At cluster level the processor elements can send messages directly to each other via the Wishbone bus as well as exchange data via the common shared memory. In this way, applications can be parallelized and processed cooperatively by the N-Core subsystems. At the top level of hierarchy, several clusters are connected through switch boxes over the network-on-chip (NoC). The NoC communication structure is described in Puttmann et al. (2007) and features packet-switched wormhole routing with up to 42 GBit/s data throughput.

3.3 Coupling of Hardware Accelerators

Depending on the requirements of the application scenario, hardware accelerators can be coupled at different levels of hierarchy, resulting in several resource utilization variants. As depicted in the top right cluster of Fig. 3, hardware accelerators in terms of instruction set extensions can be added to the N-Core subsystem. This integration means a very close coupling to the processor with little hardware overhead. In the bottom right cluster of Fig. 3, the accelerator is connected to the local on-chip bus of the cluster. In this way, each processor of the corresponding cluster can access the hardware accelerator. Moreover, the direct coupling to a switch box (see bottom left cluster in Fig. 3) results in high data throughput due to the close connection to the network-on-chip. Here, every processor of the whole system can access the hardware accelerator. In order to easily integrate the same hardware unit at different levels, we developed wrapper modules that automatically provide an interface for the desired coupling (Niemann et al., 2007).

3.4 Rapid Prototyping Environment

Depending on the software program under test, the simulation of a complex multiprocessor system-on-chip can take very long. Our approach to speed up the system verification is based on fast hardware emulation instead of comparable slow software simulation. For this purpose, we have developed an FPGA-based rapid prototyping environment called RAPTOR2000 (Kalte et al., 2002). As depicted in Fig. 4, the prototyping system consists of the RAPTOR2000

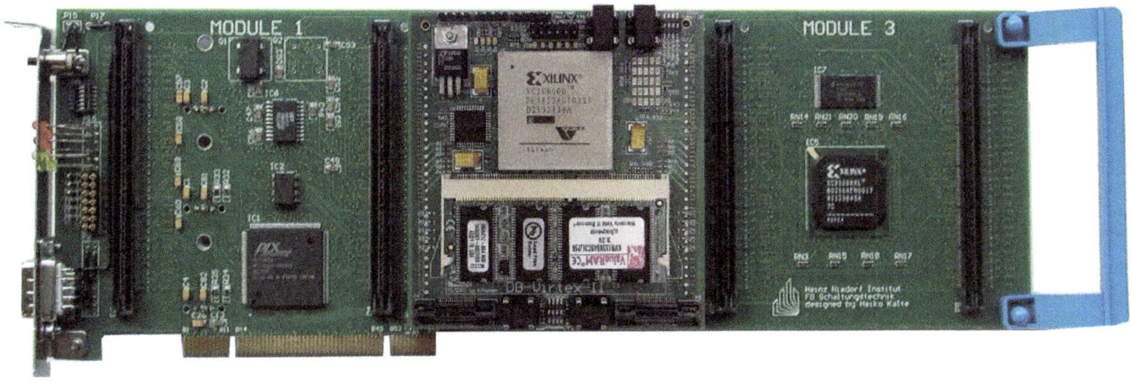

Fig. 4. Fast emulation of new hardware accelerators by using our FPGA-based rapid prototyping system RAPTOR2000.

Table 1. Synthesis results for a 65 nm standard cell technology and a Xilinx Virtex-II FPGA implementation, respectively.

	ASIC		FPGA	
	area	*power*	*utilization*	
	[mm²]	[mW]	[slices]	[RAM16]
N-Core processor	0.078	17.20	3206	0
N-Core subsystem[1]	0.613	45.16	3662	16
Cluster[2]	3.443	357.14	15362	80
Switch box	0.449	315.12	14133	0

[1] N-Core processor, 32 KB local memory, interrupt controller, timer module
[2] 4 N-Core subsystems, 32 KB shared memory, on-chip wishbone bus

motherboard, which can be equipped with up to six application specific daughterboard modules. Based on FPGA daughterboard modules, RAPTOR2000 is able to emulate circuits with a complexity of more than 200 million transistor gates. The host computer can communicate via a PCI bus interface with the RAPTOR2000 board and each attached daughterboard, respectively. Additionally, we have developed daughterboard modules that provide various communication standards. For example, an ethernet daughterboard can be used to stimulate the design under test with real world network traffic.

The implementation result of the MPSoC core components are shown in Table 1. We mapped the system to a 65 nm technology for ASIC fabrication as well as to a Xilinx XC2V8000-4 FPGA for rapid prototyping. The ASIC implementation is based on a low power (regular threshold voltage) standard cell library under typical operating conditions. The N-Core subsystem runs at a frequency of 400 MHz, which is also the operating frequency of a cluster. However, the network-on-chip as well as the switch box achieves a maximum clock frequency of 718 MHz. Each XC2V8000-4 FPGA can host one cluster including 4 N-Core subsystems and a switch box, which runs at 12.5 MHz. The system can be easily scaled by using more than one FPGA daughterboard (Niemann et al., 2007).

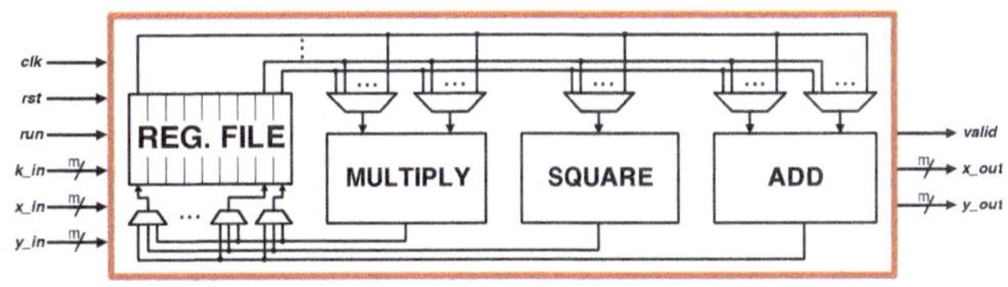

Fig. 5. Architecture of the high performance hardware accelerator for scalar multiplication.

4 Hardware Accelerators

In this section, various hardware accelerators for elliptic curve cryptography are presented. As explained in Sect. 2, we concentrate on the scalar multiplication in binary fields. Throughout the rest of this paper, we refer to the binary field $\mathbb{F}_{2^{233}}$, which is one of the recommended extension fields by the National Institute of Standards and Technology (2000). The hierarchy level for integration into the evaluation platform is chosen in respect to the application scenario. Furthermore, a metic for analyzing the resource efficiency is proposed and exemplarily applied to select the best hardware accelerator for a given application scenario.

4.1 High Performance Scenario

High performance hardware accelerators are required for systems such as internet servers, which have to handle many secure data transactions at one time, e.g. for online-banking applications. Fast computation and high data throughput are more important in this application scenario than power consumption and costs in terms of chip area.

We have developed a hardware accelerator for this application scenario, whose architecture is shown in Fig. 5. The hardware accelerator provides a modular structure comprising arithmetic units for finite field multiplication, squaring and addition (cf. Fig. 1). Addition as well as squaring is performed in one clock cycle, whereas multiplication

Table 2. Resource utilization for 65 nm standard cell technology at 200 MHz clock frequency and typical operating conditions.

	chip area		power consumption		execution time	
	absolute [μm²]	relative [%]	absolute [mW]	relative [%]	absolute [ms]	relative [%]
N-Core	57190	0.00	9.23	0.00	57.59	0.00
N-Core (ISE1)	57519	0.58	9.49	2.82	50.92	-11.58
N-Core (ISE2)	59290	3.67	7.35	-20.37	26.42	-54.12
N-Core (ISE3)	64734	13.19	12.00	30.01	20.19	-64.94

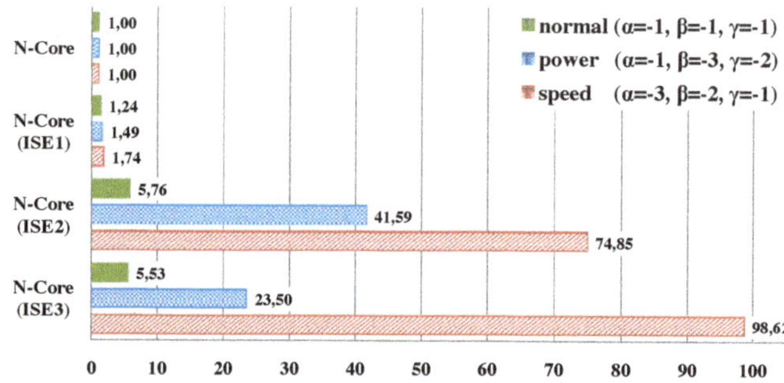

Fig. 6. Resource efficiency of different types of instruction set extensions.

based on the Karatsuba method requires 3 clock cycles in a pipelined fashion. Furthermore, the register file can store 7 finite field elements, which is sufficient for calculating the scalar multiplication. Due to the generic architecture, the hardware accelerator can easily be adapted to different field sizes. The scalar multiplication is calculated using the montgomery ladder algorithm as explained in Sect. 2. Moreover, the hardware accelerator handles coordinate transformation, i.e. affine coordinates are supported as input and output format, while internally projective coordinates are used for calculation.

The proposed hardware accelerator is coupled to a switch box port at the highest hierarchy level of the evaluation platform (cf. Fig. 3) for several reasons. At this level, the highest data throughput is achieved due to the close coupling to the network-on-chip. Furthermore, the hardware accelerator can operate autonomously without the need for an attached microprocessor.

Again, the hardware accelerator was synthesized using a 65 nm standard cell technology with typical operating conditions. A total chip area of 0.279 mm^2 is required to support the binary field $\mathbb{F}_{2^{233}}$. The hardware accelerator achieves a clock frequency of 625 MHz, i.e. the complete scalar multiplication is calculated in 7.2 μs and consumes 72.8 mW power. Respectively, 15365 slices are needed on a Xilinx XC2V8000-4 FPGA, which runs with 50 MHz clock frequency, resulting in 90 μs calculation time.

4.2 Low Power Scenario

In contrast to the high performance scenario, we now consider low power devices like smartcards. These are commonly used in mobile devices with limited energy resources and restricted chip area. Instead of high performance, the motivation for hardware accelerators here is to save energy with little increase of chip area. In order to reach this aim, we analyzed several types of instruction set extensions (ISE). Regarding the evaluation platform, instruction set extensions are directly implemented to the N-Core processor and hence coupled at the lowest hierarchy level. For a smartcard scenario, the evaluation platform is scaled down to a minimum of only one single N-Core comprising instruction set extensions.

In Puttmann et al. (2008), we present a two-stage framework for automatic instruction set extension. As a result, we implemented three different instruction set extensions. The first type of ISE combines already existing instructions, which are frequently used, to new *super instructions*. The analysis of the scalar multiplication showed that the instruction set of the N-Core processor can benefit from combinations of *shift* and *xor*. Therefore, we added three super instructions to the N-Core, namely `LSRIxor`, `LSLIxor` and `LSRIandadd`. This modification is in the following referred to as *ISE1*. Whereas ISE1 only combines existing instructions to new super instructions, the second type of instruction set extension (referred to as *ISE2*) introduces a new instruction, called `MULGF2`. This instruction multiplies two 32-bit binary field polynomials by using existing functions of the arithmetic logical unit (ALU). Multiplication with ISE2 is realized bit by bit utilizing the existing shift unit and xor function of the ALU. The third type of instruction set extension (*ISE3*) also supports this `MULGF2` instruction, but a dedicated hardware unit for binary field multiplication is implemented instead of using existing ALU functions.

Table 2 shows the absolute resource utilization as well as the relative changes compared to an original N-Core processor. The synthesis results are based on high threshold voltage standard cells, since we are considering a low power scenario. Therefore, the clock frequency of the N-Core is reduced to 200 MHz and only a chip area of 0.057 mm^2 is required. The execution time in Table 2 refers again to a scalar multiplication in $\mathbb{F}_{2^{233}}$.

4.3 Resource Efficiency

In order to select the best hardware accelerator for a given application scenario, a metric for resource efficiency is introduced in this section. The resource efficiency of the three variants of instruction set extensions is exemplarily analyzed. Therefore, we define resource efficiency (RE) as

$$RE = T^{\alpha} \times E^{\beta} \times A^{\gamma} \tag{4}$$

with *execution time* (T), *energy consumption* (E) and *chip area* (A). Each resource component can be weighted by an exponent (α, β, γ) to characterize the focus of the considered application scenario.

Figure 6 shows the resource efficiency of the three ISE variants (cf. Table 2), which are normalized to the resource efficiency of the original N-Core processor. We characterized three resource weightings, namely *normal*, *power* and *speed*. The exponents α, β and γ are all negative, so that a better utilization of any resource results in a better resource efficiency. The solid bars depict the resource efficiency with each component equally weighted, i.e. chip area, energy consumption and execution time are of the same importance. However, for the smartcard application scenario, we focus on low power implementations with little hardware overhead. Therefore, the checkered bars depict the resource efficiency, where energy consumption ($\beta = -3$) is more important than chip area ($\gamma = -2$), which in turn is more important than execution time ($\alpha = -1$). Therefore, ISE2 would be the best suited implementation for this application scenario, since it achieves the highest resource efficiency with emphasis on power. In contrast, the striped bars show the resource efficiency when focusing on speed. Consequently, the execution time ($\alpha = -3$) is here more important than the energy consumption ($\beta = -2$), which again is more important than the chip area ($\gamma = -1$). In this case, ISE3 represents the best suited hardware accelerator, because it achieves the highest resource efficiency with respect to execution speed.

5 Conclusions

In this paper we have presented various hardware accelerators for cryptography based on elliptic curves. The proposed scalable multiprocessor system-on-chip evaluation platform allows an easy integration of hardware accelerators at different levels of hierarchy. In order to speed up the verification of novel hardware accelerators, the evaluation system can be emulated by using our FPGA-based rapid prototyping environment RAPTOR2000. A high performance hardware accelerator was developed, which calculates a scalar multiplication in $\mathbb{F}_{2^{233}}$ in 7.2 μs. Moreover, different types of instruction set extensions were introduced that provide energy savings up to 63%. Finally, a metric for analyzing the resource efficiency was defined, which helps to select the best suitable hardware accelerator for a given application scenario.

Acknowledgements. The research described in this paper was funded in part by the Federal Ministry of Education and Research (Bundesministerium für Bildung und Forschung – BMBF), registered there under grant number 01AK065F (NGN-PlaNetS), and by the German Research Foundation (Deutsche Forschungsgemeinschaft – DFG) under project RU 477/8.

References

Hankerson, D., Menezes, A. J., and Vanstone, S. A.: Guide to Elliptic Curve Cryptography, Springer Professional Computing, Springer, Berlin, 2004.

Kalte, H., Porrmann, M., and Rückert, U.: A Prototyping Platform for Dynamically Reconfigurable System-on-Chip Designs, in: Proceedings of the IEEE Workshop Heterogeneous reconfigurable Systems-on-Chip (SoC), Hamburg, Germany, http://www.raptor2000.de, 2002.

Karatsuba, A. A. and Ofman, Y.: Multiplication of Multidigit Numbers on Automata, Soviet Physics Doklady, 7, 595-596, 1963.

López, J. and Dahab, R.: FastMultiplication on Elliptic Curves over GF(2m) without Precomputation, in: CHES 99: Proceedings of the First InternationalWorkshop on Cryptographic Hardware and Embedded Systems, Springer-Verlag, London, UK, 316-327, 1999.

Montgomery, P. L.: Speeding the Pollard and Elliptic Curve Methods of Factorization, Math. Comput., 48, 243-264, 1987.

National Institute of Standards and Technology (NIST): Digital Signature Standard (DSS), U.S. Department Of Commerce, chap. Recommended Elliptic Curves for Federal Government Use, FIPS 186-2, 24-48, 2000.

Niemann, J.-C., Puttmann, C., Porrmann, M., and Rückert, U.: Resource Efficiency of the GigaNetIC Chip Multiprocessor Architecture, J. Syst. Architect., special issue on Architectural premises for pervasive computing, 53, 285-299, 2007.

Puttmann, C., Niemann, J.-C., Porrmann, M., and Rückert, U.: GigaNoC - A Hierarchical Network-on-Chip for Scalable Chip-Multiprocessors, in: Proceedings of the 10th EUROMICRO Conference on Digital System Design (DSD), 29–31 August, Lübeck, Germany, IEEE Computer Society, 495–502, 2007.

Puttmann, C., Shokrollahi, J., and Porrmann, M.: Resource Efficiency of Instruction Set Extension for Elliptic Curve Cryptography, in: Proceedings of the 5th International Conference on Information Technology: New Generation (ITNG), 7–9 April, Las Vegas, Nevada, USA, IEEE Computer Society, 131–136, 2008.

6

Realtime multiprocessor for mobile ad hoc networks

T. Jungeblut, M. Grünewald, M. Porrmann, and U. Rückert

System and Circuit Technology, Heinz Nixdorf Institute, University of Paderborn, Germany

Abstract. This paper introduces a real-time Multiprocessor System-On-Chip (MPSoC) for low power wireless applications. The multiprocessor is based on eight 32bit RISC processors that are connected via an Network-On-Chip (NoC). The NoC follows a novel approach with guaranteed bandwidth to the application that meets hard realtime requirements. At a clock frequency of 100 MHz the total power consumption of the MPSoC that has been fabricated in 180 nm UMC standard cell technology is 772 mW.

Fig. 1. Transmission power control and directional communication.

1 Introduction

Portable electronic devices like PDAs, mobile phones and notebooks are increasingly equipped with wireless communication technologies, providing higher degrees of mobility and ease of use. Mobile ad hoc networks (MANETs) are a special type of wireless networks that do not require any infrastructure and whose topology can change spontaneously by the movement of participating nodes. To evaluate the performance and energy efficiency of new routing algorithms, especially including directional communication (Grünewald et al., 2005b) and transmission power control (Xu et al., 2005), we use the network simulator SAHNE (Volbert, 2002), see Fig. 1. This environment emulates the packet processing of each participating node. Simulations have shown that communicating in eigth directions can increase the total thoughput in a mobile ad hoc network by the factor of 2.5.

Efficient system components for medium access and routing are required for processing of the transmitted data packets. In our work, we evaluate Multiprocessor System-on-Chips (MPSoC) instead of single processors with higher clock frequency to achieve the required performance and resource efficiency. The MPSoC comes with an application specific "Network-on-Chip (NoC)" that offers a high bandwidth and guranteed latency (Grünewald et al., 2005a). The protocol functions of the application are automatically mapped to the processing elements via our software tool. The first MPSoC, which has been developed in our group, consists of eight processing elements (PEs), connected via the NoC. The communication with the peripherals is done via physical ports that are connected to air-to-air interfaces for the directional communication. Running a CSMA/CA-protocol, which we enhanced by the functionality of directional communication, the SoC can guarantee an external port data rate of 2.6 MBit/s and a total throughput of 21.6 Mbit/s on a system with eight physical ports, 16 processing elements and 16 switch boxes.

2 Architecture of the multiprocessor

With a high number of processing elements no common bus-based structures can be used. We use a homogeneous System-On-Chip (SoC) with a flexible NoC that is described in Sect. 2.1 in detail. The NoC connects the processing elements (PE) and the switch boxes (SB) to form the SoC. The physical ports (PP) are used for off-chip-communication. As our SoC is described in VHDL, generic and scalable, different architectures can be implemented. The number of the switch-box can be adopted to the needed number of PEs or NoC-links and through a generic uplink- and downlink interface different softcore processing elements can be evaluated. Additionally the interface of the physical ports can be adjusted to the requirements of the surrounding system. One possible architecture of the SoC is shown in Fig. 2.

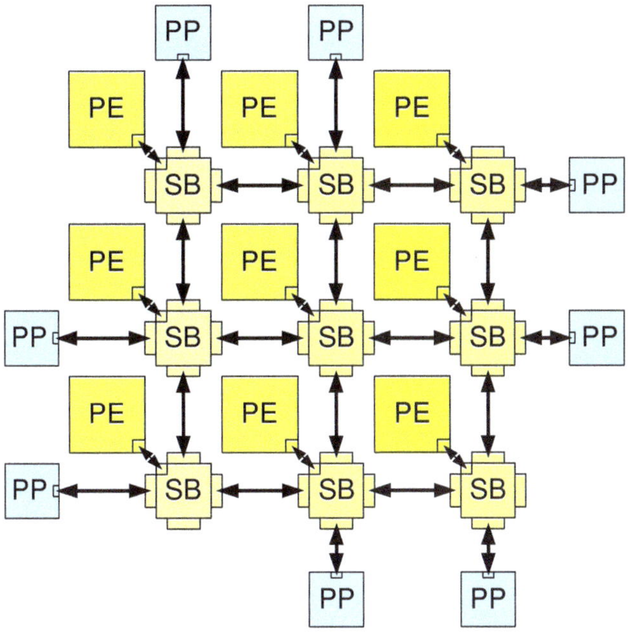

Fig. 2. Possible Architecture variant of the SoC.

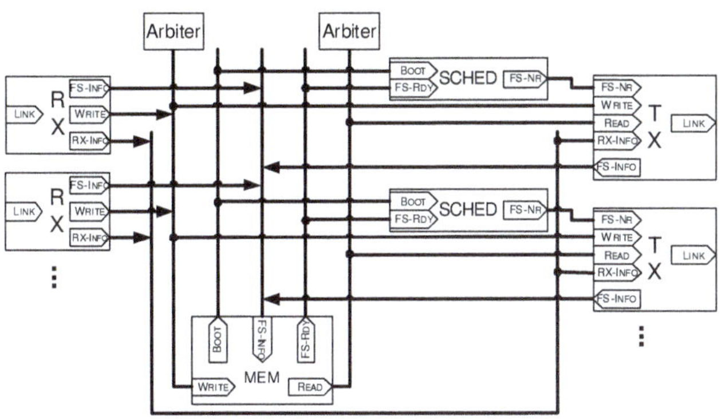

Fig. 3. Architecture of the switch box.

2.1 The Network-on-Chip

A specific feature of the multiprocessor is the predictability of its performance. Packet based communication is used instead of simple time multiplexing, enabling a high bandwidth utilization. In Grünewald et al. (2005a) methods have been proposed to assign the protocol functions to processors and to estimate the resource consumption of the final mapping. During the mapping either delay or energy consumption per packet can be minimized. An essential requirement for this methodology is that the upper bound for the latency of a packet can be calculated. As usual in NoCs, the packets are divided into data units of fixed size, called flits.

Figure 3 shows the switch box of the NoC. The SB consists of receivers (RX), transmitters (TX), schedulers (SCHED), and flit-memory. Receiver and transmitter write/read flits to/from a shared memory which is called shared memory switching. For latency reduction, the TX-, SCHED-, and RX-units are working in parallel while receiving a flit. With the given number of links of the SB (N_{SB}), the total number of execution cycles $EC_{\mathrm{SB,rx}}$ for one reception cycle is:

$$1 \leq EC_{\mathrm{SB,rx}} \leq N_{\mathrm{SB}} \qquad (1)$$

The transmit cycle starts as soon as the flow control detects an incoming flit, which has to be forwarded via the corresponding output-link. In worst case, i.e., if all transmitters detect a transmission request, the memory responds after N_{SB} cycles. With additional three clock cycles for receiving the flit from the scheduler, storing it in an internal temporary register and forwarding it to the output register, the total number of execution cycles $EC_{\mathrm{SB,tx}}$ is:

$$3 \leq EC_{\mathrm{SB,tx}} \leq N_{\mathrm{SB}} + 3 \qquad (2)$$

The first flits, the SBs receive during system start up, are boot-flits, which are used to initialize the routing tables. For communication with the surrounding system, the physical port of our MPSoC segments the flits to reduce the number of required I/O-pins. Figure 4 shows the structure of the segmented flits. The length l of a segmented flit is given by $l = q_s + q_{\mathrm{data}} + 2$, where q_s is the index of the associated flow segment, which represents a virtual connection, and q_{data} is the number of data bits.

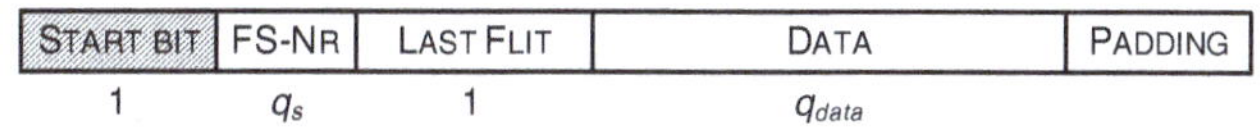

Fig. 4. Segmented flit for off-chip communication.

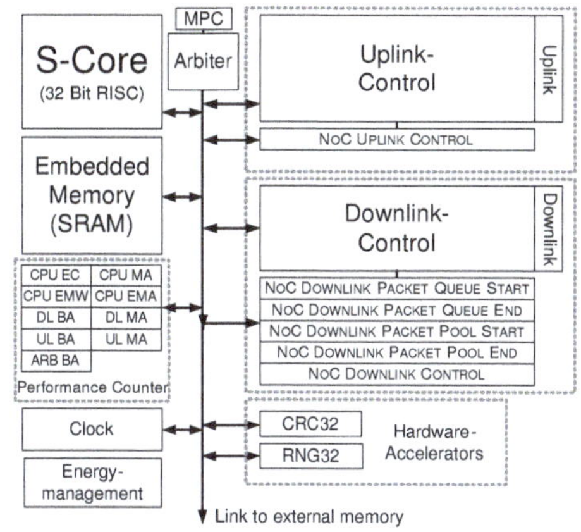

Fig. 5. Block diagram of the processing element.

2.2 The Processing Element

Figure 5 shows a block diagram of the processing element that is used in our MPSoC. Central component of the proposed architecture is the S-Core-processor (Langen et al., 2002), which has been developed in our group. S-Core is a 32 Bit-RISC-processor with a three-stage-pipeline, instruction-set compatible to the Motorola M-Core. 32 kB local static memory per PE can be used for instructions and data. Additionally, external memory can be accessed to execute memory intensive tasks. Furthermore, the PEs are equipped with CRC hardware accelerators, a timer, and a random numbers generator. Via uplink interfaces and downlink interfaces the PEs are connected to the Network-on-Chip. The number of execution cycles EC_{DL} and EC_{UL} for receiving and transmitting a flit is given by:

$$q_{\mathrm{data}}/32 + 4 \leq EC_{\mathrm{DL}} \leq q_{\mathrm{data}}/32 + 6 \tag{3}$$

and

$$q_{\mathrm{data}}/32 + 6 \leq EC_{\mathrm{UL}} \leq q_{\mathrm{data}}/32 + 8 \tag{4}$$

To determine the resource efficiency of the software and of the hardware implementation, our VHDL-based characterisation environment PERFMON is used (Grünewald et al., 2005a). PERFMON provides an infrastructure for simulation and evaluation of the whole system, including main memory, debugging units, and performance counters. Each softcore processing element can be used to substitute the currently used S-Core and to analyze its performance in the proposed multiprocessor environment. Because all parts of PERFMON are synthesizable and generic, the whole system can be mapped to a hardware technology like an ASIC or an FPGA for rapid prototyping. Target specific components are replaced automatically.

2.3 Prototyping environment

Initial implementations of the system have been mapped to FPGA architectures and have been tested in our rapid prototyping environment RAPTOR2000 (Kalte et al., 2002) (Fig. 7). This allows for fast simulation and verification in early design stages. As a proof-of-concept, the multiprocessor is integrated in the SAHNE simulation environment (Volbert, 2002), which is used to simulate the nodes of a mobile ad-hoc-network. Packet processing of one node is not simulated, but executed on the MPSoC architecture. The hardware is connected to the simulator using the hardware abstraction layer (HAL) of the packet processing library, which has been presented in Grünewald et al. (2005a). The HAL also ensures the synchronization of the hardware and the simulator. By this hardware/software co-simulation of large mobile ad-hoc-networks are possible.

2.4 The multiprocessor ASIC

After successful testing, the FPGA prototype is replaced by an fabricated ASIC. Because of the modular approach of the

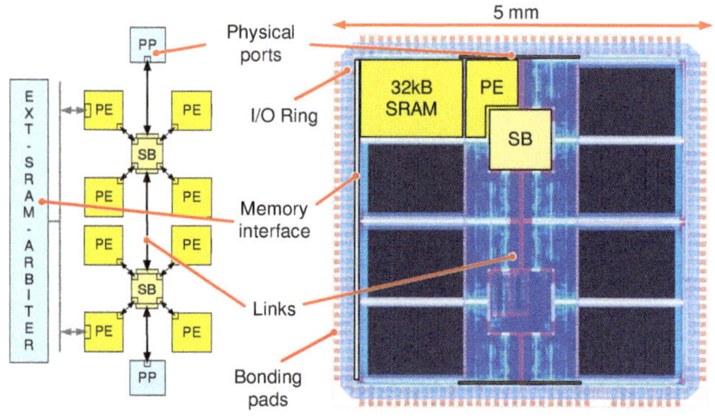

Fig. 6. Chip photo of the MPSoC.

RAPTOR2000 rapid prototyping system, the test environment can be reused as described in section 2.4.1. The hierarchical design of the ASIC shortens development time, because parts as the processing elements and the switch boxes have to be designed only once an then can be multiply instantiated. This is also an advantage of our SoCs as wire-length can be calculated better in advance and more aggressive signaling strategies can be used. In the fabricated multiprocessor (see Fig. 6) four processing enginges are connected to one switch box. Two of these processor clusters form the eight-core MPSoC. This architecture results from a design space exploration and simulation of different architectures as described in Sect. 1 and achieves a high ressource efficiency which is important for low power applications as mobile ad hoc networks. The proposed system has been manufactured in 180 nm UMC standard-cell-technology and occupies an area of 25 mm^2 using six metal layers. It embeds 2.1 MBit memory and consists of 1.6 million transistors. At a clock frequency of 100 MHz, the average power consumption is 772 mW. At this speed, a communication bandwidth of up to 2.1 Gbps is achieved for each link of the NoC. 4.2 Gbps throughput per switch box are achieved in total with all six links active which is a disadvantage of memory shared switching. The off-chip communication bandwidth via two physical ports is 500 MBit/s. A daughterboard for RAPTOR2000 has been developed, comprising the MPSoC, 4 MB external memory, and a Spartan XC3S1500 FPGA, integrating an interface to the RAPTOR2000 motherboard (see Fig. 7). The user can easily interact with the MPSoC, using the PCI bus interface of RAPTOR2000.

2.4.1 Testing the SoC

This prototyping environment is also intended to test the funcionality of new chip charges. On the host-system a monitor program controls the initialization of the processing elements and the incoming and outgoing traffic via the physical ports. Once the switch boxes are initialized, the memory images of each processing element is sent to the on-chip memory via

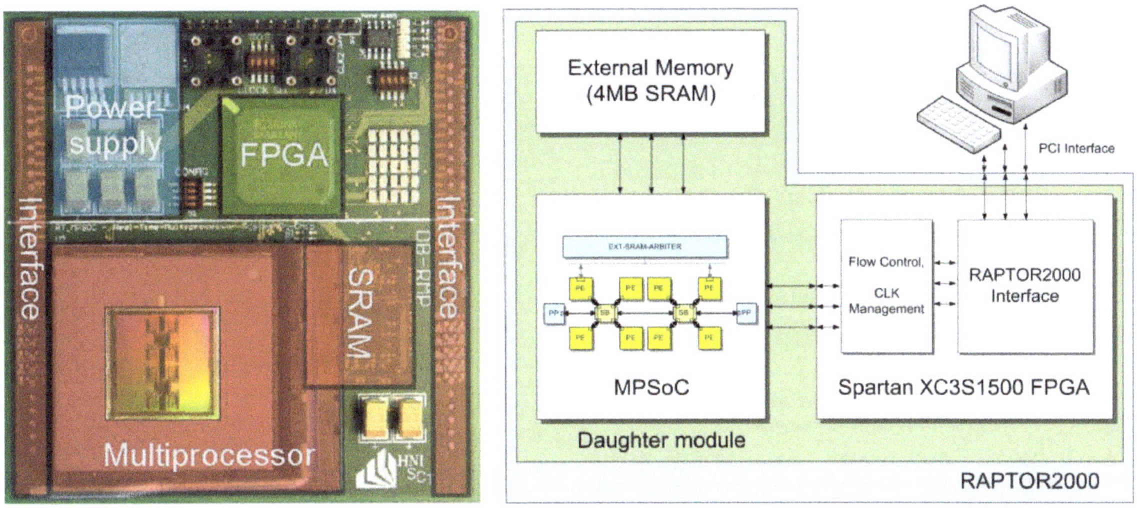

Fig. 7. Test-Environment for the MPSoC.

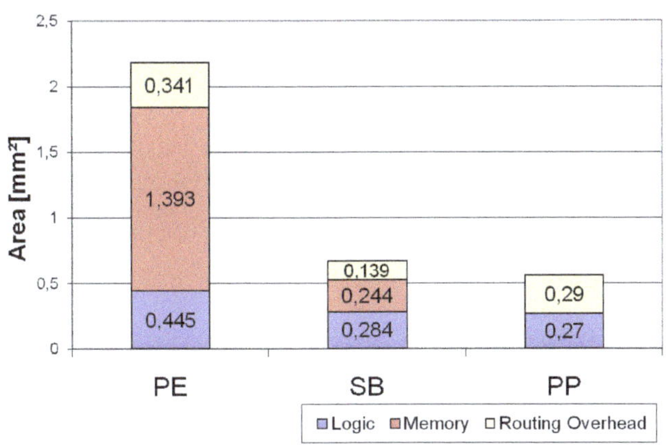

Fig. 8. Area requirements of the core components of the MPSoC.

Table 1. Power dissipation of the MPSoC [mW].

			8xPE, 2xSB, 2xPP	
	Highest Load	Idle	Highest Load	Idle
PE	74,254	35,864	594,032	286,912
SB	86,288	24,417	172,576	48,834
PP	2,6	1,451	5,2	2,902
		Total	771,808	338,648

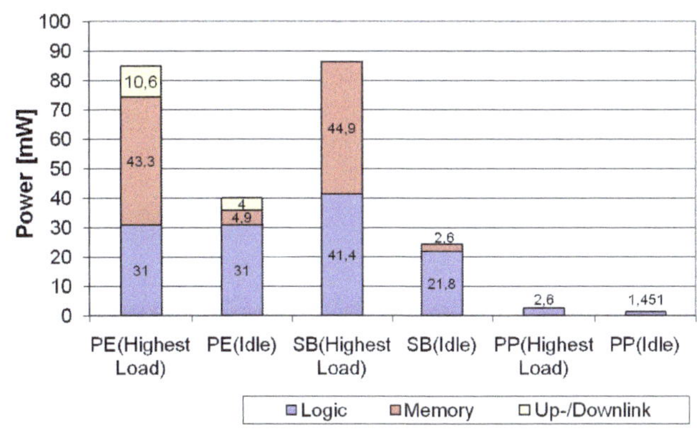

Fig. 9. Power dissipation of the core components of the MPSoC.

the NoC. We developed different test cases to test the functionality of each processing element, the on-chip memories, the switch-boxes, the interface to external memory, the NoC-links and the physical ports. To verify the correct behavior of the system, these test results can be automatically compared with those of the simulation and FPGA emulation. To determine the bottlenecks of the system or the distribution in the fabrication of the ASICs, we need to operate the components of the SoC at different clock frequencies. The interface on the additional FPGA enables the variation of the clock frequency during runtime. In this way we can operate the NoC at a specific speed while transmitting packets via the NoC and afterwards switching to a different frequency to determine the maximum performance of the processing engines.

2.5 Ressource consumption

Figure 8 shows the area consumption of the core components of the MPSoC. The largest part is the processing element ($2.18\,mm^2$), basically because of the large on-chip static memory ($1.38\,mm^2$=64%). The switch box and the physical port uses less than one third of the area of one PE.

As there are only two of each in contrast to eight PEs, their impact on the total area is insignificant in the realized architecture variant. Figure 9 shows the power dissipation for idle state and highest load. Because currently no power management is used in the processing elements, there is only an reduction of the power of the on-chip memory. With an intelligent power management, at different levels of hierarchy (Clock gating, gating of unused funtional blocks in the PEs, gating of entire PEs) energy could be saved. Table 1 shows the simulated power consumption of the entire SoC determined from Synopsys Power Compiler. The switching activities caused by the CSMA/CA protocol were annotated the get more accurate results. The measured power consuption power consumption of 470 mW is below the simulated value of 772 mW probably because of the worst case assumption of the Synopsys tools. As before, the highest impact on the total power consumption is caused by the eight processing elements.

3 Conclusions

In this work we have presented a generic, scalable architecture for Multiprocessor SoCs (MPSoC) intended for low power wireless applications as mobile ad hoc networks. Via generic uplink- and downlink interfaces, standard IP-cores can be used as processing elements. The NoC follows a novel approach to guarantee minimum bandwidth to the application to meet hard realtime requirements. An FPGA-based-prototype is used for fast hardware-software-co-simulation of mobile ad hoc networks. An eight-core MPSoC-chip prototype has been fabricated. The proposed ASIC has been manufactured in 180 nm UMC standard-cell-technology and occupies an area of 25 mm^2 at a power consumption of 772 mW.

References

Grünewald, M., Niemann, J.-C., Porrmann, M., and Rückert, U.: A framework for design space exploration of resource efficient network processing on multiprocessor SoCs, Morgan Kaufmann Publishers, 3(12), 245–277, 2005a.

Grünewald, M., Xu, F., and Rückert, U.: Increasing the Resource – Efficiency of the CSMA/CA Protocol in Directional Ad Hoc Networks, in: Proceedings of the 4th International Conference on AD-HOCNetworks & Wireless, Cancun, Mexico, 71–84, 2005b.

Kalte, H., Porrmann, M., and Rückert, U.: A Prototyping Platform for Dynamically Reconfigurable System on Chip Designs, in: Proc. of the IEEE Workshop Heterogeneous reconfigurable Systems on Chip (SoC), Hamburg, Germany, 2002.

Langen, D., Niemann, J.-C., Porrmann, M., Kalte, H., and Rückert, U.: Implementation of a RISC Processor Core for SoC Designs – FPGA Prototype vs. ASIC Implementation, in: Proceedings of the IEEE-Workshop: Heterogeneous reconfigurable Systems on Chip (SoC), Hamburg, Germany, 2002.

Volbert, K.: A Simulation Environment for Ad Hoc Networks Using Sector Subdivision, in: Proceedings of the 10th Euromicro Workshop on Parallel, Distributed and Network-based Processing, Canary Islands, Spain, 419–426, 2002.

Xu, F., Grünewald, M., and Rückert, U.: A Low Complexity Directional Scheme for Mobile Ad Hoc Networks, in: Proceedings of the 16th IEEE International Symposium on Personal Indoor and Mobile Radio Communications, 1349–1353, 2005.

7

Impact of process variations and long term degradation on 6T-SRAM cells

Th. Fischer[1], A. Olbrich[2], G. Georgakos[2], B. Lemaitre[2], and D. Schmitt-Landsiedel[1]

[1]Lehrstuhl für Technische Elektronik, Technische Universität München, München, Germany
[2]Infineon Technologies AG, München, Germany

Abstract. In modern deep-submicron CMOS technologies voltage scaling can not keep up with the scaling of the dimensions of transistors. Therefore the electrical fields inside the transistors are not constant anymore, while scaling down the device area. The rising electrical fields bring up reliability problems, such as hot carrier injection. Also other long term degradation mechanisms like Negative Bias Temperature Instability (NBTI) come into the focus of circuit design.

Along with process device parameter variations (threshold voltage, mobility), variations due to the degradation of devices form a big challenge for designers to build circuits that both yield high under the influence of process variations and remain functional with respect to long term device drift.

In this work we present the influence of long term degradation and process variations on the performance of SRAM core-cells and parametric yield of SRAM arrays. For different use cases we show the performance degradation depending on temperature and supply voltage.

1 Introduction

Increasing parametric variations due to the ongoing scaling of the semiconductor device dimensions determine more and more the circuit design in modern ultra deep-submicron (UDSM) CMOS technologies. Multiple variation mechanisms sum up to large deviations in the device characteristics.

On the one side, global variations introduced by the process lithographical width and length variation already decease the device performance, on the other side the statistical dopant distribution under the gate area and process induced local variations like line edge roughness lead to local mismatch of neighboring devices, which further reduce the device performance in sub 100 nm technologies.

Additionally the circuit designer has to consider deviations in the device parameters due to long term variations that occur during the use of a product. Most of these long term effects happen under the influence of large electrical fields. Due to the scaling of the semiconductor devices and the under-proportional scaling of the supply voltage (ITRS , 2005) the electrical fields in UDSM technologies rise. Hence aging effects such as Hot Carrier degradation and Bias Temperature Stress (BTS) play an increasing role in CMOS circuit design.

SRAM arrays consume a large portion of the chip area in today's embedded systems, hence

The influence of long term degradation effects on the SRAM cell performance plays an important role in the design of embedded systems mainly due to two factors: a) SRAM arrays consume a large portion of the chip area in todays embedded systems and highly contribute to the system yield and b) SRAM devices use minimal feature sizes and therefore already drifts are accelerated.

2 Negative Bias Temperature Instability (NBTI)

A Bias Temperature Stress (BTS) is most prominent when a p-Type MOSFET is biased in inversion. This Negative Bias Temperature Instability (NBTI) results in a higher absolute threshold voltage $|V_{th}|$ of the stressed p-MOSFET thus making the transistor harder to turn on. While the effect is not understood in full detail, it is widely accepted that positive charges are trapped near the Si/SiO_2 interface and induced interface states build up under high V_{gs} and high temperature (Ogawa, 1995). The stress induced damage is distributed homogeneously over the whole active area and the degradation on the threshold voltage is worst for a drain-source voltage of V_{ds}=0 while it shows lower degradation with higher V_{ds} (Schlünder, 2005) . The stress conditions are shown in Fig. 1.

There is also an Biased Temperature Stress effect on positive biased n-type MOSFETs, but is shows only a minor change in the device characteristics.

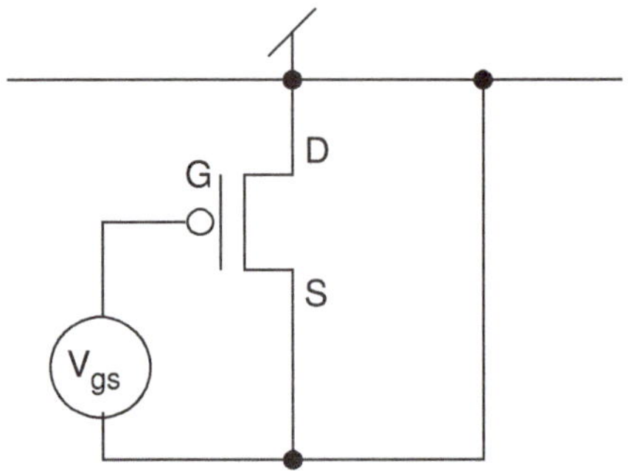

Fig. 1. Bias conditions for the Negative Bias Temperature Stress.

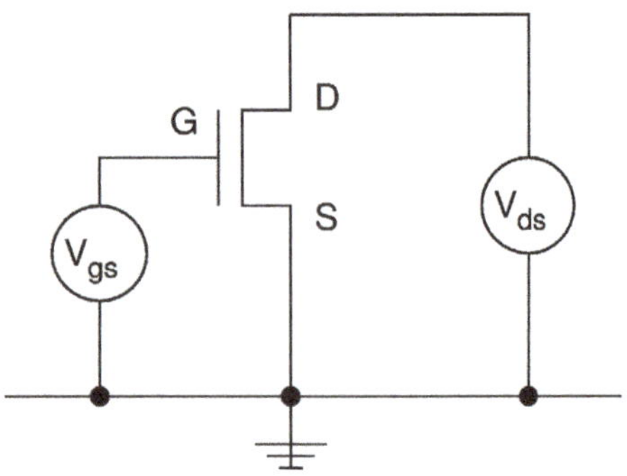

Fig. 2. Bias conditions for the Hot Carrier Stress.

The V_{th} shift due to NBTS partially recovers after device stress and AC operation has some annealing effects (Alam, 2003).

In UDSM technologies the degradation due to NBTI is becoming more critical due to the higher electrical fields across the gate resulting from the continuos size scaling without the proper voltage scaling (ITRS , 2005).

3 Hot Carrier Stress

Hot Carrier Stress is a long known hot electron induced degradation mechanism in CMOS semiconductors (Hu, 1985). Hot electrons coming from the high lateral electrical field cause electron-hole pairs by impact ionization. They can cause stress induced interface states and decrease the drain current I_d. The generated electrons can surmount the gate oxide barrier and cause the generation of trapped charges resulting in an increased threshold voltage V_{th}. The worst case operating conditions for Hot Carrier Stress are at drain-source voltages V_{ds}>0 and a gate-source voltage of $V_{gs} \simeq V_{ds}/2$ for n-type MOSFETs and $V_{gs} \simeq V_{ds}/5$ for p-type MOSFETs (Bravaix, 1999). Figure 2 shows the stress conditions for HCS.

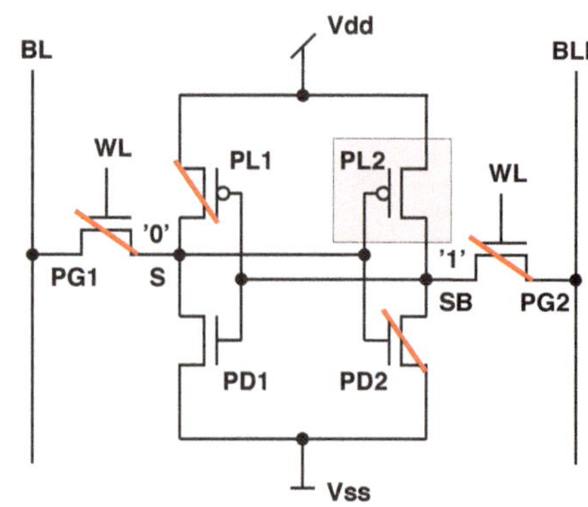

Fig. 3. SRAM core cell in hold operation. Transistor PL2 is subject to NBTS.

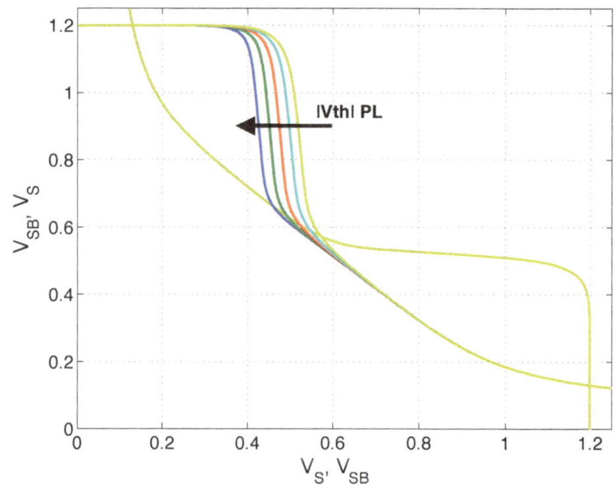

Fig. 4. Butterfly-curve for an SRAM core cell with activated wordline (read stability). The arrow shows the increase in $|V_{th}|$ of the p-type load transistor PL.

4 SRAM core cell stability degradation

An important key parameter of SRAM core cell circuits is the read stability. It describes the ability to read the information stored inside a memory cell without destroying the information. Measuring the Static Noise Margin (SNM) (Seevinck, 1987) on an SRAM cell with conducting access transistors gives a metric of the read stability.

When investigating the influence of device degradation on the performance of SRAM core cells the stress conditions of the different SRAM operating conditions have to be considered. In the following stress scenarios and the influence of the device degradation on the SNM performance of SRAM cells are investigated for data hold, read operation, and write operation.

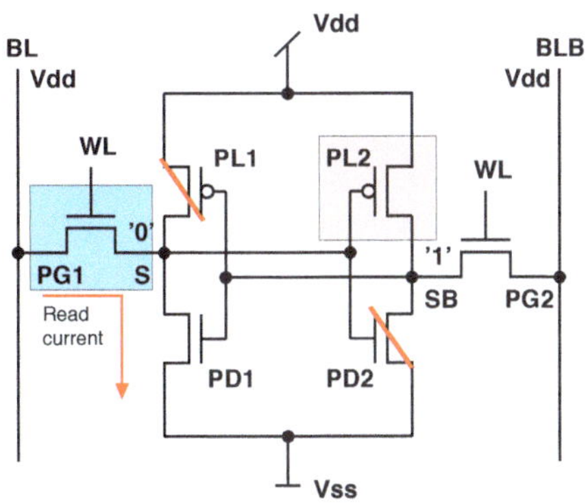

Fig. 5. SRAM core cell in read operation. Transistor PG1 is subject to HCS and Transistor PL2 to NBTS.

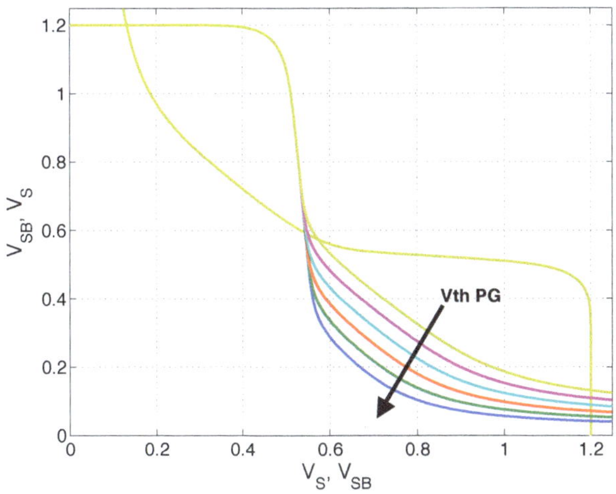

Fig. 6. Butterfly-curve for an SRAM core cell with activated word-line (read stability). The arrow shows the increase in V_{th} of the n-type access transistor PG1.

4.1 Data hold

In Fig. 3 the SRAM core cell is shown in hold operation. The two access transistors PG1 and PG2 are non-conducting and the information inside the memory cell is retained. The p-type load transistor PL2 on the "HIGH" side of the memory cell is exposed to a NBTS condition with a high field across the gate and no drain to source voltage. Whereas the p-type load transistor PG1 is non-conducting and sees no stress. In this operation mode of the core cell no Hot Carrier Stress can be expected due to the static nature of the hold state.

The NBTS on the PL2 transistor increases the threshold voltage $|V_{th}|$ of the p-MOSFET. This shifts the switching level of the inverter on the BLB side of the inverter to lower input voltages, thus reducing the eye opening of the butterfly curve that defines the read stability of the SRAM circuit (Fig. 4)

The stress to the p-type load transistor occurs whenever a

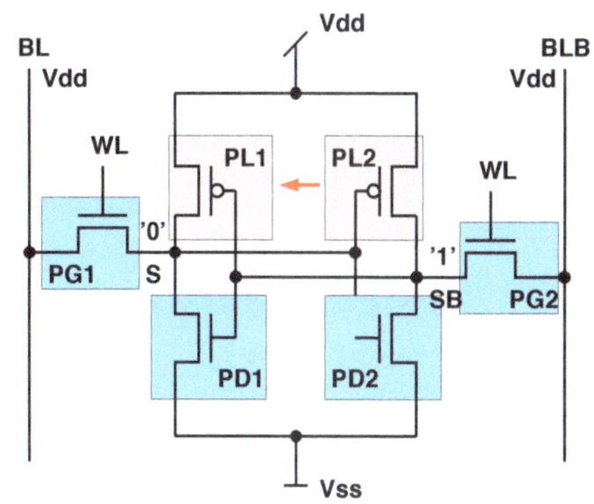

Fig. 7. SRAM core cell in write operation. The n-type transistors are subject to HCS, and NBTS changes from PL2 to PL1.

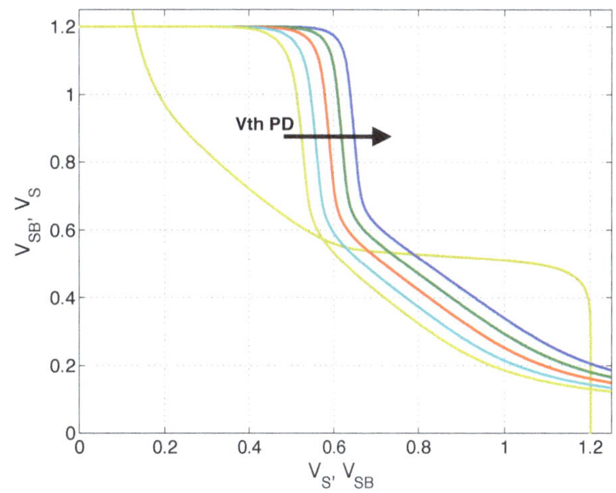

Fig. 8. Butterfly-curve for an SRAM core cell with open word-line (read stability). The arrow shows the increase in V_{th} of the n-type pull-down transistor PD.

"0" is stored on the storage node SB. In the worst case, when the data does not change, this stress is applied 100% of the operating time. The best case is at high switching activity, when the information is changed frequently so that the NBTS is present 50% of the time at each node and recovery and annealing effects also decrease the effect of the NBTS.

4.2 Data read

In the read operation the n-type word-line transistors are conducting and the two bit-lines, BL and BLB, are pre-charged to Vdd. A read current I_{read} flows through the transistors PG1 and PD1. Due to the read current and the high gate voltage the transistor PG1 is subject to HCS during the read operation. The aging of the access transistor PG1 increases the threshold voltage and decreases the drain current of the transistor. The p-type load device PL1 is still under NBTS,

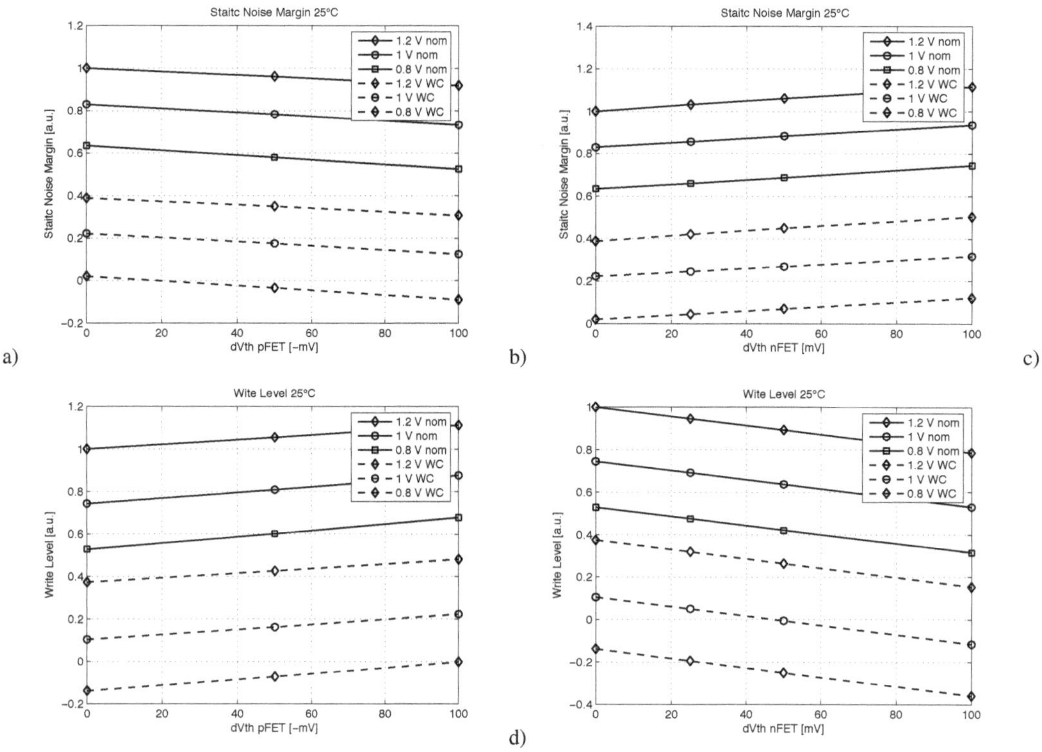

Fig. 9. Simulation results showing SRAM performance for nominal conditions (nom.) and worst case (WC) variation parameters for a 1MBit Array. **(a)** Static Noise Margin (read stability) vs. V_{th}-drift of p-FETs due to NBTS. **(b)** Static Noise Margin (read stability) vs. V_{th}-drift of n-FETs due to HCS. **(c)** Write Level (ability to write) vs. V_{th}-drift of p-FETs due to NBTS. **(d)** Write Level (ability to write) vs. V_{th}-drift of n-FETs due to HCS.

but sees a lower stress condition than in the hold state, due to the raised voltage level on node "S" (s. Fig. 5).

The influence of the threshold voltage increase of transistor PG1 on the read stability of the SRAM core cell is shown in Fig. 6. The voltage at node 'S' decreases due to the lower drain current and the eye of the butterfly curve opens.

A large number of read operation on one side of the SRAM cell stresses the access transistor, but influences the read stability in a positive way.

4.3 Data write

During the write operation the content of an SRAM core cell is changed. Therefore all transistors change their operating conditions and are stressed during the transition. The n-type access transistors and the pull-down devices are all subject to HCS due to the active current flowing in the switching process of the write operation. After the switching event the NBTS changes from transistor PL2 to PL1.

Figure 8 shows the influence of the threshold voltage increase of the transistor PD due to HCS. The increase in V_{th} changes the switching level of the core cell to higher input voltages. The voltage level defined by the voltage divider PG1/PD1 on the node "S" increases due to the weaker transistor PD1. This results in an smaller eye-opening of the butterfly-curve and a smaller read stability.

A large number of write operations stresses the core cell most due to the high number of switching events. This decreases the read stability.

5 Simulations

The impacts of the above described aging mechanisms on the performances of an SRAM array was simulated using a worst case distance simulation (Fischer, 2005) (Antreich, 1994). This simulation method allows to simulate the worst case performance for an SRAM array with a given array-size considering process variations and local mismatch. The simulations were performed for the Static Noise Margin (SNM) with conducting access transistors to obtain the read stability and for the Write Level that measures the ability to write an SRAM cell.

Figure 9a shows the degradation of the SNM on nominal SRAM core cells and on cells with worst case variations due to process and mismatch. The V_{th} shift of the threshold voltage of the p-type transistors results from NBTS. The read stability decreases with the V_{th} degradation. On the other hand

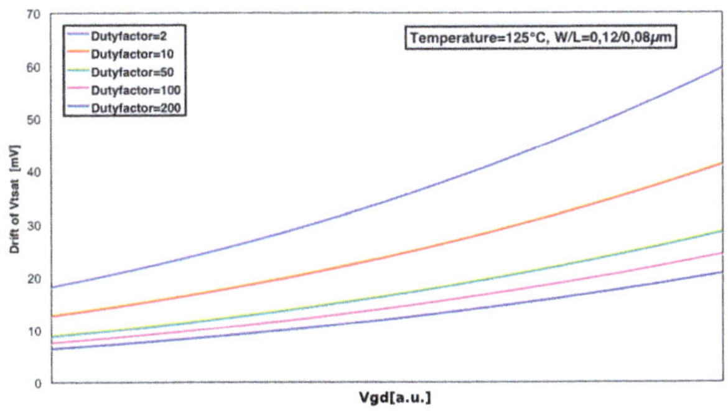

Fig. 10. The interrelation of stress voltage, duty factors, and the drift of NBTS induced threshold voltage.

the SNM increases with the a shift in the threshold voltage of the n-type transistors affected by HCS (Fig. 9b).

The ability to write, measured by the Write Level, shows the opposite behavior. When the core cell is subject to NBTS the Write Level rises with increasing V_{th} degradation and decreases with n-FET aging caused by the HCS (s. Figs. 9c and d).

It is therefore necessary to know the aging parameters and the worst case use scenarios to quantify the effect of NBTS and HCI on the SRAM performances. Figure 10 exemplary shows the interrelation of stress voltages and duty cycles and the drift of NBTS induces threshold voltage.

6 Yield considerations

The simulated Worst Case Distances, as a measure of the variation induced yield, show an increase of the total variation of 0.25 sigma for a 25 mV NBTS V_{th} shift of the p-type load transistors. The yield of an SRAM array is calculated using the simulated Worst Case Distance (WCD) and array size N (Fischer, 2005).

$$\text{Yield} = (\frac{1}{2}(1 + erf(\frac{WCD}{\sqrt{2}})))^N \quad (1)$$

Figure 11 shows the yield for a 256 k SRAM calculated form the specification of the read stability. A 25 mV NBTS induced threshold voltage shift of the p-type transistor lowers minimum supply voltage by 50 mV.

7 Conclusions

We have shown the qualitative influence of Hot Carrier Stress and Negative Bias Temperature Stress on the read stability of SRAM core cells for different operating conditions. The differing impact on the aging mechanisms depending on the operation mode of the SRAM cell needs a thorough analysis of the use case of the SRAM array. With the drift values for the individual use cases simulations show the quantitative impact

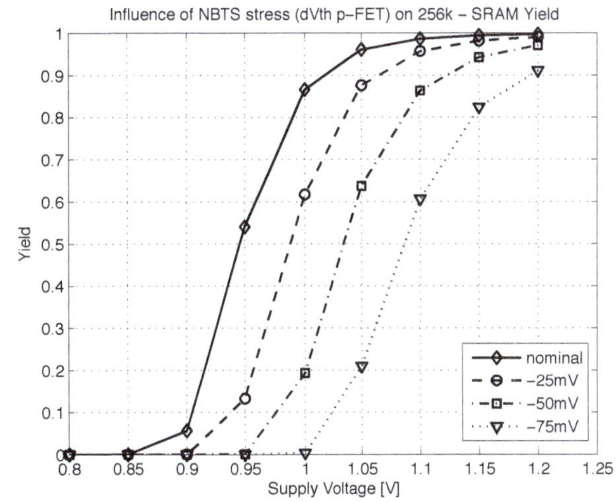

Fig. 11. The impact of NBTS induced V_{th} degradation on the read stability limited yield of a 256 k SRAM array.

on read and write performance for the total variation, including process variation, mismatch and stress degradation. The yield can be calculated and its influence on system specifications such as minimum supply voltage can be determined.

References

ITRS, International Technology Roadmap for Semiconductors 2005 Edition, http://public.itrs.net/, 2005.

Ogawa, S., Shimaya, M., and Shiono, N.: Interface-trap generation at ultrathin SiO_2 (4–6 mm)-Si Interfaces during negative-bias temperature aging, J. Appl. Phys., 77(3), 1137–1148, 1995.

Schlunder, C., Brederlow, R., Ankele, B., Lill, A., Goser, K., and Thewes, R.: On the Degradation of P-MOSFETs in Analog and RF Circuits under Inhomogeneous Negative Bias Temperature Stress, 41st Annual Intl. Reliability Physics Symposium 2003, p 5–10, 2005.

Alam, M. A.: A Critical Examination of the Mechanics of Dynamic NBTI for PMOSFETs, IEEE Intl. Electron Devices Meeting 2003, p 345, 2003.

Hu, C. M., Tam, S. C., HSU, F., Ko, P., Chan, T., and Terrill, K.: Hot-Electron-Induced MOSFET Degradation - Model, Monitor, and Imprpvement, IEEE Trans. Electron Devices, 32(2), 375–385, 1985.

Bravaix, A.: Hot Carrier Degradation Evolution in Deep Submicrometer CMOS Technologies, IEEE IRW, 174–183, 1999.

Seevinck, E., List, F., and Lohstroh, J.: Static Noise Margin Analysis of MOS SRAM Cells, IEEE J. Solid State Circuits, 22(5), 525–536, 1987.

Fischer, Th., Nirschl, T., Lemaitre, B., and Schmitt-Landsiedel, D.: Modelling of the parametric Yield in decnanometer SRAM-Arrays, Adv. Radio Sci., 4, 281–285, 2006, http://www.adv-radio-sci.net/4/281/2006/.

Antreich, K., Graeb, H., and Wieser, C.: Circuit Analysis and Optimization Driven by Worst Case Distances, IEEE Transactions on Computer-Aided Design of Integrated Circuits and Systems, 13(1), 57–71, 1994.

A low-power and high-quality implementation of the discrete cosine transformation

B. Heyne and J. Götze

University of Dortmund, Information Processing Lab, Otto-Hahn-Str. 4, 44227 Dortmund, Germany

Abstract. In this paper a computationally efficient and high-quality preserving DCT architecture is presented. It is obtained by optimizing the Loeffler DCT based on the Cordic algorithm. The computational complexity is reduced from 11 multiply and 29 add operations (Loeffler DCT) to 38 add and 16 shift operations (which is similar to the complexity of the binDCT). The experimental results show that the proposed DCT algorithm not only reduces the computational complexity significantly, but also retains the good transformation quality of the Loeffler DCT. Therefore, the proposed Cordic based Loeffler DCT is especially suited for low-power and high-quality CODECs in battery-based systems.

1 Introduction

Recently, many kinds of digital image processing and video compression techniques have been proposed in the literature, such as JPEG, Digital Watermark, MPEG and H.263 (Conzalez and Woods, 2001; Richardson, 2002). All the above standards require the Discrete Cosine Transform (DCT) Conzalez and Woods (2001) to aid image/video compression. Therefore, the DCT has become more and more important in today's image/video processing designs.

In the past few years, much research has been done on low power DCT designs (Li and Lu, 1996; Hsiao et al., 2005; August and Ha, 2004; Jeong et al., 2004; Shams et al., 2002; Fanucci and Saponara, 2002). One of the most popular ways to realize the fast DCT (FDCT) is to use the Flow-Graph Algorithm (FGA) for VLSI-implementation (Chen et al., 1977; Wang, 1984). Loeffler et al. (1989) has proposed a low-complexity FDCT/IDCT algorithm based on FGA that requires only 11 multiply and 29 add operations. However, the multiplications consume about 40% of the power and almost 45% of the total area (Sung et al., 2005). In this regard, Tran (2000) proposed the binDCT which approximates multiplications with add and shift operations. It only consumes about 38% of the power of the Loeffler DCT. However, it also looses about 3 dB in PSNR compared to the Loeffler DCT (Sung et al., 2005).

Jeong et al. (2004) has proposed a Cordic based implementation of the DCT. COordinate Rotation DIgital Computer (Cordic) is an algorithm that can be used for the evaluation of various functions in signal processing (Volder, 1959; Walther, 1971). In addition, the Cordic algorithm is highly suited for VLSI-implementation.

In this paper we propose a computationally efficient and high-quality Cordic based Loeffler DCT architecture, which is optimized by taking advantage of certain properties of the Cordic algorithm and its implementations (Goetze and Hekstra, 1995). It only requires 38 add and 16 shift operations. The resulting DCT algorithm not only reduces computational complexity significantly, but also retains the good transformation quality of the Loeffler DCT. Therefore, the presented Cordic based Loeffler DCT implementation is especially suited for low-power and high-quality CODECs.

This paper is organized as follows. Section 2 briefly introduces the algorithms of the DCT, Loeffler DCT and Cordic based DCT. In Sect. 3, we will present the proposed Cordic based Loeffer DCT algorithm. The experimental results are shown in Sect. 4, while Sect. 5 concludes this paper.

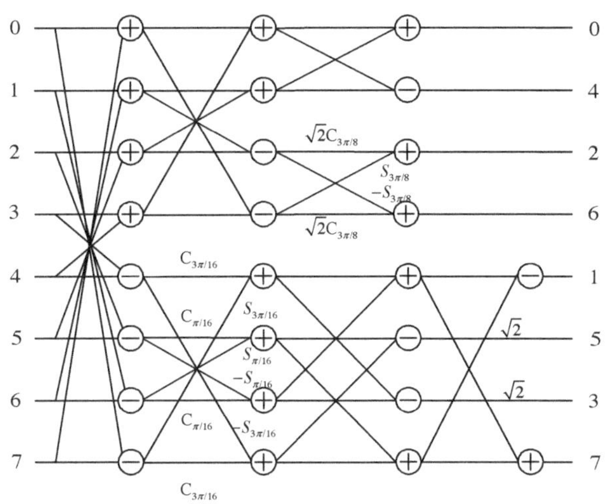

Fig. 1. Flow graph of an 8-point Loeffler DCT.

2 DCT algorithms

2.1 The DCT background

The two dimensional DCT in Eq. (1) transforms an 8×8 block sample from spatial domain $f(x, y)$ into frequency domain $F(k, l)$.

$$F(k,l) = \tfrac{1}{4}C(k)C(l)\sum_{x=0}^{7}\sum_{y=0}^{7} f(x,y) \cdot \cos[\tfrac{(2x+1)k\pi}{16}]\cos[\tfrac{(2y+1)l\pi}{16}] \quad (1)$$

$$C(m) = \begin{cases} \frac{1}{\sqrt{2}} & \text{if } m = 0 \\ 1 & \text{otherwise.} \end{cases}$$

Since computing the above 2-D DCT by using matrix multiplication requires 8^4 multiplications, a commonly used approach in hardware designs to reduce the computational complexity is row-column decomposition. The decomposition performs row-wise one-dimensional (1-D) transform followed by column-wise 1-D transform with intermediate transposition. An 8-point 1-D DCT can be expressed as follows:

$$F(k) = \tfrac{1}{2}C(k)\sum_{x=0}^{7} f(x)\cos[\tfrac{(2x+1)k\pi}{16}] \quad (2)$$

$$C(k) = \begin{cases} \frac{1}{\sqrt{2}} & \text{if k = 0} \\ 1 & \text{otherwise.} \end{cases}$$

This decomposition approach has two advantages. Firstly the number of operations is significantly reduced. Secondly, with regard to the implementation, the original 1-D DCT can be replaced easier by more efficient DCT algorithms.

2.2 MAC based Loeffler DCT

Many 1-D flow graph algorithms have been reported in the literature (Chen et al., 1977; Wang, 1984). The Loeffer 1-D 8-point DCT algorithm (Loeffler et al., 1989) requires 11 multiplications and 29 additions as shown in Table 3. The flow graph of the Loeffler DCT is illustrated in Fig. 1, with $C_x = \cos(x)$ and $S_x = \sin(x)$. One of its variations is adopted by the Independent JPEG Group (JPE, 1998) for their implementation of the popular JPEG image coding standard. Note that this factorization requires a uniform scaling factor of $\frac{1}{2\sqrt{2}}$ at the end of the flow graph to obtain the original DCT coefficients. In the 2-D transform this scaling factor becomes $\frac{1}{8}$ which can be easily implemented by a shift operation. Although the Loeffler DCT requires multipliers, which will result in larger power dissipation and area, it offers better rate distortion than the other approaches. Therefore it is especially useful for high-quality CODECs.

2.3 Cordic based DCT

The Loeffler DCT achieves good quality transformation results, but on the other hand it needs multiplications which are computationally intensive in both software and hardware implementation. In this regard, one of the popular ways to implement a fast multiplierless approximation of the DCT is using the Cordic algorithm (Jeong et al., 2004; Hsiao et al., 2005). The Cordic has a very regular structure suitable for VLSI design. Figure 2 shows the flow graph of an 8-point Cordic based DCT using six Cordics (Jeong et al., 2004), requiring 104 additions and 84 shift operations as shown in Table 3.

In order to realize a vector rotation for the Cordic algorithm, that is rotating a vector (x, y) by an angle θ, the circular rotation angle is described as

$$\theta = \sum_i \sigma_i \cdot \tan^{-1}(2^{-i}) \quad \text{with } \sigma_i = 1, -1. \quad (3)$$

Then, the vector rotation can be performed using the iterative equation given in (Mariatos et al., 1994; Jeong et al., 2004):

$$\begin{aligned} x_{i+1} &= x_i - \sigma_i \cdot y_i \cdot 2^{-i} \\ y_{i+1} &= y_i + \sigma_i \cdot x_i \cdot 2^{-i}. \end{aligned} \quad (4)$$

In Eq. (4), only shift and add operations are required in digital hardware. Next, the results of the rotation iterations need to be compensated (scaled) by a compensation factor s. This is also done using an iterative approach:

$$\begin{aligned} x_{i+1} &= x_i(1 + \gamma_i \cdot F_i) \\ y_{i+1} &= y_i(1 + \gamma_i \cdot F_i) \end{aligned}$$
$$\text{with } \prod_i (1 + \gamma_i \cdot F_i) \cong s \text{ and } \gamma_i = (0, 1, -1),\ F_i = 2^{-i}. \quad (5)$$

Table 1. Parameters for the DCT based on Jeong et al. (2004).

Angle	$\frac{\pi}{4}$	$\frac{3\pi}{8}$	$\frac{7\pi}{16}$	$\frac{3\pi}{16}$
Rotation iterations $[\sigma_i, i]$ according to Eq. (4)				
1	$-1, 0$	$-1, 2$	$+1, 0$	$-1, 1$
2	–	$-1, 3$	$+1, 1$	$-1, 3$
3	–	$-1, 6$	$+1, 3$	$-1, 10$
4	–	$-1, 7$	$+1, 10$	$-1, 14$
Compensation iterations $[1 + \gamma_i \cdot F_i]$ according to Eq. (5)				
1	$1-\frac{1}{4}$	$1+\frac{1}{32}$	$\frac{1}{2}+\frac{1}{8}$	$1-\frac{1}{8}$
2	$1-\frac{1}{16}$	$1+\frac{1}{128}$	$1+\frac{1}{256}$	$1+\frac{1}{64}$
3	$1+\frac{1}{256}$	$1+\frac{1}{1024}$	$1+\frac{1}{4096}$	$1+\frac{1}{1024}$
4	$1+\frac{1}{512}$	$1+\frac{1}{4096}$	–	$1+\frac{1}{4096}$
5	$1+\frac{1}{4096}$	–	–	–

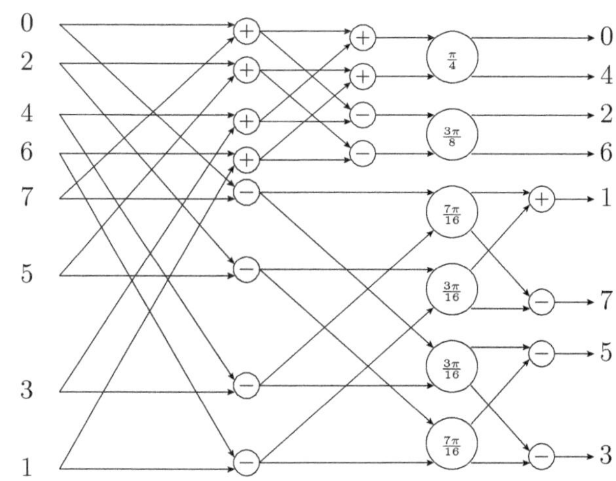

Fig. 2. Flow graph of an 8-point Cordic based DCT (Jeong et al., 2004).

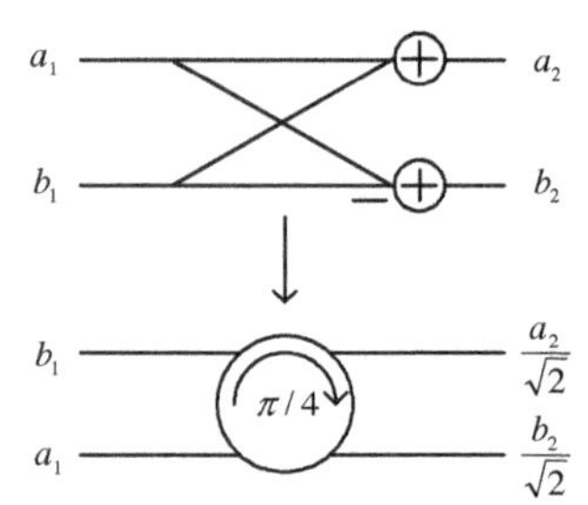

Fig. 3. Cordic rotation of the $\pi/4$ angle.

When using the Cordic to replace the multiplications of the 8-point DCT the angles θ_x are fixed. Therefore, we can skip some unnecessary iterations without losing accuracy. Table 1 shows the detailed number of iterations and compensations for the Cordic based algorithm (Jeong et al., 2004). Although the Cordic based DCT can reduce the number of computations in image/video compression, it still needs more operations than the binDCT does (Tran, 2000).

3 Cordic based Loeffler DCT

Based on our previous work about Cordic based FFTs (Heyne and Goetze, 2004; Heyne et al., 2004), we now propose an optimized Cordic based Loeffler DCT in this paper. This implementation requires only 38 add and 16 shift operations. We have taken the original Loeffler DCT as the starting point for our optimization, because the theoretical lower bound of the number of multiplications required for the 1-D 8-point DCT had been proven to be 11 (Duhamel and H'Mida, 1987).

In order to derive the proposed algorithm, we first consider the butterfly at the beginning of Loeffler's flow graph as shown in Fig. 1. In this case the butterfly can be expressed as:

$$\begin{bmatrix} a_2 \\ b_2 \end{bmatrix} = \begin{bmatrix} 1 & 1 \\ -1 & 1 \end{bmatrix} \cdot \begin{bmatrix} b_1 \\ a_1 \end{bmatrix} \tag{6}$$

The first matrix can then be decomposed to

$$\begin{bmatrix} 1 & 1 \\ -1 & 1 \end{bmatrix} = \sqrt{2} \cdot \begin{bmatrix} \frac{1}{\sqrt{2}} & \frac{1}{\sqrt{2}} \\ -\frac{1}{\sqrt{2}} & \frac{1}{\sqrt{2}} \end{bmatrix} \tag{7}$$

which equals a Cordic rotating the input values by $\pi/4$, followed by a scaling of $\sqrt{2}$ as shown in Fig. 3.

The scaled butterflies with scaling factors $3\pi/8$, $1\pi/16$ and $3\pi/16$ can also be replaced by Cordics using $\theta=3\pi/8$, $1\pi/16$ and $3\pi/16$ respectively. Hence, we can replace all butterflies in the Loeffler DCT to derive the pure Cordic based Loeffler DCT as shown in Fig. 4.

The most commonly used DCT-based CODECs for signal processing are usually followed by a quantizer. In this regard we can skip some Cordic iterations without losing visual quality, and shift the compensation steps to the quantization table without using additional hardware.

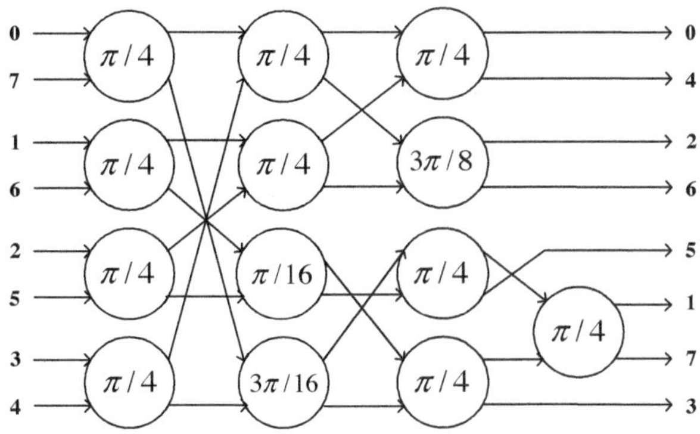

Fig. 4. Pure Cordic based Loeffler DCT architecture.

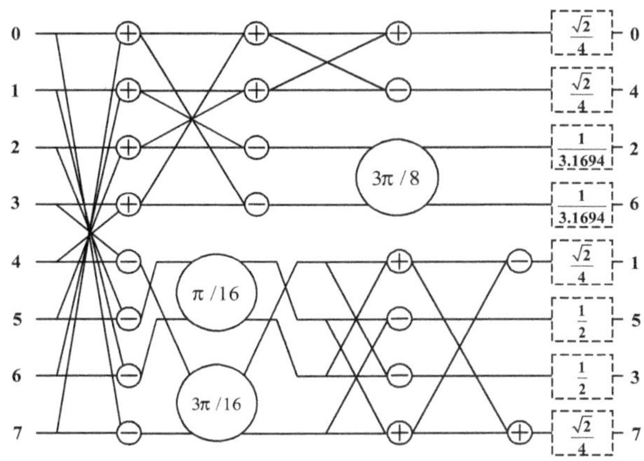

Fig. 5. Flow graph of an 8-point Cordic based Loeffler DCT architecture.

Next, we will start to optimize each rotation angle and reduce the computational complexity.

At first,due to the special structure of the Loeffler DCT, the scaling of $\sqrt{2}$ and the five needed compensation steps as shown in Table 1 can be performed at the end of the flow graph for the angle $\theta = \pi/4$. In other words, the $\pi/4$ rotation only needs two add operations to carry out the Cordic rotation.

Secondly, for the angle $\theta = 3\pi/8$ we reduce the number of iterations to three and also shift all compensation steps to the quantizer. Although the optimized $3\pi/8$ rotation will decrease the quality of the results, the influences are not noticeable in video sequence streams or image compression.

Thirdly, when we take a close look at the angle $\theta = \pi/16$, it can be easily observed that the needed compensation of the $\pi/16$ rotation is very close to one. Thus, we can ignore the compensation steps of the $\pi/16$ rotation. Therefore, it only needs two iterations in the Cordic calculation. Unfortunately we can not shift any compensation steps of the $3\pi/16$ rotation to the end of the graph, due to the data correlation

Table 2. Cordic based Loeffler DCT – Cordic parameters.

Angle	$\frac{\pi}{4}$	$\frac{3\pi}{8}$	$\frac{\pi}{16}$	$\frac{3\pi}{16}$
Rotation iterations $[\sigma_i, i]$ according to Eq. (4)				
1	$-1, 0$	$-1, 0$	$-1, 3$	$-1, 1$
2	-	$-1, 1$	$-1, 4$	$-1, 3$
3	-	$+1, 4$	-	-
Compensation iterations $[1 + \gamma_i \cdot F_i]$ according to Eq. (5)				
1	–	–	–	$1 - \frac{1}{8}$
2	–	–	–	$1 + \frac{1}{64}$

between the following stages of the $\pi/16$ and $3\pi/16$ rotations. However, we still can ignore some unnoticeable iterations and compensation steps to reduce the computational complexity of the angle $\theta = 3\pi/16$.

Table 2 shows the summary of Cordic iterations and compensation steps for the proposed Cordic based Loeffler DCT. In Fig. 5 the optimized flow graph is shown, including the scaling factors incorporated into the quantization table.

It only requires 38 add and 16 shift operations to realize the DCT transformation. In short, we try to ignore some unnoticeable iterations and shift the compensation steps of each angle to the quantizer to derive the optimized Cordic based Loeffler DCT. Moreover, the proposed DCT algorithm not only reduces the computational complexity significantly, but also keeps the high transformation quality as well as the original Loeffler DCT does.

Therefore, the proposed DCT algorithm has the same computational complexity as the binDCT, but as shown in the next Section it can perform as well as the Loeffler DCT in quality.

4 Experimental results

In our experiments we have used different criteria to evaluate four architectures: Loeffler DCT, Cordic based DCT, binDCT-C5 and Cordic based Loeffler DCT. Table 3 summarizes the number of operations of each DCT architecture. It can be easily observed from Table 3 that the proposed DCT reduces the computational complexity significantly compared to the original Loeffler DCT and Cordic based DCT.

Table 3. Complexity of different DCT architectures.

DCT type \ Operation	Mult	Add	Shift
Loeffler	11	29	0
Cordic [Jeong et al. (2004)]	0	104	82
Cordic Loeffler	0	38	16
binDCT-C5 [Tran (1999)]	0	36	17

In order to demonstrate the quality features of the proposed DCT algorithm, we have also applied it to the video coding standard MPEG 4, by using a publicly available XVID CODEC software (XIV, 2005). The DCT in the CODEC of the selected XVID implementation is based on Loeffler's factorization with floating-point multiplications. In this part we have applied each DCT algorithm to the XVID software, and simulated with some well-known video sequences to show the ability of the proposed approach. Figure 6 shows the average PSNR results of four DCT algorithms from low to high compression ratio (i.e. quantization steps from 1 to 10) with the "Foreman" video sequence.

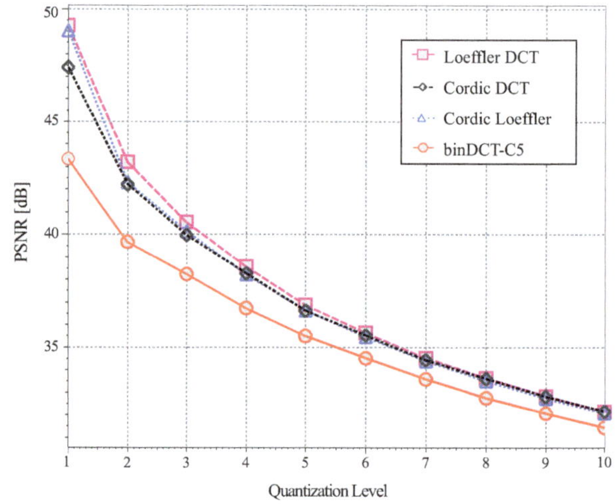

Fig. 6. The average PSNR of the "Foreman" sequence from low to high compression ratio (XVID CODEC).

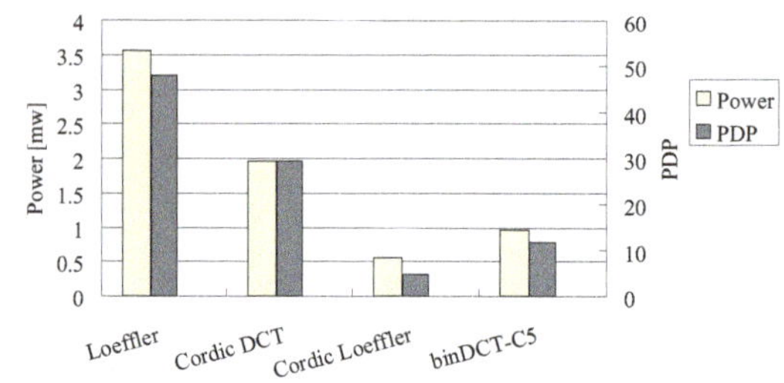

Fig. 7. Experimental results of the Power and PDP.

It can be seen from the PSNR simulation results, that the proposed algorithm performs as well as the Loeffler DCT does. Also, the average PSNR is about 2 dB higher than that of the binDCT-C5. In summary, the overall performance of the Cordic based Loeffler DCT is very similar to the Loeffler DCT in terms of video quality. However, it only requires the same computational complexity as the binDCT-C5 does.

To analyze the performance and the energy consumption of the proposed Cordic based Loeffler DCT, we have modeled the four different DCT architectures as RTL. After synthesizing with Synopsys Design Compiler, we have used Synopsys PrimePower to estimate the power consumption at gate-level.

The simulation results are shown in Table 4. Some important points can be observed easily. Firstly, the proposed DCT architecture only consumed 19% of the area and about 16% of the power of the original Loeffler DCT. Secondly, the proposed DCT architecture occupied the same area as the binDCT-C5. However, it has only about 59% of the power dissipation and half the delay time of the binDCT-C5.

To further display the low-power features of the proposed algorithm, we have analyzed the Power, Power-Delay Product (PDP), Energy-Delay Product (EDP) and Energy-Delay-Delay Product (EDDP). The PDP is the average energy consumed per DCT transformation. A lower PDP means that the power consumption is better translated into speed of each operation and the EDP represents that one can trade increased delay for lower energy of each operation. Finally, the EDDP represents whether or not it results in a voltage-invariant efficiency metric.

Figure 7 illustrates the experimental results of the power consumption and the PDP. As illustrated for the PDP the proposed DCT algorithm only consumes about 10% of the PDP of the Loeffler DCT and 16% of the PDP of the Cordic based DCT respectively. It also reduces the PDP by about 59% compared to the binDCT-C5. Figure 8 shows the EDP and the EDDP. As shown, the performance of the proposed DCT algorithm is far superior to the other DCT algorithms, especially in the EDDP.

Hence, the proposed DCT not only reduces the computational complexity significantly, but also achieves the best performance in all criteria.

5 Conclusions

In this paper a low complexity and high quality DCT transformation based on the Cordic algorithm is presented. The

Table 4. Power, Area and Time delay simulation results at gate-level.

Measures \ DCT Arch.	Loeffler	Cordic based (Jeong et al., 2004)	Cordic based Loeffler	binDCT-C5 (Tran, 1999)
Power (mW)	3.557	1.954	0.5616	0.9604
Area $(GateCount)$	15.06 K	6.66 K	2.81 K	2.83 K
Delay (ns)	13.49	15.08	8.37	12.17

TSMC 0.13-μm at 1.2 V without pipelining.

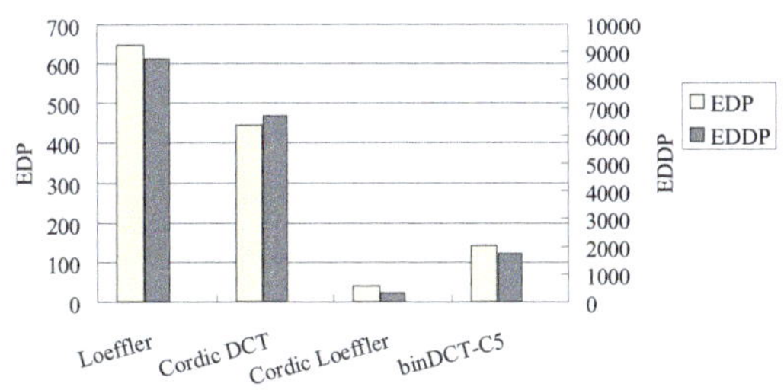

Fig. 8. Experimental results of the EDP and EDDP.

proposed Cordic based Loeffler DCT architecture only requires 38 add and 16 shift operations to carry out the DCT transformation, which is about the same complexity as the binDCT-C5's. The proposed algorithm not only reduces the computational complexity significantly compared to the original Loeffler DCT, it also keeps the good quality transformation result. In this regard, the proposed DCT algorithm is very suitable for low-power and high quality CODECs.

References

The JPEG-6b Website, http://www.ijg.org/, 1998.

The XVID Website, http://www.xvid.org/, 2005.

August, N. J. and Ha, D. S.: Low Power Design of DCT and IDCT for Low Bit Rate Video Codecs, IEEE Transactions on Multimedia, 6, 414–422, 2004.

Chen, W. H., Smith, C., and Fralick, S.: A Fast Computational Algorithm for the Discrete Cosine Transform, 25, 1004–1009, 1977.

Conzalez, R. C. and Woods, R. E.: Digial Image Processing, Prentice-Hall, Inc., New Jersey, 2001.

Duhamel, P. and H'Mida, H.: New 2^n DCT Algorithms Suitable for VLSI Implementation, in: IEEE International Conference on ICASSP, 12, 1805–1808, 1987.

Fanucci, L. and Saponara, S.: Data Driven VLSI Computation For Low-Power DCT-Based Video Coding, in: International Conf. on Electronics, Circuits and Systems, 541–544, 2002.

Goetze, J. and Hekstra, G.: An Algorithm and Architecture Based on Orthonormal Micro-Rotations for Computing the Symmetric EVD, in: Integration – The VLSI Journal, 20, 21–39, 1995.

Heyne, B. and Goetze, J.: A Pure Cordic Based FFT for Reconfigurable Digital Signal Processing, in: 12th European Signal Processing Conference, 2004.

Heyne, B., Buecker, M., and Goetze, J.: Implementation of a Cordic Based FFT on a Reconfigurable Hardware Accelerator, in: 3rd Karlsruhe Workshop on Software Radios, 2004.

Hsiao, S., Hu, Y., Juang, T., and Lee, C.: Efficient VLSI Implementations of Fast Multiplierless Approximated DCT Using Parameterized Hardware Modules for Silicon Intellectual Property Design, 52, 1568–1579, 2005.

Jeong, H., Kim, J., and Cho, W. K.: Low-Power Multiplierless DCT Architecture Using Image Correlation, 50, 262–267, 2004.

Li, J. and Lu, S. L.: Low Power Design of Two-Dimensional DCT, in: IEEE Conf. on ASIC and Exhibit, 309–312, 1996.

Loeffler, C., Lightenberg, A., and Moschytz, G. S.: Practical Fast 1-D DCT Algorithms With 11-Multiplications, in: Proc. ICASSP, 2, 988–991, Glasgow,UK, 1989.

Mariatos, E., Metafas, D., Hallas, J., and Goutis, C.: A Fast DCT Processor, Based on Special Purpose CORDIC Rotators, in: IEEE International Symposium on Circuits and Systems, 4, 271–274, 1994.

Richardson, I. E. G.: Video Codec Design, John Wiley & Sons Ltd, Atrium, England, 2002.

Shams, A., Pan, W., Chidanandan, A., and Bayoumi, M. A.: A Low-Power High Performance Distributed DCT Architecture, in: IEEE Computer Society Annual Symposium on VLSI, 21–27, 2002.

Sung, C. C., Ruan, S. J., Lin, B. Y., and Shie, M. C.: Quality and Power Effcient Architecture for the Discrete Cosine Transform, IEICE Transactions on Fundamentals Special Section on VLSI Design and CAD Algorithms, 2005.

Tran, T. D.: A Fast Multiplierless Block Transform for Image and Video Compression, in: International Conf. on Image Processing, 822–826, 1999.

Tran, T. D.: The BinDCT: Fast Multiplierless Approximation of the DCT, 7, 141–144, 2000.

Volder, J.: The CORDIC Trigonometric Computing Technique, IRE Trans. Electron. Comput., EC-8, 330–334, 1959.

Walther, J.: A Unified Algorithm for Elementary Functions, in: Proc. Spring Joint Comput. Conf., 38, 379–385, 1971.

Wang, Z.: Fast Algorithms for the Discrete W Transform and for the Discrete Fourier Transform, 32, 803–816, 1984.

9

Physical IC debug – backside approach and nanoscale challenge

C. Boit[1], R. Schlangen[1], A. Glowacki[1], U. Kindereit[1], T. Kiyan[1], U. Kerst[1], T. Lundquist[*], S. Kasapi[**], and H. Suzuki[***]

[1]TUB Berlin University of Technology, Berlin, Germany
[*]DCG Systems, Fremont, CA, USA
[**]NVIDIA, Santa Clara, CA, USA
[***]Hamamatsu Photonics KK, Japan

Abstract. Physical analysis for IC functionality in submicron technologies requires access through chip backside. Based upon typical global backside preparation with 50–100 μm moderate silicon thickness remaining, a state of the art of the analysis techniques available for this purpose is presented and evaluated for functional analysis and layout pattern resolution potential. A circuit edit technique valid for nano technology ICs, is also presented that is based upon the formation of local trenches using the bottom of Shallow Trench Isolation (STI) as endpoint for Focused Ion Beam (FIB) milling. As a derivative from this process, a locally ultra thin silicon device can be processed, creating a back surface as work bench for breakthrough applications of nanoscale analysis techniques to a fully functional circuit through chip backside. Several applications demonstrate the power and potential of this new approach.

1 Introduction

It has become common practice with recent IC technologies that the physical techniques to support functional circuit analysis have to access the active information through chip backside (see Fig. 1) (Barton et al., 2004). For this purpose, the device receives typically a global polishing preparation. In most cases, mechanical stability and / or heat sink requirements limit the minimum remaining silicon substrate thickness to 550–100 μm, which will be categorized here as moderate silicon thickness.

The penetration depth of particle beams is orders of magnitude smaller, so the previously very successful electron or ion beam microscopy derivatives using voltage contrast like E Beam Probing are not applicable under these circumstances. But bulk silicon is transparent to light of photon energies smaller than the Si band gap of 1.1eV, i.e. the near infrared (IR) regime. A survey of the analysis techniques available for backside access of moderate silicon thickness is presented in chapter 2.

Fig. 1. Flip chip package. Structured side of IC (die) is turned to substrate (image from [1]).

Circuit edit also needs to be performed through chip backside. The creation and optimization of FIB-based processes for this purpose is subject of the following chapter 3. The key of backside circuit edit is the milling of local trenches to the bottom level of the STI, leaving an SOI (Silicon on Insulator) – like ultra thin layer of silicon (ca. 300 nm) for active device function. This preparation technique, expanded to planes of up to 0.03 mm^2, offers a new breakthrough approach of physical functional analysis techniques: While the circuit will be remaining fully functional, the active volume then is small enough to interact with high energy photons or particle beams applied through chip backside. The 4th chapter will discuss the opportunities of such a concept for nanoscale resolution of functional analysis techniques and demonstrate how powerful it is in use.

2 Backside access functional analysis on devices of moderate bulk silicon thickness

Global polishing of bulk silicon is a high yield process as long as a moderate thickness of 50–100 μm remains. The preferred interactive medium of functional analysis techniques in this case is near IR optics with photon energies smaller than silicon energy band gap of 1.1eV, corresponding to wavelengths of 1.1 μm and higher. Three main dynamic approaches have been established: Time resolved photon emission (TRE), laser voltage probing (LVP) and laser

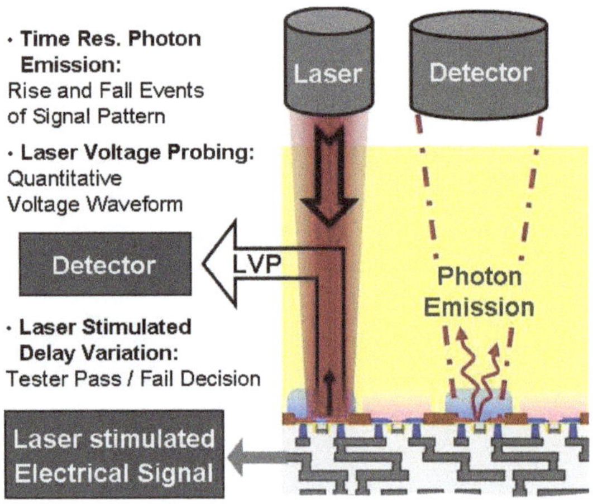

Fig. 2. Optical probing and stimulation techniques.

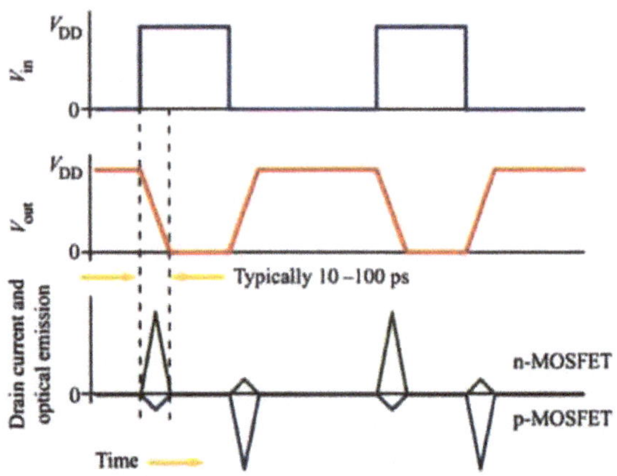

Fig. 3. TRE and signal pattern (from [24]).

stimulation (LS) techniques (Fig. 2). Time resolution and signal pattern response characteristics of these techniques will be discussed shortly in this chapter, before the critical resolution potential of the general approach using moderate silicon thickness will be assessed.

2.1 Time resolved photon emission (TRE)

Photon Emission is the standard functional analysis technique of semiconductor devices as it occurs in a MOS Transistor without any further stimulus when a considerable number of carriers gain kinetic energy high enough to relax by emitting photons (Boit, 2004). This is happening in CMOS digital circuits mainly during rise and fall time of the signal pattern (Fig. 3). Although this dynamic light signal is faint, TRE is a very powerful, reliable and easy to interpret technique with low risk of artifacts and time resolution in 10 ps regime (Vallett, 2004). Due to the low signal level, sampling of the signal is necessary. As the waveform is not reproduced in full but the signal only represents the short switching pulses, trigger stability and jitter are critical (Kasapi and Woods, 2006). TRE emission intensity decreases nonlinearly with supply voltage (Rowlette and Eiles, 2003). The resulting SNR challenge of recent IC technologies may not be overly severe as recent studies indicate (Tosi et al., 2006).

Spectral information of the emission has so far only marginally been used in analysis. Detailed understanding of device physics and the mechanisms of photon emission may change this in the future.

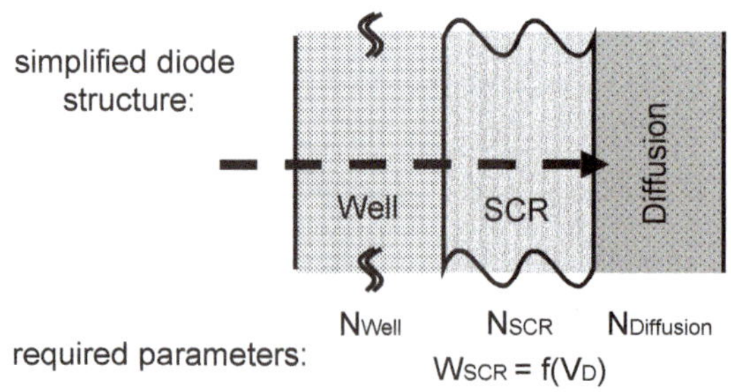

Fig. 4. Simplified diode structure for simulation (from [8])

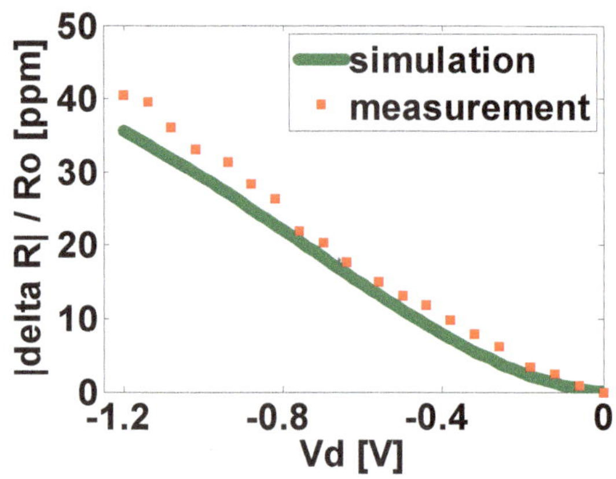

Fig. 5. Comparison of simulation results from model in Fig. 4 with measurement on PFET drain area.

2.2 Laser Voltage Probing (LVP)

When an IR laser illuminates the active devices through chip backside, the reflected light is modulated by active device operation in the ppm (parts per million) range. This signal has been proven to be linearly correlated to the voltage applied to the device and the waveform of a signal pattern passing the node can be reproduced, very similar to E Beam Probing (EBP)(Paniccia et al., 1998). Originally, the signal has been attributed to electrical fields in the device due to free carrier dynamics (Franz-Keldysh-Effect). Recent investigations using device simulation succeeded to correlate the signal to the dynamics of absorption and refractive index due to the presence or absence of free carriers by space charge modulation (Kindereit et al. 2007), which allow simpler modeling

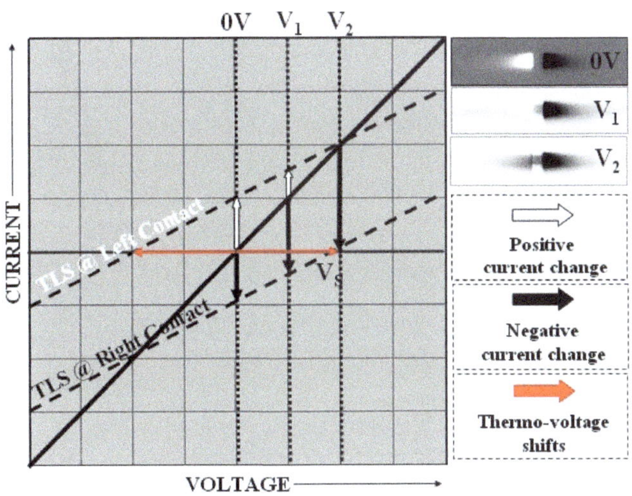

Fig. 6. Illustration of SEI and OBIRCH influence on I/V characteristic (from[10]).

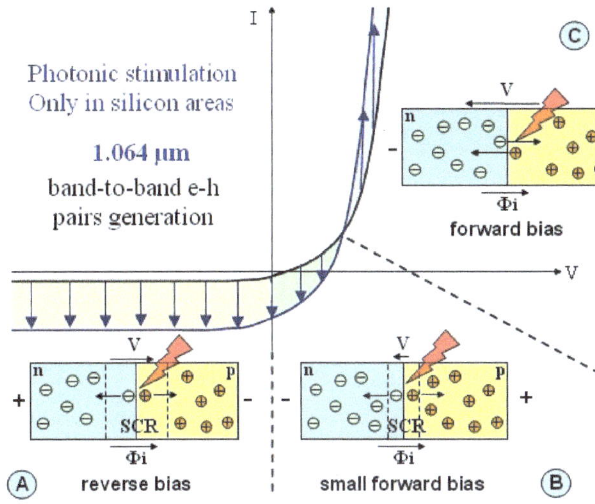

Fig. 8. p-n junction IV characteristic at PLS (from[10]).

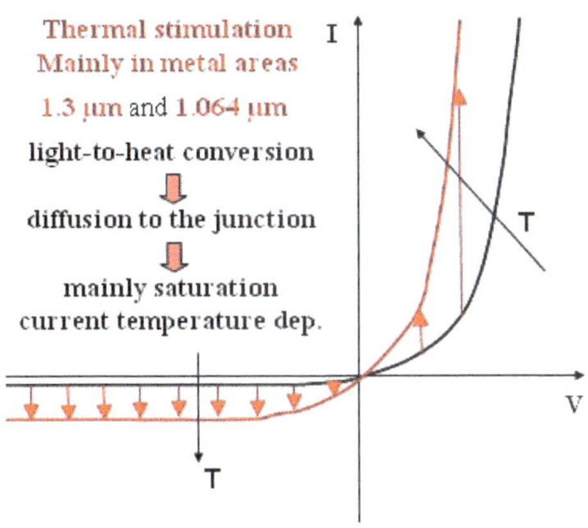

Fig. 7. p-n junction IV characteristic at TLS (from[10]).

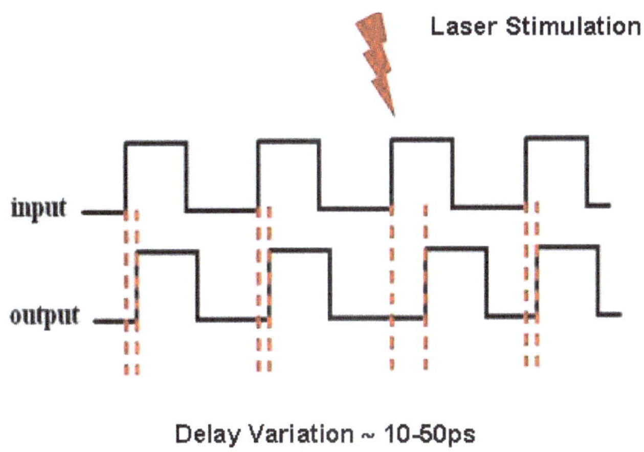

Fig. 9. Delay variation with laser stimulation.

of the effect (see Fig. 4 and 5). In this example, a qualitative and even a quantitative fit has been successful using well doping level of 10^{17} cm^{-3} and a diffusion doping level of 10^{21} cm^{-3}.

Time resolution is in the same range as TRE. Due to the waveform reproduction, the signal pattern information can still be obtained when trigger stability is problematic.

Probe placement may not be trivial when local interference is affecting the signal level.

Although available for a long time, in recent past LVP stood in the shadow of TRE. But the power of this technique has been improved by the mentioned in depth investigation of the interaction mechanisms, and a new product generation using polarized light, and seems to become increasingly attractive again, especially as it offers some robustness to IC clock instabilities.

2.3 Laser stimulation

Scanning IR lasers with several hundred mW can stimulate electrical signal alteration either by local heat conversion of the absorbed light modulating electrical parameters (TLS = Thermal LS, Beaudoin, 2004), or by photocurrent induced into the signal path if photon energies reach silicon band gap level and electron hole (e-h) pairs are generated (PLS=Photoelectric LS).

As the electrical circuit behavior can be altered in a wide variety, many techniques have been invented that are based on the LS principle. Table 1 gives an overview of the most common abbreviations used in LS of statically and dynamically driven circuits.

The transformation of the stimulating effect into an electrical signal that can be read out over IC terminals occurs on the four paths resistivity change, thermoelectric voltage generation, temperature coefficient of pn junction and photocarrier generation. In the Figs. 6–8, the corresponding influence

Table 1. Laser Stimulation techniques.

Stim.	Effect	Technique	Device Analysis
1300 nm	**TLS** Thermal Laser Stim.	**OBIRCH** Optical Beam Induced Resitivity Change **TIVA** Thermally Induced Voltage Alteration	Identify areas of high local resistance
		RIL Resisitive Interconnect Localization **SDL** Soft Defect Localization	Test vectors: Identify area of highest pass-fail sensitivity (ShmooVariation)
		SEI Seebeck (Thermoelectric) Effect Imaging	Identify sensitive material interfaces
1064 nm	**PLS** Photo -electric Laser Stim., parasitic TLS	**OBIC** Optical Beam Induced Current	Localize p-n junctions – Space Charge Region (SCR)
		LIVA Light Induced Voltage Alteration	Read out signals sensitive to logical state
		LADA Laser Assisted Device Alteration, **Dynamic LS**	Signal Pattern in IC: Delay Variation
		OBIRCH, SEI	As parasitic effect

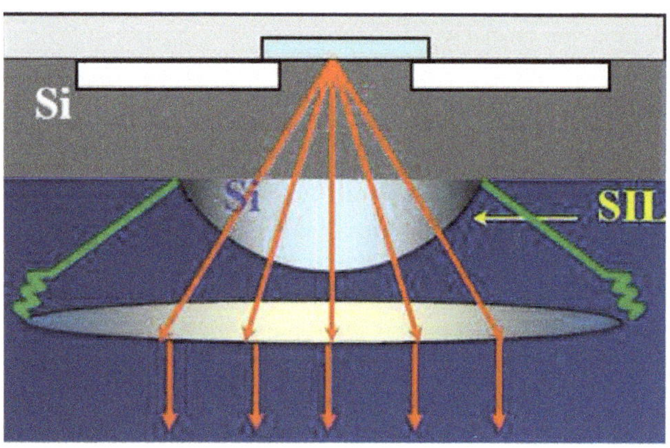

Fig. 10. Principle of Solid Immersion Lens (from [23]).

of the stimulation on equivalent device i-v characteristics is presented. All static and dynamic properties of devices and circuits can be derived from these models. In case of TLS in an FET, typically the channel conductivity will be modulated like OBIRCH, but in two competitive mechanisms, change of

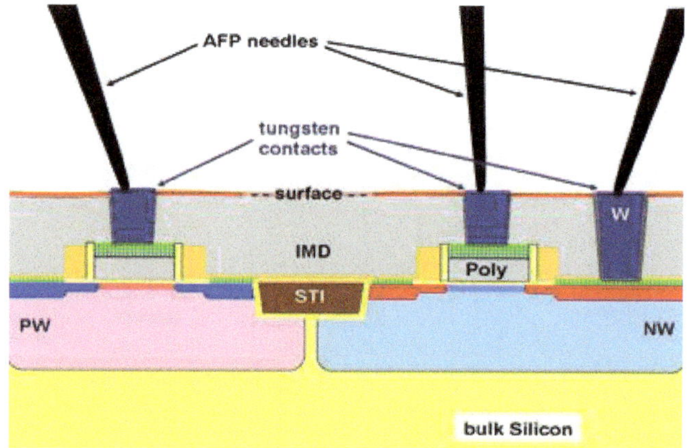

Fig. 11. Principle of Nanoscale Atomic Force Probing (AFP) on parallel polished sample.

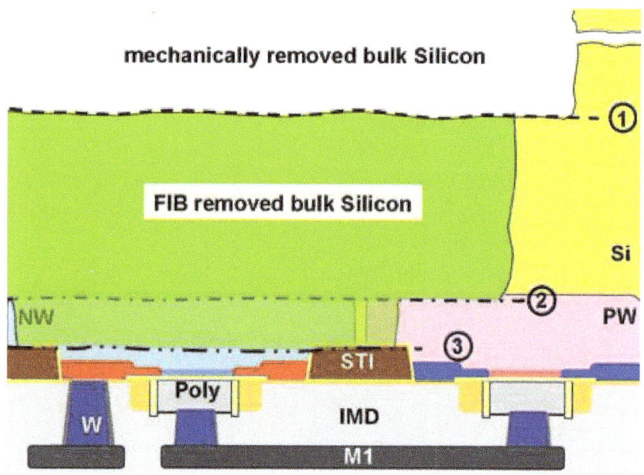

Fig. 12. FIB preparation for backside circuit edit.

mobility and threshold voltage. But to understand the readout properly, several side effects like local conversion efficiencies, heat or carrier diffusions and distant Seebeck voltages may contribute to the final signal. A detailed description of all those effects is given in (Glowacki, 2007).

In a dynamically driven IC, LS will influence the timing. The results of dynamic LS can be read out either by pass/fail condition of the tester under stimulation (Shmoo variation) or direct evaluation of the signal pattern variation (Fig. 9). A systematic overview on the LS at dynamic circuit operation like in (Bruce et al., 2002) for static techniques is nor yet available, so in Table 1 some of the techniques seem to be overlapping and some possible options have not been published yet.

A very effective analysis uses an already established test signal path of a soft failing device that is driven below a critical temperature, and the laser is raising the illuminated area above the critical level. Such a Shmoo variation using TLS is RIL if the soft defect is located in interconnect levels or SDL

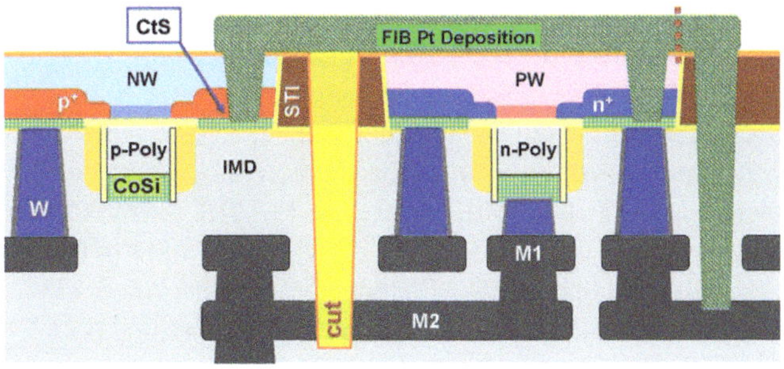

Fig. 13. Variety of FIB contacts using UTS.

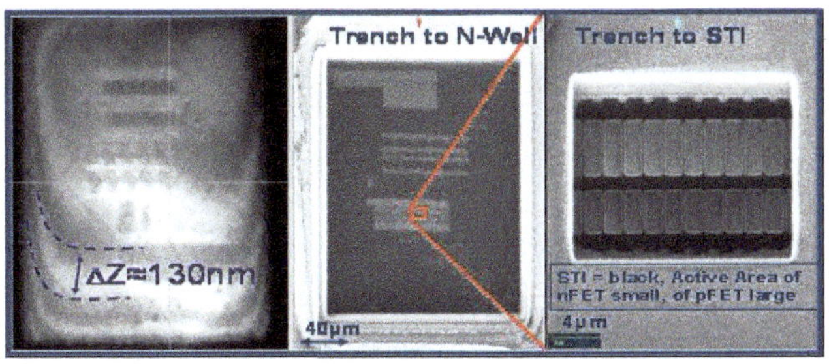

Fig. 14. Image of FIB trench location (left: IR-optics, middle: FIB n-well end point, right: FIB STI opening).

(Bruce et al., 2002), when active devices are affected. PLS causes similar effects.

The other group of techniques read out the signal pattern directly like LADA or Delay Variation (Sanchez et al., 2006a,b). In a transistor, the effects of PLS and TLS on the output current must be studied in detail for correct evaluation (Glowaki et al., 2007). In general, if PLS is enabled by selection of the proper laser wavelength, it usually prevails over TLS.

Time resolution and required trigger quality of LS is very much a question of the signal extraction procedure.

As in LS a large variety of stimulation and readout opportunities can be applied, the overall potential of this approach is by far not exhausted yet.

2.4 Nanoscale potential of IR analysis techniques

Maximum microscopic resolution is the light wavelength referring to the length of a repeating structure, the so called pitch. A pitch consists of minimum 2 features, so the minimum resolved feature is 2× smaller. Realistic pitch resolution is ca. 800 nm, respectively 400 nm feature size. But feature sizes are rapidly shrinking with technology innovation. Solid Immersion Lenses (SIL, Fig. 10) have been introduced, making bulk silicon a part of the microscope and increasing resolution theoretically by the refraction index of silicon n_{Si}=3,5. A resolution of of 200 nm has been achieved.

As technologies progress further into the nanometer range, IR techniques even with SIL can localize only the area of in-

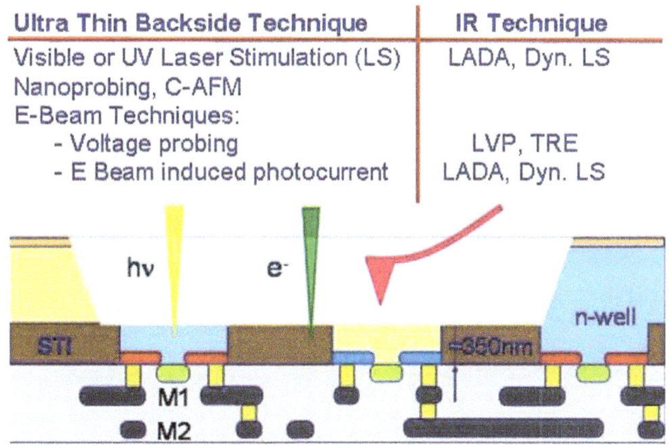

Fig. 15. Probing on Ultra Thin Silicon.

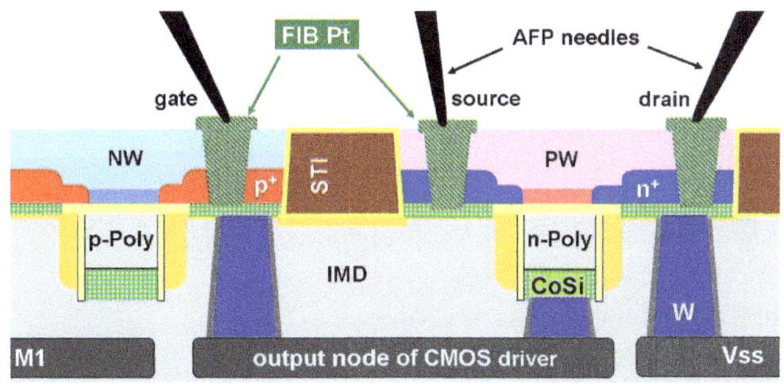

Fig. 16. Nanoprobing on FIB edited contacts to Silicide.

terest but not anymore the exact device. In addition, scanning probe techniques and derivatives are increasingly inevitable to identify the critical node in nanoscale regime. Nanoprobing makes use of the contacts to the active structures under investigation (Fig. 11) (Lin et al., 2006; Mulder et al., 2006). Planar polishing techniques need to be applied from chip frontside, finally removing all interconnect layers to reveal the tungsten contacts of multiple devices in the suspected fail area.

This process is time consuming and full of risks: If the failing transistor cannot be found in the final measurement the problem might have been removed with the backend, leaving no second chance for other FA techniques.

3 Circuit edit through chip backside

The challenge for focused ion beam (FIB) based circuit edit (CE) today is access through chip backside, requiring circuit cut and paste operations much closer to the active layers than global preparation techniques can provide. A trenching process using a FIB with coaxial IR optical microscope has been established that removes bulk silicon down to the bottom level of shallow trench isolation (STI). This process is performed in three steps as indicated in Fig. 12: After a global

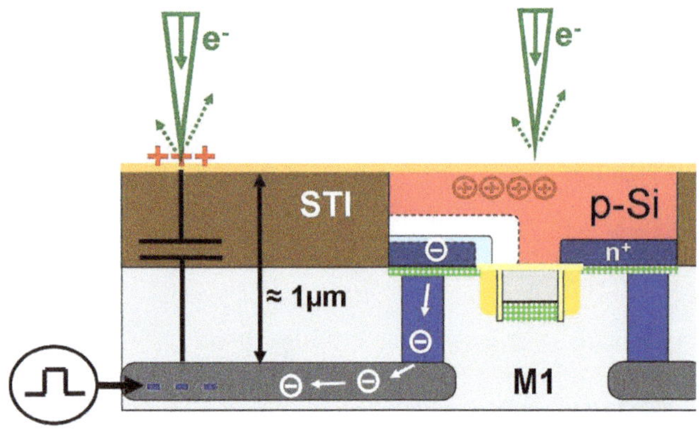

Fig. 17. Backside E-Beam Probing targeting M1 or directly measuring on the FET floating body.

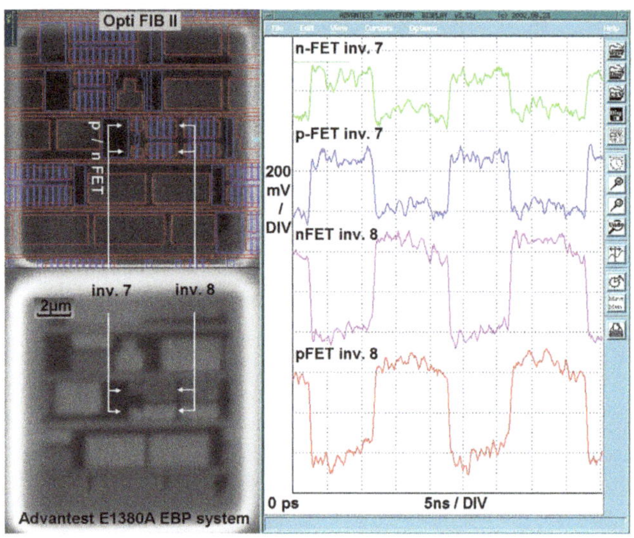

Fig. 18. EBP results measured directly of FET body of 90 nm bulk CMOS after FIB backside prep down to STI level.

polishing technique as applied in chapter 2, the region of interest is identified by the IR microscope. Then, after a special FIB cleaning procedure, a 200×200 μm^2 wide trench is milled until the operator stops on n-well level. The endpoint becomes visible because of a voltage contrast between the n-wells and the substrate due to the built-in potential of the pn-junction (Boit et al., 2003). Next, a smaller local trench will be milled until the STI becomes visible as imaging contrast - lower secondary electron emission of SiO_2 – and the preparation is ended.

By this approach, the active device material is reduced to 300 to 400 nm (the depth of the STI trenches) and the chip is ready for the actual edit process, the milling of narrow trenches for the cuts and contact holes (Boit et al., 2005).

This backside FIB process is offering several more opportunities to edit the circuit than the frontside process: The technique to endpoint the contact trenches on the Silicide interface, which is only 300 nm below the trench to STI, is meanwhile established (Schlangen et al., 2006a), allowing to contact any source/drain area on the chip, even those that hove not been contacted on IC interconnect level. All the functional nodes are represented on the active device level. A typical backside CE is illustrated in Fig. 13. Any CE contact can be performed using the contact to silicide (CtS) process, only cut operations still have to be performed by trenching deeper into metal interconnects. The lower aspect ratio for CtS compared to contacts to metal lines (right part of Fig. 13) is very fortunate for the contact resistance, being the critical parameter for nanoscale application. With current platinum deposition, contacts on silicide have resistivities below $60\Omega/\mu m^2$. Typical contact sizes of $1\times0.2\ \mu m^2$ end up with 300Ω resistance per contact, considerably lower than FET- channel resistances of several kΩ (Schlangen et al., 2007a) and acceptable for most CE and probing tasks. When SOI technologies are considered, the contact height will be reduced by a factor of 3, reducing resistance further. The utilization of advanced conductor deposition materials as Molybdenum may offer further improvement potential.

Taking into account state of the art FIB resolution performance, this backside CE technique can be rated as fully suitable for technologies scaling well into the nanometer regime.

An invasiveness study of backside CE processing using intentionally enlarged trench sizes down to STI level (Fig. 14), resulted in almost unchanged single device- and circuitry-performance at regular processing (Boit et al., 2005).

This study showed also that a wide FIB trench to n-well level can be performed with remarkable co-planarity to the chip levels. In Fig. 14, the coaxial IR microscope image of the trench shows Newton rings indicating the slope of silicon thickness. A planarity mismatch less than 200 nm on 250 μm trench expansion or 0.08% deviation is state of the art. With such a process, a new base plane can be created for the application of analysis techniques through the backside of ultra thin silicon.

4 Functional IC analysis through the backside of Ultra Thin Silicon (UTS)

The FIB preparation routines developed for circuit edit through chip backside (chapter 3) produce a back surface of an ultra thin active device volume maintaining full IC functionality. The back surface of the remaining ultra thin active volume can now be regarded as a completely new work bench for physical techniques to access fully functional circuits. This innovative concept opens a very short path to device interaction for analysis techniques with improved resolution potential (Fig. 15), which is summarized in Table 2.

Nanoprobing can now be performed on a fully functional circuit, avoiding the risks of parallel polishing from IC frontside that are pointed out in chapter 2. Particle beam techniques like E Beam Probing with the resolution potential of an electron microscope, can be reconsidered for IC analysis on a higher technical level again. Optical techniques may be shifted to much smaller wavelengths with the respective resolution improvement as no bulk silicon has to be passed anymore.

In case of substantial power consumption in the IC's area of interest, solutions for proper heat dissipation paths need to be developed.

4.1 Nanoprobing on UTS

Nanoprobing on UTS can be done on edited contacts because of their low resistance, taking advantage of the full circuit function (Fig. 16). For characterization of single devices, some cuts might be necessary to separate the transistors from their environment (Schlangen et al., 2007a), but full circuit functionality can be reestablished easily afterwards by simple conductor deposition.

4.2 E-Beam Probing on UTS

The interaction opportunities through UTS are expanded to particle beams as well. Then, high resolution probing or stimulation with electron beams return to the IC debug toolkit. In (Schlangen etal., 2006b) it is demonstrated, that switching information on lower metal lines is transferred to the back surface (capacitive coupled voltage contrast, CCVC). Additionally, the full active device area can be used to pick up node waveforms (Fig. 17), based on the modulation of the floating body potential. In this way, the spot size of the pulsed e-beam can be as big as the full transistor body (always by a factor of 3–5 bigger than the smallest metal lines) and the resolution requirements become more relaxed. Furthermore, this method offers all advantages frontside EBP did: fast signal acquisition, linear Vdd signal scaling and superior signal to noise ratio (Jain et al., 2007).

Figure 18 shows first results measured with a 10 year old EBP, clearly resolving 90 nm technology (Schlangen et al., 2007b). Regarding latest improvements in low keV SEM development, this technique should be suitable for scaling further into nanometer regime.

4.3 Optical Techniques on UTS

On UTS, visible and even UV light may be used to stimulate the devices. This way, optical resolution would be instantly improved even without SIL. In addition, short wavelength SIL solutions are within reach. Efficient PLS in small active volumes like SOI devices requires light of short absorption ranges. Only UTS provides the condition for such a stimulation.

Another problem is the stimulation character in multi-material active volumes as it is used more and more in strain technologies. Stimulation as performed today with photon energies near the Si band gap, has hardly predictable absorption properties for Ge compounds used for straining and mobility enhancement. With shorter wavelengths, absorption properties are clearly defined independent of the semiconductor material. If laser stimulation can be blanked over actives and only performed over STI, it may be possible to suppress photocurrent generation and thermal stimulation could then be accomplished with high resolution.

To gain resolution beyond all optical limits though, backside E-Beam Probing and E-Beam stimulation seem to be most promising.

Table 2. Comparison of Feature resolution potential of Functional Analysis Techniques though moderate and ultra thin silicon.

Technique	Resolution	Potential	Status
(IR) Optical bulk Si	400 nm	200 nm (SIL)	standard technique
(IR) Optical UTS	200 nm	100 nm (SIL)	future concept
UV UTS	100 nm	<50 nm (SIL)	future concept
E Beam UTS	100 nm	10–20 nm	demonstrated
Nanoprobing UTS	50 nm	10 nm	demonstrated

5 Summary

The IR analysis techniques for IC debug are very powerful and offer a huge potential to localize circuit problems which is not fully exhausted yet. The only drastic disadvantage is the increasing optical resolution mismatch in nanoscale technologies.

For circuit edit (CE), a process for access through chip backside is established as well with a FIB preparation technique processing towards STI level.

The back surface of the ultra thin silicon (UTS) produced for backside CE is full of opportunities for functional analysis of nanoscale devices because this preparation technique offers for analysis techniques with nanoscale resolution the unique option to keep the Integrated Circuit fully functional. UTS analysis potential matches the requirements for several technology generations to come. The utilization has just begun.

Acknowledgements. The authors thank S. Schömann and P. Egger of Failure Analysis Munich at Infineon Technologies for the electronic devices, and A. Eckert at TU Berlin for excellent backside preparation.

References

Barton, D., Cole, Jr. E., and Bernhard-Höfer, K.: Flip Chip and "Backside" Sample Preparation Techniques, Microelectronics Failure Analysis Desk Reference, ASM International, 5th Edition, 42–48, Ohio, USA, 2004.

Boit, C.: Fundamentals of Photon Emission (PEM) in Silicon-Electroluminescence for, Analysis of Electronic Circuit and Device Functionality, Microelectronics Failure Analysis Desk Reference, ASM International, 5th Edition, 356–368, Ohio, USA, 2004.

Vallett, D.: Picosecond Imaging Circuit Analysis – PICA, Microelectronics Failure Analysis Desk Reference, ASM International, 5th Edition, 369–377, Ohio, USA, 2004.

Kasapi, S. and Woods, G.: Voltage Noise and Jitter Measurement Using Time Resolved Emission, Proc. 32nd EDFAS ISTFA International Symposium for Testing and Failure Analysis, Austin, TX, USA, 438–443, 2006.

Rowlette, J. A. and Eiles, T.: Critical Timing Analysis in Microprocessors Using Near-IR Laser Assisted Device Alteration (LADA), Proc. IEEE ITC International Test Conference, Charlotte NC, USA, 264–273, 2003.

Tosi, A., Stellari, F., Pigozzi, A., Marchesi, G., and Zappa, F.: Hot Carrier Photon Emission in Scaled CMOS Technologies – A Callenge for Emission Based Testing and Diagnosis, Proc. 44th IEEE IRPS International Reliability Physics Symposium San Jose 2006, 595–601, 2006.

Paniccia, M., Eiles, T., Rao, V. R. M., et al.: Novel optical probing technique for flip-chip packaged microprocessors, Proc. IEEE ITC International Test Conference, Paper 30.2, 740–747, 1998.

Kindereit, U., Woods, G., Tian, J., Kerst, U., Leihkauf, R., and Boit, C.: Quantitative Investigation of Laser Beam Modulation in Electrically Active Devices as used in Laser Voltage Probing, IEEE TDMR Transactions on Device and Materials Reliability, 7, 1, 19–30, 2007.

Beaudoin, F., Desplats, R., Perdu, P., and Boit, C.: Principles of Thermal Laser Stimulation Techniques, (), Microelectronics Failure Analysis Desk Reference, ASM International, 5th Edition, 417–425, Ohio, USA, 2004.

Glowacki, A. S., Brahma, H., Suzuki, C., and Boit, C.: Systematic Characterization of Integrated Circuit Standard Components as Stimulated by Scanning Laser Beam, IEEE TDMR Transactions on Device and Materials Reliability, 7, 1, 31–49, 2007.

Bruce, M., Bruce, V., Eppes, D., Wilcox, J., Cole, Jr. E., Tangyunyong, P., and Hawkins, C. F.: Soft Defect Localization (SDL) on ICs , Proc. 28th EDFAS ISTFA International Symposium for Testing and Failure Analysis, Phoenix, AZ, USA, 21–28, 2002.

Sanchez, K., Desplats, R., Beaudoin, F., Perdu, P., Lewis, D., Vedagarbha, P., and Woods, G.: Delay Variation Mapping Induced by Dynamic Laser Stimulation, Proc. 43th IEEE IRPS International Reliability Physics Symposium San Jose 2005, 305–311, 2006a.

Sanchez, K., Desplats, R., Beaudoin, F., Perdu, P., Dudit, S., and Vallet, M.: Dynamic Thermal Laser Stimulation Theory and Applications, Proc. 44th IEEE IRPS International Reliability Physics Symposium San Jose 2006, 574–584, 2006b.

Lin, K., Zhang, H., and Lu, S.-S.: Conductive Atomic Force Microscopy Application for Semiconductor Failure Analysis in Advanced Nanometer Process, Proc. 32nd EDFAS ISTFA International Symposium for Testing and Failure Analysis, Austin TX, USA, 178–181, 2006.

Mulder, R., Subramanian, S., and Chrastecky, T.: Case Studies in Atomic Force Probe Analysis, Proc. 32nd EDFAS ISTFA International Symposium for Testing and Failure Analysis, Austin TX, USA, 153–162, 2006.

Boit, C., Wirth, K., and LeRoy, E.: Voltage Contrast like N-Well Imaging, Proc. 29th EDFAS ISTFA International Symposium for Testing and Failure Analysis, Santa Clara, CA, USA, 331 pp., 2003.

Boit, C., Kerst, U., Schlangen, R., Kabakow, A., Le Roy, E., Lundquist, T., and Pauthner, S.: Impact of Back Side Circuit Edit on Active Device Performance in Bulk Silicon ICs, Proc. IEEE ITC International Test Conference, Austin TX, paper 48.2, 1–9, 2005.

Schlangen, R., Sadewater, P., Kerst, U., and Boit, C.: Contact to Contacts or Silicide by Use of Backside FIB Circuit Edit Allowing to Approach Every Active Circuit Node,Proc. 17th ESREF European Symposium on Reliability of Electron Devices, Failure Physics and Analysis, Wuppertal, Germany, 1498–1503, 2006a.

Schlangen, R., Kerst, U., Boit, C., Schömann, S., Krüger, B., Jain, R., Malik, T., and Lundquist, T.: FIB Backside Circuit Modification on Device Level Allowing to Access Every Circuit Node with Minimum Impact on Device Performance by Use of Atomic Force Probing, accepted at 33rd EDFAS ISTFA International Symposium for Testing and Failure Analysis, San Jose, USA, 2007a.

Schlangen, R., Leihkauf, R., Kerst, U., and Boit, C.: Functional IC Analysis through Chip Backside with Nano Scale Resolution – E-Beam Probing in FIB Trenches to STI Level, Proc. 32nd EDFAS ISTFA International Symposium for Testing and Failure Analysis, Austin TX, USA, 376–381, 2006b.

Jain, R., Malik, T., Lundquist, T., Schlangen, R., Leihkauf, R., Kerst, U., and Boit, C.: Novel Flip-Chip Probing Methodology Using Electron Beam Probing, Proc. 14th IEEE IPFA International Symposium on the Physical and Failure Analysis of Integrated Circuits, Bangalore, India 39–42, 2007.

Schlangen, R., Leihkauf, R., Kerst, U., Boit, C., Jain, R., Malik, T., Wilsher, K., Lundquist, T., and Krüger, B.: Backside E-Beam Probing on Nano Scale Devices, accepted at IEEE ITC International Test Conference, Santa Clara, CA, USA, 2007b.

Vickers, J. S., Ispasoiu, R., Cotton, D., Frank, J., Lee, B., and Kasapi, S.: Time-Resolved Photon Counting System Based on a Geiger-Mode InGaAs/InP APD and a Solid Immersion Lens, Proc. 16th Annual Meeting of the IEEE LEOS, 2, 600–601, 2003.

Kash, J., Tsang, J., Knebel, D., and Vallett, D.: Non-Invasive Backside Failure Analysis of Integrated Circuits by Time-Dependent Light Emission: Picosecond Imaging Circuit Analysis, Proc. 24th ISTFA International Symposium for Testing and Failure Dallas, TX, USA, 483–488, 1998.

10

ESD full chip simulation: HBM and CDM requirements and simulation approach

E. Franell[1], S. Drueen[2], H. Gossner[2], and D. Schmitt-Landsiedel[1]

[1]Institute for Technical Electronics, Technical University Munich, Germany
[2]Infineon Technologies AG, Munich, Germany

Abstract. Verification of ESD safety on full chip level is a major challenge for IC design. Especially phenomena with their origin in the overall product setup are posing a hurdle on the way to ESD safe products. For stress according to the Charged Device Model (CDM), a stumbling stone for a simulation based analysis is the complex current distribution among a huge number of internal nodes leading to hardly predictable voltage drops inside the circuits.

This paper describes an methodology for Human Body Model (HBM) simulations with an improved ESD-failure coverage and a novel methodology to replace capacitive nodes within a resistive network by current sources for CDM simulation. This enables a highly efficient DC simulation clearly marking CDM relevant design weaknesses allowing for application of this software both during product development and for product verification.

1 Introduction

Robustness of ICs against electrostatic discharge (ESD) is one of the leading topics in actual chip design flows due to the shrinking of devices and especially gate oxide thicknesses in modern CMOS processes. ESD on chip level results in current discharge pulses (1A–10A) through a not mounted IC.

ESD protection conducts the discharge current through the chip on a safe path. Provided that the discharge path itself is safe, remaining possible failure sites may be found inside the core in elements which are not close to the discharge current path. Voltage differences on cells which are connected to more than one voltage domain are prone to cause this kind of problems.

The design optimization for HBM and especially CDM safety is still an evolutionary approach involving failure analysis and re-designs. This is expensive and may delay time-to-market of a product tremendously. Thus software based verification is required which predicts weaknesses of the IC design before tape-out.

Device simulation has established as powerful tool during the development of ESD protection concepts (Esmark, 2000), but due to the high numerical effort, it can only be applied for analysis of small structures, e.g. protection devices or parasitic structures. Some further efforts have been done searching for possible current paths through the IC (Liu, 2006). Furthermore, a layout-based verification of ESD-rules has been proposed (Zhan, 2005).

This paper will present ESD simulation methodologies for the verification of ULSI microchips, different from all established variants of ESD-simulation. The second section shows the requirements for a verification-oriented ESD simulation. Section 3 presents the simulation approach and a way to comply with the shown requirements. In Sect. 4 the functionality of the tool and the requirements for such software are demonstrated with a product example. The extension of this approach to CDM verification is discussed in Sect. 5. A conclusion is given in Sect. 6.

2 Full chip ESD-simulation – requirements

For ESD issues a complex netlist from circuit design contains too much information about non-relevant parts of the IC. Therefore the basic steps for an efficient ESD simulation are the reduction of complexity to an amount which allows a huge number of stress combinations for highly complex ICs. As it has been shown for a HBM type, 2 pin discharge with externally forced current a DC simulation of simplified netlists together with special simulation models is sufficient

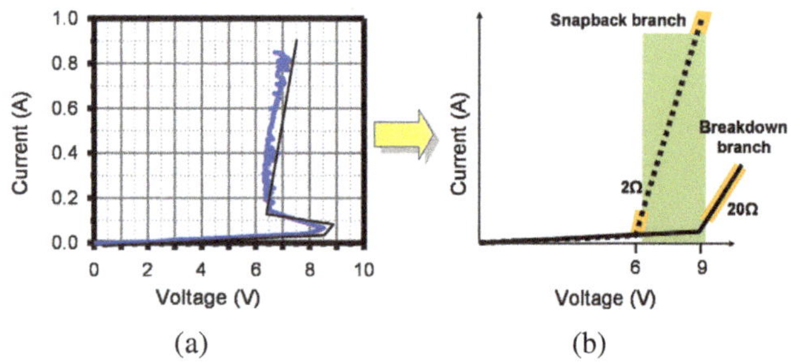

Fig. 1. Measured IV-curve and simplified model for ESD-protection elements with snapback behavior.

to detect design weaknesses in the ESD-protection (Drueen, 2007).

Waveforms and failures could be reproduced in transient ESD simulations, but failure prediction was not satisfying (Etherton, 2006). A truly predictive and useful simulation approach needs to be based on the (huge) input database of a design project without major manual modification. Our approach to facilitate the verification is that a highly accurate transient simulation is typically not required.

2.1 Requirements

The demand for a predictive simulation in the product concept phase implicates that the ESD netlist can not be based on extraction from complete schematics and layouts. A second request is the product ESD-review for the final check of all ESD-rules. At this moment the final database for the product already exists and often achieves a complexity which makes manual checking very time consuming.

The device behaviour during an ESD pulse differs from normal operation conditions. At typical ESD current levels of several 100mA to several amperes MOS and bipolar devices typically expose snapback behaviour that must be handled by the ESD simulation. An example for a typical snapback IV characteristic is given in Fig. 1.

The characteristics of a snapback are the trigger voltage, here V_{t1}=9 V, and the snapback voltage V_{SP}=6 V. In the green zone, both characteristics may be valid depending on history. The yellow zones represents the invalid parts of the pwl descriptions.

2.2 Simulation concept

Our particular ESD netlist building and simulation process is based on the following five considerations:

1. In most cases a new design will employ a precharacterized I/O cell library. The ESD view has to provide all major ESD current paths through the I/O cell. Therefore, a much simplified ESD view for the library cells will be sufficient for chip level ESD simulation.

2. The setup of a proper ESD protection concept avoids destructive current paths through the IO cell logic or the chip core. Correspondingly simulation will still be correct when these blocks are excluded from the ESD netlist. An extremely simplified but sufficiently accurate ESD netlist can be built by plugging together the ESD views of the individual I/O cells in the pad ring to a pad ring netlist, adding the chip core supply mesh, the bondout, and merging it with the package netlist.

3. One of the main sources of ESD failures is the division of the IC into a certain number of electrical supply voltage domains or ground nets on one chip. This division requires additional protection elements and package concepts, which must be included in the simulation netlist.

4. The actual ESD standards provide opportunities to decrease the amount of stress combinations by exploiting the particular chip topology (EOS/ESD Association, 2001). The proposed simulation tool is able to build a stress plan which contains every relevant pin combination to yield full test coverage and limit the simulation time as much as possible.

5. The simulation runs generate a huge amount of output data which have to be evaluated by the software. An additional postprocessor is used to find critical stress cases and sort them by significance. This software part has to deliver a summary of results and find possible failures or incorrect runs.

3 Simulation approach

3.1 Simplified models

The modeling of the snapback curve's negative differential resistance (NDR) usually is associated with severe problems in convergence, numerical stability and simulation time (Zhang, 2001). A good example are 2 protection elements in parallel with a voltage drop around the triggering voltage. This state will forces a normal simulator to decide which single element will jump into the snapback, and the decision may be arbitrary especially taking device variations into account.

The key to our method is the replacement of the compact model by a piecewise linear model of the protection element's IV characteristic with two or more independent model parts. For devices with diode-like character this description contains a Zener-Diode-like description with the corresponding resistances. Devices with a snapback will be represented by two Zener-diode-like devices in one model. Thus, each device has to be considered in its two possible states: breakdown (BD – device not triggered) and snapback (SP – device is triggered).

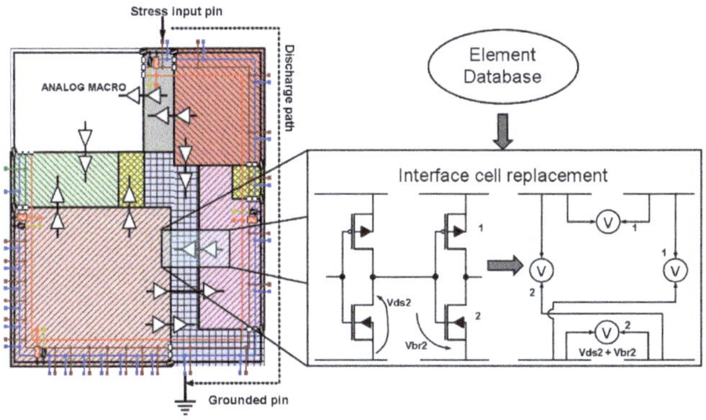

Fig. 2. Transformation of interface elements to combined voltage sensors in the corresponding critical paths.

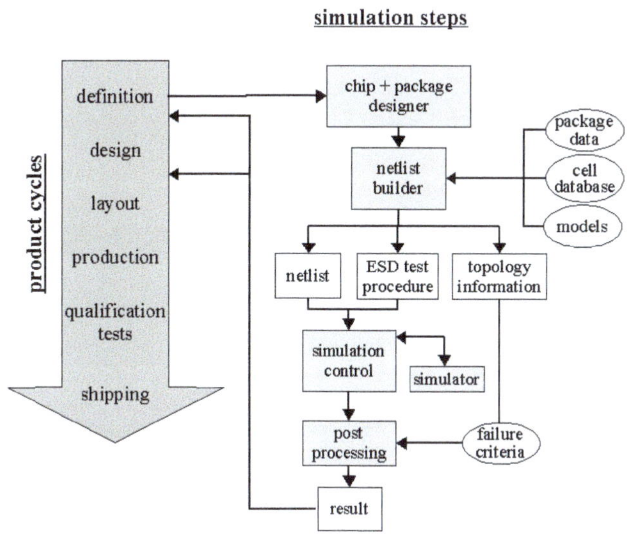

Fig. 3. Simulation flow and influence on development flow.

As shown in Fig. 1 the model contains some physical impossible zones which are blocked by the parameters for the validity range in the model. Because there are two parts of these linear descriptions and starting the simulations it is not known in which state the device will be, all combinations of states have to be regarded.

3.2 Permutational approach – implementation

The permutational approach takes into account that devices could have two possible states and covers every combination of single device states. This simulation method will be necessary if more than one protection device reaches the trigger voltage and the DC simulation is used.

Thus, for finding every critical constellation the only safe solution is to try out every possible circuit state. The jump into snapback is only a very short temporary condition, which does not have to be modeled for finding the critical condition.

In the first simulation run every element is assumed to be in a non-triggered high ohmic state in order to detect all possible discharge paths. Afterwards the elements which exceed the breakdown voltage of the breakdown branch will be marked as permutation elements. The software will now perform additional runs with a permutational variation of the states of all marked devices. Physically impossible solutions arise e.g. when a voltage higher than the BD voltage arises at a device in BD state. They are rejected during validity check. Introducing a valid-criterion and a fail-voltage together with the basic circuit elements in the models it is possible to use every circuit simulator for the calculations.

3.3 Possible hot spots in the core

Possible victims for damage are interface devices between two or more voltage domains (Huh, 2005). In Fig. 2 a simplified product example is drawn with interfaces between different supply blocks. If this were a general problem it would be covered during the place&route step by special rules. But such a scenario only happens with a certain small probability, especially together with adverse influence of the package, and is dependent on the interface devices.

In a worst case a sensitive gate oxide in such an interface device will receive the full voltage difference between two domains. This total voltage difference drops over the sender's channel (Vds2) and the receiver's gate oxide (Vbr2). By using voltage monitor cells with technology specific borders according to the device types the simulation tool can detect critical conditions, Fig. 2.

To track these constellations a special step is performed during the source data conversion which searches for interface elements between domains and checks their device types. For ESD simulations only one representative element of every device combination between two domains is sufficient to predict the ESD robustness. Equal combinations of devices between the same blocks will be considered as one kind of interface element.

3.4 Stressplan generation and simulation flow

International standards for ESD-qualification give strict guidelines for ESD-tests and how devices are to be stressed. A basic rule for two-pin stress is that every pin combination has to be stressed. In detail there are some differences between existing standards, but they resemble in the amount and order of discharges per combination. To reduce the testing efforts pins are sorted into groups and will be stressed depending on their group. Furthermore, for simulation issues it is not needed to simulate every stress combination, e.g. if the input or output pin is part of a low-ohmic package network with more than one pin.

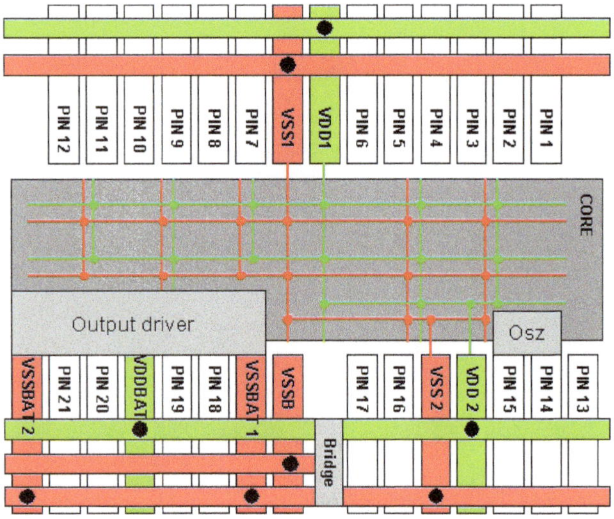

Fig. 4. Product example – layout and principle of pin configuration.

The whole simulation framework is divided into three separate parts (see Fig. 3). The first part generates the netlist and additional data for simulation control. The second main part contains the calculation routines for internal simulation, the control and the interface to the external simulator. The third part is used for the result interpretation.

The topology and design information from a chip/package co-design tool is combined in the ESD netlist builder with data taken from a library cell database. The netlist builder will provide the chip level ESD netlist, an ESD stress plan, i.e. a simulation job file containing all ESD test events as required by current test standards, and a topology data file needed for automated post-simulation analysis. The simulation driver will step through the stress plan and take care of the permutational simulation runs. The user can choose between the implemented circuit simulator and an external circuit simulator, e.g. SPICE.

3.5 Result processing

This simulation generates a huge amount of data which contains the voltage distribution and partial currents of every element along the current path and for every stress combination. A first and rough simulation result analysis can by done in the simulation driver based on the voltage sensors in the ESD netlist. Full data evaluation is performed by the simulation data post-processor. Based on the topology data file voltage drops over functional blocks are monitored and assessed. Current and voltage distributions in the I/O cell ring can be compiled, organized in various graphical views and displayed to the user.

4 Application example

As a proof of concept a simulation run on a real product was performed without additional extractions or reports from the design group.

4.1 Source of netlist information

Within the first step an existing pinlist for ESD-review is used to generate an ESD-netlist, a full stress plan and a file with topology information for the result processing. The subcircuits for padcells are created during library generation and reused for this simulation. The model files with the piecewise linear description of the protection elements are made for this technology and can be used for all products on this technology platform. The input data contains only padring data and abstract information on the core circuitry as is available in the earliest design steps.

4.2 Product description

This sample was a fully digital design with a custom-built output line driver block and an oscillator circuit. Figure 4 shows the simplified floorplan and power connectivity without signal lines.

This 28 pin IC uses a 90 nm standard CMOS process and has two voltage domains. As an additional challenge this chip does not have a closed padring and uses the core power mesh as connection between the two pad stripes. The core mesh for power and ground has to be added to the simulation netlist since it is part of the protection network. Every signal pin has protection elements and is connected to the shown rails for power and ground. The circuit has several control lines with additional protection elements from the digital core to the output driver block which are identified as interface elements and added to the simulation netlist. The bridge between the domains is required by the package. It contains additional protection elements and separates the domains.

4.3 Interface elements

The full IC netlist is used to find every network which is connected to two or more devices in different domains and replace them by voltage sensors. The fail voltage levels are calculated by the sum of the gate oxide breakdown voltage Vbr2 of the receiver and the threshold voltage Vds2 of the sender, as showed in Fig. 2. The search routine for interface elements found 10 common networks between the two domains. These networks are connected to 19 elements which could be simplified to two interface elements due to identical device types. The elements are added to the input pinlist as subcircuits including voltage monitors.

4.4 Netlist generation and simulation

The simulation netlist is generated in two steps by the software suite. First step is the extraction of interface elements and the generation of special subcircuit cells. The whole core mesh is divided into 22 subcircuits as equivalent circuits for the VDD- and VSS-metallization. The division of the core blocks follows product specific and geometrical aspects and leads to a combined subcircuit cell for power and ground network. Between the VDD- and VSS-networks there are high resistances as a substitute for currents during an ESD-pulse caused by capacitance charging. The simulation is controlled from a platform independent graphical user interface.

4.5 Simulation results

The full stressplan for the example product contains 538 stress combinations. A full simulation run at nominal stress conditions (HBM: 2 kV – 1.3 A) consumed about 4 h on a UNIX workstation cluster and generated no fail condition of any used device. Two additional simulation runs were performed for testing the applicability of this software to typical cases in product development. Based on the results of the real product ESD-test a new stressplan was generated. The new stressplan contains test combinations of a selected signal pin (Pin 21) against each power-, ground- and one signal-pin of both domains. During the ESD-test the IC was stressed accordant to the JEDEC-standard which stresses the IC first and performs current measurements afterwards.

4.5.1 Increased stress level

This simulation should check the product behavior for higher stress level. The ICs were subject experimentally to a HBM test with increased stress level (3 kV – 2 A) resulting in leakage current on output driver's signal pins and increased power consumption within the output driver. This implies a fail in a protection element of the signal pin.

The new simulation result showed some failures relating to the signal pins protection elements and the voltage monitors between the two domains. In Fig. 5 the voltage distribution within the VDD mesh is drawn and shows a significant voltage drop between the domains. The fail condition appear after stress combinations which cause a current path over the domain borders.

The failing protection element results in a hard error on an element between VDD_BAT and the stressed signal pin. A reason for this fail in the protection element might be a too weak protection element and a too high resistance in the VDD-VDD_BAT bridge. The resulting error would cause increased leakage current, as was measured in the real IC.

Furthermore the voltage monitor reports an overvoltage which indicates a possible gate oxide gate breakdown at the interface between the domains.

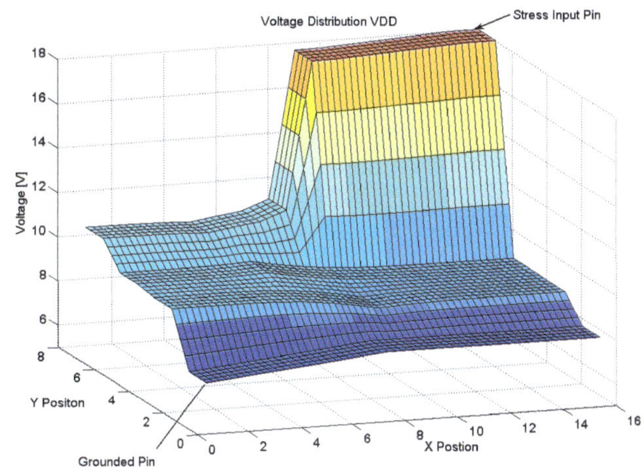

Fig. 5. Voltage distribution VDD mesh.

4.5.2 Required interface protection

A second simulation run with a modified input netlist and a HBM stress level of 2 kV was performed. In this version the additional protection elements at the domain border were removed. These protection elements are mandatory only in non-common ground packages. This simulation run emulates the situation if a design for non-common ground packages were used together with a common ground package.

Voltage monitors were retained at the positions of the removed interface protection elements. They showed warnings in several stress cases. The failure locations indicate a possible breakdown at the interface between domain VDD and VDD_BAT domains. The monitors produced the same voltage warning levels as before. The removed protection elements limited the voltage difference between the power networks in this location. By removing these elements the voltage difference was increased by about 2.5 V at this interface. This would cause a gate oxide breakdown of the involved transistors.

5 CDM simulation methodology

In contrast to specifically designed test circuits, an IC design on full chip level challenges by its high complexity. The proposed "CDM-alike" method adapts the combinational simulation method to extract the voltage levels of a worst case quasi static (DC) state.

The total chip area is normally densely filled by circuit parts. As the p-substrate silicon is connected to VSS and nwells are connected to VDD, the charge present before the CDM discharge can be assumed to be on the metal layers.

CDM discharge is strongly dependent on package type and size (Brodbeck, 1998), as package represents an essential part of the discharge path and a charge source. The package

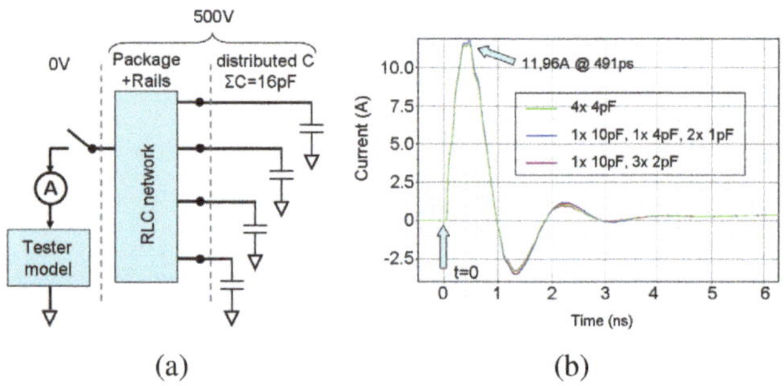

Fig. 6. Simulation setup for transient CDM simulation.

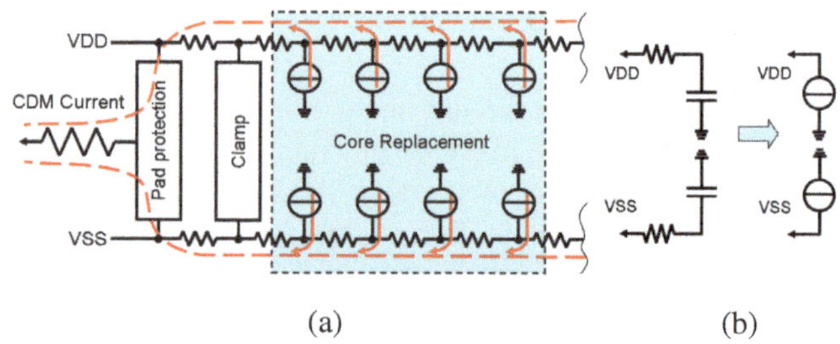

Fig. 7. Simulation setup for DC simulation **(a)** with transformation of capacitors to current sources **(b)**.

together with the ESD-test set-up determines the peak level and the shape of the discharge current to a large extent. The peak CDM current is generated by the distributed capacitors between the IC and the ground plane of the CDM-tester and flows through the testpin.

For transient simulations the distributed capacitors and the capacitances of the package model were charged to 500 V and discharged through the tester, modeled as shown in Fig. 6a. The total capacitance of 16 pF was divided into different combinations of 4 capacitors. The transient results, Fig. 6b, show no differences in peak level and shape of the discharge current through the testpin.

These results suggest a transformation of the CDM charge sources to constant current sources like shown in Fig. 7b. The known total peak current for a specific package is divided into several current sources. The strength of a single current source corresponds to the fraction of the external capacitance contribution of this point.

The final simulation netlist Fig. 7a includes the package and tester models and the ESD-relevant elements of the core area. This core netlist covers the IO-ring with pwl-models, a resistance network for the power and ground-meshes and voltage monitors on typical places. The capacitors within core and package are replaced by constant current sources, Fig. 7b. The sum of all current sources is equivalent to the known peak current of the used package. The voltage monitors will issue an alert if the voltage limit is exceeded.

The biggest advantage of this transformation is the fast and simple simulation setup and the concentration on results relevant for the product verification. The input data can be either abstract information available during floor-planning or extractions from the final layout. In the early design stages both the package and the IO-ring are already known.

The core mesh will be represented by typical resistance networks with a size conformal to the domain size. The voltage monitor limits depend on technology and already known topology of the IC. For simulations on the final layout the extracted netlist of power and ground meshes will be used for the core representation. The package model and the IO-ring model will be the same in both cases.

6 Conclusions

Chip level ESD simulation has been shown to be a vital need in the design flow of ULSI circuits. It will assist in the ESD compliant setup of the overall ESD protection concept mainly determined by the correct layout of I/O cell ring, bond-out and package. A simulation flow and framework was presented that reduces modelling extraction and simulation efforts. The simulation covers possible failure locations and helps to assure the ESD-robustness of upcoming products. The presented ESD simulation methodology can be applied from the product concept phase until the final ESD-review. Three HBM simulation runs showed that the simulation software is able to reproduce realistic errors in the ESD-protection concept.

The presented "CDM-alike" approach satisfies the requirements for a CDM verification procedure during the IC design phase. With the transformation of capacitances to current sources it is possible to set up a fast and meaningful CDM simulation which delivers sufficient information to determine the CDM weaknesses of an IC design reliably. Even when only presented for simple examples these considerations show the direction a DC-simulation based verification for complex system on chip design can go.

References

Brodbeck, T. and Kagerer, A.: Influence of the device package on the results of CDM tests-consequences for tester characterization and test procedure, Proc. EOS/ESD Symposium, p. 320–327, 1998.

Drueen, S.: Virtual ESD Test – An ESD Analysis Methodology at Chip Level, PhD thesis, Technical University of Munich, Germany, Shaker Verlag, 2007.

EOS/ESD Association Inc.: 7900 Turin Road, Rome, NY 13440, USA, Human Body Model (HBM) Component Level – ESD-STM5.1-2001, 2001.

Esmark, K., Stadler, W., Wendel, M., Gossner, H., Guggenmos, X., and Fichtner, W.: Advanced 2D/3D ESD device simulation-a powerful tool already used in a pre-Si phase, Proc. EOS/ESD Symposium, p. 420–429, 2000.

Etherton, M.: Charged Device Model (CDM) ESD in ICs, PhD thesis, ETH Zurich, Switzerland, Hartung-Gorre Verlag, 2006.

Huh, Y., Bendix, P., Min, K., Chen, J., Narayan, R., Johnson, L. D., and Voldman, S. H.: ESD-Induced Internal Core Device Failure: New Failure Modes in System-on-Chip (SoC) Designs, in: Proc. 5th Workshop on SoC For Real-Time Applications (Iwsoc'05), p. 47–53, 2005.

Liu, H.-Y., Lin, C.-W., Chou, S.-J., Tu, W.-T., Chang, Y.-W., and Kuo, S.-Y.: Current path analysis for electrostatic discharge protection, Proc. of ICCAD, p. 510–515, 2006.

Zhan, R., Xie, H., Feng, H., and Wang, A.: ESDZapper: a new layout-level verification tool for finding critical discharging path under ESD stress, Proc. of ASP-DAC, pp. 79–82, 2005.

Zhang, X. Y.: Modeling and Characterezation of Substrate Resistance for Deep Submicron ESD Protection Devices, PhD thesis, Department of Electrical Engineering of Stantford University, 2001.

11

Design issues of arithmetic structures in adiabatic logic

Ph. Teichmann, J. Fischer, F. Chouard, and D. Schmitt-Landsiedel

Institute for Technical Electronics, Technical University Munich, Munich, Germany

Abstract. Since adiabatic logic uses a supply that incorporates both supply voltage and clock signal in one line, adiabatic logic systems have a built-in micro-pipelined architecture. Considering this fact, different design constraints have to be observed compared to static CMOS designs. Complex arithmetic building blocks, like multipliers, mainly consist of adders. Therefore, a comparison of adder structures is performed. Based on these results, multipliers and complex systems can be built. A Discrete Cosine Transformation (DCT) is taken as example for an arithmetic system. Comparing an adiabatic logic implementation of a DCT to its static CMOS counterpart, a significant saving factor of more than 10 can be achieved with the adiabatic system.

1 Introduction

Adiabatic logic circuits are known for offering an energy dissipation less than the $E=\frac{1}{2}CV_{DD}^2$ limit in static CMOS circuits. The Positive Feedback Adiabatic Logic (PFAL) (Vetuli et al., 1996) family has proven to be a reasonable choice for todays VLSI integration (Amirante, 2004; Fischer, 2006) as it is ultra low-power and robust against parameter variations. A chain of PFAL gates is operated via a four-phase clock which acts as power supply and clocking line for the gates, and it is therefore called power-clock. As consequence, cascaded gates form a micropipeline which is an inherent property of adiabatic logic. This fact has to be observed if complex systems are investigated.

Arithmetic structures are basic building blocks in a variety of digital signal processing tasks. E. g. adders are used to build more complex arithmetic functions like multipliers, filters, CORDICs etc. Many proposals for adders focusing on different design goals like power, area and performance have been presented for static CMOS circuits (Koren, 2002; Zimmermann, 1997; Beaumont-Smith and Lim, 2001). But PFAL's inherent properties make it desirable to analyze the known arithmetic structures with respect to their applicability in adiabatic logic.

This work will first introduce the idea of adiabatic logic and the micropipeline in Sect. 2 followed by the results gained from the considerations on adders and multipliers in Sect. 3 followed by a case study of a DCT in Sect. 4. The results are concluded in Sect. 5.

2 Brief introduction of adiabatic logic

In static CMOS the fundamental limit for the energy dissipation per switching event of $E=\frac{1}{2}CV_{DD}^2$ can only be lowered by reducing the capacitance C or by scaling of the supply voltage V_{DD}. To overcome this limit adiabatic circuits have been proposed, that trade frequency for energy. By lowering the operating frequency, theoretically an asymptotic convergence to zero dissipation can be reached. The energy dissipation in a PFAL adiabatic logic gate is described by $E=\frac{RC}{T}CV_{DD}^2$, where R is the path resistance, C is the capacitance at the output, T is the rise/fall time of the power-clock signal. A low threshold voltage V_{th} on the one hand reduces the path resistance R due to an improved gate overdrive voltage, on the other hand it increases the leakage currents which limit the energy dissipation in the mid- and low-frequency range. PFAL circuits show a minimum in energy dissipation at the crossing point of the adiabatic losses ($\propto f$) and the leakage losses ($\propto \frac{1}{f}$). For a 130 nm CMOS process this minimum is retrieved around 100 MHz.

The PFAL logic family uses a four-phase power-clock signal, that acts as the supply voltage and the clocking signal (see Fig. 1), and inherently imposes the circuits beeing operated in a pipelined fashion. The pipeline style gives a first conception how a system in adiabatic logic should be implemented, as overhead due to synchronization degrades the savings gained through the application of adiabatic logic, furthermore pipelining leads to a rise in latency.

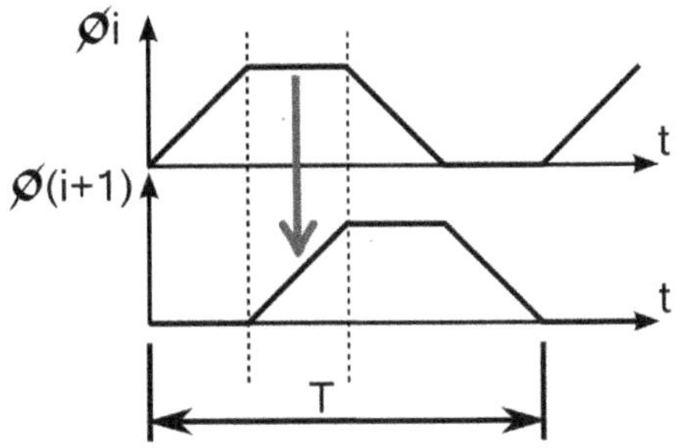

Fig. 1. Two preceeding phases Φi and $\Phi(i+1)$ of an adiabatic four-phase power-clock are shown. When Φi is in its stable high phase, the data is evaluated in the succeeding stage, indicated by the arrow.

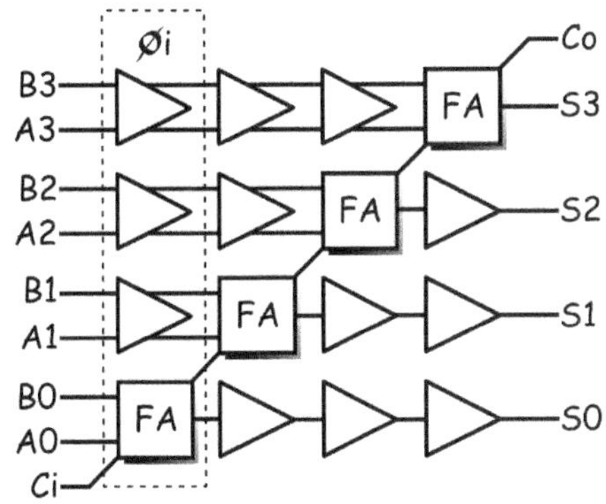

Fig. 2. A N=4 ripple-carry adder in adiabatic logic. Notice the overhead due to the synchronization buffers. Each input bit of A and B has to be buffered, leading to two buffers per bit position. The dashed box shows a clock domain of the power clock Φi.

3 Binary adder structures

The inherent properties of adiabatic logic must be taken into account in the design of adiabatic arithmetic structures. Large overheads in structures result in suboptimal designs.

One advantage of PFAL is its dual-rail signal representation, so a signal is inverted without additional gates. Subtractors are easily obtained from adders, by swapping the signal rails of the subtrahend and feeding a logic one into the carry-in signal of the adder. The results for the adders gained in this section are valid for the subtractors as well.

Basically two groups of adders exist, namely the carry-propagate and the parallel-prefix adders (Sklansky, 1960), (Kogge and Stone, 1973), (Ladner and Fischer, 1980), (Brent and Kung, 1982) and (Han and Carlson, 1987). The ripple-carry adder (RCA) is the simplest implementation of a carry-propagate adder, using N full-adder (FA) cells in static CMOS, where N is the input width of the data words. But for adiabatic logic, the rippling of the carry signal leads to an overhead of synchronizing buffer

stages of $\mathcal{O}(N^2)$, making the ripple-carry structure in adiabatic logic (Fig. 2) an inproper choice if we talk about high bit width N. Additionally, the adiabatic ripple-carry adder (RCA) utilizes $N/4$ clock cycles, leading to a relatively fast rising latency. Only for several suqsequent addition operations, the RCA can be used in a nested way, as pictured in Fig. 3, can be applied efficiently, as the absolute overhead of synchronizing buffers per arithmetic operations remains almost constant. Especially for butterfly structures, as used in the Discrete Cosine Transformation (DCT) in Sect. 4, the nested RCA is advantageous. Other approaches try to reduce the critical path, i.e. the path of the rippling carry, to gain speed in static CMOS designs, respectively to reduce the latency in adiabatic logic. As we are talking about synchronous, pipelined designs, approaches like the carry-bypass adder cannot be applied to adiabatic logic. The carry-select scheme splits the input words into smaller groups, i.e. 2 of size $N/2$ each and pre-calculates the sums

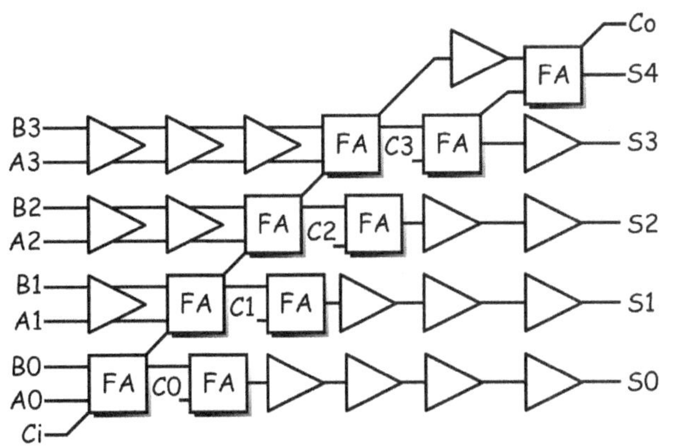

Fig. 3. For a nested RCA (here with two suqsequent addition operations) structure, the relative overhead due to synchronization is reduced. Please note, that inputs C also have to be delayed via input buffers.

for both input carry alternatives. The incoming carry from the preceeding group selects the appropriate output via a multiplexer. Such a design trades area, respectiveley energy, against speed in static CMOS and latency in adiabatic logic respectively. Additionally an overhead from the multiplexers arises for such an adder. If we take the RCA from Fig. 2 and split the adder in two groups of 2 bits, the design uses 3 blocks consisting of 2 full-adder cells and three buffers each, one 6:3 multiplexer and buffers that synchronize the output bits of the lower block to the outputs of the multiplexer. In Fig. 4 for the arrangement with a multiplexer of logic depth equal to 1 it can be seen, that this design uses more full-adders than the RCA in Fig. 2, but less buffers. If the RCA and the carry-select adders are compared for high bit width N it can be seen that the RCA suffers from a higher energy consumption. This estimation does not account for the energy

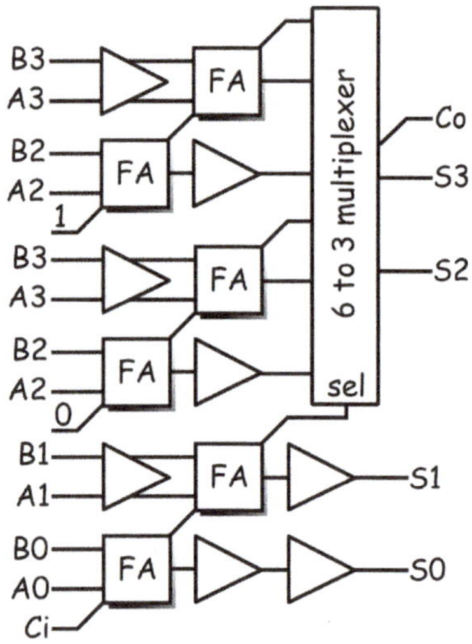

Fig. 4. A 4 bit RCA carry-select adder using a multiplexer with logic depth of 1.

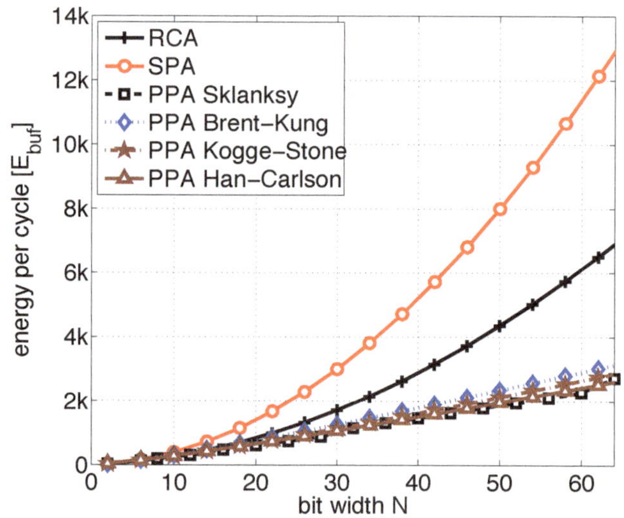

Fig. 5. Comparison of various adder structures up to 64bits. The parallel-prefix adders show the lowest energy consumption.

dissipation caused in the multiplexer. But as only $N/2$ XOR gates are used for the multiplexer, the energy consumed by the multiplexer is negligible. To avoid the synchronizing overhead observed for the RCA, parallel-prefix schemes are investigated. The parallel-prefix adders (PPA) allow to calculate the output of the m-th bit via logical combination of the inputs 0 to $mu-1$. Therefore, theoretically each output of the adder can be calculated using one complex gate. Because of practical reasons in stacking transistor devices, only a maximum of 3 or 4 inputs is combined at a time. In literature different schemes for the parallel prefix algorithm are reported, differing in logical depth and fanout. Here a variety of PPA structures (Sklansky, Brent-Kung, Kogge-Stone, Han-Carlson) are estimated to see, which adder structure is desirable in adiabatic logic. E.g. the Sklansky structure has the lowest logical depth, but the layout is supposed to be irregular. The Han-Carlson structure applies a more regular layout for the price of a higher logical depth. Additionally, the RCA structure and a serial prefix adder (SPA) are estimated in respect on their energy consumption. Estimations are performed by counting the number X_i of each class of gates i (e.g. INV, FA) and multiplying it with its corresponding mean energy value E_i. The whole system's energy dissipation E is then calculated by

$$E = \sum_i X_i E_i. \tag{1}$$

Gates are characterized by means of SPICE simulation; industrial 130nm CMOS process parameters of low-V_{th} devices are used. The gates are simulated at a frequency of 100MHz and a supply voltage of V_{DD}=1.2V. In Fig. 5 the investigated adder structures are presented. The RCA rises in energy consumption quadratically with the bit width N. The parallel-prefix adders present the lowest energy consumption. The serial-prefix adder consumes the largest amount of energy and thus is no alternative. In the detail view in Fig. 6 only the RCA and the most promising PPA structures are presented. Up to 8bit input width, the RCA is the best choice.

For the decision between Sklansky and Han-Carlson different properties have to be taken into consideration. As said before, the Han-Carlson structure has a more regular design, thus saving effort in layouting. Additionally the Han-Carlson adder has a fixed maximum fan-out of 2. For the Sklansky structure, the effort for layout will be higher, and the fanout rises with $N/2$. So the decision, which design is a suitable low-power adder structure will depend on the post-layout extraction, taking into account the rise of the energy consumption due to parasitic capacitances.

4 Discrete cosine transformation: a case study

In image processing the Discrete Cosine Transformation (DCT) is used to compress pictures. A DCT with minimal hardware amount is presented in (Heyne et al., 2006). As the DCT is a strongly parallel algorithm, that can be efficiently implemented using the adiabatic micropipeline, it is a good demonstrator to show the energy savings gained by using adiabatic logic in comparison to static CMOS. The DCT is built of adders, subtractors and buffers as can be seen in Fig. 7.

Different static CMOS implementations have been compared. Flip-flops are power-consuming components in a synchronous design in static CMOS. Different levels of pipelining have been investigated and been opposed to the adiabatic logic DCT implementation. The adiabatic reference design

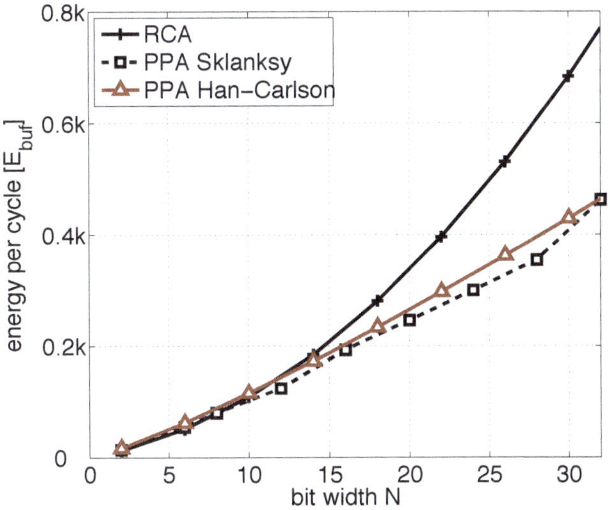

Fig. 6. Comparison of RCA and most promising PPA adders, Han-Carlson and Sklansky PPA, up to 32bits. The RCA exhibits the lowest energy consumption up to 8bit.

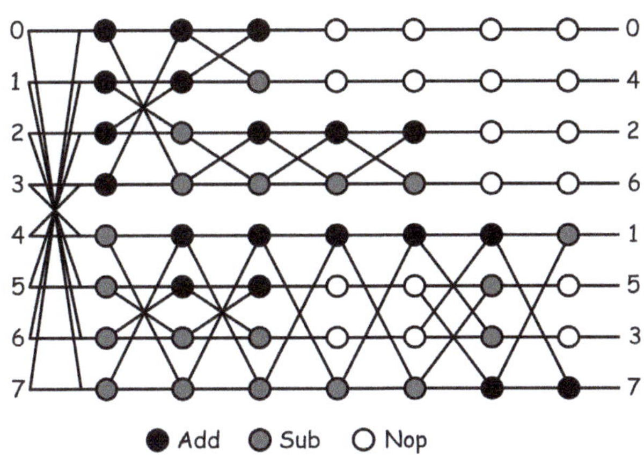

Fig. 7. Scheme of the DCT, where a Nop (No Operation) is a synchronizing buffer stage.

is implemented as nested RCA structure, thus a minimum number of FA cells are used and the overhead due to synchronization is reduced due to nesting. For static CMOS a 2D-pipelined carry-propagate adder (CPA) has been estimated first. To reduce the number of flip-flops, in **V1** to **V3** a CSA has been used to avoid skewing and deskewing at the inputs, respectively at the outputs. In **V1** a flip-flop is introduced after each CSA stage, in **V2** after every second CSA stage, and in **V3** no pipelining was introduced at all. Results are calculated according to

$$E_{AL} = X_{FA,AL} * E_{FA,AL} + X_{BUF,AL} * E_{BUF,AL} \quad (2)$$

$$E_{CMOS} = X_{FA,CMOS} * E_{FA,CMOS} \quad (3)$$

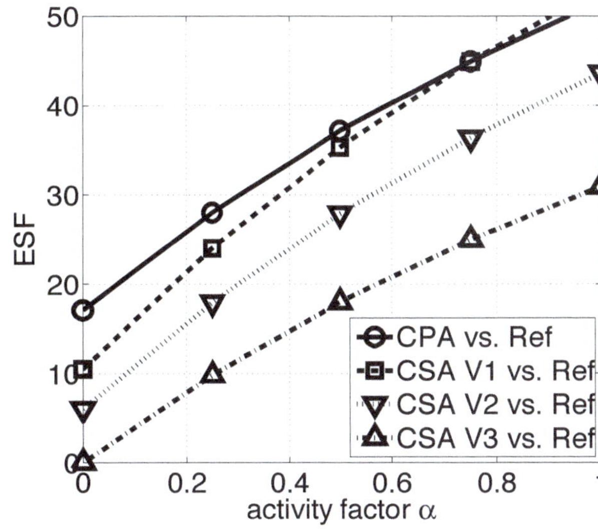

Fig. 8. Comparing the differing static CMOS DCT implementations to the adiabatic DCT. For static CMOS, first a 2D-pipelined carry-propagate adder (CPA) has been implemented and **V1** to **V3** show carry-save array (CSA) structures with different degrees of pipelining, where **V3** has no pipelining at all.

$$+ X_{INV,CMOS} * E_{INV,CMOS}$$

$$+ X_{FF,CMOS} * E_{FF,CMOS}$$

where the index AL stands for adiabatic logic, FA for full-adder, BUF for a buffer, INV for an inverter, and FF for a flip-flop.

Especially for static CMOS an important parameter for energy dissipation is the activity α, whereas in PFAL activity has a minor impact on the dissipation. To allow a fair comparison, activity has to be considered in the estimations via

$$E(\alpha) = E_{\alpha=0} + \alpha \left(E_{\alpha=1} - E_{\alpha=0}\right), \quad (4)$$

where $E_{\alpha=0}$ and $E_{\alpha=1}$ are the respective mean energy dissipation values for activity α=0 and α=1.

The results of the comparison are given in Fig. 8. All estimations are based on the simulations in an industrial 130nm CMOS process, at 100 MHz and with a V_{DD}=1.2V. The mean energy dissipation values for the used gates are summarized in Table 1 for α=0 and α=1.

In Fig. 8 we see that the best implementation for static CMOS is **V3**, that uses no pipelining. There the energy saving factor

$$\mathrm{ESF} = \frac{E_{\mathrm{diss,CMOS}}}{E_{\mathrm{diss,AL}}} \quad (5)$$

reaches its minimum value. Depending on the activity the ESF reaches up to a factor of 30 for the highest activity of the system. Applying a realistic activity value of α=30% (Ye and Roy, 2001) an ESF of factor 10 can be gained. The

dissipated energy per cycle [J]					
family	AL		CMOS		
gate	FA	BUF	FA	INV	FF
α=0	2.10f	0.24f	3.48a	1.59a	6.46f
α=1	2.67f	0.33f	37.6f	2.66f	19.0f

Table 1. Simulated energy per cycle dissipation values for adiabatic and static CMOS gates. All estimations are performed for α=0 and α=1.

generation of the four-phase clock itself is lossy, the ESF must take into account the efficiency of the oscillator. Oscillators with an efficiency of 50% have been reported. There is still room for further improvements of the oscillator efficiency. For a system level comparison the losses on the clock net in an static CMOS system have to be considered here as well. This requires detailed information on the topologie of the clock net. Predictions for high-performance systems show that almost as much energy in the clock distribution is used as in the logic itself. At least flip-flops are used at the inputs of the structure and the outputs of the structure, leading to a slight increase of the dissipation even for the adder **V3** on system level. Thus the factor of 10 presented here should also be a good indication for the savings in a complete system achievable with adiabatic logic.

5 Conclusions

In this paper the inherent properties of adiabatic logic have been analyzed for arithmetic building blocks. Adder structures were investigated and compared for the first time showing that due to the synchronization overhead of $\mathcal{O}(N^2)$ the ripple-carry adder is only suitable for bit sizes $N \leq 8$. Parallel-prefix adders are a good choice, as the reduced depth of the adder also reduces the area and the synchronization overhead and thus the power consumption. The Han-Carlson and the Sklansky architecture show the lowest energy consumption of all parallel-prefix schemes. The Discrete Cosine Transformation is an arithmetic structure making use of adiabatic logic's inherent pipelining structure. The nested ripple-carry adder scheme allows major savings against comparable designs in static CMOS. For an activity factor of 30%, savings of around factor 10 can be achieved.

Adiabatic logic is therefore a circuit family that allows the implementation of ultra low-power digital signal-processing tasks.

References

Amirante, E.: Adiabatic Logic in Sub-Quartermicron CMOS Technologies, vol. 13 of Selected Topics of Electronics and Micromechatronics, Shaker, 2004.

Beaumont-Smith, A. and Lim, C.-C.: Parallel prefix adder design, in: 15th IEEE Symposium on Computer Arithmetic, 218–225, 2001.

Brent, R. P. and Kung, H. T.: A regular layout for parallel adders, IEEE Transactions on Computers, C-31, 260–264, 1982.

Fischer, J.: Adiabatische Schaltungen und Systeme in Deep-Submicron-CMOS-Technologien, vol. 24 of Selected Topics of Electronics and Micromechatronics, Shaker, 2006.

Han, T. and Carlson, D. A.: Fast area-efficient VLSI adders, in: Proceedings of the 8th Symp. Comp. Arithmetic, 49–56, 1987.

Heyne, B., Sun, C.-C., Goetze, J., and Ruan, S.-J.: A Computationally Efficient High-Quality CORDIC based Loefller DCT, in: 14th European Signal Processing Conference (EUSIPCO), 2006.

Kogge, P. M. and Stone, H. S.: A parallel algorithm for the efficient solution of a general class of recurrence equations, IEEE Transactions on Computers, C-22, 786–793, 1973.

Koren, I.: Computer Arithmetic Algorithms, A K Peters, second edn., 2002.

Ladner, R. E. and Fischer, M. J.: Parallel Prefix Computation, J. Assoc. Computing Machinery, 27, 831–838, 1980.

Sklansky, J.: Conditional-sum addition logic, IRE Transactions, EC-9, 226–231, 1960.

Vetuli, A., Pascoli, S. D., and Reyneri, L. M.: Positive feedback in adiabatic logic, Electronics Lett., 32, 1867–1869, 1996.

Ye, Y. and Roy, K.: QSERL: Quasi-Static Energy Recovery Logic, IEEE Journal of Solid-State Circuits, 36, 239–248, 2001.

Zimmermann, R.: Binary Adder Architectures for Cell-Based VLSI and their Synthesis, Ph.D. thesis, ETH Zurich, 1997.

12

Area- and energy-efficient CORDIC accelerators in deep sub-micron CMOS technologies

U. Vishnoi and T. G. Noll

Chair of Electrical Engineering and Computer Systems, RWTH Aachen University, Aachen, Germany

Correspondence to: U. Vishnoi (vishnoi@eecs.rwth-aachen.de)

Abstract. The COordinate Rotate DIgital Computer (CORDIC) algorithm is a well known versatile approach and is widely applied in today's SoCs for especially but not restricted to digital communications. Dedicated CORDIC blocks can be implemented in deep sub-micron CMOS technologies at very low area and energy costs and are attractive to be used as hardware accelerators for Application Specific Instruction Processors (ASIPs). Thereby, overcoming the well known energy vs. flexibility conflict. Optimizing Global Navigation Satellite System (GNSS) receivers to reduce the hardware complexity is an important research topic at present. In such receivers CORDIC accelerators can be used for digital baseband processing (fixed-point) and in Position-Velocity-Time estimation (floating-point). A micro architecture well suited to such applications is presented. This architecture is parameterized according to the wordlengths as well as the number of iterations and can be easily extended for floating point data format. Moreover, area can be traded for throughput by partially or even fully unrolling the iterations, whereby the degree of pipelining is organized with one CORDIC iteration per cycle. From the architectural description, the macro layout can be generated fully automatically using an in-house datapath generator tool. Since the adders and shifters play an important role in optimizing the CORDIC block, they must be carefully optimized for high area and energy efficiency in the underlying technology. So, for this purpose carry-select adders and logarithmic shifters have been chosen. Device dimensioning was automatically optimized with respect to dynamic and static power, area and performance using the in-house tool. The fully sequential CORDIC block for fixed-point digital baseband processing features a wordlength of 16 bits, requires 5232 transistors, which is implemented in a 40-nm CMOS technology and occupies a silicon area of 1560 μm^2 only. Maximum clock frequency from circuit simulation of extracted netlist is 768 MHz under typical, and 463 MHz under worst case technology and application corner conditions, respectively. Simulated dynamic power dissipation is 0.24 uW MHz^{-1} at 0.9 V; static power is 38 uW in slow corner, 65 uW in typical corner and 518 uW in fast corner, respectively. The latter can be reduced by 43 % in a 40-nm CMOS technology using 0.5 V reverse-backbias. These features are compared with the results from different design styles as well as with an implementation in 28-nm CMOS technology. It is interesting that in the latter case area scales as expected, but worst case performance and energy do not scale well anymore.

1 Introduction

The CORDIC-approach is a well known versatile algorithm, which can be used to implement a large variety of arithmetic functions being quite complex to implement on standard processor architectures. Consequently, it is widely applied in Digital Signal Processing (DSP) on today's SoCs for especially but not restricted to digital communications. Typical tasks comprise (Numerically Controlled Oscillator) NCO, phase rotating, filtering, modulation and demodulation to name a few. A very attractive application of this is to use dedicated CORDIC macros as accelerators for certain arithmetic functions to Application Specific Instruction Processors (ASIPs). If the CORDIC macro is properly optimized for very low area and energy costs, the overhead becomes very small and by that a remedy to the well known "flexibility vs. energy" conflict can be achieved.

In this contribution, CORDIC macros are considered for the use in highly area- and energy efficient digital GNSS receivers. It will be shown in the later sections of the paper that CORDIC can be used in the baseband signal processing part as well as in the so-called Position Velocity Time (PVT) estimation unit. In the future, further applications of CORDIC building blocks can be expected. For example, it can be used in Singular Value Decomposition- (SVD) based jammer or interferer supression in multisystem GNSS receivers.

In order to cover the different specifications for these GNSS receiver applications and also to ensure re-usability in other application domains, a highly flexible, parameterized CORDIC architecture was elaborated. Thereby, a major goal was to achieve an architecture being well suited for implementation in today's deep sub-micron CMOS technologies.

This paper is divided into 5 sections. A brief description of CORDIC algorithm is presented in Sect. 2. Section 3 refers to the proposed flexible CORDIC architecture template and the building blocks required for its implementation are described in detail. Implementation of an exemplary CORDIC macro and results also are presented in the end of Sect. 3. In Sect. 4, as an exemplary application, a CORDIC macro well-suited as a floating-point co-processor to an ASIP is discussed.

2 CORDIC algorithm

The Coordinate Rotation Digital Computer (CORDIC) algorithm was first introduced by Volder (1959) for the computation of trigonometric functions. Later on, it was generalized to hyperbolic functions, multiplication, division etc. by Walther (1971). The algorithm allows for a hardware efficient implementation of arithmetic and trigonometric functions by rotating a vector in a two-dimensional plane using only shift- and addition/subtraction operations being well suited for VLSI-CMOS realization. Thereby, a given input vector $(\boldsymbol{x}_{\text{in}}, \boldsymbol{y}_{\text{in}})$ is rotated in an iterative approach by a series of predetermined angles in each step; the direction of each elementary rotation is determined similarly to non-restoring division.

In more detail, after an initialization

$$x_0 = \boldsymbol{x}_{\text{in}}, y_0 = \boldsymbol{y}_{\text{in}}, z_0 = z_{\text{in}} \tag{1}$$

the vector is iteratively rotated by applying the CORDIC equations

$$\begin{aligned} x_{i+1} &= x_i - m \cdot y_i \cdot d_i \cdot 2^{-i} \\ y_{i+1} &= y_i + x_i \cdot d_i \cdot 2^{-i} \\ z_{i+1} &= z_i - d_i \cdot e_i, \end{aligned} \tag{2}$$

where, $e_i = \tan^{-1}(2^{-i})$ is the i-th element of the predefined step angle sequence. The parameter m specifies whether the rotations are performed on a circle $(m = 1)$, on a straight line $(m = 0)$, or on a hyperbola $(m = -1)$. There are two modes of CORDIC operations for each of the above defined rotations: the "rotate mode" and the "vectoring mode". For this application it is sufficient to consider the case of $(m = 1)$ with the two possible CORDIC operations. In the rotate mode, the vector $(\boldsymbol{x}_{\text{in}}, \boldsymbol{y}_{\text{in}})$ has to be rotated by a given angle z_{in} and in order to converge z_i to zero the direction d_i of rotation in each individual iteration is determined by the sign of the remaining angle z_i

$$d_i = \text{sgn}(z_i). \tag{3}$$

In the vectoring mode, the vector has to be rotated into the x-axis and therefore, in order to converge y_i to zero the direction d_i of rotation is determined by the sign of the remaining angle z_i

$$d_i = -\text{sgn}(y_i). \tag{4}$$

As Eq. (2) actually perform so-called pseudo-rotations only, at the very end, after n iterations the x- and y-components have to be corrected by a constant K .

$$\begin{aligned} x_{\text{out}} &= K \cdot x_n \\ y_{\text{out}} &= K \cdot y_n \\ z_{\text{out}} &= K \cdot z_n. \end{aligned} \tag{5}$$

This constant is determined by the series of step angles only and especially being independent from the individual directions of rotation d_i. The series of step angles can be optimized by repeating individual elements towards a correction constant K featuring a small number of non-zero digits in canonical signed digit representation easing these final multiplications.

3 CORDIC architecture template

As can be seen from Eq. (2), only additions/subtractions and shift operations are required for CORDIC iterations. In addition to that, registers are required in pipelined CORDIC architectures and multiplexers are required in the feedback time-shared architectures.

Figure 1 shows the block diagram of a fully iterative CORDIC macro performing rotations on a circle $(m = 1)$. The input $rot/vect$ is used to switch between rotate and vectoring mode and Initialization is controlled by the input $init$. A single CORDIC stage is multiplexed in time i.e. for each iteration the intermediate results are fed back. The multiplication by 2^{-i} is implemented using shifters which are controlled by the iteration index i. A look-up table (not shown in the figure) feeds the step-angle values e_i, 2's complement subtraction is implemented by using XOR stages and proper LSB 2's complement correction. Final multiplication by the correction constant K, can be implemented by simple shift and add operations (not shown in the Fig. 1).

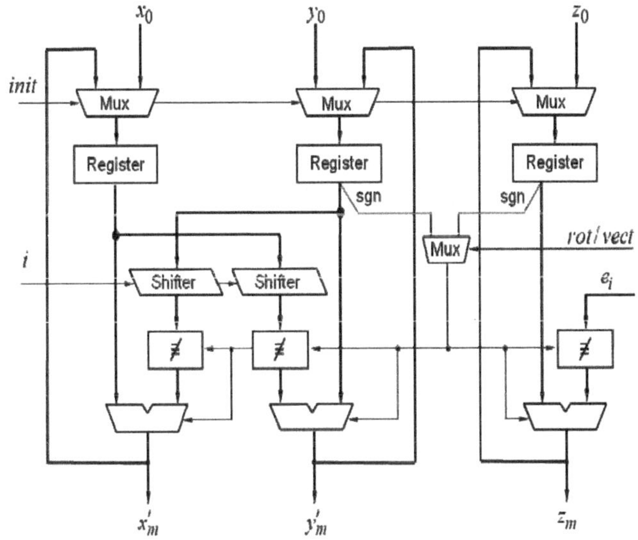

Fig. 1. Basic structure of a fully iterative CORDIC macro.

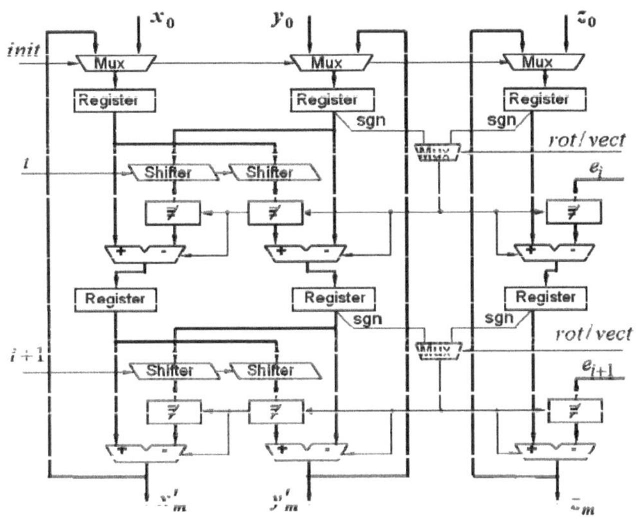

Fig. 2. Partially unrolled CORDIC block diagram (unrolling factor $k = 2$).

As shown in Fig. 2, area can be traded for throughput by partially unrolling the iterative CORDIC structure. Applying two (unrolling factor $k = 2$) pipelined stages increases the area by a factor of about two, but halves the number of clock cycles per CORDIC operation and improves the throughput by a factor of about two. The unrolling can be extended to the case where the unrolling factor k becomes equal to the number of iterations n. With increasing k, the requirements to the programmable shifters reduce. Finally for the fully unrolled structure no shifters are required at all anymore, just the dedicated interconnects are enough. In conclusion, the parameters of this architecture template consist of wordlengths, number of iterations and unrolling factor.

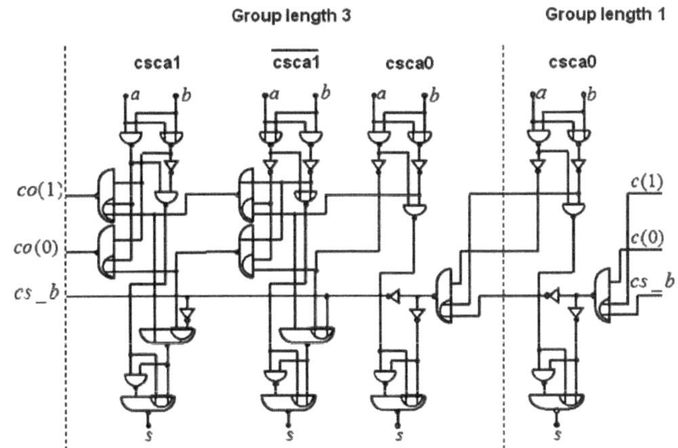

Fig. 3a. The basic cells of carry select adder.

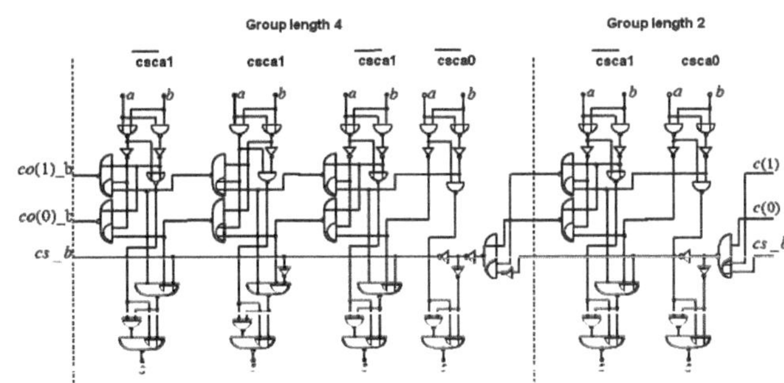

Fig. 3b. The basic cells of carry select adder.

3.1 Building blocks

The major building blocks of the template architecture are the adders and shifters. For the envisaged applications with wordlengths in the range of 16 to 32 bits, carry select adders offer a high area- and energy efficiency (i.e. performance over cost ratio). Figure 3a shows the cell arrangement in the LSBs of carry select adder following the simple arithmetic progression of group lengths 1, 3, 5, 7... and Fig. 3b shows the cell arrangement for the progression of 2, 4, 6, 8... In conclusion, from only four basic cells, delay optimal carry select adders with wordlengths of 9, 12, 16, 20, 25, 30, 36 can be generated and for other wordlengths the MSB group has to be pruned.

The basic cells were implemented in 40-nm CMOS technology; Fig. 3c shows the layouts. To improve the Area-Throughput-Energy (ATE) efficiency, device sizes of the basic cells were optimized automatically by using an in-house tool. All devices belonging to the critical path are carefully optimized for the best possible timing. The tool starts from minimum sized n- and p- devices and in every iteration all the possible permutation of increasing the transistor widths are analyzed. If no further improvement can be achieved the iterations are stopped. Finally from a set of dimensioning candidates the one with lowest ATE complexity was selected. By

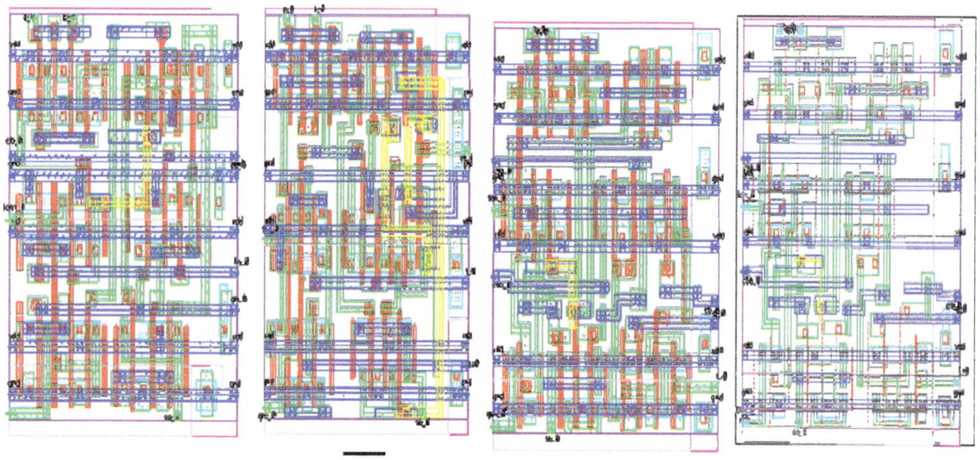

Fig. 3c. Layouts of the basic cells of the carry select adder.

applying this procedure to the adder basic cells, typically a performance improvement of about 25 % was achieved, without sacrificing the efficiency. Moreover, on the positive note, the ATE complexity could even be reduced by typically 5 %.

For the shifter, a logarithmic shifter structure was chosen because it offers high efficiency for the range of above mentioned wordlengths. Only one basic cell namely a one-bit multiplexer had to be implemented in layout. The multiplexer cells as well as the other cells (XOR, register) were optimized in the same way as was described for the carry select adder cells.

3.2 Implementation and results

The CORDIC template architecture was described as a parameterized signal flow graph. With an in-house datapath generator tool, a complete CORDIC macro layout could be generated automatically from only five basic cells for different sets of parameters.

CORDIC macros for two exemplary cases in a digital GNSS receiver were implemented and analyzed in detail.

The first one is used in the baseband processing unit and requires 16 bits wordlength in integer data format and 10 iterations (Kappen, 2011). Figure 4 shows the layout of the CORDIC macro ($k = 1$) without including the Read Only Memory- (ROM) based step angle look-up table in it which can be placed into the blank space. In 40-nm CMOS technology, it contains 5232 transistors in total and occupies a silicon area of 1560 μm^2 only. The maximum clock frequency from extracted netlist simulation is 768 MHz under typical, and 463 MHz under worst case technology and application corner conditions, respectively. Total power dissipation at 500 MHz and 0.9 V supply voltage is 185 uW only.

As already mentioned, one of the architecture parameters is the unrolling factors k. Figure 5 shows the Area-Throughput (AT) complexity diagram for CORDIC blocks with $n = 10$ iterations and for different unrolling factors k. For partial unrolling, silicon area increases with a factor of less than two because one level of the shifters can be saved. In the case of fully unrolled macro silicon area is significantly smaller than would be expected from ideal scaling as the shifters can be completely replaced by dedicated interconnects. A fully unrolled CORDIC macro ($k = 10$) allows for a minimum cycle time of 1.2 ns under typical and 2ns under worst case conditions.

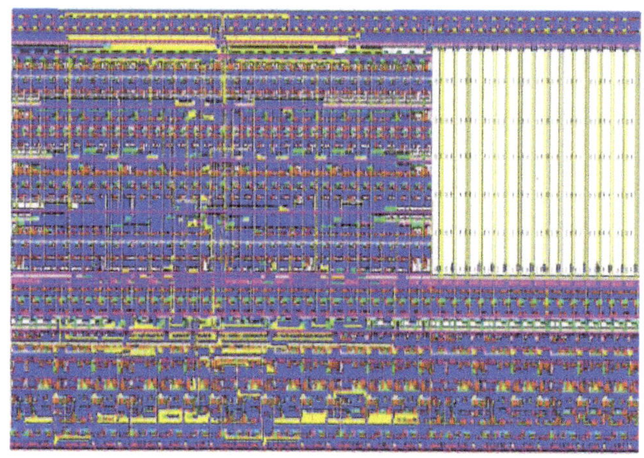

Fig. 4. Layout of CORDIC macro in 40-nm CMOS technology.

The fully iterative macro ($k = 1$) was also extracted and simulated in 28-nm CMOS technology, leading to 56 % of saving in area and a maximum clock frequency is 876 MHz under typical, and 555 MHz under worst case conditions, respectively. Total power dissipation at 500MHz and 0.9 V supply voltage is 134 uW.

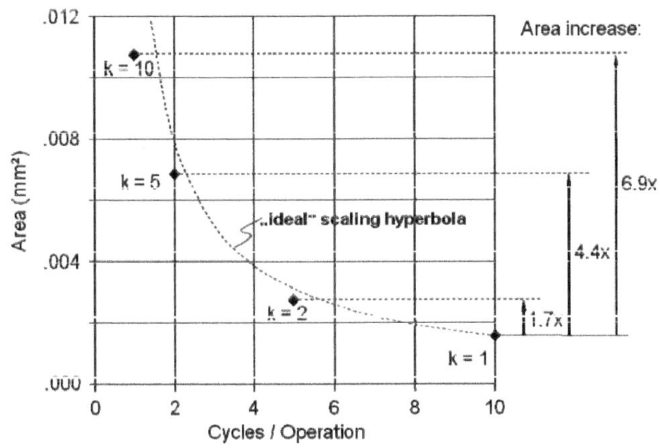

Fig. 5. AT complexity diagram for different unrolling factors k.

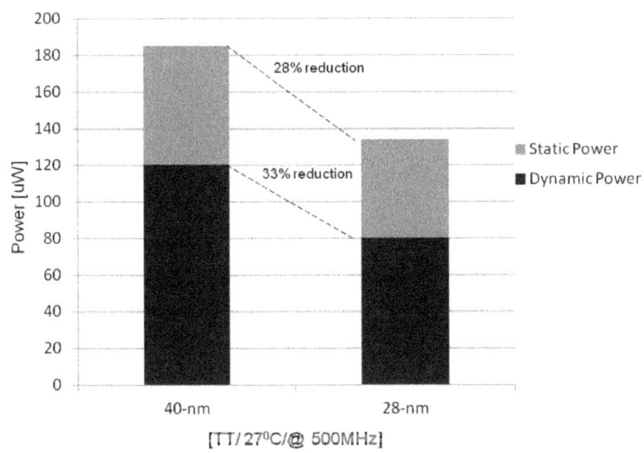

Fig. 6a. Power breakdown at 500 MHz [TT/27 °C/500 MHz].

Figure 6a shows the power breakdown under typical conditions for 40- and 28-nm CMOS technologies. In the fast corner at 125 °C as shown in Fig. 6b static power dominates by far for both technologies.

And if a reverse back bias voltage of 0.5 V is applied, 43 % of the total power is saved in 40 nm and 34 % of the total power is saved in 28-nm CMOS technology.

4 CORDIC macro as an ASIP co-processor

The second CORDIC macro is used as a co-processor to an ASIP, which in the digital GNSS receiver is mainly used to perform the PVT estimation (Kappen, 2011). In this case floating point data format with 28 bits mantissa wordlength and a number of $n = 30$ iterations is required. As long as rotations are performed only on circles, integer CORDIC macros can be used for floating point data format too with only some simple pre-processing as shown in Fig. 7a. As the length of the vector is kept constant during rotation (except for the correction factor) an initial normalization

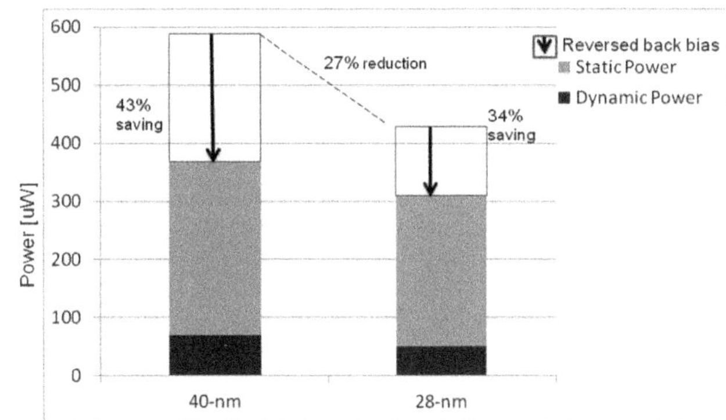

Fig. 6b. CORDIC macro power dissipation.

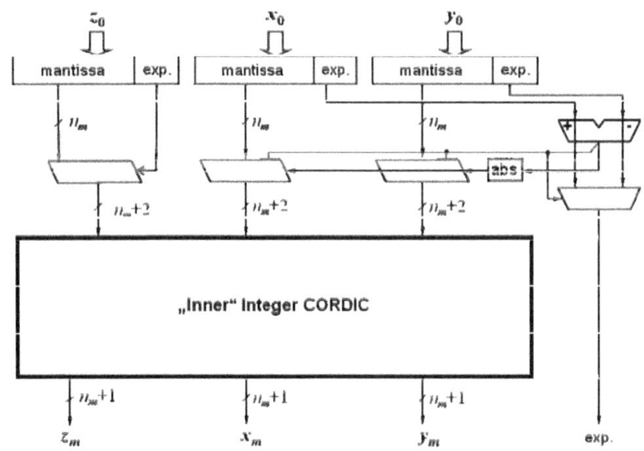

Fig. 7a. Schematic of the Pre-processing for a floating point CORDIC.

of x_0 and y_0 has to be performed so that the exponents get aligned. This ensures that in all the successive iterations the mantissa words can be added/subtracted without further normalization.

So, after this pre-processing the integer CORDIC can be used, if provided two extra front bits. One extra bit is required because the x- and y-components can be enlarged by a factor of $\sqrt{2}$ during rotation and another extra bit is required because the correction by K is done at the very end. This pre-processing can be implemented on dedicated hardware. In case the CORDIC is used as a co-processor to an ASIP an attractive alternative is to get the normalization done by the ASIP itself. Same is true for the post-processing, i.e. the multiplication of the output values with the correction factor K.

Figure 7b shows the block diagram of the coupling between the ASIP and the co-processor. In principle, the co-processor can be tightly or loosely coupled to the ASIP. For the application in PVT estimation a tightly coupled co-processor is the preferred alternative.

Table 1 contains the features of the GNSS-ASIP, of non-optimized synthesized floating point CORDIC and of the

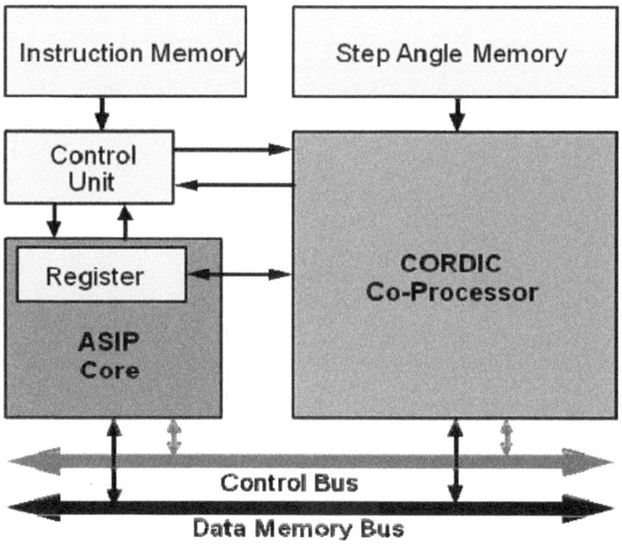

Fig. 7b. Block diagram of the ASIP/CORDIC co-processor structure.

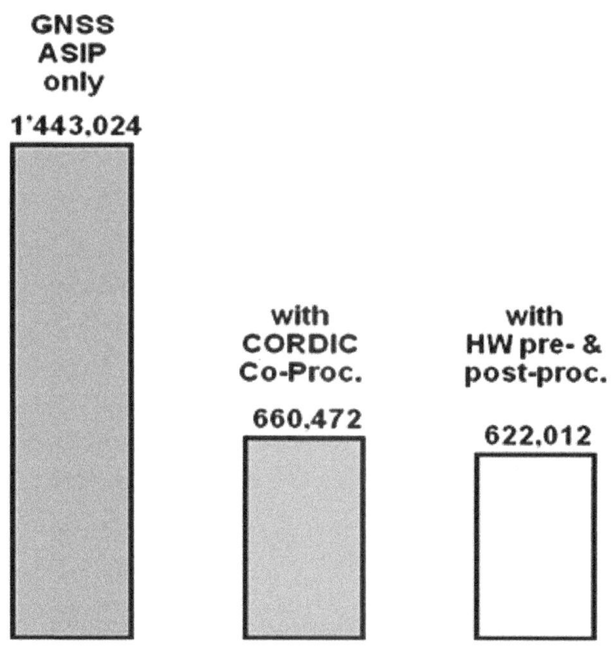

Fig. 7c. Clock cycle per PVT calculation (scenario with 7 satellites).

implementation presented here. The features for the ASIP as well as for the synthesized CORDIC were derived from a 90-nm CMOS implementation (Kappen, 2011) and properly scaled down to 40-nm CMOS technology. As can be seen from the table, the non-optimized CORDIC operated significantly slower than the ASIP itself and the optimized CORDIC would be well-suited to accelerate a faster version of the ASIP. As an alternative with a modified coupling the

Table 1. GNSS-ASIP/Co-Processor features (40 nm).

	Synthesized GNSS ASIP IP	Not optimized Synthesized FP CORDIC	30-bit Optimized CORDIC
Silicon area	0.02 mm^2	0.013 mm^2	0.0033 mm^2
Max. clock Frequency (typ.)	96 MHz	35 MHz	> 500 MHz
Energy per clock cycle (typ. at fmax)	12 pJ	9 pJ	0.78 fJ

co-processor could be operated at 500 MHz while the ASIP is operated at about 100 MHz. Moreover, the silicon area and energy per clock cycle of the optimized CORDIC macro are improved by orders of magnitude in comparison to the non-optimized one and are absolutely negligible in comparison to the ASIP.

Figure 7c shows a comparison of the number of clock cycles required per PVT calculation (for constellation with 7 satellites). The ASIP without co-processor requires 1 443 024 clock cycles. The ASIP with co-processor and with the pre- and post-processing also being done by the ASIP requires 660 472 clock cycles. If the floating point pre- and post-processing would be realized in hardware too, then the number of clock cycles would be further reduced to 622 012. Clearly, the most attractive compromise is to let the pre- and post- processing be done by the ASIP which shows an improvement by a factor of 2.18 in cycle count per PVT estimation.

This improvement can even be increased by letting the ASIP continue to execute instructions concurrently to the multi-cycle operations of the co-processor, i.e. the ASIP is not halted during the CORDIC executions.

5 Conclusions

A micro architecture for highly flexible, scalable, area and energy efficient CORDIC macro being well-suited for implementation in deep sub-micron CMOS technologies has been elaborated. Exemplary macros were designed in 40-nm and 28-nm CMOS technologies for application in a digital GNSS receiver. These macros feature very attractive performance, area and power figures. As to be expected for those technologies worst case power dissipation is dominated by static power. But even in the most advanced 28-nm CMOS technology static power dissipation can be significantly reduced by reverse back biasing.

It has also been shown that coupling of the CORDIC macros to an ASIP usage as a co-processor is a very attractive application. For the case of requirement of floating point data format it has been shown that mapping the proper pre-

and post-processing to the ASIP and applying only an integer CORDIC macro yields an attractive solution.

Acknowledgements. The authors thank their colleague Götz Kappen for discussion and for features of the GNSS ASIP.

References

Kappen, G.: Konzeption und Entwurf Flächen- und Energie-effizienter digitaler GNSS-Empfänger, PhD thesis, RWTH Aachen University, June, 2011.

Kappen, G., Kurz, L., Priebe, O., and Noll, T. G.: Design Space Exploration for an ASIP/Co-Processor Architecture used in GNSS Receivers, J. Signal Process. Sys., 1, 41–51, 2008.

Volder, J. E.: The CORDIC trigonometric Computing technique, IRE Trans. Electronic Computers, EC-8, 330–334, September, 1959.

Walther, J. S.: A unified algorithm for elementary functions, AFIPS Spring Joint Computer Conference, 38, 379–85, 1971.

13

A CMOS integrated impedance-to-frequency converter for sensing cellular adhesion

A. Mucha[1,2]**, M. Schienle**[1]**, and D. Schmitt-Landsiedel**[2]

[1]Corporate Research and Technologies, Siemens AG, Munich, Germany
[2]Lehrstuhl für Technische Elektronik, Technische Universität München, Munich, Germany

Abstract. Sensing cellular adhesion via impedance measurements provides a versatile and easily accessible means for monitoring in-vitro cell cultures. Previous works used external electronics connected via cables to microelectrodes to achieve this goal, thus incurring parasitic impedance, electromagnetic interference, and bulky measurement setups. In this work we present a CMOS impedance-to-frequency converter integrated with biocompatible planar surface electrodes to make a compact and robust sensor chip for in-vitro cell monitoring. The system features an 8×8 array of individually addressable electrodes connected to four impedance-to-frequency converter circuits with externally adjustable biasing and square wave output. We present first measurement results obtained with the integrated electronics that demonstrate the successful operation of the system and show good agreement with models of the electrode and cell impedances.

1 Introduction

Quick and reliable information about the presence of bioactive substances in water can be vital in many situations. From developing countries, where infrastructure for drinking water might be nonexistent or unreliable, to modern waste water recycling facilities, often the question is not primarily which specific substances are in the water, but rather simply if it is safe to consume, i.e. if it has any negative effects on the human organism. Although chemical analysis technology has now reached a maturity which allows the detection of a wide range of specific contaminants with high sensitivity, it is impractical to test for all relevant substances. Chemical analyses by their nature also cannot give information on the biological activity of substances. This is the strength of biosensors, which combine biological elements with an electronic readout device to measure the reaction of living systems to a stimulus. Cells cultured "in-vitro" react to stimuli much the same way as in the whole organism ("in-vivo"). Thus whole cell biosensors using mammalian or human cell lines are a promising avenue to create a water monitoring system with a high validity as an indicator of bioactive contaminants. Cells can be cultured on sensor chips equipped with transducers for various physical and chemical parameters such as impedance, oxygen concentration and acidity to monitor cell health and metabolism (Bousse, 1996). In this paper, we focus on the measurement of impedance as an indicator of cell morphology and adhesion. It is a valuable parameter for the assessment of cell reactions because it is very fast and many substances cause morphological changes in cells (Wegener et al., 2000).

Previously published whole cell sensor systems used passive chips connected to external electronics. This is problematic because the signals generated by the transducers are very weak, making the systems susceptible to noise and interference. The use of CMOS technology to integrate the transducers with on-chip electronics offers a great advantage in this regard because the reduced physical distance between the sensing elements and the front end amplifier circuitry translates into reduced parasitic impedance, interference and noise. On-chip analog-to-digital conversion and data multiplexing also reduce the number of external connections and thus further improve the robustness.

2 Principle of cell impedance measurement

Animal cells have a phospholipid bilayer membrane to separate the interior from the exterior of the cell. Electrically, this membrane is mainly an insulator, as it does not allow the passage of charge except through ion channels embedded in it. This results in an electrical behavior which can

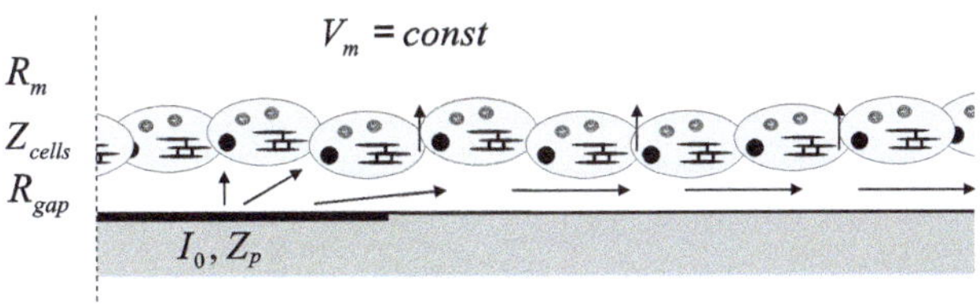

Fig. 1. Schematic cross section of a cell monolayer, having a distributed impedance Z_{cells}, growing on the sensor. A current I_0 is injected into the sensing electrode with polarization impedance Z_p on the left and spreads in the space under the cells and between the cells into the bulk medium, which is assumed to be at a fixed potential V_m. The dashed line at the left indicates the axis of symmetry.

be modeled as a parallel RC circuit. Published values of the cell membrane impedance are around 1–2 kΩcm^2 and 1–4 μFcm^{-2} (Ritter et al., 1990; Ehret et al., 1997; Lo et al., 1995). Culture media used to grow cells contain various ions in concentrations similar to those in the original organism, giving them a relatively low resistivity of around 65 Ωcm.

If a small current is applied to the medium by means of electrodes on the surface of the sensor chip, a low resistance (voltage drop) will consequently be measured in the absence of cells. If cells are however present on the chip or even the electrodes themselves, a larger voltage drop will arise due to their higher impedance. This effect is amplified when the cells are growing as a confluent monolayer, as the current is then mostly constrained to a narrow gap under the cell layer. In this case, even small morphological changes of the cells can result in a large change of the collective impedance of the cell layer as the tightness of the junctions between the cells varies and current can escape into the bulk medium above them more or less easily. The various paths the current can take are illustrated in Fig. 1.

2.1 Electrode impedance

If one of the electrodes has a much larger area than the others, it acts as a counter electrode which keeps the bulk electrolyte at a fixed potential. Its impedance can be neglected in this case. The other electrodes with the smaller area, which we will call sensing electrodes, however present a significant impedance which is added to the useful signal represented by the cell impedance. In Fig. 2, the components of the electrode impedance are shown: The charge transfer resistance R_{ct} represents the passage of charge carriers between the metal and the electrolyte. As the redox reactions involved in this process need some activation energy, its value depends on the applied voltage to the electrode. In the case of a cell impedance sensor, electrochemical reactions are undesired as they corrupt the electrodes. Thus only very small excitations can be applied to the electrodes and the charge transfer resistance can be considered infinite.

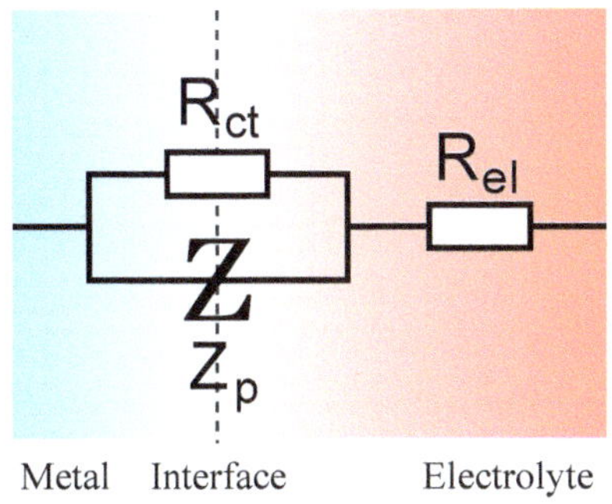

Fig. 2. Equivalent circuit diagram of a metal electrode in contact with an electrolyte.

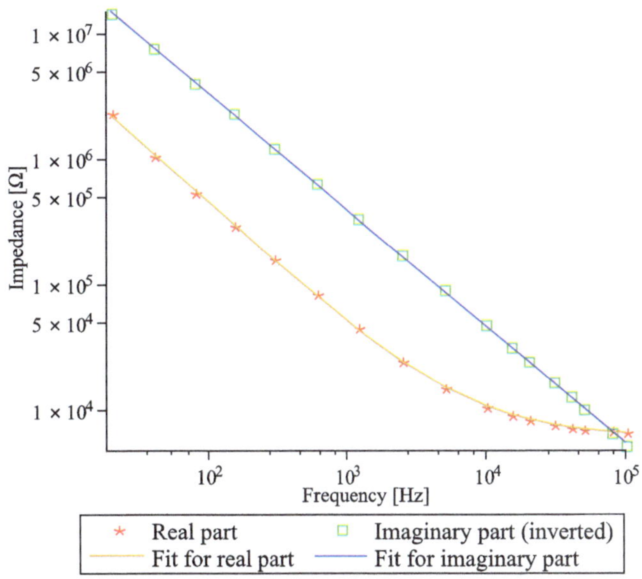

Fig. 3. Typical impedance spectrum of a gold electrode in cell culture medium.

This leaves the polarization impedance Z_p as the only element between metal and electrolyte. It can be modeled phenomenologically by the equation

$$Z_p = Af^{-\alpha} - jBf^{-\beta} \tag{1}$$

with the complex impedance Z_p in ohms and A, B, α and β positive constants (Ragheb and Geddes, 1991). As the frequency f approaches zero, the real and imaginary parts of the polarization impedance go towards positive and negative infinity, respectively, so no DC current can pass through it. In series with the polarization impedance is the spreading resistance through the medium R_{el}, which must thus be added to the total impedance (Franks et al., 2005).

In Fig. 3, a typical impedance spectrum of one of the electrodes on our chip is shown. The real part of the impedance is represented by the asterisk symbols, and the imaginary part (with inverted sign) by the squares. Also shown is our best fit of the Eq. (1) to the data, with the addition of a real constant term for the spreading resistance. Its value was extracted

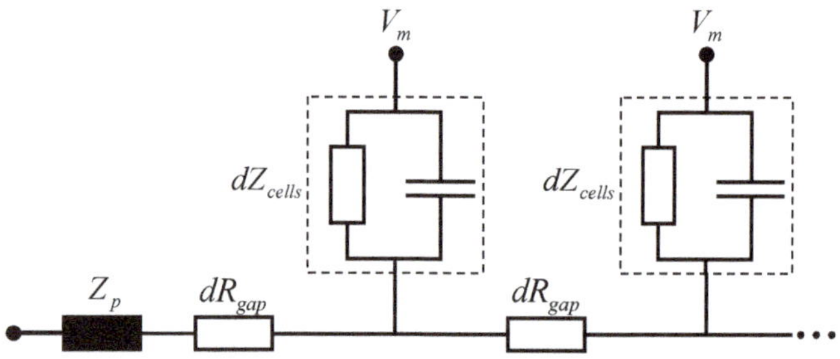

Fig. 4. Simplified schematic of distributed cell impedance.

as 6 kΩ. The exponents for the real and imaginary parts of the polarization impedance are $\alpha = 0.99$ and $\beta = 0.93$, respectively, and their constant factors $A = 4.23 \times 10^7$ and $B = 2.43 \times 10^8$. Since the imaginary part is dominant in Z_p and its frequency dependence is close to that of an ideal capacitance, i.e. $\beta = -1$, the polarization impedance can be approximated to first order as a capacitance.

2.2 Cell layer impedance

The impedance of a cell layer growing on the chip is unfortunately not as easy to model. A simplified schematic of the cell impedance, corresponding to the geometry in Fig. 1, is depicted in Fig. 4. The constriction resistance R_{gap} of the space under the cells and the cells themselves form a distributed impedance which in the one-dimensional case resembles a transmission line structure. A complete model of the cell impedance Z_{cells} has to take into account both the capacitive behavior of the cell membrane and the resistive paths, which can be either through ion channels and the cytosol or through gaps between the cells. However, the capacitive coupling through the cells in general only becomes noticeable at frequencies significantly above 10 kHz, so the cell impedance can be approximated as a pure resistance R_{cells} at lower frequencies.

For the one-dimensional case, i.e. restricting the analysis to geometries with axial symmetry and using some assumptions, analytical models can be derived (Lo et al., 1995; Urdapilleta et al., 2006). For other geometries, numerical methods such as finite elements modeling (FEM) have to be used. We thus used ANSYS finite elements software to evaluate various electrode layouts and to optimize the electrode sizes. Values for the cell layer resistance and typical heights of the cell-substrate gap were obtained from published literature (ibid.). Especially the gap height h is a critical parameter, since it directly influences the constriction resistance of the cell layer and thus the useful signal. The published values in excess of 1 μm that were extracted using the previously mentioned analytical models however seem greatly exaggerated compared to actual measurements (Braun and Fromherz, 1998), so we explored the relationship between h and the expected impedance in a series of simulations.

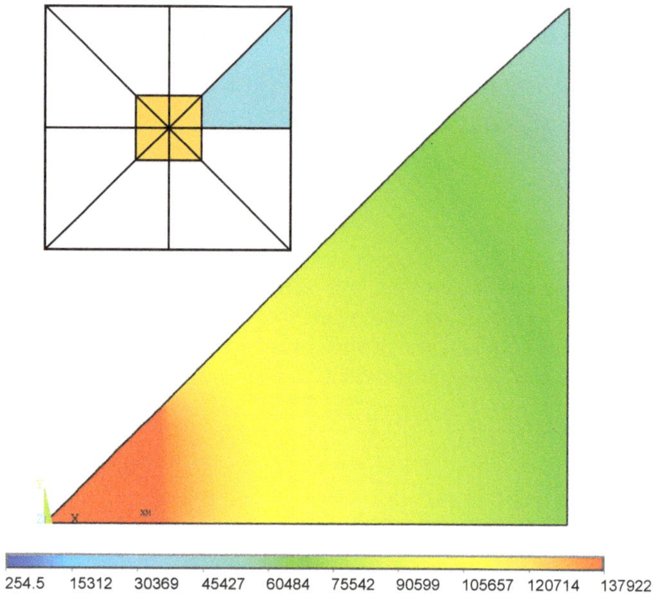

Fig. 5. Top view of voltage distribution in the gap under the cells ($h = 2$ μm) simulated for an imposed current of 1 A. The insert in the upper left shows the segmentation of the finite elements model into eight identical sectors.

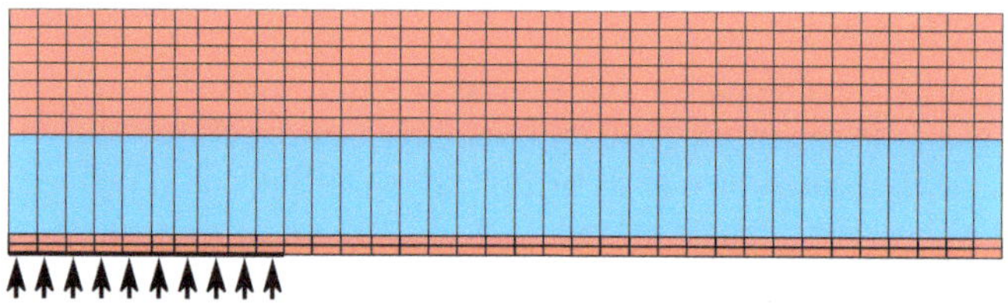

Fig. 6. Cross section of finite elements model along XZ-plane. Blue elements represent the cell layer, red elements cell culture medium.

The simulated geometry in top view is depicted in Fig. 5. A square electrode (modeled as an equipotential region) is located in the center of the simulation space, which has a boundary condition for far field decay imposed, thus simulating an infinitely large counter electrode at infinite distance. By exploiting the symmetries of the geometry, only a sector of 1/8 of the space needs to be simulated. A cross section of the model, showing the gap, the cell layer and the bulk medium above it, is depicted in Fig. 6. The arrows on the left indicate the area where the electrode current is impressed.

Examination of the simulated current distribution revealed that the current indeed spreads mostly in the gap under the cells. One could thus assume that the spreading resistance can be described by an equation of the form

$$R_{\text{gap}}(h) = F_{\text{geom}} \frac{\rho}{h} \tag{2}$$

with F_{geom} as a factor describing the geometry (e.g. electrode area and shape), gap height h and the medium resistivity ρ. A plot of R_{gap} over $1/h$ should thus yield a straight line. However, a significant spreading resistance R_0 of several kΩ was

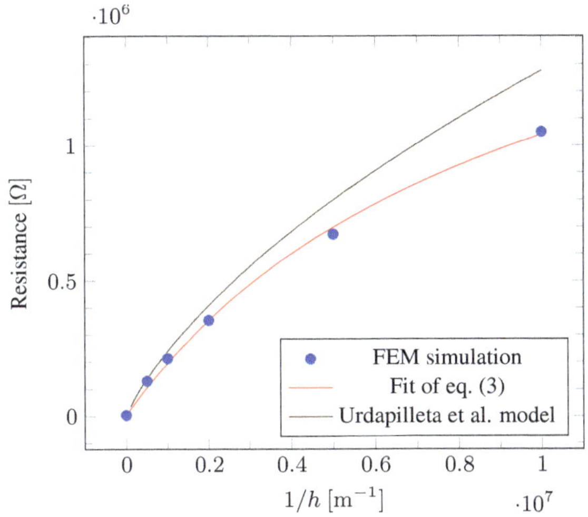

Fig. 7. Plot of cell impedance versus inverse gap height: FEM simulation, fitted curve and analytical model by Urdapilleta et al. for a center electrode of $55 \times 55\ \mu m^2$.

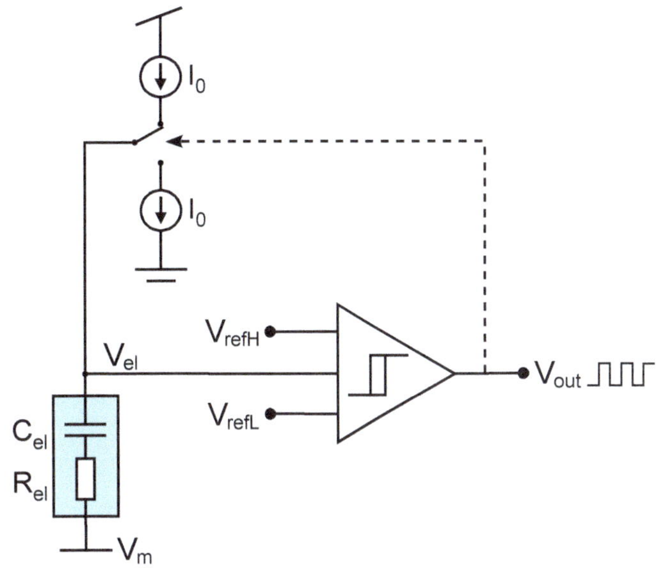

Fig. 8. Simplified block diagram of impedance-to-frequency converter.

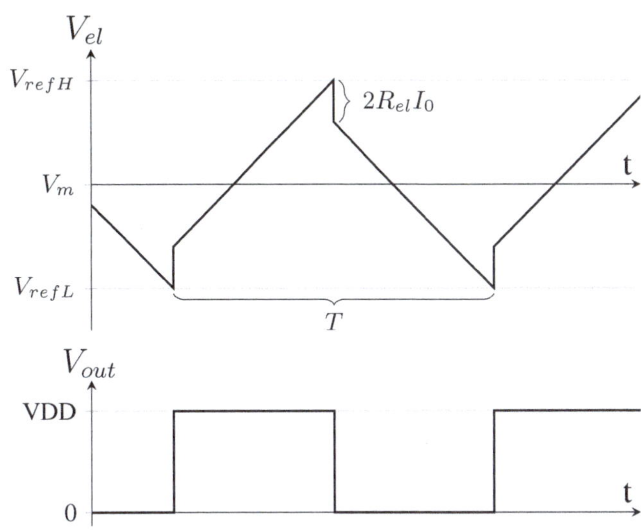

Fig. 9. Idealized plot of the electrode voltage and digital output over time.

evident in our preliminary measurements even in the absence of cells (see Fig. 3). Some current also does cross the cell layer, mainly directly above the electrode, so the total behavior is better described by the gap resistance in parallel with another resistive term for this direct crossing, i.e. as

$$R_{\text{total}} = R_{\text{gap}}(h)||R_{\text{cells}} + R_0 \tag{3}$$

The plot of the simulated values in Fig. 7 with a superposed fit of Eq. (3) confirms this model. For comparison, the values yielded by application of the Urdapilleta et al. model with the same parameters to a circular electrode of equal area are also plotted.

Summarizing the simulation results, even in the assumed worst case ($h = 2\ \mu$m), the cell layer would present a resistive impedance in excess of 100 kΩ, a significant increase over the spreading resistance without cells. To separate this useful signal from the background of the electrode impedance, the measurement has to be performed at frequencies around 10 kHz, where the polarization impedance has sufficiently decreased.

3 Impedance-to-frequency converter

Our modeling thus yielded a first-order approximation of the impedance to be measured as a capacitance for the electrode-electrolyte interface in series with a resistance for the cell layer and the medium. Based on this model we developed an integrated circuit to measure the cell impedance. A key aim in the design process was to enable analog-to-digital conversion of cell adhesion data. Thus, an impedance-to-frequency converter was ideally suited, since it only needs an additional counter to directly produce digital values with a wide dynamic range.

3.1 Principle of operation

The measurement principle of our circuit is illustrated by the simplified block diagram in Fig. 8: asurface electrode in contact with the electrolyte is connected to a comparator and a switchable current source. In the shown switch state, the interface capacitance is charged by a constant current I_0 while the comparator senses the electrode potential. As soon as the upper reference voltage V_{refH} is reached, the comparator output V_{out} changes from high to low, which in turn triggers the switch so the electrode is now discharged by $-I_0$. When V_{el} reaches the lower reference voltage V_{refL}, the comparator switches again, completing the cycle. The series resistance

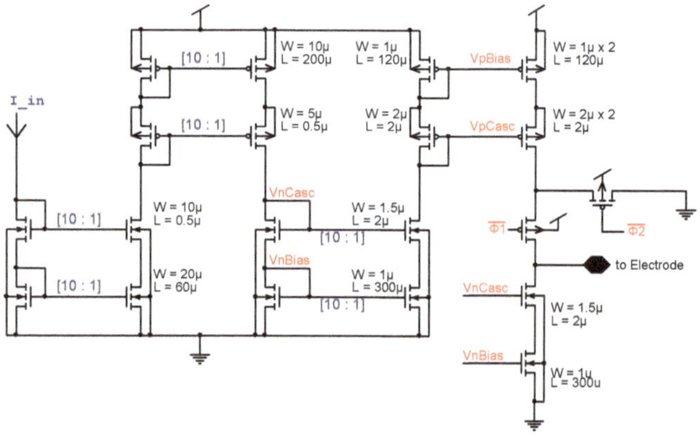

Fig. 10. Schematic of electrode current generator circuit.

R_{el}, constituted either by the spreading resistance or the cell layer, causes an ohmic voltage drop of $R_{el}I_0$ which changes sign at each transition, effectively reducing the voltage window by $2R_{el}I_0$. The period of one charge-discharge cycle as illustrated in Fig. 9 is thus given by

$$T = 2\frac{C_{el}}{I_0}(\Delta V - 2R_{el}I_0) = 2C_{el}\left(\frac{\Delta V}{I_0} - 2R_{el}\right) \tag{4}$$

with $\Delta V = V_{\mathrm{refH}} - V_{\mathrm{refL}}$. When cells cover the electrode, increasing R_{el}, the frequency of V_{out} also increases. The output of the comparator thus serves as a digital frequencymodulated signal containing information about cell adhesion.

The same circuit can be connected to an array of sensing electrodes via a switching matrix, accomplishing a spatially resolved measurement of cell adhesion. In order to maximize the long-term stability of the electrodes, redox reactions have to be avoided under all circumstances. This leads to the requirement to limit the polarization voltage of the electrodes to few tens of millivolts. It is also advisable to minimize the electrode current in order to avoid effects like depolarization of the cell membrane, although no exact limits to the current density that cells can be exposed to without harm are known. Surveying published literature about cell impedance measurements led us to the conclusion that a current density of few milliamperes per square centimeter should not be exceeded (Wegener et al., 2000).

These limitations had to be considered in the design of the integrated circuit. They translate into the need for high gain of the comparator to switch at sub-millivolt accuracy while taking steps to mitigate the resulting sensitivity to noise, such as decoupling the reference and supply voltages. The choice of the electrode area needs to strike a balance between the desired high sensitivity to cell adhesion, requiring small electrodes, and the technological limitations as the allowed current decreases with electrode area. With these factors in mind, we designed our sensor for a value of $\Delta V = 20$ mV and the current I_0 adjustable in a range of $10 - 100$ nA.

Accurately generating currents of this magnitude and reliably transporting them to the electrode requires consideration

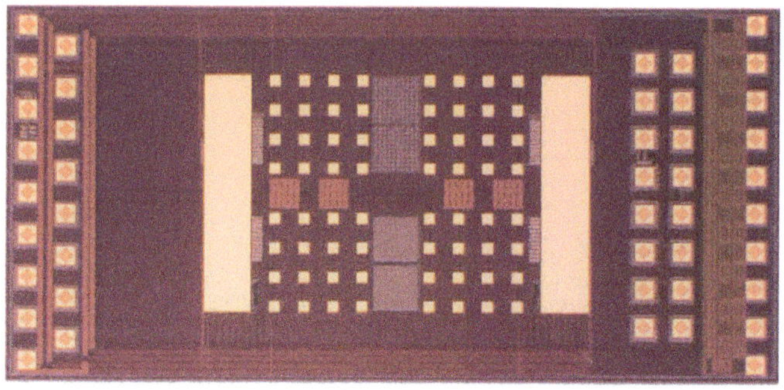

Fig. 11. Micrograph of the impedance-to-frequency converter test chip, showing the four electrode arrays with two large counter electrodes to both sides. The leftmost row of pads are supply and reference connections, the pads on the right hand side are the address and control bits and the digital output. Also visible are additional pads for various test structures.

of leakage in the switching matrix and the statistical mismatches introduced in the manufacturing process. The unavoidable variations of transistor parameters such as threshold voltage generally scale with the inverse of the transistor dimensions and become more pronounced in the weak inversion regime. In order to limit these variations, all transistors need to operate in the saturation region. To ensure this we designed the electrode current source as a chain of three cascode current mirrors, each with a scaling factor of 10:1, implementing a downscaling of a reference current by a factor of 1000 (see Fig. 10). Non-overlapping signals generated from the comparator output control the switches between current source and electrode.

3.2 Test chip

A test chip for the impedance-to-frequency converter was produced in a standard 0.35 μm 3.3 V CMOS technology (Austria Micro Systems). An additional lift-off process was performed in-house to process the surface electrodes consisting of a Ti-Pt-Au stack. Based on our simulations and the considerations outlined above, we chose an electrode size of 55×55 $\mu\mathrm{m}^2$ and an inter-electrode distance of 100 μm. In total 64 sensing electrodes are located on the chip, organized in four arrays of 4×4, with the electrode closest to the chip center of each array accessible via a pad for testing purposes. Each of the electrodes can be addressed individually through a digital control interface. Two large counter electrodes keep the electrolyte at an externally supplied fixed potential. A chip photograph is shown in Fig. 11.

4 Measurement results

For the measurements with cells, the chip was mounted in a CPGA package and the bond wires were sealed with an epoxy resin. A small polystyrene dish was then mounted on top to act as a liquid reservoir. The biasing of the chip was

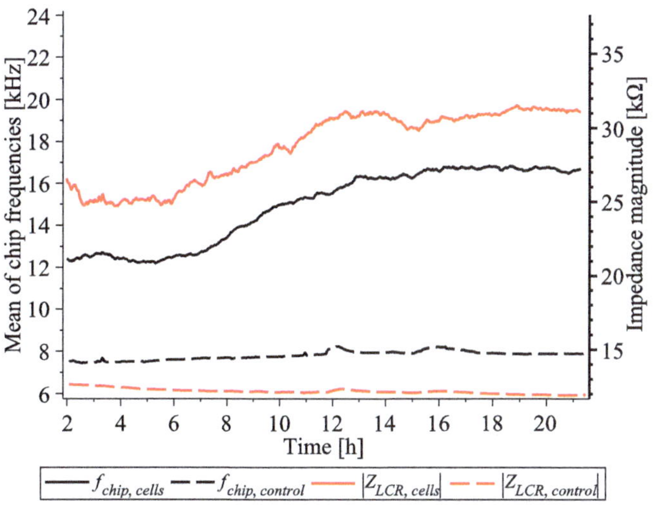

Fig. 12. Cell growth on the sensor chip captured with our impedance-to-frequency converter (black solid line, left axis) and actual impedance at 10 kHz measured simultaneously via a LCR meter connected to on-chip electrodes (red solid line, right axis). Also shown are data from a control experiment without cells (dashed lines).

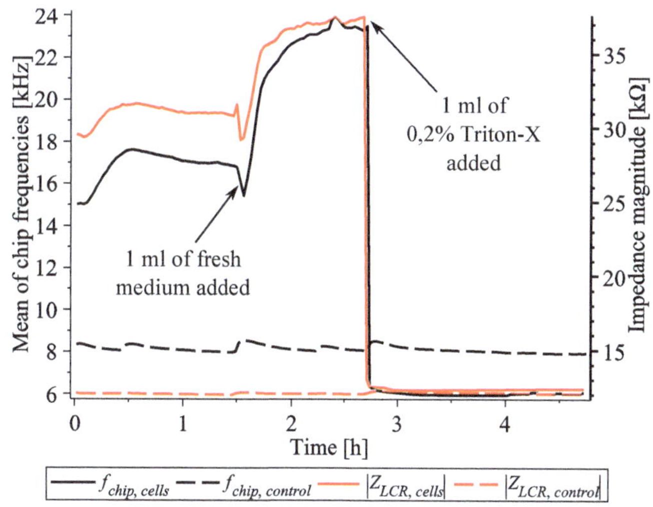

Fig. 13. Continuation of the experiment shown in Fig 12. Cell reactions to the addition of fresh medium and later to the addition of the detergens Triton-X 100 are clearly visible.

adjusted for an electrode current of 70 nA. For comparison to the output of our impedance-to-frequency converter, we also connected a Hewlett Packard 4284A LCR meter to the chip, allowing us to record the actual impedance of the four center electrodes and at the same time measure the remaining 60 electrodes with the on-chip circuitry. A multifunction I/O box attached to a PC was used to output the digital address bits, supply power to the chip and to evaluate the frequency-modulated digital output.

To verify the functionality of the sensor system, we seeded cells on our test chip (1 million V79 cells in 2 ml of Dulbecco's Modified Eagle Medium), placed the chip with its support PCB in an incubator at 37°C and recorded the output frequencies of the impedance-to-frequency converters for each electrode, along with the impedance measured by the LCR meter on the four center electrodes.

In Fig. 12, the mean output frequency of our circuit for all sensing electrodes during this experiment is represented by the black curve. The red curve shows the impedance magnitude of the center electrodes as measured by the LCR meter. At the beginning of the plot, the cells have just settled on the chip from suspension, but have not yet attached to the substrate. In the following hours, they spread out and connect to neighboring cells, forming a confluent monolayer (Wegener et al., 2000). This is accompanied by an increase in impedance, which both curves show in good agreement.

The next day, the measurement was briefly interrupted and the formation of a tight cell coverage confirmed visually. The chip was then placed in the incubator again and the measurement continued. As can be seen in Fig. 13, the cells were not disturbed by this interruption and the signals stabilized at their previous values within one hour. We then exchanged 1 ml of the used medium with an equal amount of fresh medium at room temperature. This caused a short dip of both curves, which we interpret as a temperature shock reaction, followed by a longer rise, which could be explained by the supply of oxygen and fresh nutrients by the new medium. When we repeated the identical experiment without cells as a control (dashed lines), no significant reaction could be observed, confirming that the effect was caused by the cells.

After allowing for the signals to stabilize, we again exchanged 1 ml of the used medium with 1 ml of medium containing 0.2 % Triton-X 100, a detergent which kills cells by destroying their membrane. An immediate drop of the impedance down to the base level of the control experiment followed as the cells died off and detached from the chip.

During the whole experiment, the changes of the mean chip frequency and the impedance measured by the LCR meter show high correlation, confirming the viability of our impedance-to-frequency converter for cell adhesion measurement. The slight discrepancies can be sufficiently explained by the different physical measurement locations on the chip and the nonlinearity of the relation between impedance magnitude and chip output frequency.

5 Conclusions

The evaluation of cell adhesion by impedance measurements can give valuable information about cell reactions and is thus an important element for a water quality biosensor. To harness the advantages of CMOS integration for such a sensor, we developed a model of the impedance under study, comprising contributions of the electrode and the cells. Using a combination of analytical and numerical techniques, this allowed us to predict the nature and magnitude of the cellular impedance. An integrated impedance-to-frequency converter circuit was developed based on these results, allowing for a straightforward digital readout of cell adhesion data while taking into account the electrochemical limitations of the allowed currents and voltages. Experiments with cells proved the functionality of the measurement system.

References

Bousse, L.: Whole cell biosensors, Sensors and Actuators B, 34, 270–275, doi:10.1016/S0925-4005(96)01906-5, 1996.

Braun, D. and Fromherz, P.: Fluorescence Interferometry of Neuronal Cell Adhesion on Microstructured Silicon, Physical Review Letters, 81, 5241, doi:10.1103/PhysRevLett.81.5241, 1998.

Ehret, R., Baumann, W., Brischwein, M., Schwinde, A., Stegbauer, K., and Wolf, B.: Monitoring of cellular behaviour by impedance measurements on interdigitated electrode structures, Biosensors and Bioelectronics, 12, 29–41, doi:10.1016/0956-5663(96)89087-7, 1997.

Franks, W., Schenker, I., Schmutz, P., and Hierlemann, A.: Impedance characterization and modeling of electrodes for biomedical applications, Biomedical Engineering, IEEE Transactions on, 52, 1295–1302, doi:10.1109/TBME.2005.847523, 2005.

Lo, C. M., Keese, C. R., and Giaever, I.: Impedance analysis of MDCK cells measured by electric cell-substrate impedance sensing., Biophys. J., 69, 2800–2807, 1995.

Ragheb, T. and Geddes, L.: The polarization impedance of common electrode metals operated at low current density, Annals of Biomedical Engineering, 19, 151–163, doi:10.1007/BF02368466, 1991.

Ritter, M., Lang, F., Grübl, G., and Embacher, H. G.: Determination of cell membrane resistance in cultured renal epithelioid (MDCK) cells: effects of cadmium and mercury ions, Pflügers Archiv European Journal of Physiology, 417, 29–36, doi:10.1007/BF00370765, 1990.

Urdapilleta, E., Bellotti, M., and Bonetto, F. J.: Impedance analysis of cultured cells: A mean-field electrical response model for electric cell-substrate impedance sensing technique, Physical Review E (Statistical, Nonlinear, and Soft Matter Physics), 74, 041908–11, doi:10.1103/PhysRevE.74.041908, 2006.

Wegener, J., Keese, C. R., and Giaever, I.: Electric Cell-Substrate Impedance Sensing (ECIS) as a Noninvasive Means to Monitor the Kinetics of Cell Spreading to Artificial Surfaces, Experimental Cell Research, 259, 158–166, doi:10.1006/excr.2000.4919, 2000.

14

Fully-integrated LDO voltage regulator for digital circuits

M. Lüders[1], **B. Eversmann**[2], **D. Schmitt-Landsiedel**[1], **and R. Brederlow**[2]

[1]Munich University of Technology, Munich, Germany
[2]Texas Instruments Deutschland, Freising, Germany

Abstract. Low-dropout (LDO) voltage regulators are widely used to supply low-voltage digital circuits. For recent ultra-low-power microcontroller systems, a fully-integrated LDO without any external capacitance is preferred in order to achieve a fast and energy-efficient wake-up.

Commonly, an LDO is specified, designed and verified for DC load currents. In contrast, a digital load creates large current spikes. As an LDO designed for low quiescent current is too slow to react on fast current spikes, a minimum on-chip capacitance is required to keep the supply voltage within a certain error window. Different fully-integrated LDO topologies are investigated regarding their suitability to supply low-voltage digital circuits. The any-load stable LDO topology is selected and implemented on a 0.13 μm test-chip. The LDO is able to provide a maximum load current of 2.5mA while consuming a quiescent current of 17 μA.

1 System overview: ultra-low-power microcontroller

Ultra-low-power microcontrollers (MCU) are used in many small scale battery-powered applications today. Typical examples are smart meters, alarm systems and medical monitoring devices. These applications spend most of their time in sleep mode consuming only a few microamperes and waking up periodically to perform a certain task. When activated their current consumption abruptly increases usually by more than three decades. Assuming a rather typical 1/1000 active/sleep ratio both modes consume half of the available energy and need to be optimized to maximize the battery lifetime (Zwerg, 2011).

While the external supply voltage of an ultra-low-power MCU system is determined by battery voltage over lifetime, its digital core requires a lower sub-regulated supply voltage. It is therefore supplied by an integrated LDO as depicted in Fig. 1. Keeping the LDO and consequently the digital core always powered causes digital leakage current as well as LDO quiescent current flowing even in sleep mode. To minimize LDO quiescent current in sleep mode, some ultra-low-power MCU systems use different LDOs, each optimized for a dedicated power mode, but with a complex and slow transition between them (Kristjansson, 2006).

As in modern CMOS technologies leakage current increases, the battery lifetime for typical ultra-low-power applications becomes more and more limited by leakage current in sleep mode. The most effective approach to minimize leakage current in sleep mode is to switch-off the digital core by disabling the LDO. Thereby, the pass-transistor of the LDO acts as power switch. Only some digital control and interrupt circuitry in the battery voltage domain remain active in sleep mode.

From an energy perspective, a minimum sleep time can be defined for which it is useful to switch-off the digital core. At each transition into sleep mode, the capacitance at the LDO output is discharged and consequently has to be recharged at wake-up. This results in a trade-off between saved leakage energy on the one hand and additional energy required for recharging the load capacitance on the other hand, as depicted in Fig. 2. To achieve fast and energy-efficient transitions between power modes, lowest capacitance at the LDO output is obviously preferred. Consequently, the large external capacitance at the LDO output, which is typically in the range of some 100 nF, is removed and only a much smaller on-chip capacitance remains. A welcome side-effect is a higher system integration resulting in lower system costs. However, as an LDO typically use the external capacitance for compensation, its absence greatly affects the LDO design.

In this paper, the fundamental design considerations for a fully-integrated LDO supplying the digital core of an ultra-low-power MCU are presented. In Sect. 2, the supply voltage requirements of the digital core and the interferences with the LDO are discussed. In Sect. 3, different LDO topologies are investigated regarding their suitability to supply the digital core. The any-load stable LDO topology is selected and implemented on a 0.13 μm test-chip. The measurement results are presented in Sect. 4.

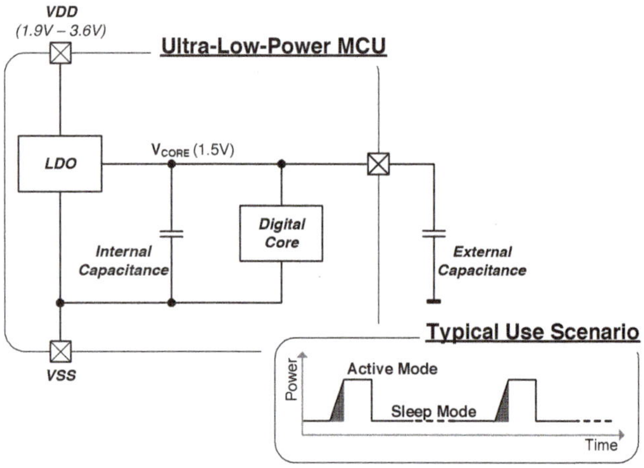

Fig. 1. Power supply system and typical use scenario of an ultra-low-power MCU.

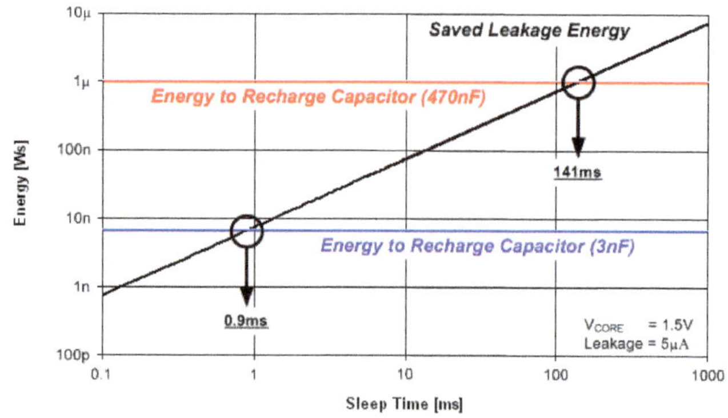

Fig. 2. The minimum sleep time for which it is useful to switch the digital core off is proportional to the LDO load capacitance.

2 LDO operating environment

To guarantee fault-free operation of the digital core, the core voltage must remain within a certain error window. This window is determined by reliability constraints of the digital core on the upper boundary side and speed requirements on the lower boundary side. To minimize current consumption of the digital core, the core voltage in an ultra-low-power MCU system is typically set to the lower boundary with some safety margin. This safety margin is determined by several tolerances within the supply system, which accumulate:

- Static inaccuracy of the reference voltage due to limited accuracy of the bandgap reference.
- Static inaccuracy due to limited LDO loop gain and LDO offset.
- Dynamic inaccuracy in response to fast line and load transient conditions due to limited LDO bandwidth.
- Supply noise within the on-chip power distribution network due to IR-drop and di/dt-drop.

The tolerances are budgeted during the system specification such that the core voltage always remains in the specified error window, which in the present case is $1.5\,\text{V} \pm 0.1\,\text{V}$.

The digital core operates synchronous to a system clock; it draws current by charging the various load capacitances whenever they are switched. Thereby, as depicted in Fig. 3a, the digital core creates large current spikes at each clock edge. The average current consumption of the digital core, which is equal to the LDO load current, can be expressed as:

$$I_{\text{LOAD}} = \sum_{i=1}^{k} (\alpha_i C_i)\, f_{\text{CLK}} V_{\text{CORE}} \tag{1}$$

where α_i is the activity rate and C_i is the load capacitance of a single digital logic gate, f_{CLK} is the system clock and V_{CORE} is the core voltage. The current consumption of the digital core can instantly change from zero to maximum (and back) when the system clock is gated. An LDO supplying the digital core must therefore have a large bandwidth in order to react as fast as clock speed determines.

To ensure the power integrity of the digital core, capacitance is allocated and distributed on-chip (Popovich, 2007). This capacitance acts at the same time as load capacitance for the fully-integrated LDO. As an LDO designed for low quiescent current is too slow to react on single current spikes produced by the digital core, there is a certain minimum on-chip capacitance required. Each current spike results in a voltage ripple ΔV_{CORE}, which is determined by charge sharing:

$$V_{\text{CORE}} C_{\text{LOAD}} = (V_{\text{CORE}} + \Delta V_{\text{CORE}}) \left(C_{\text{LOAD}} + \sum_{i=1}^{k} (\alpha_i C_i) \right)$$

$$\Delta V_{\text{CORE}} = -\frac{\sum_{i=1}^{k} (\alpha_i C_i)}{C_{\text{LOAD}} + \sum_{i=1}^{k} (\alpha_i C_i)} V_{\text{CORE}} \tag{2}$$

where C_{LOAD} is the on-chip capacitance. Inserting Eq. (1) into (2) and assuming that the load capacitance is sized much larger than the effective switching capacitance results in:

$$\Delta V_{\text{CORE}} \approx -\frac{I_{\text{LOAD}}}{f_{\text{CLK}} C_{\text{LOAD}}} \tag{3}$$

Thereby, the digital core is assumed to have the same effective switching capacitance at each clock edge. The resulting voltage ripple adds to the overall LDO inaccuracy. To limit this voltage ripple to a sufficient level of some 10 mV, a minimum on-chip capacitance is required, which depends on the maximum load current and is in the range of some nanofarads in case of an ultra-low-power MCU system.

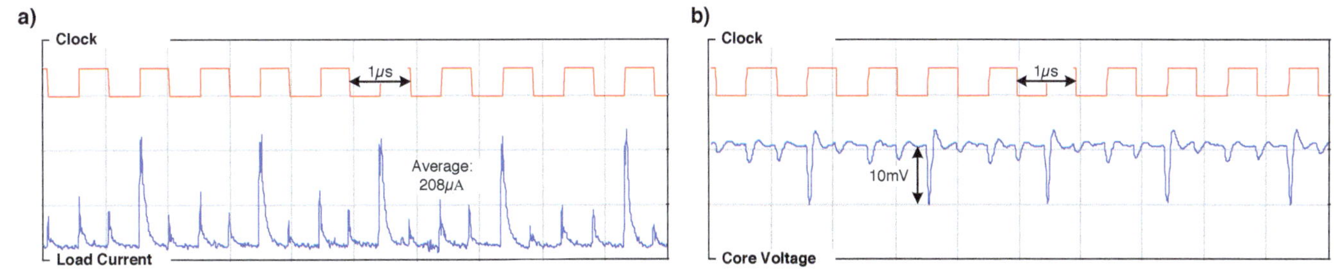

Fig. 3. (**a**) Simulated transient load current profile of the digital core when operating at 1 MHz, and (**b**) the resulting core voltage when the load current profile is presented to the any-load stable LDO with an on-chip capacitance of 3 nF.

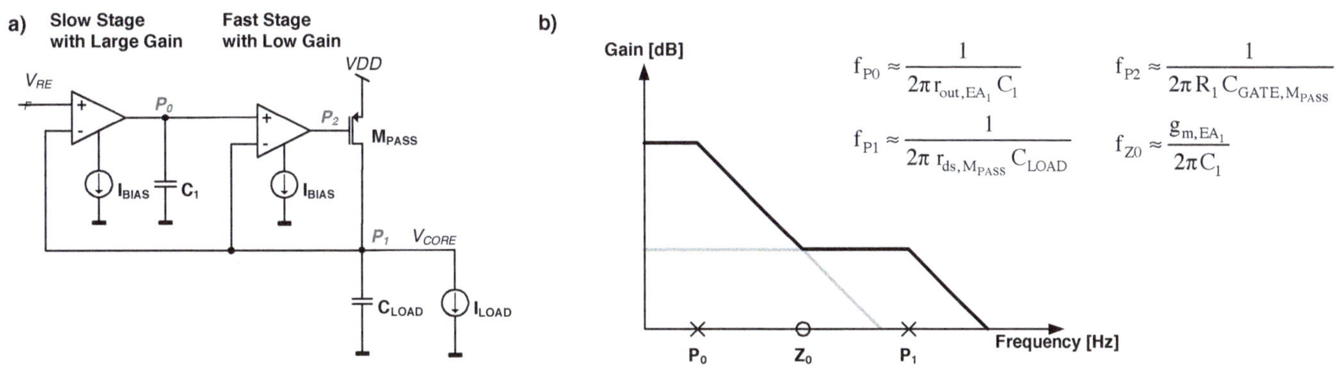

Fig. 4. (**a**) Block diagram of the any-load stable LDO, and (**b**) its open-loop gain over frequency: The any-load stable LDO topology is compensated by an active feed-forward compensation scheme.

Figure 3b shows the simulated core voltage when the transient load current profile of the digital core is presented to the fully-integrated any-load stable LDO with an on-chip capacitance of 3 nF.

3 Fully-integrated LDO

An LDO can be considered as a voltage buffer with a certain current drive capability. Ideally, the voltage buffer should have high gain for an accurate final value and large bandwidth for fast settling while loop stability is guaranteed under all operating conditions. In practice, it is not easy to combine these requirements. This is especially true for a fully-integrated LDO without a large (external) capacitance.

The design of fully-integrated LDOs has received a lot of attention in literature. While conventional LDO topologies are compensated by a large (external) capacitance with the dominant pole being at the LDO output, most of the presented, fully-integrated LDO topologies use some form of miller compensation to establish an internal, dominant pole (Guo, 2010; Leung, 2003). These LDO topologies in principle suffer from stability issues in light load conditions. Thereby, the required minimum load current increases for a larger capacitance at the LDO output. For example, the LDO topology presented in (Guo, 2010) requires a minimum load current of 3 mA while the capacitance is limited to 50 pF in order to achieve loop stability. Hence, these topologies are not best suited to supply the digital core of an ultra-low-power MCU system.

An alternative LDO topology, called any-load stable LDO, is presented in (Ivanov, 2009). This LDO topology achieves high gain and large bandwidth by combining two stages in two control loops as shown in Fig. 4a. The LDO topology is compensated by an active feed-forward compensation scheme (Thandri, 2003). As depicted in Fig. 4b, this compensation scheme uses the positive phase shift of a left-half-plane zero caused by the fast control loop to cancel the negative phase shift of the first dominant pole P_0. A second dominant pole P_1 is located at the LDO output. This pole widely moves with load current due to changing output impedance of the pass-transistor.

The transistor-level implementation of the any-load stable LDO topology is shown in Fig. 5a. While the slow stage is implemented as a folded-cascode amplifier, the fast stage is based on a cascoded flipped voltage follower (FVF) (Ramirez-Angulo, 2005). The cascoded FVF is formed by the common-gate transistors M_{CG1} and M_{CG2}, the current

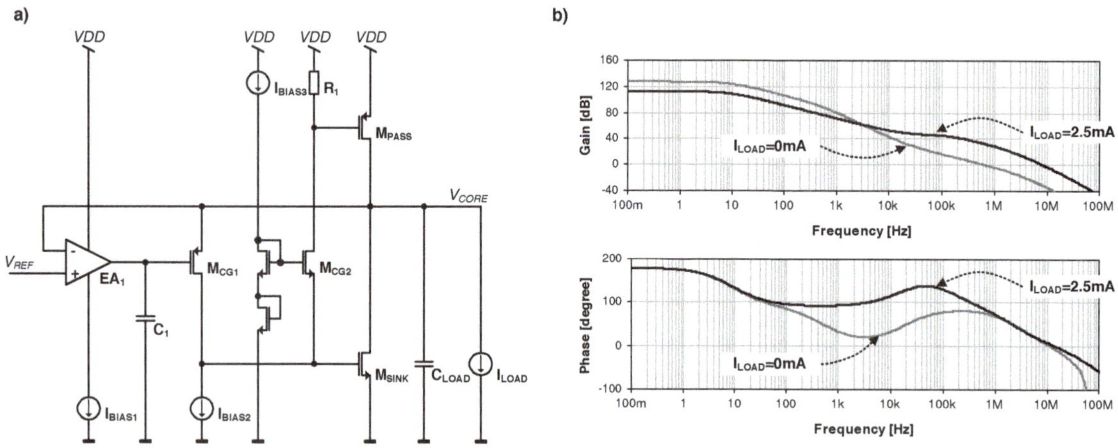

Fig. 5. **(a)** Basic circuit diagram of the any-load stable LDO topology, and **(b)** the simulated Bode plots for minimum and maximum load current at VDD = 3.0 V and C_{LOAD} = 3 nF.

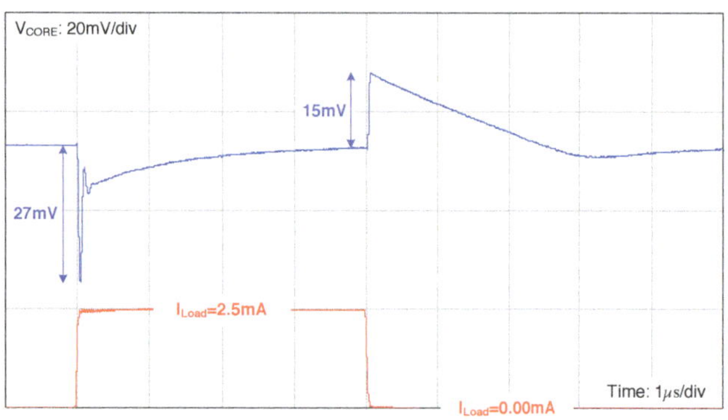

Fig. 6. Measured LDO transient behavior when applying a load current step from 0 mA to 2.5 mA under worst case conditions (VDD = 1.9 V, T = −40°C).

source I_{BIAS2}, the resistor R_1 and the pass-transistor M_{PASS}. The two control loops are combined at M_{CG1}.

The first non-dominant pole P_2 of the cascoded FVF is defined by R_1 and the gate capacitance of M_{PASS}. The cascoded FVF can be easily stabilized for any capacitance at the LDO output by adapting the resistance R_1. Thereby, loop stability for smaller capacitances requires an increased LDO quiescent current.

The simulated Bode plots of the any-load stable LDO topology in minimum and maximum load current condition are depicted in Fig. 5b. By employing the active feed-forward compensation scheme, loop stability is easily achieved over the full load current range without any limitations on the on-chip capacitance. The fully integrated any-load stable LDO is therefore well suited to supply the digital core of an ultra-low-power MCU system.

4 Measurement results

The any-load stable LDO has been implemented as part of a complete ultra-low-power MCU in a 0.13 μm digital CMOS process. The supported supply voltage range is determined by battery voltage over lifetime (1.9 V–3.6 V); the nominal output voltage is 1.5 V. Reference voltage and bias current are provided by a reference circuit not discussed here. The on-chip capacitance coming with the digital core is 3 nF. For test purposes, a DC load current profile can be generated by an on-chip circuit resulting in rise and fall times of 1 ns.

The LDO is able to provide a maximum load current of 2.5 mA while consuming a quiescent current of 17 μA. Fig. 6 shows the measured LDO transient behavior when applying a load current step from 0mA to 2.5 mA under worst case conditions (VDD = 1.9 V, T = −40°C). The resulting transient voltage errors are −27 mV and +15 mV, respectively. The effect of the multi-loop LDO topology with its slow and fast control loop is well noticeable from the transient response.

5 Conclusions

In this paper, the fundamental design requirements for a fully-integrated LDO supplying the digital core of an ultra-low-power MCU are summarized. Among other presented LDO topologies, the any-load stable LDO is selected and implemented on a test-chip. The LDO is able to provide a maximum load current of 2.5 mA while consuming a quiescent current of 17 μA. Its loop stability over the full load current range is proven by measurement results. The any-load stable LDO is therefore well suited to supply digital circuits.

References

Zwerg, M., Baumann, A., Kuhn, R., Arnold, M., Nerlich, R., Herzog, M., Ledwa, R., Sichert, C., Rzehak, V., Thanigai, P., and Eversmann, B.: An 82 μA/MHz Microcontroller with Embedded FeRAM for Energy-Harvesting Applications, IEEE ISSCC 2011, 334–335, 2011.

Kristjansson, E.: ARM7 Low Power Design, White Paper www.microcontroller.com/ARM7_Low_Power_Design_-_White_Paper.htm, November 2006.

Popovich, M., Friedman, E., Secareanu, R., and Hartin, O.: Efficient placement of distributed on-chip decoupling capacitors in nanoscale ICs, IEEE ICCAD 2007, 811–816, 2007.

Guo, J. and Leung, K.: A 6 μW Chip-Area-Efficient Output-Capacitorless LDO in 90-nm CMOS Technology, IEEE JSSC, 45(9), 1896–1905, 2010.

Leung, K. and Mok, P.: A capacitor-free CMOS low-dropout regulator with damping-factor-control frequency compensation, IEEE JSSC, 38(10), 1691–1702, 2003.

Ivanov, V.: Design Methodology and Circuit Techniques for Any-Load Stable LDOs with Instant Load Regulation and Low Noise, Analog Circuit Design, 339–358, 2009.

Thandri B. and Silva-Martinez, J.: A robust feedforward compensation scheme for multistage operational trans-conductance amplifiers with no Miller capacitors, IEEE JSSC, 38(2), 237–243, 2003.

Ramirez-Angulo, J., Gupta, S., Padilla, I., Carvajal, R., Torralba, A., Jimenez, M., and Munoz, F.: Comparison of conventional and new flipped voltage structures with increased input/output signal swing and current sourcing/sinking capabilities, IEEE MWSCAS 2005, 2, 1151–1154, 2005.

15

An 8 bit current steering DAC for offset compensation purposes in sensor arrays

G. Bertotti[1,2], A. Laifi[1], E. Di Gioia[1], M. Masoumi[1], N. Dodel[1], E. F. Scarselli[2], and R. Thewes[1]

[1]Chair of Sensor and Actuator Systems, TU Berlin, Berlin, Germany
[2]ARCES, University of Bologna, Bologna, Italy

Correspondence to: G. Bertotti (gabriel.bertotti@tu-berlin.de)

Abstract. An 8 bit segmented current steering DAC is presented for the compensation of mismatch of sensors with current output arranged in a large arrays. The DAC is implemented in a 1.8 V supply voltage 180 nm standard CMOS technology. Post layout simulations reveal that the design target concerning a sampling frequency of 2.6 MHz is exceeded, worst-case settling time equals 60.6 ns. The output current range is 0–10 μA, which translates into an LSB of 40 nA. Good linearity is achieved, INL < 0.5 LSB and DNL < 0.4 LSB, respectively. Static power consumption with the outputs operated at a voltage of 0.9 V is approximately 10 μW. Dynamic power, mainly consumed by switching activity of the digital circuit parts, amounts to 100 μW at 2.6 MHz operation frequency. Total area is $38.6 \times 2933.0\,\mu m^2$.

1 Introduction

Sensor arrays implemented on CMOS chips with thousands up to millions of sensor sites repeated in a regular manner have found many applications in today's world: Whereas the most prominent example may be the imaging chip in CMOS cameras from consumer to professional applications (http://isscc.org/doc/2011/2011_Trends.pdf), other applications are found in space industry (Lin et al., 2009), security (Perenzoni et al., 2011), biomedical research (Eversmann et al., 2003; Frey et al., 2010), and further fields. There, the output signal of the sensor sites can be given in the voltage domain – which usually applies to the pixels of CMOS imaging chips – or in the current domain. Frequently, the output date received from the respective sensor sites are prone to biasing offsets and random parameter variations.

In this work, we consider sensor sites with current output, i.e. the sensors can be modeled as current sources, whose output currents consist of a bias value (I_{BIAS}), which is prone to variations (ΔI_{BIAS}), and a signal dependent portion (I_{SIGNAL}). Under worst-case conditions, the following relation holds:

$$I_{BIAS} \gg \Delta I_{BIAS} \gg I_{SIGNAL} \qquad (1)$$

In such cases, it can be advantageous to implement compensation mechanisms for the deviation ΔI_{BIAS}. In this work, a digital-to-analog converter (DAC) is presented, operated at the sensor array border and used for the compensation of the bias current deviations. The required data patterns used for compensation are derived in an initial calibration phase.

For our purposes – aiming for an application in the biomedical domain, the following design specifications have been identified: full-scale current range = 10 μA, resolution = 8 bit, operating frequency should at least amount to 2.6 MHz and, moreover, the layout must not measure more than 40 μm in one direction.

Many different solutions implementing current steering DACs are available in the literature, e.g. Bastos et al. (1996), Bastos et al. (1998), Miki et al. (1986), Chen and Gielen (2007), Lei Luo (2008) and many more. However, these solutions are designed for full scale output current ranges of order 10 mA and cannot simply be adjusted to our design targets with 10 μA full scale range. Moreover, all state of the art DACs achieve their performances thank to the application of symmetrical layouts, which cannot be adapted to our aspect ratio of order 100.

The paper is organized as follows: in Sect. 2 the fundamental concepts of current steering DACs are introduced. Section 3 presents mismatch considerations whose understanding is necessary for the design. Section 4 briefly comments on the transistor-level implementation, and Sect. 5 focuses on layout issues. In Sect. 6, post-layout simulation results are presented, and finally, in Sect. 7, the paper is concluded.

2 Current steering DAC topologies

Current steering digital-to-analog converters are based on the principle of providing a digitally controlled sum of currents. Depending on the manner how respective current sources are tailored and summed, Binary Weighted (BW) and Unary Weighted (UW) topologies are distinguished (Baker et al., 1998). In the latter case, all current sources provide a unity current value (associated with the Least Significant Bit, LSB). The number of current sources equals the number of quantization levels, and their connection to the DAC's output is controlled by a thermometer code signal derived from the input word. The Binary Weighted architecture uses only N current sources with N being the resolution in bits. Consequently, they are binary weighted and directly controlled by the N input signals.

Both architectures have advantages and drawbacks: for instance, the simplicity of the BW approach allows to save area and power but leads to worse accuracy, in the worst-case even to non-monotonic behavior. On the other hand, for high resolutions UW converters require huge binary-to-thermometer decoders, which increase latency, limit the operating frequency, and lead to very expensive solutions in terms of area. However, monotonic behavior is always guaranteed and a linear dependence between output glitch and transition value is obtained (Lin and Bult, 1998).

A mixture of UW and BW topologies leads to the so-called segmented current steering topology, where the advantages of both approaches are merged by concurrently converting the more significant bits through a UW DAC and the less significant bits by means of a BW converter (Maloberti, 2008).

3 Mismatch considerations

As shown in Cong and Geiger (2002), Differential Non Linearity (DNL) and Integral Non Linearity (INL) errors of unary weighted current steering DACs can be analytically related to the matching properties of the unity current sources. Moreover, since binary scaled current sources consist of ensembles of unity current sources, it can be shown that the worst INL error occurs at midcode. This applies to UW, BW, and segmented current steering topologies, and it amounts to:

$$E[\mathrm{INL}^2_{\mathrm{WORST}}] \simeq \frac{\sigma^2_{I,\mathrm{Rel}}}{4}\left(2^N - 1 - \frac{1}{2^N-1}\right) \quad (2)$$

There, $E[\mathrm{INL}^2_{\mathrm{WORST}}]$ is the variance of the INL error at midcode, N is the resolution in bits, i.e. $(2^N - 1)$ is the total number of unity current sources. The parameter $\sigma_{I,\mathrm{Rel}}$ is the standard deviation of the relative error of the unity current sources' output current. This mismatch is mainly dependent on the dimensions of the transistor responsible for the current generation and on its operating point (Pelgrom et al., 1989; Sansen, 2008; Bastos et al., 1996).

On the basis of similar considerations it can be shown that, if the DAC is segmented or binary weighted, the worst-case DNL code is given by the code which shares the smallest number of current sources with its previous one: in a BW-DAC that is the midcode, while in a segmented architecture it is every code, that makes the BW part convert a zero. If m is the number of bits with binary weighted conversion, we achieve:

$$E[\mathrm{DNL}^2_{\mathrm{WORST}}] = \left(2^{m+1} - 1 - \frac{1}{2^N-1}\right)\sigma^2_{I,\mathrm{Rel}} \quad (3)$$

From Eqs. (2) and (3) the optimum percentage of segmentation and the required precision of the unity current source can be evaluated, once maximum INL and DNL are defined. For this work, aiming for an 8 bit input signal and a monotonic behavior with high probability (99.9 %), a maximum $\sigma_{I,\mathrm{Rel}}$ of 1.904 % and an m-value of ≤ 5 are calculated, respectively. Consequently, in the next paragraphs the related blocks of a current steering DAC with a segmentation of 50 % and unity current sources with a target current of 40 nA and an absolute standard deviation of less than 760 pA are presented.

4 Transistor level design

As shown in Fig. 1, the binary weighted part of the converter consists of four scaled current sources which generate 1×LSB, 2×LSB, 4×LSB, and 8×LSB currents, respectively. They are controlled by the four inputs for the less significant bits. The 4 more significant bits are fed into a binary-to-thermometer decoder. Its fifteen outputs control the 16×LSB current sources which are used in the DAC's unary weighted part.

4.1 Binary-to-thermometer decoder

Figure 2 shows the small decoder that translates the four input signals into the 15 thermometer outputs. The clear hierarchical structure with its exponential dependence on the number of bits to be converted is the issue to be faced when designing high-precision high-bit unary weighted DACs. Some sophisticated solutions have been developed to bypass this challenge taking advantage of the unary weighted topology (Miki et al., 1986; Bastos et al., 1998): There, use of a row-column selection architecture reduces the area demand on the cost of additional digital gates in every current source unit. Another solution has been reported by Jeong et al. (2010), where sub-arrays are used. However, since in this work we are aiming to interface the DAC with a small-pitch sensor array, harvesting of any opportunity to save area is mandatory. Thus, we have chosen not to implement any sophisticated strategy but to accept a slight linearity degradation due to the segmented topology.

4.2 LSB current cell

If the LSB current source is operated in strong inversion the following equation applies (Pelgrom et al., 1989):

$$\sigma^2_{I,\mathrm{Rel}} = \frac{A^2_\beta}{WL} + \frac{4A^2_{V\mathrm{TH}}}{WL(V_{\mathrm{GS}} - V_{\mathrm{TH}})^2} \tag{4}$$

On this basis the following considerations can be drawn:

1. Given the mismatch related parameters A_β and $A_{V_{\mathrm{TH}}}$ with A_β being the transistor constant related matching constant in [% μm] and $A_{V_{\mathrm{TH}}}$ being the threshold voltage related matching constant in [mV μm], the required standard deviation $\sigma_{I,\mathrm{Rel}}$ can be achieved by choosing an adequate channel area *WL* and/or by choosing an appropriate effective gate voltage $V_{G,\mathrm{EFF}} = (V_{\mathrm{GS}} - V_{\mathrm{TH}})$.

2. Consequently, for a given standard deviation, the application of a higher effective gate voltage always translates into the option to decrease the active transistor area.

3. However, given a defined current to be generated by the MOSFET, the only way to increase $V_{G,\mathrm{EFF}}$ is to make the ratio *W*/*L* smaller.

A comparison with the mismatch behavior for devices operated in the subthreshold region yields:

$$\sigma^2_{I,\mathrm{Rel}} = \left(\frac{q}{nkT}\right)^2 \frac{A^2_{V\mathrm{TH}}}{WL} \tag{5}$$

There, q is the elementary electronic charge, k the Boltzmann constant, T the absolute temperature, and n the so-called non-ideality factor. For a chosen standard deviation and a fixed current, transistors operated in strong inversion at sufficiently high effective gate voltage are smaller than transistors operated in subthreshold region.

Given the boundary conditions in this work, this, however, would translate into an extremely high *L*/*W* ratio as the required current is small. This is not convenient from the layout related point of view. As a consequence, the LSB current cell in this work is based on an NMOS transistor operated in subthreshold region, with more reasonable values for W and L at an active transistor area of roughly 60 μm^2.

The entire schematic of the LSB current source is shown in Fig. 3. Two output nodes are used to ensure an optimum transient response by never switching off the current sources (Baker et al., 1998; Sansen, 2008). A cascode transistor is added in order to minimize the parasitic capacitance at the source of the switch transistors, and thus to optimize the settling-time of the circuit. For the same reason the switches have minimum dimensions and dummy switches are avoided as the use of dummies would imply doubling the amount of switch transistors and related parasitics.

As shown in Fig. 4, the effect of using a cascode transistor on current source mismatch is negligible. Associated to

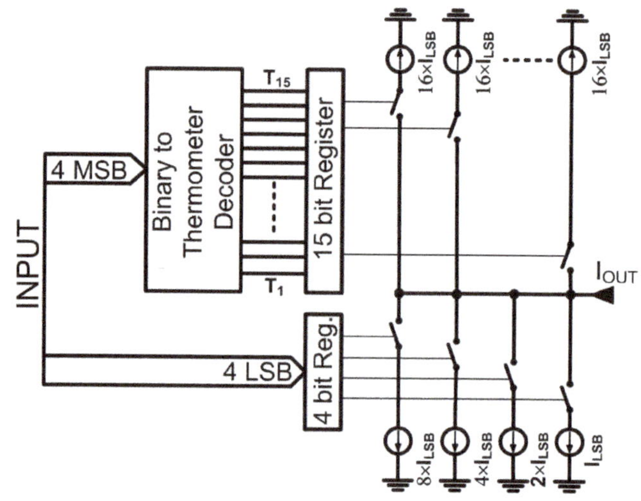

Fig. 1. Block diagram.

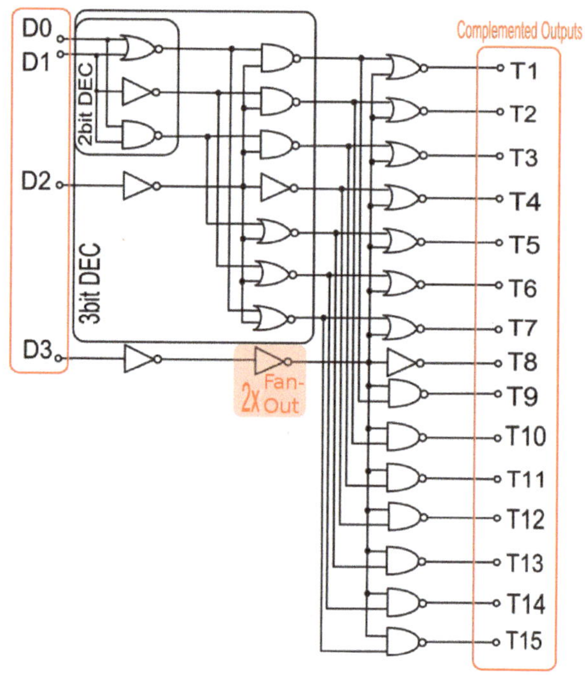

Fig. 2. Binary-to-thermometer decoder.

the switchable current source is also a small driver (Fig. 3). By means of that circuit the control signals SW and SWn are generated: their voltage swing is limited to the window from 0.7 V to 1.2 V in order to reduce the clock feedthrough (CFT) to the output while still maintaining proper switching functionality. Moreover, a slower transient drive capability caused by the minimum dimensions of the output stage of the driver is advantageous in the sense that CFT is further reduced. Thus, the voltage error induced on node Vs of the current source is minimized, which is also advantageous for the overall settling time.

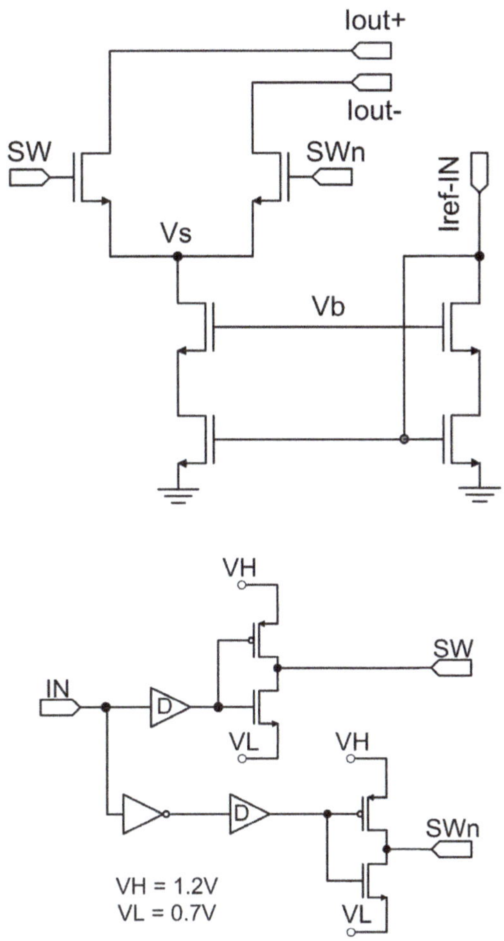

Fig. 3. LSB current source with related driver. Bias transistors connected to input "Iref-IN" are shared by the entire array.

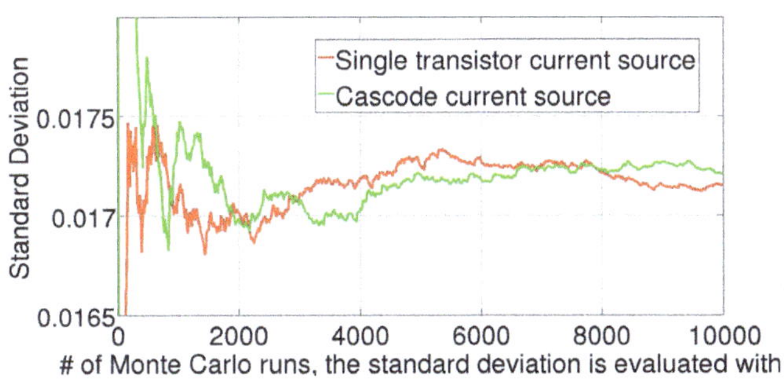

Fig. 4. Relative output current standard deviation of a single transistor based and of a cascode current source.

4.3 Scaled current cells

The current sources required for the unary weighted part of the DAC and for the other bits of the binary weighted part can be directly derived from the circuit presented in Sect. 4.2. The generic weighted cells are implemented as replica cells of the LSB current cell (i.e. LSB current source with related driver) since the small driver in Fig. 3 is optimized only for one LSB current source. Although the area consumption consequently slightly increases, the layout regularity due to the replication of the entire cell has a positive impact on the linearity and suppresses the sensitivity to on-chip parameter gradients.

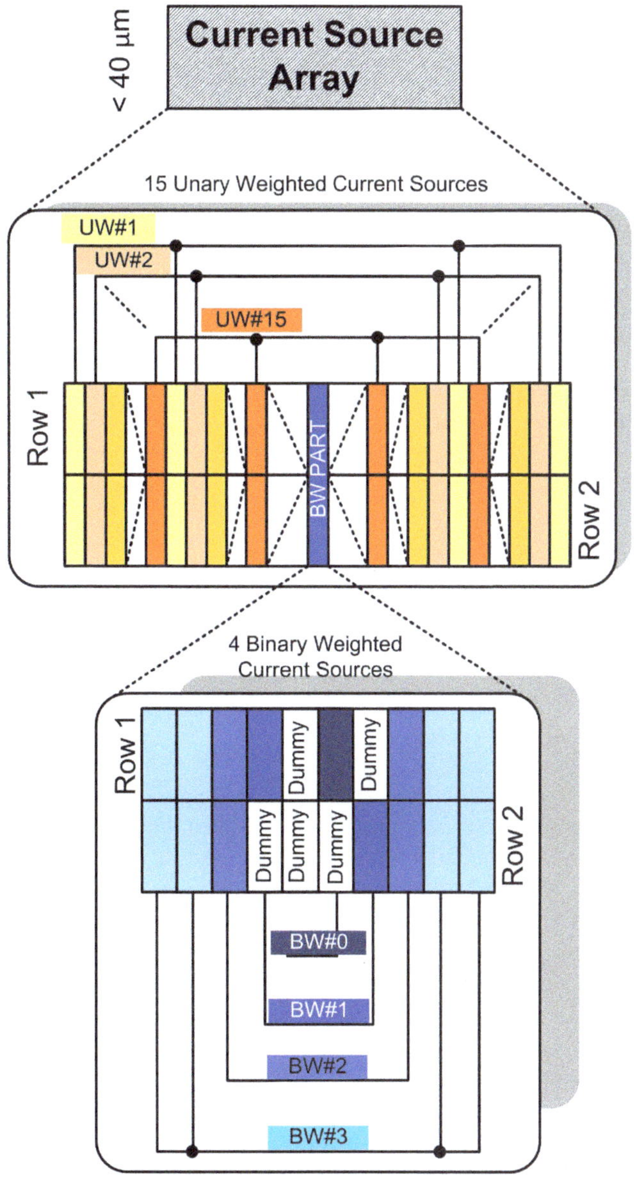

Fig. 5. Schematic plot of the realized layout.

5 Layout

As e.g. emphasized by Bastos et al. (1998), the performance of digital-to-analog converters is strictly related to their layout implementation. As the array of current sources can demand a huge area depending on the number of bits, precision,

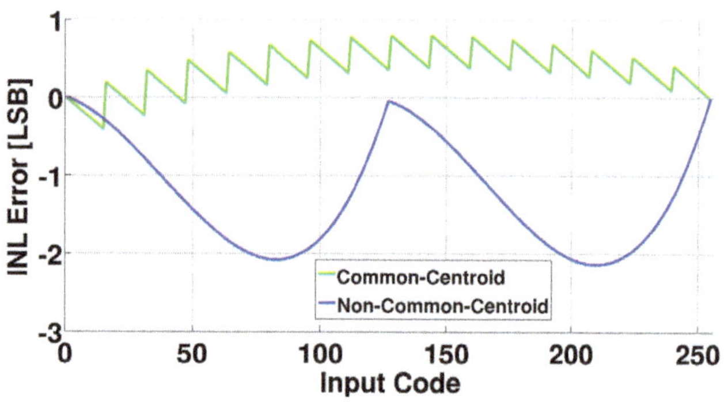

Fig. 6. Example for the impact of a systematic error applied to the array of current sources for a Common-Centroid and a Non-Common-Centroid layout. Details concerning the assumed error function see Eq. (6).

and full-scale current, CMOS process parameter gradients on the chip cannot be neglected. A systematic error of the current sources' parameters would indeed determine a distortion of the static characteristics and might have a severe negative impact on DNL and INL. For that reason a Common Centroid layout is used. Some further techniques may additionally give rise to even better results (Van der Plas et al., 1999; Nakamura et al., 1991), however, as they also require to invest a huge amount of extra area, layout constraints do not allow for their implementation in this work.

In Fig. 5, details are depicted of the Common Centroid layout used here. The current sources of the binary weighted part are placed in the center of the array in a symmetrical manner supported by the use of dummy LSB cells. The current cells of the unary weighted part form the rest of the array. The 15 UW cells, each one consisting of 16 LSB cells connected together, are not arranged side by side in space but symmetrically distributed with respect to the binary weighted part within the entire array. The effects of such a local distribution are exemplarily shown in Fig. 6, where for instance a parabolic error as a function of the current source location is considered following the relation

$$\Delta I(x,y) = I_{\text{LSB}}\left[K_X(x-x_0)^2 + K_Y(y-y_0)^2\right] \quad (6)$$

with $K_X = 5\times10^{-8}\,\mu\text{m}^{-2}$, $K_Y = 1\times10^{-7}\,\mu\text{m}^{-2}$, $x_0 = -500\,\mu\text{m}$, $y_0 = 100\,\mu\text{m}$, $x \in [-1311;1311]\,\mu\text{m}$, $y \in [-2.3;2.3]\,\mu\text{m}$.

6 Results

Figures 7 and 8 provide post-layout simulation results of DNL and INL, respectively. For these simulations, a low-ohmic 0.9 V (=VDD/2) voltage source is connected as a load to the DAC's output. The shown results are based on 5000 Monte-Carlo runs: the data are generated by means of a Matlab script, which computes the transfer characteristics on the basis of DC simulations of 255 equal independent LSB cells. With the DNL being always smaller than half of an LSB, the Integral Non Linearity error is ≤ 0.5 LSB with a probability of 99.82 %. The slight deviation from the targeted 99.9 % may originate from the limitations of the model used in Eq. (5), from the impact of the cascode transistor, but also from the finite number of Monte-Carlo runs. Anyway, the result is considered to be acceptable.

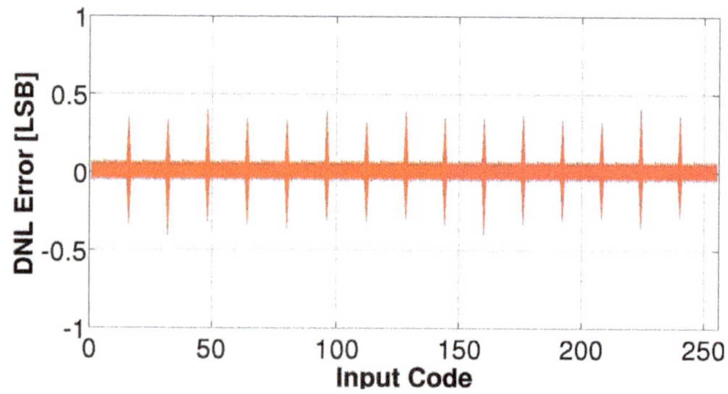

Fig. 7. DNL error vs. input code for 5000 Monte-Carlo simulations.

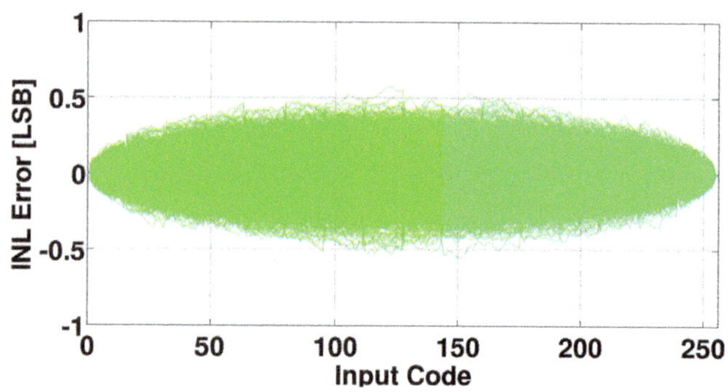

Fig. 8. INL error vs. input code for 5000 Monte-Carlo simulations.

Using transient simulations and the same output configuration as above, the overall correct functionality, settling time, glitch values, and the dynamic power consumption are considered. The results are summarized in Table 1. Since the settling time at 5 % of LSB level in a full scale transition is much smaller than required for the specified 2.6 MHz operating frequency, a maximum operating frequency calculated as reciprocal of the 3-sigma worst-case Monte-Carlo settling time appears to be 10.2 MHz. At 1.8 V supply voltage a dynamic power of 100 μW is dissipated at the operating frequency of 2.6 MHz, and a DC power of 10 μW is absorbed mainly by the low-ohmic load.

7 Conclusions

An 8 bit, 2.6 MHz, 10 μA full scale CMOS current steering DAC with high linearity, low noise, and small area has been presented. Using a standard 180 nm CMOS technology,

Table 1. Summary of simulation results.

Parameter	Value
LSB Current	40 nA
DNL	< 0.5 LSB
INL	< 0.5 LSB
Settling Time (nominal, 5 %-LSB)	60.6 ns
Settling Time (Monte-Carlo, 5 %-LSB)	$\mu = 68.2$ ns $\sigma = 9.98$ ns
Output Noise [20 Hz; 2.6 MHz]	2.75 nA
Average Glitch Current (1 LSB transition)	14 pA
DC Power Consumption	10 μW
Dynamic Power Consumption (2.6 MHz Full Scale Transition)	100 μW
Maximum Operating Frequency	10.2 MHz

it dissipates 100 μW from a single power supply voltage of 1.8 V at 2.6 MHz operating frequency. Monte-Carlo simulations reveal DNL and INL of $\leq$0.4 LSBs and $\leq$0.5 LSB, respectively. Maximum operating frequency is 10.2 MHz. Following the target application, a layout with total dimensions $38.6 \times 2933\ (\mu\text{m})^2 = 0.113\ \text{mm}^2$ is provided. Common Centroid layout techniques are applied to the crucial blocks in order to minimize the impact of possible on-chip gradients.

Acknowledgements. This work was supported by the "Bundesministerium für Bildung und Forschung" (BMBF), reference 0315636B.

References

Baker, R. J.: CMOS Circuit Design, Layout, and Simulation, 800–802, IEEE Press, 1st edn., 1998.

Bastos, J., Steyaert, M. and Sansen, W.: A High Yield 12-bit 250-MS/s CMOS D/A Converter, IEEE Cust. Integr. Cir., 431–434, 1996.

Bastos, J., Marques, A. M., Steyaert, M. S. J. and Sansen W.: A 12-Bit Intrinsic Accuracy High-Speed CMOS DAC, IEEE J. Solid.-St. Circ., 33, 1959–1969, 1998.

Chen, T. and Gielen, G. G. E.: A 14-bit 200-MHz current-steering DAC with switching-sequence post-adjustment calibration, IEEE J. Solid.-St. Circ., 42, 2386–2394, 2007.

Cong, Y. and Geiger, R. L.: Formulation of INL and DNL yield estimation in current-steering D/A converters, Proc. Iscas., 3, 149–152, 2002.

Eversmann, B., Jenkner, M., Hofmann, F., Paulus, C., Brederlow, R., Holzapfl, B., Fromherz, P., Merz, M., Brenner, M., Matthias, S., Gabl, R., Plehnert, K., Steinhauser, M., Eckstein, G., Schmitt-Landsiedel, D. and Thewes, R.: A 128 × 128 CMOS Biosensor Array for Extracellular Recording of Neural Activity, IEEE J. Solid.-St. Circ., 38, 2306–2317, 2003.

Frey, U., Sedivy, J., Heer, F., Pedron, R., Ballini, M., Mueller, J., Bakkum, D., Hafizovic, S., Faraci, F. D., Greve, F., Kirstein, K. and Hierlemann, A.: Switch-Matrix-Based High-Density Microelectrode Array in CMOS Technology, IEEE J. Solid.-St. Circ., 45, 467–482, 2010.

Jeong, S., Kim, K., Jung, K., Park, K., Cho, S.: A Novel Current Steering Cell Matrix DAC Architecture with Reduced Decoder Area, Ele. Com. Eng.-CISST, 10, 167–170, 2010.

Lei Luo, J. Y. R.: An 8-bit 700 Mss current-steering DAC, Solid-State and Integrated-Circuit Technology, Int. Conf. on, 1925–1928, Beijing, 20-23 Oct. 2008.

Lin, C.-H. and Bult, K.: A 10-b, 500-MSample/s CMOS DAC in 0.6 mm^2, IEEE J. Solid.-St. Circ., 33, 1948–1958, 1998.

Lin, J.-F., Chang, S.-J, Chiu, C.-F., Tsai, H.-H and Wang, J.-J.: Low-Power and Wide-Bandwidth Cyclic ADC with Capacitor and Opamp Reuse Techniques for CMOS Image Sensor Application, IEEE Sens. J., 9, 2044–2054, 2009.

Maloberti, F.: Data Converters, Springer, Dordrecht, The Netherlands, 1st edn., 124–129, 2008.

Miki, T., Nakamura, Y., Nakaya, M., Asai, S., Akasaka, Y. and Horiba, Y.: An 80-MHz 8-bit CMOS D/A Converter, IEEE J. Solid.-St. Circ., 21, 983–988, 1986.

Nakamura, Y., Miki, T., Maeda, A., Kondoh, H. and Yazawa, N.: A 10-b 70-MS/s CMOS D/A Converter, IEEE J. Solid.-St. Circ., 26, 637–642, 1991.

Pelgrom, M. J. M., Duinmaijer, A. C. J. and Welbers, A. P. G. : Matching Properties of MOS Transistors, IEEE J. Solid.-St. Circ., 24, 1433–1440, 1989.

Perenzoni, M., Massari, N., Stoppa, D., Pancheri, L., Malfatti, M. and Gonzo, L. : A 160 × 120-Pixels Range Camera With In-Pixel Correlated Double Sampling and Fixed-Pattern Noise Correction, IEEE J. Solid.-St. Circ., 46, 1672–1681, 2011.

Sansen, W.: Analog Design Essential, 614–615, Springer, Dordrecht, The Netherlands, 1st edn., 2008.

Van der Plas, G. A. M., Vandenbussche, J., Sansen, W., Steyaert, M. S. J., Gielen, G. G. E.: A 14-bit Intrinsic Accuracy Q^2 Random Walk CMOS DAC, IEEE J. Solid.-St. Circ., 34, 1708–1718, 1999.

16

Design and quantitative analysis of parametrisable eFPGA-architectures for arithmetic

B. Neumann, T. von Sydow, H. Blume, and T. G. Noll

Chair of Electrical Engineering and Computer Systems, RWTH Aachen University, 52062 Aachen, Germany

Abstract. Future SoCs will feature embedded FPGAs (eFPGAs) to enable flexible and efficient implementations of high-throughput digital signal processing applications. Current research projects on and emerging products containing FPGAs are mainly based on "standard FPGA"-architectures that are optimised for a very wide range of applications. The implementation costs of these FPGAs are dominated by a very complex interconnect network. This paper presents a method to improve the efficiency of eFPGAs by tailoring them for a certain application domain using a parametrisable architecture template derived from the results of a systematic evaluation of the requirements of the application domain.

Two different architectures are discussed, a reference architecture to illustrate the methodology and possible optimisation measures as well as a specialised arithmetic-oriented eFPGA for applications like correlators, decoders, and filters. For the arithmetic-oriented architecture, a novel logic element (LE) and a special interconnect architecture that was designed with respect to the connectivity characteristics of regular datapaths, are presented. For both architecture templates, physically optimised implementations based on an automatic design approach have been created.

As a first cost comparison of these implementations with standard FPGAs, the LE-density (number of logic elements per mm^2) is evaluated. For the arithmetic-oriented architecture, the LE-density could be increased by an order of magnitude compared to standard architectures.

1 Introduction

The algorithmic complexity of current and future digital signal processing systems leads to an ever growing demand for performance, while short product cycles, mobility and cost-restrictions make the use of dedicated integrated circuits unfeasible. Fig. 1 illustrates the design space in terms of power efficiency (in mW/MOPS) and area efficiency (in MOPS/mm^2) for frequently used digital signal processing tasks implemented on different architectures like general purpose processors, Field Programmable Gate Arrays (FPGAs), physically optimised macros etc. As is well known from Fig. 1, programmable processor-architectures provide the lowest power- and area efficiency but feature highest flexibility. In contrast, physically optimised macros lead to the highest efficiency, but are inflexible. FPGAs offer an attractive compromise between these two extremes, as they allow for highly parallelised implementations while preserving in-system reconfigurability.

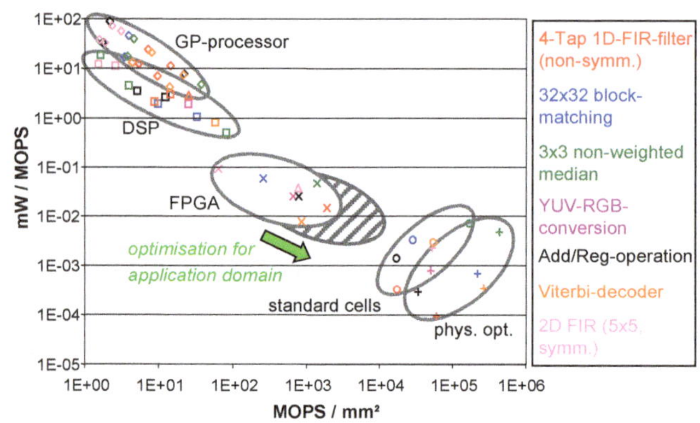

Fig. 1. Design space for different architectures.

To meet the contrary demands of today's and future systems, different architecture blocks are integrated on a single chip building a heterogeneous System-on-a-Chip (SoC). A so-called SoC platform addresses a set of applications with similar requirements and characteristics (application domain). This offers the possibility to optimise the included components for this application domain. In the same way

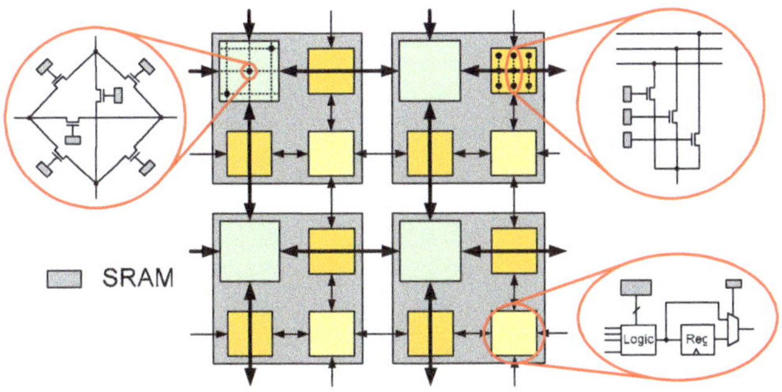

Fig. 2. General island-style FPGA-architecture.

the optimisation for an application domain improves the efficiency of a Digital Signal Processor (DSP) over a general purpose CPU, the architecture of an embedded FPGA (eFPGA) can be optimised by designing it for a dedicated application domain. The potential of improvement in efficiency for application domain specific eFPGAs was demonstrated for example in Neumann et al. (2003) and Leijten-Nowak et al. (2003). This paper focuses on arithmetic-oriented applications like correlators, decoders and filters, as they promise a high optimisation potential concerning all structural elements of an eFPGA-architecture.

The paper is structured as follows: In Sect. 2, a commonly applied architecture for FPGAs and its characteristics concerning area and power dissipation costs is sketched. The properties of eFPGAs specified for an arithmetic-oriented application domain are described in Sect. 3. In Sect. 4, the concept of parametrisable eFPGA-architectures is explained and two architecture templates with corresponding implementations are demonstrated. Conclusions are given in Sect. 5.

2 FPGA-architectures

Most common FPGA-architectures, as described e.g. in Brown et al. (1996), consist of three basic building blocks: Logic Elements (LE), Routing Switches (RS) and Connection Boxes (CB). Figure 2 shows the architecture of a so-called island-style FPGA-architecture as it was used in the first commercial devices. Most state-of-the-art FPGAs are still based on a modified island-style architectural concept.

An LE typically implements basic boolean functionality on bit-level. In most commercial FPGAs, the architecture of an LE is based on one or more lookup-tables (LUTs). A LUT with N inputs shall be called LUT-N in the following. Connection Boxes allow to connect interconnect lines either as in- or outputs of the corresponding LEs. The routing switch provides configurable connections between all its terminals to route signals through the FPGA. Most modern FPGAs use SRAMs to store the configuration information. This allows fast reconfiguration (and in principle dynamic reconfigurability).

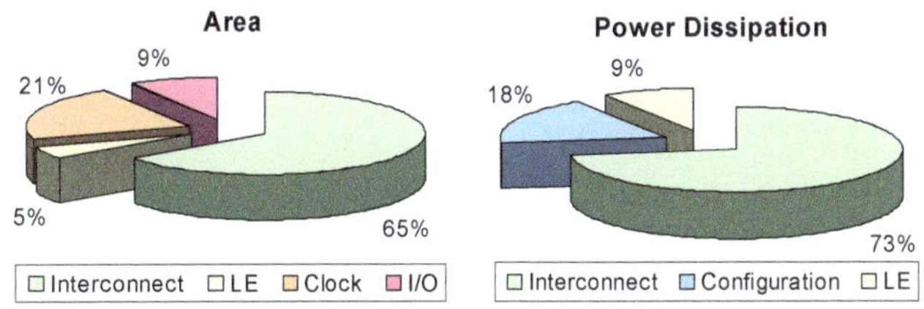

Fig. 3. FPGA area- and power breakdown.

To relax the global interconnect requirements, several logic elements can be combined in so-called clusters that share a central connection box. Typically, a cluster consists of about ten LEs. As delays for signals travelling long distances on the FPGA can get significant since they have to pass several routing switches, it is a common technique to provide segmented interconnects with routing lines of different lengths. In this case, certain routing channels skip several routing switches before connecting to another RS.

The programmable interconnect features the highest contribution to area, power dissipation and delay times of an FPGA. Figure 3 shows an area- and power breakdown taken from Kusse et al. (1998) and George et al. (2001). The figures show that the highest optimisation potential can be expected from the interconnect architecture. It is therefore indispensable to analyse the specific interconnect requirements of the given application domain and hence design the appropriate eFPGA-architecture.

3 Arithmetic-oriented eFPGAs

Commercial FPGA components are optimised for a wide range of applications, as they are intended for universal use. Significant research has been conducted in the optimisation of these universal FPGAs, using benchmark circuits from different application domains (e.g. Betz et al., 1999). An analysis of the logic- and interconnect requirements of arithmetic-datapaths reveals that the architectural requirements differ significantly from irregular logic. In this contribution, an eFPGA-architecture and the corresponding structural elements have been tailored to an arithmetic-oriented application domain. Considering the interconnect requirements of different applications mapped to FPGAs, it is useful to consider the histogram of the lengths of allocated interconnections. Fig. 4 shows such a qualitative histogram of the allocated connections between logic elements. For irregular logic, the number of connections between the LEs decreases with the connection length L. Arithmetic-datapaths, however, have a very high locality, hence exposing a peak in the distribution for short connections. Only few lines of intermediate length are required, while some long connections, usually representing broadcast lines, are used.

To confirm this qualitative histogram, an investigation of the allocated routing resources in a commercial FPGA de-

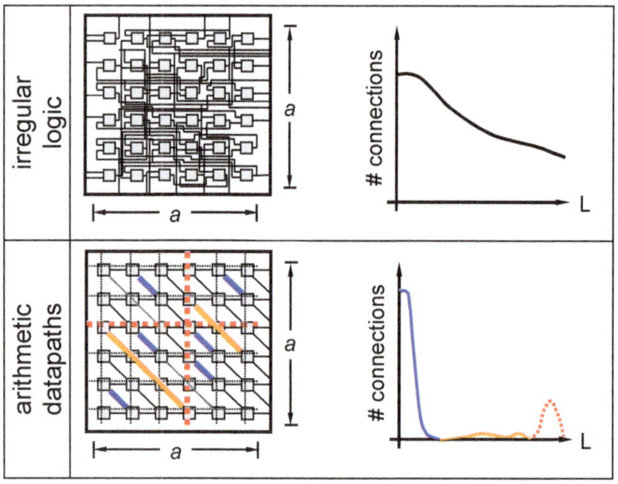

Fig. 4. Application domain specific routing requirements.

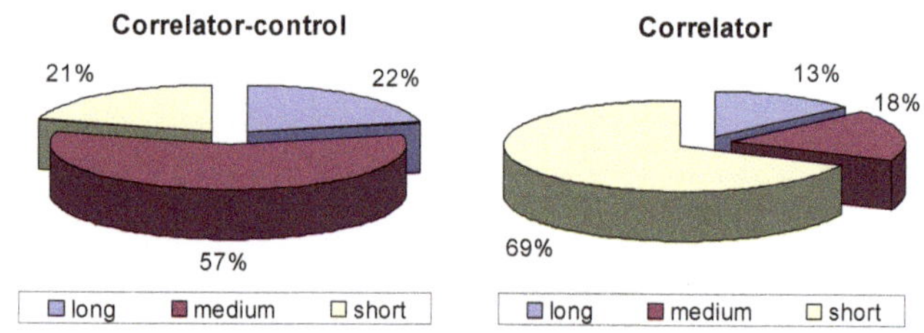

Fig. 5. Allocated interconnect lengths for irregular logic and arithmetic.

Table 1. Exemplary architectural parameters.

LE	no. of in- /outputs amount of dedicated logic
Cluster	no. and alignment of LEs no. of in- /outputs
Interconnect	no. of hierarchy levels size of routing channels segmentation of routing channels flexibility
Configuration	no. of parallel configuration sets no. of shared configuration sets
Others	granularity of supply and clock domains

vice has been conducted. For this purpose, key components of a GPS-receiver (Global Positioning System) have been implemented on an Altera Stratix device (Kappen et al. 2006). Using the corresponding design software, the number of allocated routing resources according to their segment length has been determined. Figure 5 shows the results for two exemplary components, the correlator and the corresponding control logic. While the irregular control logic has a high fraction of intermediate segments and considerably less short and long segments, the correlator logic itself has a significant amount of direct nearest-neighbour connections. Since the overhead in the interconnect architecture of today's FPGAs is mainly due to their high flexibility, a considerable reduction of this overhead can be expected if the architecture is tailored to this application domain.

A further reduction can be achieved by carefully designing the logic elements to preserve the communication locality of arithmetic datapaths. In order to relax the interconnect requirements, appropriate dedicated logic and connections are provided in the LEs. A basic example is the use of dedicated carry-chains in commercial FPGAs. Direct connections between the LEs enable the use of adders with no additional routing involved in the carry-propagation.

Additional optimisation can be achieved if the LEs are designed such that frequently used basic operations of typical datapaths are directly mapped to one LE or LE cluster respectively. An example of optimising the LE architecture in that sense is to provide the possibility of realising a gated full-addition efficiently within one LE, as gated full-addition is a common part of many arithmetic-oriented applications (multiply, etc.). In the same way, other dedicated logic and dedicated interconnect can be employed to improve the efficiency of the architecture. For example, while typical logic elements feature one register, arithmetic datapaths with extensive pipelining require significantly more registers per LE.

4 Parametrisable eFPGA-architectures

In the following section, the design flow applied in this contribution is presented. This flow is based on an architecture template of the eFPGA with adjustable parameters. From this template, physically optimised realisations of the macro are created.

4.1 Layout automation and parameters

The basic design concept is based on generic architecture templates which are parametrisable in a wide range and a couple of physically optimised so-called leafcells. An automatic layout generation using a flexible datapath generator as described in (Weiss et al. 2001) is used to generate the layout of the eFPGA-macro. Table 1 lists some exemplary parameters influencing the basic structural elements of an eFPGA-architecture.

These parameters feature strong interdependencies. For example, the architecture of the LE and the number of in- and outputs have a strong impact on the interconnect architecture.

In the present work, the design of a physically optimised layout for a parametrisable architecture is based on the following steps:

An architecture template has been conceived and realised in form of a generic structural description. A small set of key components like logic elements and switch points used for

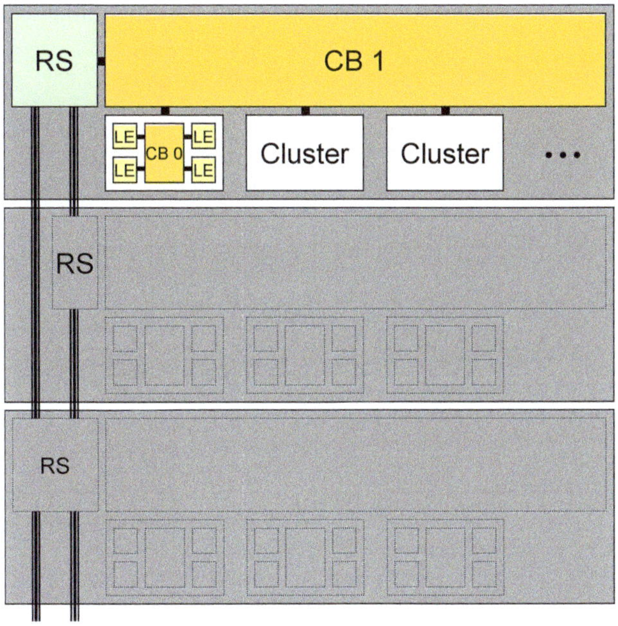

Fig. 6. Hierarchical reference architecture.

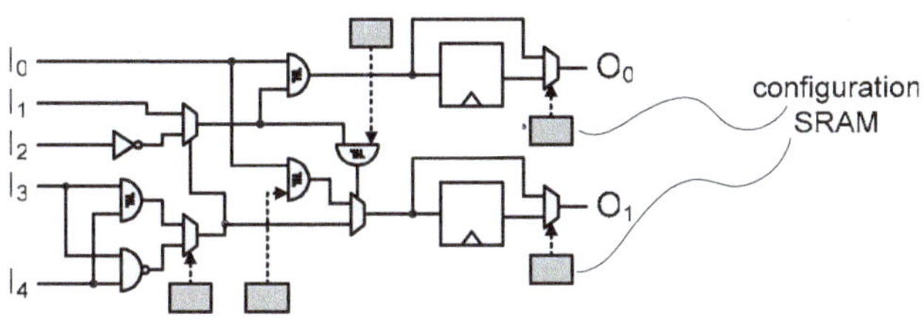

Fig. 7. LE of reference architecture.

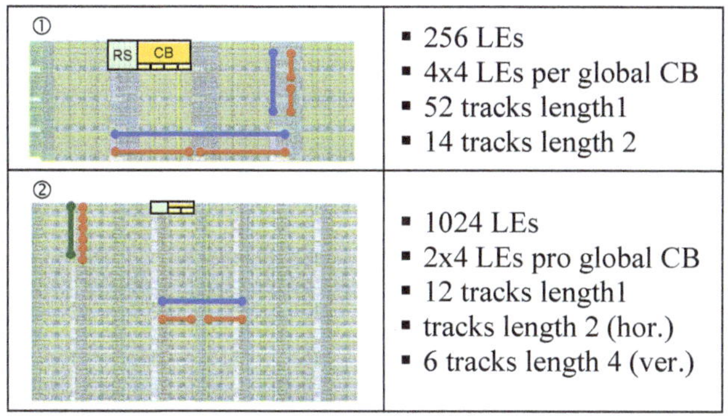

Fig. 8. Reference architecture macros and corresponding parameter settings.

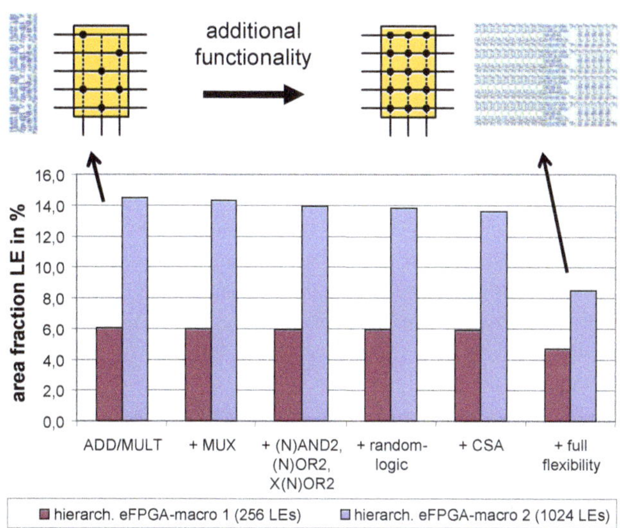

Fig. 9. Interrelation of flexibility and area within CBO.

routing switches has been designed. After adjusting all required parameters of the architecture template, a physically optimised macro is generated automatically applying the datapath generator. After the netlist extraction of this macro and setting up the simulation environment, a detailed timing and power analysis can be performed. If design constraints are not met, parameters can be easily adapted within the generic description and the macro generation process can be initiated again. The cycle time of adjusting parameters and acquiring cost and performance values depends on the complexity of the target macro.

4.2 Reference architecture

The first architecture template which has been realised serves as reference (Fig. 6). The interconnect architecture of this reference is based on the interconnect structure of modern commercial devices (Website Altera, Website Xilinx). A local connection box (CB 0 in Fig. 6) provides the connection between LEs arranged as clusters and the global connection box (CB 1 in Fig. 6). The global CB provides the connection to the global routing lines. Routing of signals coming from the global connection lines is executed by the RS. There are two types of routing switches: the first types enables regular vertical and horizontal routing between neighbouring RS, the second type is enhanced and offers the possibility of segmented routing for certain lines. Adaptable parameters of the corresponding architecture template are for example the number of routing tracks, the number of clusters per CB and the length of interconnect lines for the segmented routing.

The LE architecture unlike commercial devices has been designed for arithmetic-oriented applications and is based on dedicated logic without employing LUTs (Fig. 7). Gated full-addition, add-compare-select and other arithmetic operations can be implemented very efficiently. In particular, the second register allows for efficient implementation of pipelined datapaths. The logic complexity of the LE is smaller than for a LUT with the same number of inputs. The number of SRAM-cells and hence overall silicon area required for this LE is much smaller than a LUT-based solution.

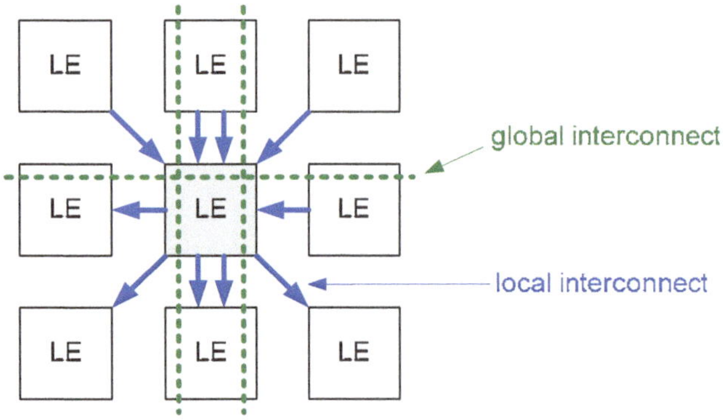

Fig. 10. Local and broadcast-interconnect.

Applying the design flow mentioned above, several macros with different parameter settings have been designed in a 180 nm-technology. Two exemplary macros featuring different cluster sizes, different number of LEs per macro and different number and segmentation of routing tracks are depicted in Fig. 8.

For the local CB (CB0 in Fig. 6) a detailed analysis of flexibility requirements has been conducted using a variety of exemplary basic operations. Fig. 9 shows the area fraction of the LEs proportional to the complete macro for different levels of connection box flexibility. The flexibility of the complete architecture is determined by the number of possible connections within a CB. At first, the interconnect of the local CB has been designed for simple arithmetic datapaths (e.g. gated full addition, carry-ripple addition etc.). This means that only those connections are realised which are required for ADD/MULT operations. Such a restricted CB would mean that other types of operations cannot be performed with this CB configuration. In the next steps, the interconnect structure has been enhanced in a way that multiplexer functionality and randomised logic can be implemented (step 2, 3 and 4).

In the following step, more complex arithmetic datapaths have been included (e.g. carry-save arithmetic, CSA). The reference point in this depiction corresponds to the full flexible solution (rightmost bars). Above the diagram, the according layouts of the CB, which are created automatically, are shown. The area fraction of the LEs is directly influenced by the routing parameters described above. Macro 1 includes more global routing tracks. Therefore, this solution features a higher flexibility than macro 2 which has got a more restricted routing architecture. The area of the LE and the local CB is the same for both macros.

For this first architecture template, the routing architecture is still the dominating factor in terms of area, timing and power consumption. Using the design methodology (see Sect. 4.1) established with the reference architecture, an eFPGA-architecture tailored for an arithmetic-oriented application domain has been developed.

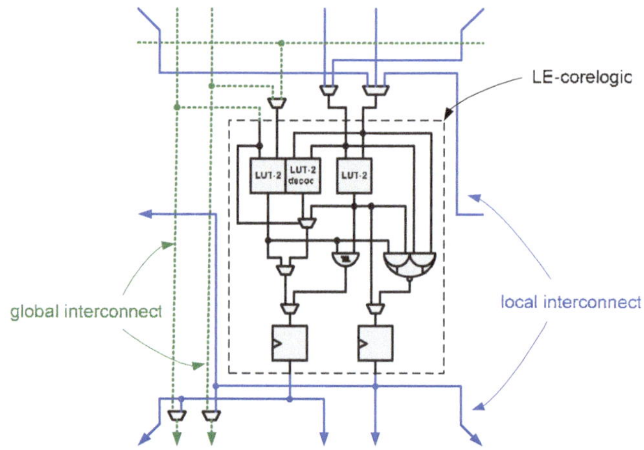

Fig. 11. Arithmetic-oriented LE.

4.3 Arithmetic-oriented eFPGA-architecture

The requirements of arithmetic-oriented datapaths have been addressed by the eFPGA architecture described in the following. The architecture uses arithmetic-oriented LEs that are connected as two-dimensional clusters, reflecting the typical orientation of processing elements in filters and multipliers. The LEs are connected with a directional nearest-neighbour interconnect. Rather than providing random connections to all surrounding neighbours, this structure reflects the typical nature of arithmetic-oriented datapaths featuring a preferred data flow direction. In this architecture, a typical data flow from top to bottom and right to left is assumed. Diagonal connections are provided in both left- and right-downwards direction to allow an efficient implementation of both right- and left-shift-operations. For efficient mapping of datapaths with arbitrary wordlengths, local connections are also provided across cluster borders. Signals from the global interconnect, typically representing operands in datapaths, are distributed along the LEs inside a cluster as so-called broadcast lines. This is in contrast to the highly flexible (and large) connection boxes in standard FPGAs that allow each LE to access global routing resources. Fig. 10 illustrates this data flow oriented interconnect.

The logic element used in the arithmetic-oriented eFPGA is depicted in Fig. 11. The core logic inside the dashed line consists of two lookup-tables with two inputs each, dedicated sum- and carry-logic and two latches.

One LUT-2 can be configured as partial-product gate using the global inputs to enable a direct mapping of array-multiplier-based structures to the eFPGA. In addition, the two lookup-tables can be combined to a single LUT with three inputs (LUT-3). Dedicated multiplexers allow to connect the outputs of an LE to the broadcast-lines. The use of two latches per LE enables efficient pipelining of array-based datapaths. Rather than using complete registers, small transmission-gate-based latches reduce the implementation costs.

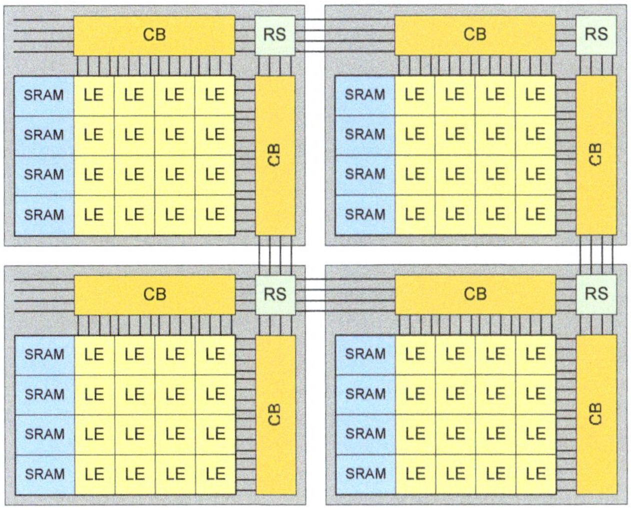

Fig. 12. Arithmetic-oriented architecture template.

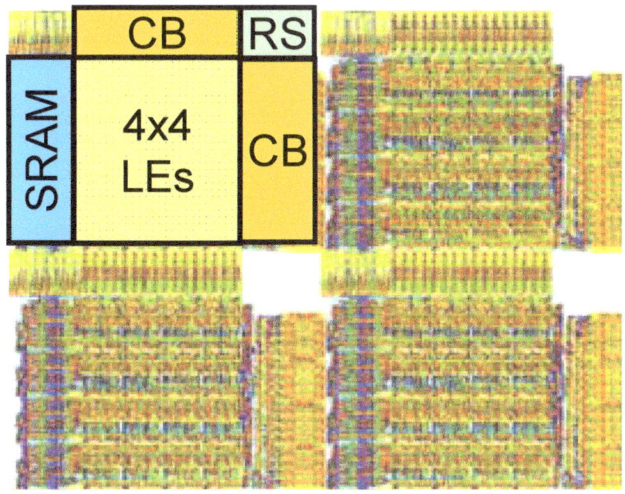

Fig. 13. Layout of arithmetic-oriented eFPGA.

The regularity and the typical wordlengths of arithmetic datapaths allow the shared use of configuration information, as for example presented in Ye et al., 2003). In this case, several logic elements inside a cluster use the same configuration bits. As the SRAMs storing the configuration typically amount for about half the area of a logic element, significant area-reductions can be achieved by choosing the appropriate granularity for the eFPGA-architecture. The same concept is applied to the global interconnect lines, thus reducing the overhead of the costly interconnect architecture. The overall architecture is shown in Fig. 12.

An according eFPGA-macro was designed in a 130 nm-technology using a cluster size of 4×4 and a configuration granularity of four. The global interconnect provides only 16 tracks, as the communication in arithmetic-oriented datapaths is mainly based on local connections as shown before. A layout of four complete clusters with configuration blocks, CBs and RSs is shown in Fig. 13.

As can be seen from the layout, the area of the macro is dominated by the logic elements, while the area required by the global interconnect could be reduced significantly. The area fraction of the LE for the arithmetic-oriented eFPGA is 40%, compared to less than 10% in typical standard FPGAs.

Figure 14 illustrates the LE-density (number of logic elements per mm^2) for different (e)FPGA-architectures. It compares commercial standard FPGAs like the Altera APEX 20K (Website Altera) or the M2000 FlexEOS-eFPGA (Website M2000) as well as an exemplary research project, the Chimaera-architecture (Hauck et al. 2004) with the eFPGA-macros presented in this paper. Also, a basic island-style architecture as described in Fig. 2 is considered. As expected, the reference macro 1 has an LE-density similar to that of the APEX 20K, while reference macro 2 achieves a higher density due to the reduced number of routing tracks. Both the Chimaera-architecture and the commercial Flex-EOS show a significant improvement compared to the discrete standard FPGA.

The arithmetic-oriented eFPGA presented in this paper achieves an LE-density that is one order of magnitude higher than for standard FPGAs. This comes at the cost of reduced flexibility when mapping arbitrary, irregular logic. Mapping irregular logic to this architecture would lead to losses in efficiency due to misalignments in data flow, inefficient usage of shared configurations and insufficient functional complexity on LE-level. However, for arithmetic-oriented applications the functional complexity of the LEs is equivalent to LE-complexities in standard FPGAs (i.e. a similar number of LEs is required to map a certain function to either a standard- or the arithmetic-oriented eFPGA). Hence, the area efficiency of arithmetic-datapaths can be improved by as much as an order of magnitude with the present architecture.

5 Conclusion

In this paper, an architecture template based design-methodology for domain specific eFPGAs has been presented. Based on the results of an exhaustive analysis of routing-, functionality- and regularity requirements of arithmetic-oriented applications, several architectural optimisations have been applied. The parametrisable template enables systematic optimisations according to the requirements of an application domain. Based on an analysis of arithmetic-oriented applications like filters and correlators, a novel eFPGA-architecture was presented. For this architecture, an area reduction of one order of magnitude compared to common FPGA-architectures was achieved.

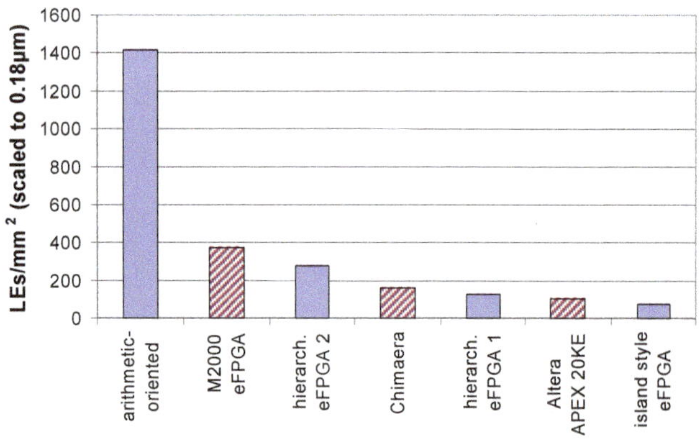

Fig. 14. LE-density for different eFPGA-architectures.

References

Betz, V., Rose, J. and Marquardt, A.: Architecture and CAD for Deep-Submicron FPGAs, Kluwer International Series in Engineering and Computer Science, ISBN 0-7923-8460-1, 1999.

Brown, S. and Rose, J.: FPGA and CLPD architectures: A tutorial, IEEE Design and Test of Computers, 13, 42–57, 1996.

George, V. and Rabaey, J. M.: Low Energy FPGAs – Architecture and Design, Kluwer International Series in Engineering and Computer Science, 2001.

Hauck, S., Fry, T. W., Hosler, M. M., and Kao, J. P.: The Chimaera Reconfigurable Unit, IEEE Trans. on VLSI Systems, 12, 206–217, 2004.

Kappen, G. and Noll, T. G.: Application Specific Instruction Processor Based Implementation of a GNSS Receiver on an FPGA, accepted for IEEE DATE, 2006.

Kusse, E. and Rabaey, J.: Low-energy embedded FPGA structures, Symp. on Low Power Electronics and Design (IEEE/ACM), 155–160, 1998.

Leijten-Nowak, K. and van Meerbergen, J. L.: An FPGA Architecture with Enhanced Datapath Functionality, Proc. FPGA '03, 194–204, 2003.

Neumann, B., Blume, H., Feldkämper, H., and Noll, T. G.: Embedded FPGAs für Multimedia-Applikationen, Proceedings ITG-Fachtagung "Elektronische Medien", Dortmund, 147–152, 2003 (in german).

Website Altera: http://www.altera.com

Website M2000: http://www.m2000.fr

Website Xilinx: http://www.xilinx.com

Weiss, O., Gansen, M., and Noll, T. G.: A flexible Datapath Generator for Physical Oriented Design, Proceedings ESSCIRC, 408–411, 2001.

Ye, A., Rose, J., and Lewis, D.: Architecture of Datapath-Oriented Coarse-Grain Logic and Routing for FPGAs, IEEE CICC, 61–64, 2003.

17

Modelling of the parametric yield in decananometer SRAM-Arrays

Th. Fischer[1], **T. Nirschl**[2], **B. Lemaitre**[2], **and D. Schmitt-Landsiedel**[1]

[1]Lehrstuhl für Technische Elektronik, Technische Universität München, München, Germany
[2]Infineon Technologies AG, München, Germany

Abstract. In today's decananometer (90 nm, 65 nm, ...), CMOS technologies variations of device parameters play an ever more important role. Due to the demand for low leakage systems, supply voltage is decreased on one hand and the transistor threshold voltage is increased on the other hand. This reduces the overdrive voltage of the transistors and leads to decreasing read and write security margins in static memories (SRAM). In addition, smaller dimensions of the devices lead to increasing variations of the device parameters, thus mismatch effects increase. It can be shown that local variations of the transistor parameters limit the functionality of circuits stronger than variations on a global scale or hard defects.

We show a method to predict the yield for a large number of SRAM devices without time consuming Monte Carlo simulations in dependence of various parameters (Vdd, temperature, technology options, transistor dimensions, ...). This helps the designer to predict the yield for various system options and transistor dimensions, to choose the optimal solution for a specific product.

1 Introduction

With the scaling of modern VLSI circuits according to Moore's law and the ITRS (ITRS , 2005) new problems arise. The smaller dimensions of transistors in advanced CMOS technologies allow the circuits to consume smaller chip area and to operate at higher clock rates. But on the other hand the standby power consumption is rising due to higher leakage currents. Short channel effects increase the sub-threshold leakage and with the thinner gate oxides of scaled-down transistors the gate leakage is also increasing.

Therefore, today many systems operate at lower supply voltages and the threshold voltages of the transistors increase, resulting in a lower power dissipation. However, the functionality of many circuits depends on this decreasing overdrive voltage, i.e. $V_{DD} - V_{th}$. Hence the read and the write security margins of static memory devices (SRAM) are decreasing.

On the other side the variations of the device characteristics (e.g. transistor length and width, threshold voltage) are ever more increasing with proceeding technologies. Printing these small devices gets harder due to the subwavelength lithography used in todays semiconductor industry. Also the statistical variation of the dopant concentration under the gate of a MOSFET transistor increases with smaller dimensions.

With emerging technologies the share of static random access memory (SRAM) gets larger with every generation. Due to the large demand of memory, SRAM cells must be designed as small as possible, and the cell area is an important benchmark for the quality of a design shrink. The large number of SRAM cells on a System on a Chip (SOC) and the usage of minimum feature size transistors makes the SRAM cell an ideal device for the characterization of variation effects and a good vehicle for yield testing.

In the following section, local and global variations are described. In Sect. 3 the read and write margins of SRAMs are introduced. Sections 4 and 5 present a method to model the parametric yield using a worst case distance analysis. Section 6 shows some results of the yield analysis and is followed by a summary.

2 Local and global variations

Variations of transistor parameters occur on various scales, lot to lot, wafer to wafer, die to die and device to device. They can be divided into variations on a more global scale and a local scale. Lot to lot, wafer to wafer and die to die variation are of global nature. These variations are common

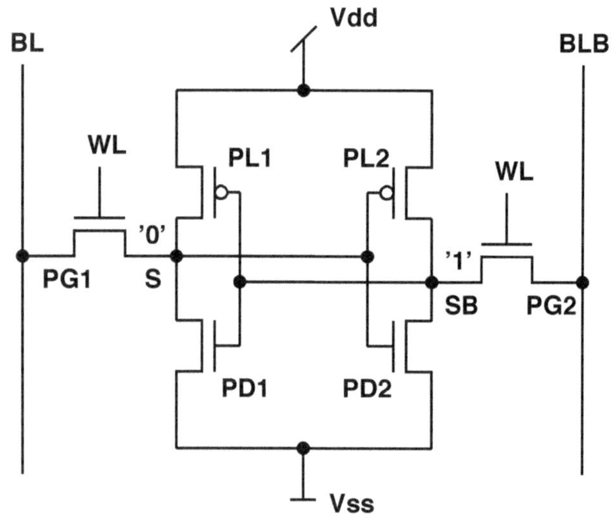

Fig. 1. A 6T SRAM core cell with the two pass gate n-FET transistors (PG1,2), the two pull down n-FET transistors (PD1,2) and the two p-FET pull up transistors (PL1,2).

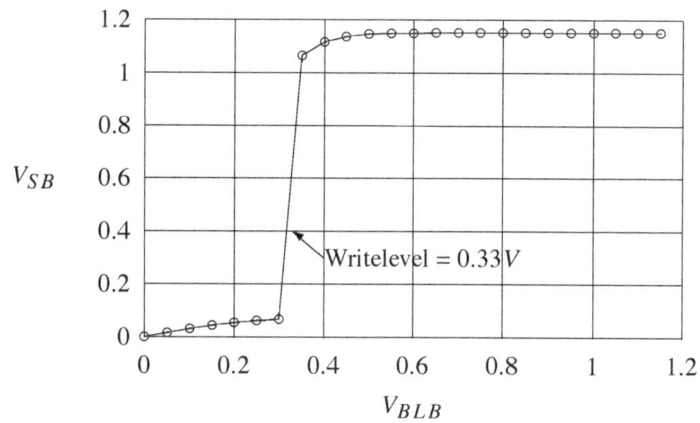

Fig. 2. Voltage at the storage node V_{SB} while lowering V_{BLB}. The core cell has a Write Level of 0.33 V.

to all devices of a die. An example for this type of variations is the oxide thickness t_{ox} or a variation of the width W and length L of the devices due to a global misalignment of the masks. Also the threshold voltage V_{th} of all the transistors on a die can be shifted by a constant value. These variations are of systematic nature. They can be improved during production with a tighter process control, and the impact of the global variations can be attenuated during design with layout techniques such as symmetry of important devices.

Local variations rom device to device are known as device mismatch. This are stochastic variations, such as the number of dopants under the gate area of a MOSFET device. These variations increase with smaller area of a device. Pelgrom et al. (1989) shows that the variation of the threshold voltage V_{th} obeys the following equation:

$$\sigma V_{th} = \frac{A_{V_{th}}}{\sqrt{WL}} \tag{1}$$

where W is the width and L is the length of the transistor. $A_{V_{th}}$ is the matching constant for the threshold voltage. The variaton of the transconductance factor $\beta = \mu C_{ox} W/L$ is modeled in a similar way:

$$\frac{\sigma_\beta}{\beta} = \frac{A_\beta}{\sqrt{WL}} \tag{2}$$

In this case A_β, is the process related matching constant for the transconductance factor.

The only way a designer can improve the matching of the devices is to improve the area.

3 Read and write margin of SRAM cells

Figure 1 shows a 6-transistor SRAM core cell. During read access the bitlines BL and BLB are pre-charged to a pre-charge level – usually V_{DD}. For further discussions we assume a '0' stored in the storage node S at the BL side. When opening the access devices PG1 and PG2, the voltage level on the storage node S at the bitline side rises due to the voltage-divider formed by the transistors PG1 and PD1. The cell ratio CR is defined as

$$CR = \frac{W_{PD}/L_{PD}}{W_{PG}/L_{PG}} \tag{3}$$

and sets the voltage V_S on the storage node. When this voltage V_S rises above the switching level (SWL) of the cross coupled inverters of the SRAM core cell, the cell looses its data while reading the cell. This is called destructive read. In this work, the static noise margin (Seevinck et al., 1987) is used to measure the stability of the SRAM core cell during read access. A destructive read occurs when the SNM drops to zero or below. Another performance measure of the SRAM core cell is the read current I_{read} through the transistors PG and PD.

When writing a '0' to the storage node SB on the BLB side of the core cell, the voltage on BLB is lowered to 0. Here the voltage V_{SB} on the storage node SB is defined by the voltage-divider formed by the transistors PL2 and PG2. Analog to the read case a pull-up ratio PR can be defined as

$$PR = \frac{W_{PU}/L_{PU}}{W_{PG}/L_{PG}}. \tag{4}$$

While lowering the voltage on BLB, the write level is defined as the value of the BLB when the cell flips (see Fig. 2). If the write level is larger than 0 the cell can be written.

The static noise margin, the write level and the read current are the three performances to define the functionality of a core cell in this work.

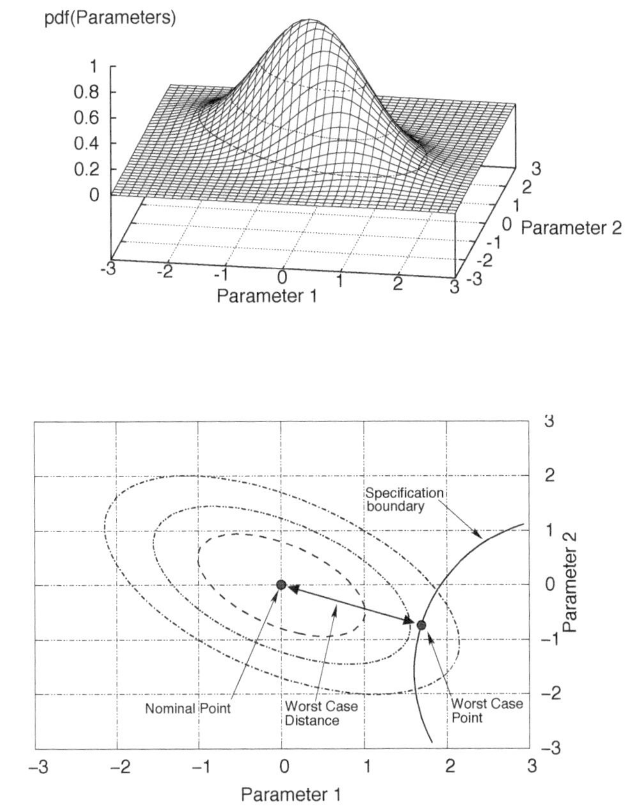

Fig. 3. Schematic probability density of two transistor parameters (top) and worst case distance and worst case point with specification boundary (bottom).

4 The worst case analysis

In order to define the yield of an SRAM core cell array the Worst Case Distance Analysis is used (Antreich et al., 1994). Simply spoken, the Worst Case Distance is a probability measure for the most likely violation of a performance specification. In Fig. 3 the parameter space of a circuit is symbolized. For example, the parameter space can be the local threshold voltage and mobility variations of all six transistors of an SRAM core cell. Here only two parameters are shown in the $x-y$-plane. The z-axes shows the probability density of the parameter space.

For every parameter-set a performance can be simulated with the according parameters as input for the circuit simulator. Now a specification for a circuit performance can be defined. On the top of Fig. 3 we see the top view of the probability density function. The solid line represents a specification boundary. This is given by all parameters sets that result in the same specified performance value. Everything right of the specification boundary violates the specification and results in an non-functional SRAM core cell.

The smallest distance between the nominal point, i.e. the parameter set with the highest probability, and the specification border is the worst case distance WCD. The point on the specification border that is closest to the nominal point has the highest probability that a parametric fault occurs.

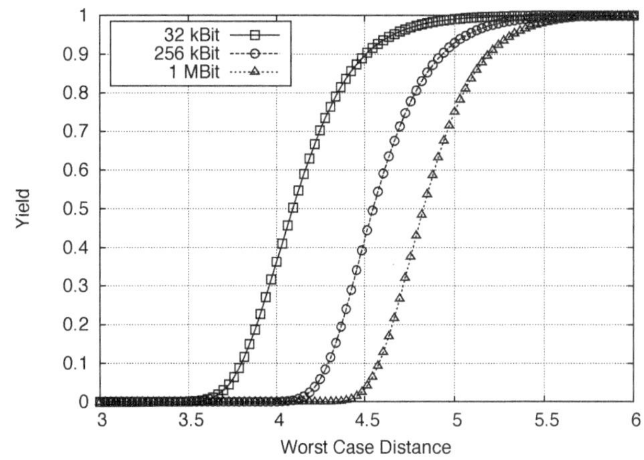

Fig. 4. Yield vs. Worst Case Distance of local variations for different SRAM core cell array sizes.

The WCD can be translated into a probability of a functional core cell array:

$$\text{Yield} = \left(\frac{1}{2}\left(1 + erf\left(\frac{\text{WCD}}{\sqrt{2}}\right)\right)\right)^{N} \tag{5}$$

where N is the number of core cells in an array. In order to achieve a yield of 90% for a 1MBit SRAM array a WCD of 5.2 (see Fig. 4) must be reached or redundancy must be used.

5 First global than local

In order to find the yield of an SRAM array, global and local variations must be simulated independently. First a set of global parameters must be found that represent the worst die in the production process — not including mismatch. Therefore only parameters that define global variations are considered for analysis, e.g. the oxide thickness t_{ox}, the global fluctuations of the n-FET and p-FET threshold voltage V_{thn} and V_{thp} and the length and width variations of the device x_L and x_W. For a given statistical variation of a process, the algorithm looks for the set of the global parameters that results in the worst performance of the core cell.

Once the parameter set for the worst die is found, an analysis of the local variations in an SRAM cell can be made. The local parameters considered in this work were the V_{th} variations and the mobility variations of the six transistors in the core cell. Hence 12 local parameters were taken into account.

In order to find the worst case distance, an optimization problem in a multidimensional space must be solved.

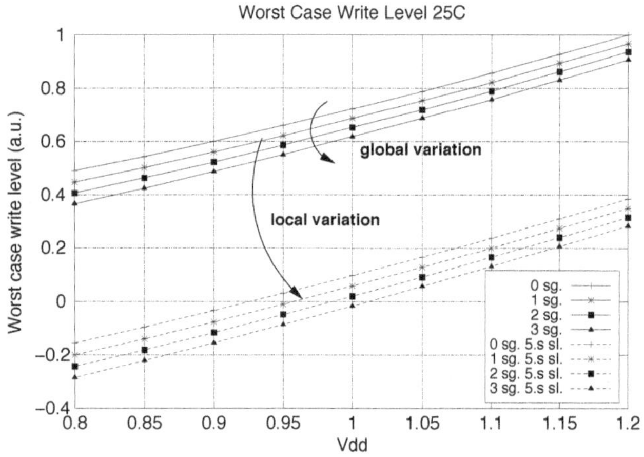

Fig. 5. Worst Case Write Level for global and local variations. Variations in sigma global (sg.) and sigma local (sl.).

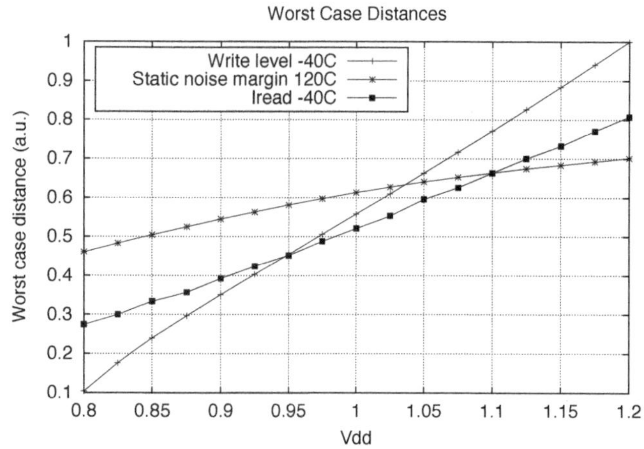

Fig. 6. Worst Case Distances for the write level, the static noise margin and the iread of a SRAM cell. All temperatures are the worst case temperatures.

For this task we used a tool called WiCkeD from Muneda (http://www.muneda.com). This tool uses an sequential quadratic programming (SQP) algorithm to find the Worst Case Distance (Antreich et al., 2000). An external TCL script was used to control the WicKeD program to run the worst case distance analysis first with just global variation, and then with local variations using the worst case parameters obtained by the global WCD analysis. A SPICE simulator simulated the performances of the SRAM core cell for a set of parameter values, and a Perl script wrote the simulation results back to WiCkeD for further evaluation.

Two modes of an analysis of the parametric yield are realized. First of all, a WCD can be found by defining specifications for each performance, like Write Level or Static Noise Margin. On the other hand it is possible to look for the worst performance in an array of a given size. For that purpose a WCD is chosen according to the SRAM array size. Then the parameter set for the worst performance with the given WCD is searched for.

6 Results and performance

For the simulation of the SRAM performances, the BSIM4 parameters of an 65 nm technology (Luo et al., 2002) were used. First of all, the influence of global and local variations on SRAM core cell performances were simulated. In Fig. 5 we see a normalized plot of the Write Level of an SRAM core cell for different supply voltages. The upper set of lines shows the relative worst case Write Level depending on global variations ranging from 0 to 3 sigma. With the worst case parameter-set found by the global simulation, a simulation for the worst case performance using just local variations such as the V_{th} and the mobility μ for the six core cell transistors was made. The local worst case distance WCD of 5.2 sigma was chosen to represent a SRAM core cell array with 1 MBit (see Fig. 4).

The local variation has a large influence on the functionality of the SRAM array due to two effects. First the large number of core cells that must be fulfill the specifications in order to get a SRAM array with 100% working cells. Second the smaller size of the transistors increase the local variations and with lower overdrive voltage result in more core cells failing the specification.

Although the local variations are more dominant, one can see form Fig. 5 that a tighter process control can result in circuits operating at a lower supply voltage.

To identify the performance that is the main yield detractor, the worst case distances of the write level, the static noise margin and the read current were simulated. The circuit is simulated with the temperature corners of the specification that results in the respective worst performance. This is -40°C for the write level and the read current and 120°C for the SNM. So the worst cases of the performances are comparable. It is shown in Fig. 6 that different performances limit the yield of the core cell. For high supply voltages the read stability (SNM) is the main yield detractor, whereas the write level limits the yield for supply voltages lower than $1.1V$, and the read current is the yield detractor for the low voltage operation.

The Worst Case Distance simulations are much faster than a comparable number of Monte Carlo simulations. To verify a yield of 5 sigma in a 95% confidence interval, 10 million Monte Carlo runs need to be simulated. These simulations take about 200 h of runtime on a single processor machine. The Worst Case Distance simulations need about 2 min of time on a comparable computer. Both, the Monte Carlo simulations and the WCD simulations can be divided into multiple threads and run on multiple processors. However, these figures show clearly that simulations of parametric yield can be simulated much faster with the Worst Case Distance analysis. Therefore multiple yield simulations with multiple parametric variations, such as temperatures, supply voltages and even different transistor sizes, can be done and used for design centering and verification considering global and local variations.

7 Conclusions

We presented a method to effectively simulate the yield and the worst case corners of several SRAM performances. This Worst Case Distance Analysis is by 6 orders of magnitudes faster than a Monte Carlo Analysis. So a six-sigma simulation can be done for several parameter alterations of a circuit. It is possible to find the main yield limiting performances for the worst case operating conditions. This gives the circuit designer a tool to simulate different transistor sizing options under the influence of global and local parameter variations.

References

ITRS, International Technology Roadmap for Semiconductors 2005 Edition,http://public.itrs.net/

Pelgrom, M., Duinmaije, A., and Welbers, A.: Matching Properties of MOS Transistors, IEEE J. Solid State Circuits, 24, 1433–1440, 1989.

Seevinck, E., List, F., and Lohstroh, J.: Static Noise Margin Analysis of MOS SRAM Cells, IEEE J. Solid State Circuits, 22, 525–536, 1987.

Antreich, K., Graeb, H., and Wieser, C.: Circuit Analysis and Optimization Driven by Worst Case Distances, IEEE Transactions on Computer-Aided Design of Integrated Circuits and Systems, 13, 57–71, 1994.

Antreich, K., Eckmueller, J., Graeb, H., Pronath, M., Schenkel, F., Schwencker, R., and Zizala, S.: WiCkeD: Analog Circuit Synthesis Incorporating Mismatch, IEEE Conf. on Custom Integrated Circuits, 511–514, 2000.

Luo, Z., et. al.: High Performance and Low Power Transistors Integrated in 65 nm Bulk CMOS Technology, Proc. IEEE IEDM, 661–664, 2004.

18

From algorithm to implementation: a case study on blind carrier synchronization

D. Schmidt, T. Brack, U. Wasenmüller, and N. Wehn

Microelectronic System Design Research Group, University of Kaiserslautern, Erwin-Schrödinger-Straße, 67663 Kaiserslautern, Germany

Abstract. Increasing chip complexities demand a higher design productivity. IP cores, which implement commonly needed operations, can help to dramatically shorten development and verification times for new designs. They often allow for a efficient mapping of algorithmic tasks to a hardware architecture. In this paper we present a novel configurable building block for blind carrier synchronization that features combined frequency and phase offset estimation and an alternative modulation removal that improves communication performance compared to state-of-the-art designs. The used design flow exploits the benefits of IP cores for rapid development times while still offering the designer the full range of optimization possibilities for a specific design. It allowed us to do an almost complete design space exploration, assuring a near-optimal solution to the given problem. The implementation platform is a XILINX Virtex II Pro FPGA.

1 Introduction

It is well known that the increasing chip complexities demand a higher design productivity. An important method for decreasing development effort and shortening development time is the use of IP cores. Many common building blocks, such as multipliers, dividers and dual port rams, can be addressed by freely available IP cores. Configurability regarding precision, finite word lengths or throughput make them suitable for a wide range of designs, thus drastically reducing engineering effort for design and verification.

This paper demonstrates the design of a configurable IP core for blind carrier synchronization with novel features (combined frequency and phase offset estimation, an advanced modulation removal technique, and a configurable windowed spectral analysis). This building block is in use in a large scale communication system. The system throughput requirements can not be fulfilled by a DSP and due to the low volume an FPGA solution is far more cost efficient than an ASIC implementation. Thus, XILINX Virtex II Pro FPGAs were selected as implementation platform. We used the freely available IP cores generated with the XILINX core generator tool. The case study shows the high level design flow from the basic algorithms of the application to the implemented building block. This design flow includes the mapping of algorithmic steps to the available IP cores and the evaluation of algorithmic variations, considering both, communications performance and implementation complexity. While designing a communication system the designer always has to trade off between these two. Our building block offers a configurable interface to accept different word lengths at the input.

Every inner receiver for wireless communication must estimate the generally unknown parameters of timing, frequency and phase offset of the received signal. All possible negative influences introduced by them have to be eliminated. As mentioned, this paper concentrates on the frequency and phase offset estimation and correction (carrier synchronization) of bursts with Phase Shift Keying modulation for different modulation indices M (MPSK), e.g. $M=2$ for BPSK and $M=4$ for QPSK.

The rest of the paper is structured as follows. In Sect. 2 an overview about the used design flow is given. Section 3 provides an overview of the problem of carrier synchronization and the basic algorithms for the synchronization task. The individual steps carried out in this case study, according to the design flow, are covered in the following sections. Section 4 shows possible optimizations of the basic algorithm as well as a way to combine the task of frequency and phase offset estimation. Section 5 deals with the examination of the implementation alternatives and the mapping of algorithmic tasks to XILINX IP cores. The resulting architecture is presented in Sect. 6, followed by the conclusions in Sect. 7.

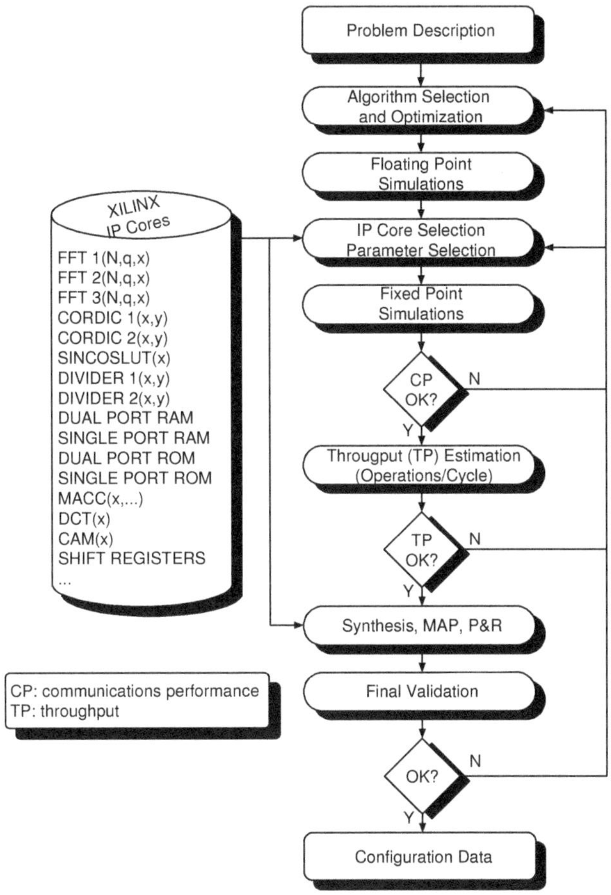

Fig. 1. The used design flow for IP core based design.

2 Design flow

Figure 1 shows the used design flow. Typically any given problem can be solved with various algorithms. Additionally, each of the algorithmic alternatives may depend on several parameters. There may also be different ways to implement an algorithm using different IP cores. Hence, all implementation alternatives form a tree, the so called design space. In the phase of design space exploration the designers task is to select the algorithm, the parameters and the IP cores to solve the problem under all given constraints. In a communication system these constraints usually regard the implementation complexity, the minimum throughput and maximum latency as well as the achievable communications performance. As the characteristics of each algorithmic alternative are not known in advance it is necessary to estimate them for each algorithm under consideration until a feasible solution is found. This is indicated by the iterative loops in the design flow diagram.

To minimize the time spent for design space exploration it is vital to exclude as many implementation alternatives as early in the process as possible. This is best done by evaluating the communications performance of each algorithm prior to the time consuming estimation of the implementation complexity and comparing it to the requirement specifications. A first estimate of the communications performance of an algorithm can be derived using a floating point model. The use of a high level modeling language, like SystemC or MATLAB, allows for faster development and shorter simulation times compared to a VHDL model. These first models are neither cycle accurate nor bit true models of the later implementation. Once an algorithm achieving a sufficient communications performance is found, the different implementation alternatives have to be reviewed. Thus a more accurate fixed point model of the hardware is needed that takes into account the effects of different quantization levels and of the IP cores used. Cores from the XILINX IP core library can be used to reduce development effort. To evaluate the impact of the decisions made on the communications performance, the simulation results of the fixed point model are compared to the results of the floating point simulation. The fixed point models can still be written in a high level language.

The core library provides IP cores for many high latency operations like FFTs or dividers. In cases where the number of clock cycles needed to perform a calculation is dominated by such complex operations, a rough throughput estimation can be done based on the specifications of the used IP cores. The glue logic does not significantly contribute to the computation complexity or the implementation complexity. Hence, to estimate the number of occupied slices, it is sufficient to instantiate and connect the main blocks of the design. Then, only for the design with the optimal trade-off a synthesizable VHDL-model has to be described and validated.

As a last step the resulting design has to be verified using the fixed point model. In the area of communication systems this can either be done with a bit true C-model or by comparison of statistical bit error rates. After synthesis, mapping and place and route have been performed, a final validation of throughput and area ensure the compliance with the given constraints.

3 Problem description

This section contains all fundamental algorithms needed for the development of our blind carrier synchronization building block. Further details and resources can be found in Meyr et al. (1997); Viterbi and Viterbi (1983); Mengali and D'Andrea (1997); van Trees (1968); Barry et al. (2004). The basic synchronization steps are estimation of the frequency offset, frequency offset correction, estimation of the phase offset and correction of the phase offset.

In an MPSK modulation scheme the set of M possible symbols s_m is given by:

$$s_m := e^{j\left(\frac{2\pi \cdot m}{M}\right)} \qquad m = 0, 1, .., M-1 \tag{1}$$

The received sample sequence r is given in the complex baseband according to Eq. (2):

$$r(l) := s(l) \cdot e^{j(2\pi f_o l T + \Phi)} + n(l) \qquad l = 0, 1, .., L-1 \qquad (2)$$

The sample sequence r with L elements is based on MPSK symbols s with one sample per symbol and symbol duration T, and is disturbed by a noise sequence n. The frequency offset f_o and phase offset Φ have to be estimated and corrected. They are considered fixed during one burst transmission.

The estimation of the unknown parameters f_o and Φ can be done by data aided or non data aided methods. Data aided methods need known symbols in the transmitted bursts. The introduced redundancy is exploited to estimate the frequency and phase offset, at the cost of decreasing bandwidth and power efficiency (Meyr et al., 1997). Hence, non data aided or blind methods become mandatory. They estimate the parameters without the help of such a training sequence. In this case especially estimation of frequency offset becomes a very critical task with high impact on communications performance. Deviation of the frequency estimation results in bursts with high bit error rates, resulting in high frame error rates after channel decoding.

For blind frequency or phase offset estimation the modulation introduced by the unknown transmitted symbol sequence s has to be removed from the received sample sequence r. State-of-the-art modulation removal is based on a power of M operation on r, where M is the modulation index.

$$\tilde{r}(l) := r(l)^M = |r(l)|^M \cdot e^{jM \cdot arg(r(l))} \qquad (3)$$
$$l = 0, 1, .., L-1$$

Note that by such a power of M operation all the MPSK symbols defined in Eq. (1) are transformed to the symbol s_0. After the modulation has been removed, an approximation of the maximum likelihood estimation of the frequency offset can be found by calculating the spectral power of the sequence $\tilde{r}$. This is done by performing a Fast Fourier Transformation (FFT) (Mengali and D'Andrea, 1997) with N points according to Eq. (4):

$$X(k) := \sum_{n=0}^{N-1} x(n) \cdot e^{-j\left(\frac{2\pi \cdot k \cdot n}{N}\right)} \qquad k = 0, 1, .., N-1 \qquad (4)$$

The sequence x with N elements corresponds to the sequence $\tilde{r}$ with $L \leq N$ elements of Eq. (4), which is padded with zeros to achieve the N elements of Eq. (4). Note that each bin $X(k)$ represents the spectral power of x in a range of $(M \cdot N \cdot T)^{-1}$ Hz, not only in one distinct frequency. This results in a loss of accuracy, depending on the number N of points. The estimated frequency offset $\tilde{f}_o$ is given by spectral analysis of the FFT output, i.e. the FFT bin corresponding to the estimated frequency offset is given by:

$$k_f, \ s.t.: \ |X(k_f)| = \max_k |X(k)| \quad k = 0, 1, .., N-1 \qquad (5)$$

After spectral analysis of the FFT bins, the sample sequence r must be corrected with the calculated frequency offset $\tilde{f}_o$. The corrected sample sequence u is then given by:

$$u(l) := r(l) \cdot e^{-j\left(\frac{2\pi \cdot k_f \cdot l}{M \cdot N}\right)} \qquad l = 0, 1, .., L-1 \qquad (6)$$

Phase offset estimation can be done using the V and V algorithm presented in Viterbi and Viterbi (1983). This algorithm requires the modulation to be removed from the sample sequence u, analogous to Eq. (4):

$$\tilde{u}(l) := u(l)^M \qquad l = 0, 1, .., L-1 \qquad (7)$$

The phase offset of the sample sequence u is then estimated to be the average phase of the symbols in the sequence $\tilde{u}$, as given in Eq. (8):

$$\tilde{\Phi} := M^{-1} \cdot \arg \sum_{l=0}^{L-1} \tilde{u}(l) \qquad (8)$$

The sequence $u(l)$ must be corrected with the estimated phase offset $\tilde{\Phi}$, which results in

$$v(l) := u(l) \cdot e^{-j\tilde{\Phi}} \qquad l = 0, 1, .., L-1 \qquad (9)$$

Equation (8) still leaves an M-time phase ambiguity. Apart from this phase ambiguity the sequence $v(l)$ gives an estimation of the transmitted symbols $s(l)$. The resolution of the phase ambiguity can be achieved using several methods, e.g. differential encoding can be used. But this strongly depends on the actual communication system. Thus this step of the synchronization process is not considered in our case study.

4 Algorithm optimization

During the design space exploration we reviewed several modifications of the above mentioned algorithms. As mentioned, we were particularly interested in their impact on implementation complexity and communications performance.

Since the power operation for modulation removal in Eqs. (4) and (7) enhances the noise in the signal super proportional, so called self noise is inferred. This results in large frequency offset estimation errors especially at medium to low signal to noise ratio ranges. Thus bursts with high bit error rate and therefore high frame error rate after channel decoding result.

To elude this problem another approach for removing the modulation, proposed in Wang et al. (2003); Viterbi and Viterbi (1983), is used. This technique for modulation removal is based on the following calculation:

$$\tilde{r}(l) := |r(l)|^k \cdot e^{jM \cdot arg(r)}$$
$$k = 0, 1, .., M; \ l = 0, 1, .., L-1 \qquad (10)$$

The parameter k is an optimization parameter. Observe that in case of $k = M$ Eq. (10) transforms into Eq. (4).

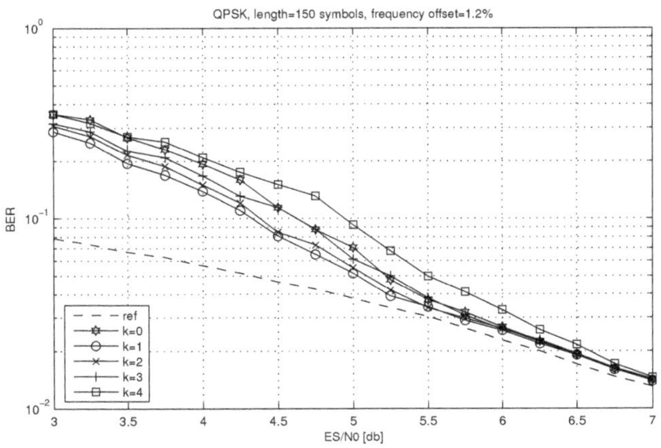

Fig. 2. BER of 150 symbols QPSK bursts for different values of k.

The calculation of the phase offset $\tilde{\Phi}$ can be simplified with further algebraic transformations. Combining Eqs. (6–8) gives an equivalent equation for calculating the estimated phase offset $\tilde{\Phi}$:

$$\tilde{\Phi} := M^{-1} \cdot \arg\left(X(k_f)\right) \tag{11}$$

Equation (11) uses the FFT result of Eq. (4) for the bin k with $k = k_f$. Thus frequency and phase offset estimation can be combined efficiently. Note that in this case phase offset estimation and frequency offset estimation implicitly use the same technique for modulation removal. The input burst has to be corrected with the estimated values of frequency and phase offset, formally described as:

$$v(l) := r(l) \cdot e^{-j\left(\frac{2\pi k_f \cdot n}{M \cdot N}\right) + \tilde{\Phi}} \qquad l = 0, 1, .., L-1 \tag{12}$$

If the range of the occurring frequency offset is known in advance, the communications performance can be improved significantly by simply limiting the spectral analysis. This windowing technique is given by substituting k_f in Eq. (12) with k_{fw} from Eq. (13):

$$k_{fw}, \; s.t.: \; |X(k_{fw})| = \max_k |X(k)|$$
$$k = 0, .., w_u, w_l, .., N-1 \tag{13}$$

The parameters w_u and w_l define the FFT bin range, where the frequency offset has to be found. Setting w_u to $N/2-1$ and w_l to $N/2$ describes the FFT bin k_f given in Eq. (5).

Evaluating the impact of the parameter k on communications performance is done by simulation. Figure 2 shows the bit error rates (BER) of QPSK bursts with a length of 150 symbols for the different modulation removal techniques. We simulated transmission of the bursts over an additive white Gaussian noise (AWGN) channel using a floating point model written in C. A reference graph ref is provided to allow comparison of the simulation results to a perfect frequency and phase synchronization reflecting the performance limits of the AWGN channel. Clearly, $k=1$ yields the best performance. Simulations with different burst lengths and BPSK bursts showed similar results. A detailed discussion of the presented algorithms regarding communications performance can be found in Brack et al. (2004).

The frequency and phase estimation can be performed in a combined manner defined by Eq. (11) or in a separate manner, defined by Eqs. (6–8). Obviously, the combined frequency and phase offset estimation reduces the overall computational complexity. Simulation results also showed that using different modulation removal techniques for frequency and phase offset estimation does not yield a better communications performance.

5 IP core selection

To obtain a first estimate of the number of slices occupied by each implementation alternative, a VHDL-model that merely instantiates and connects the needed IP cores and all major building blocks, is used. The time consuming task of describing all the needed control logic and validating the design is not yet necessary at this stage, because for computational intensive algorithms, like the ones under consideration, the control logic does not significantly contribute to the implementation complexity.

The components of the synchronization circuit are derived from the optimized algorithm and mapped to XILINX IP cores. We also used specific XILINX resources like the internal multipliers (MULT) and block RAM (BRAM) available on the Virtex II Pro FPGA.

For the FFT calculation defined in Eq. (4) three XILINX IP cores with different architectures are available (Xilinx, 2003a). The architectures differ in throughput, latency and needed resources. Due to the throughput requirements we had to choose the fastest FFT core with a throughput of one FFT bin calculation per clock cycle independent of the parameter k. The number of FFT points must be chosen to be at least twice the maximum burst length in the application Mengali and D'Andrea (1997). It also has to be considered that more points result in a higher resolution of the FFT bins and hence in a higher estimation accuracy.

For the modulation removal defined in Eq. (10) the mapping to IP cores depends on the parameter k. A mapping from the equation to the implementation requires polar coordinates. Since, however, the input samples r are given in Cartesian coordinates a transformation is necessary. The calculation of $\arg(r)$ and $|r|$ can be done with the CORDIC algorithm (Volder, 1959), which works iteratively. XILINX provides a lot of different variations of a CORDIC IP core (Xilinx, 2003b). A fully pipelined version with a throughput of one sample per clock cycle and a serial version with a throughput inverse proportional to the number of iterations are available. To fulfill high throughput requirements the pipelined IP core was chosen.

Table 1. Selected alternatives for modulation removal and their IP core and resource requirements.

Alt.	Equation	Cores	MULT	SLICES
0	$\alpha := e^{j4arg(r)}$	aCORDIC, SCL	0	273
1	$\|r\| \cdot \alpha$	CORDIC, SCL	2	331
2a	$r^4/(\Re(r)^2 + \Im(r)^2)$	DIVIDER	10	297
2b	$\alpha(\Re(r)^2 + \Im(r)^2)$	aCORDIC, SCL	4	290
2c	$\|r\|^2 \cdot \alpha$	CORDIC, SCL	3	325
3	$\|r\|^3 \cdot \alpha$	CORDIC, SCL	4	329
4	r^4		6	79

In some cases of k and M the use of a CORDIC module obsolete. For $k = M$ Eq. (10) reduces to a power of M operation, which can be implemented using 3 multipliers in the case of BPSK and 6 multipliers for QPSK modulation.

For $M = 4$ and $k = 2$ Eq. (10) equals $\tilde{r} = r^4/|r|^2$. To calculate $|r|^2 = \Re(r)^2 + \Im(r)^2$ two multipliers are needed, so the CORDIC module can be replaced by 10 Multipliers and a divider. As a third alternative for $k = 2$, we can use a simplified CORDIC module (aCORDIC) that only calculates arg(r) but not $|r|$. Then, again, we need two additional multipliers to calculate $|r|^2$. The simplified CORDIC module is also sufficient to perform the calculation of Eq. (10) in the case of $k = 0$.

After modulation removal the resulting sample sequence $\tilde{r}$ has to be transformed back to Cartesian coordinates for the FFT calculation (see Eq. (4)). This is accomplished with a sine-/cosine-look-up-table (SCL) (Xilinx, 2002), also available as an IP core, and internal multipliers to realize the e^{jx} operation.

The realization of Eq. (5) as well as the improved version, defined in Eq. (13), is not based on IP cores. The windowing feature proposed in Eq. (13) to evaluate k_{fw} is realized using a limited maximum absolute value search on the determined FFT bins. This can be implemented in a very efficient way using only counters and comparators. Since this part does not differ between two algorithmic alternatives, it does not have to be taken into account when comparing their implementation complexities.

After determining the frequency bin k_{fw}, the phase offset is estimated according to Eq. (11). Again, the argument function needed for this equation is realized with a CORDIC IP core. Because only one argument calculation per phase estimation has to be performed, we can use a rather small serial aCORDIC core without the ability to calculate $|r|$.

Rotation of the samples, as described, in Eq. (12) is accomplished with complex number multiplications and a second SCL.

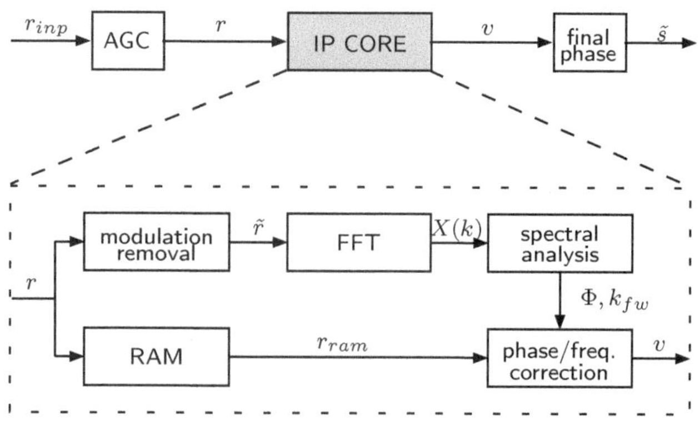

Fig. 3. Base architecture for blind carrier synchronization.

The considered implementation alternatives for modulation removal and the IP cores needed by each of them are listed in Table 1. It also shows the number of slices and other hardware resources occupied. The resources needed by the FFT and for the other building blocks (phase offset estimation, spectral analysis with windowing and combined correction) are not taken into account, because they are the same for all values of k.

6 Results

We used XILINX ISE 6.1 for synthesis and place and route with XILINX Virtex II Pro FPGA XC2VP50 as target platform. Figure 3 shows the main building blocks of the implemented architecture. We assume that the input signal is filtered using a matched filter and is sampled at one sample per symbol. As a front-end to the IP core an automatic gain control unit (AGC) is required to adapt the input samples bit width to the IP core input bit width without limiting the dynamic range. The back-end must deal with the M-times phase uncertainty introduced by Eq. (8).

The building blocks modulation removal, FFT, spectral analysis, and phase/freq. correction correspond directly to the algorithms from Sect. 3 and Sect. 4. The RAM buffer is used to guarantee constant throughput by hiding the FFT latency.

The final working and validated IP core is configurable regarding different bit widths (BW) of the input signal r and maximum burst lengths (MBL). As can be seen in Table 2 the size of the IP core grows strongly with increasing MBL due to the FFT component. The results also show that the overhead introduced by the alternative modulation removal with $k = 1$ (about 250 slices, Table 1) is very small compared to the overall resource needs of the IP. Considering the 0.5 dB gain in communications performance over the state-of-the-art technique with $k = 4$ (Fig. 2), we considered implementation of the alternative modulation removal to be worthwhile.

Table 2. Synthesis results.

BW	MBL	Slices	BRAM	MULT	MHz
7	64	2926	20	18	119.9
7	512	3636	30	27	119.4
7	4096	5018	78	36	104.0
6	64	2863	20	18	125.3
6	512	3548	30	27	123.5
6	4096	4964	78	36	116.1
5	64	2816	20	18	128.4
5	512	3505	26	27	126.3
5	4096	4931	78	36	116.0

The resulting maximum sample throughput is about half the clock speed because the FFT point size is at least twice the maximum burst length (Sect. 5). Thus the corresponding bit rate is determined by the modulation:

$$\text{throughput}[{}^{\text{bits}}/_{\text{sec}}] \approx (2 \cdot \text{cycle})^{-1} \cdot \log_2(M) \qquad (14)$$

For example, using QPSK modulation Eq. (14) evaluates to a bit rate beyond 100 Mbps for every configuration (Table 2).

We have also shown that an IP core based design flow allows not only for a rapid implementation but also for a fast and exhaustive design space exploration. As in many designs resource needs and throughput are dominated by operations that can be covered with pre-existing IP cores, it is possible to quickly and accurately estimate the performance of each implementation alternative based on high level simulations (in SystemC/MATLAB) and the specifications of the used IP cores. The time demanding step of describing and verifying a fully functional and synthesizable VHDL-model has only to be done for the final architecture. Thus, the IP core based design flow proved to combine near optimal results with very short development times.

7 Conclusions

In this paper we presented a configurable building block that utilizes advanced techniques like combined frequency and phase synchronization, improved modulation removal and a windowed spectral analysis. These features yield high throughput, compact size and best communications performance achievable for transmissions without training sequence. We used a design flow suitable for a wide range of designs. In our case study we showed how to employ it to efficiently explore the design space for a blind frequency and phase synchronization module for MPSK bursts.

References

Barry, J. R., Lee, E. A., and Messerschmitt, D. G.: Digital Communication, Kluwer Acad. Publishers, Boston/Dordrecht/London, 2004.

Brack, T., Wasenmüller, U., Schmidt, D., and Wehn, N.: Design Space Exploration for Frequency Synchronisation of BPSK/QPSK Bursts, in Kleinheubacher Berichte, 48, 337–341, 2004.

Mengali, U. and D'Andrea, A.: Synchronization Techniques for Digital Receivers, Plenum Publ. Corp., New York, 1997.

Meyr, H., Moeneclaey, M., and Fechtel, S.: Digital Communication Receivers, Wiley and Sons, Inc., New York, 1997.

van Trees, H. L.: Detection, Estimation and Modulation Theory, Wiley and Sons, Inc., New York, 1968.

Viterbi, A. J. and Viterbi, A. M.: Nonlinear Estimation of PSK Modulated Carrier Phase with Application to Burst Digital Transmission, IEEE Trans. on Inf. Theory, 32, 543–551, 1983.

Volder, J.: The CORDIC Trigonometric Computing Technique, IRE Transactions on Electronic Comp., 8, 330–334, 1959.

Wang, Y., Serpedin, E., and Ciblat, P.: Optimal Blind Carrier Recovery for MPSK Burst Transmissions, IEEE Trans. on Commun., 51, 1571–1581, 2003.

Xilinx Inc. Fast Fourier Transform v2.0, Mar. 2003a.

Xilinx Inc. CORDIC v2.0, Mar. 2003b.

Xilinx Inc. Sine/Cosine Look-Up Table v4.2, Nov. 2002.

Simultaneous large-scale reliability analysis of ultra-thin MOS gate dielectrics using an automated test system

A. Domdey, K. M. Hafkemeyer, W. H. Krautschneider, and D. Schroeder
Hamburg University of Technology, Institute of Nanoelectronics, Hamburg, Germany

Abstract. This article presents an automated test system targeting the large-scale analysis of ultra-thin MOS gate dielectric degradation. The system allows for stress tests at elevated temperatures as well as supply voltages and long-term tests of thousands of MOS devices simultaneously. The aim is to build-up large and hence significant statistics about the degradation process as a function of time.

1 Introduction

The reliability of ultra-thin MOS gate dielectrics is limited by time-dependent degradation processes inter alia. Gate dielectric thicknesses of MOS devices currently in fabrication range from 1.7 nm for a 130 nm technology down to 1.2 nm for a 65 nm technology or even below for further advanced technologies and correspond to only a few atom layers. Considerable tunnelling currents flow through these ultra-thin dielectrics and cause local rearrangements of the atomic structure. These so-called "weak spots" Lombardo (2005) attract ever more current flow leading to further restructuring on atomic level ("soft breakdown"). At the end, a crystalline Si conducting path through the gate dielectric shortens channel and gate ("hard breakdown") as illustrated in Fig. 1.

Atomistic computer simulations are expected to expose physical mechanisms originating gate dielectric degradation and to give information on time-dependent breakdown probabilities. A device simulator considering degradation effects and reproducing ageing of electronic circuits is the long-term objective of this project.

Broad and significant databases of appropriate experiments are essential to verify and calibrate the simulations mentioned above. A large-scale automated test system has been developed for this purpose. It offers the possibility to simultaneously analyse thousands of MOS devices under stress conditions (elevated temperatures and supply voltages) as well as under long-term operating conditions.

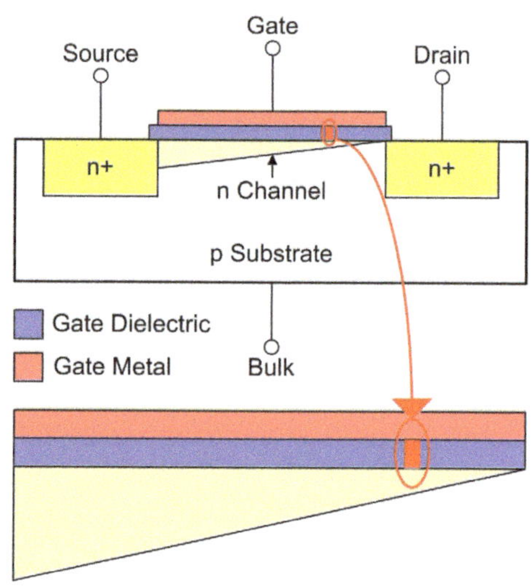

Fig. 1. Hard gate dielectric breakdown.

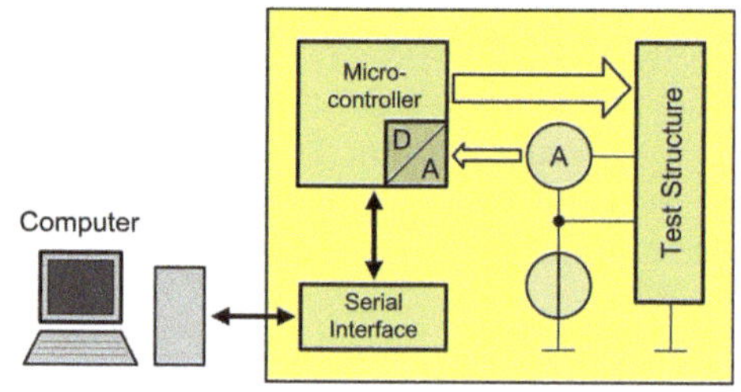

Fig. 2. Automated test system.

2 Automated test system

The automated test system is shown in Fig. 2. It consists of an integrated test structure, a current sense circuit, a voltage source, a microcontroller with A/D-converter, and a serial interface. During testing, the automated test system operates stand-alone. A computer is only required to set up tests at the beginning and to evaluate the results at the end.

3 Integrated test structure

Figure 3 depicts the block diagram of the integrated test structure. In the case shown, the MOS device under test (DUT) cells are arranged in a 32x32 matrix. Every single DUT cell can be accessed via column/row decoders and a clock signal CLK. The column/row decoders are driven by two external 5 bit busses X, Y.

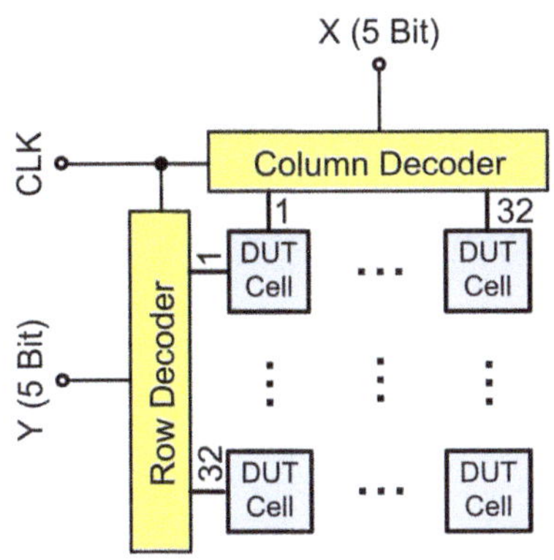

Fig. 3. Integrated test structure overview.

Details of the DUT cells are shown in Figure 4. Each of the 32x32 cells consists of the MOS DUT itself, a switch, and a logic with a memory. The memory stores the state of the switch that either connects the DUT gate to a supply line common to all cells in a row or disconnects it from that line. Its memory contents can be written via a data line when the decoders set the appropriate row and column lines. The logic then connects the data line to the memory to set the state of the switch. Every single cell of the matrix can be addressed, measured, and switched-off in case of breakdown separately in this manner.

A switch at the beginning of every row connects that row either to CSTRESS ("common stress") or to SSTRESS ("single stress"). CSTRESS is the stress voltage line. All rows are normally connected to that line during testing. Temporarily, one single row can be switched to SSTRESS in order to measure the current of that single row with the external current sense circuit. All other rows are not affected by that operation. A single gate current is measurable when all other DUTs of a row are shut off.

To ensure all DUTs are stressed under the same conditions, each source is connected to CSOURCE ("common source") and each drain is attached to CDRAIN ("common drain") respectively. With CSOURCE and CDRAIN at the same potential, it is possible to stress the DUTs as MOS capacitors. Real operating conditions can be obtained with CSOURCE and CDRAIN at different potentials.

The integration of both the DUTs and the periphery (logic, decoders) on one chip means that all devices are stressed with the same conditions and degrade similarly. This problem is solved by the usage of MOS transistors with different gate dielectric thicknesses for DUTs (thin dielectrics) and logic as well as decoders (thick dielectrics). Figure 5 shows the layout of the integrated test structure with 32x32 devices. The 130 nm technology chosen features 1.7 nm and 6.0 nm gate dielectrics.

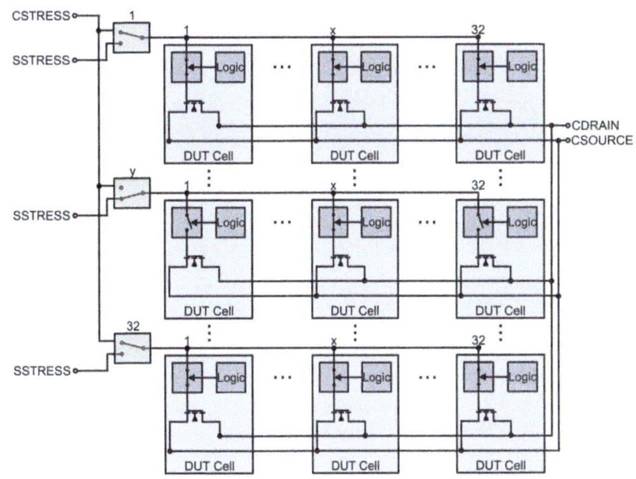

Fig. 4. Integrated test structure details.

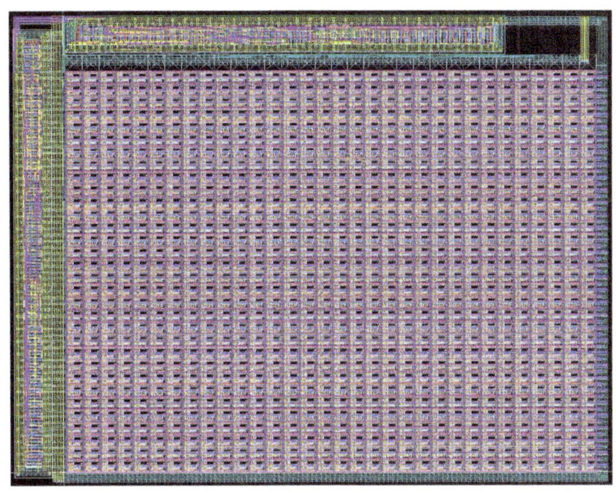

Fig. 5. Integrated test structure layout.

4 Current sense circuit

Gate dielectric degradation causes increasing gate currents. The lowest currents to be measured occur at the beginning of the DUT lifetime. These are due to direct tunnelling. In the course of time, the gate currents increase which is accompanied by noise. At a certain point, the noise affects the functionality of the DUT. This is called soft breakdown [Lombardo (2005)]. At the end of lifetime, the gate dielectric finally breaks down which means gate electrode and channel are shorted. This is the so-called hard breakdown. The currents to be measured are comparatively high in this case. The

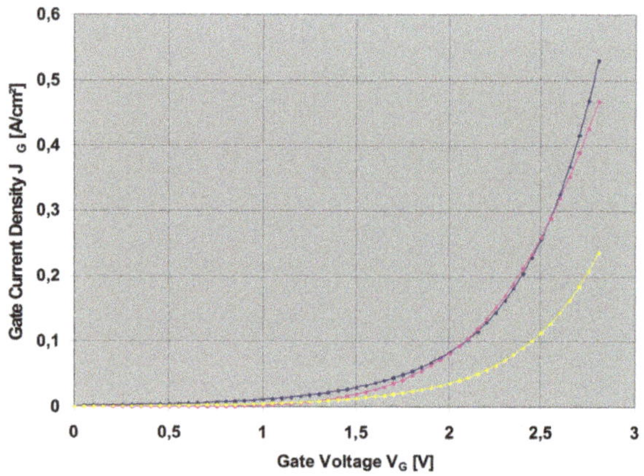

Fig. 6. Direct tunnelling current densities.

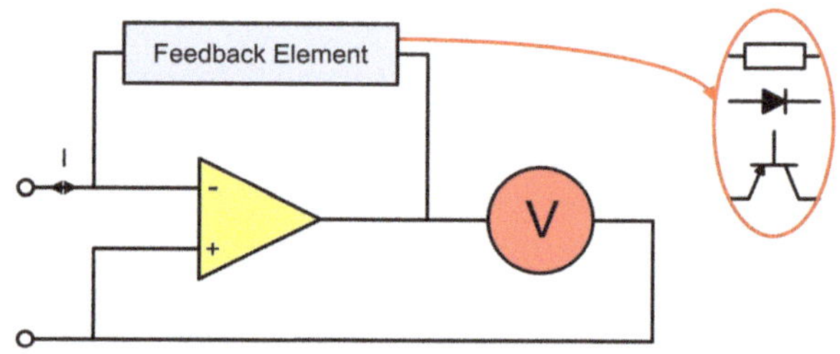

Fig. 7. Feedback current sense circuit.

gate current variation at the beginning and at the end of lifetime covers several orders of magnitude.

Direct tunnelling currents through the gates of fresh DUTs have been estimated with the aid of literature Depas (1995) and computer simulations. Figure 6 shows the results (yellow curve: rough estimation with literature values, blue curve: simulation results, magenta curve: calculated values). It can be seen that the expected direct tunnelling current density ranges from 0.04 $\frac{A}{cm^2}$ to 0.08 $\frac{A}{cm^2}$ at 1.8 V gate voltage. This means direct tunnelling currents from 300 pA to 1 nA at 1.8 V stress voltage and with the geometries chosen for the DUTs.

Research workers have investigated the gate currents of MOS devices after hard breakdown. E.g. Lombardo (2005); Avellan (2004) report an increase of the gate current density of several orders of magnitude. Breakdown gate currents up to the mA region have been measured. Both literary sources consider MOS devices with geometries comparable to those of the integrated test structure.

The current sense circuit in Fig. 2 must cover the measuring range from pA to mA. An appropriate measurement instrument for this purpose is the feedback current sense circuit as shown in Fig. 7 schematically. The circuit is based on a low leakage, high input resistance, high gain operational amplifier. The feedback element defines the measurement range. In case a resistor is used, the pA region can be measured linearly. If a diode or a bipolar transistor is applied, the feedback current sense circuit covers the range from pA to mA logarithmically.

5 Microcontroller

The microcontroller scans the test structure for broken down DUTs. It selects one row after the other, connects the current sense circuit to that row, switches off all DUTs in that row except for one, and measures the gate current of the one single DUT. In case the gate current is unacceptably high, a hard breakdown has occured and the broken device is shut off. These steps are repeated for every single DUT. The microcontroller automatically ends the test if no working DUT is left. Note that the operations in one row do not affect the other rows.

The microcontroller stores characteristic data of the gate dielectric degradation for every single DUT in its internal memory. In case of long-term tests, the microcontroller is able to capture the gate current of all DUTs every second for up to four years. If short-term stress tests are performed, an accordingly increased resolution in time is possible.

Generally, the microcontroller can be set up for test through any terminal program. A graphical user interface (GUI) under MATLAB eases the setup and the data transmission to as well as evaluation on a computer.

6 Conclusions

This paper presents an automated test system for MOS gate dielectric analysis based on a 1024 DUT integrated test structure. A version with 4096 DUTs is also available.

Acknowledgements. The authors gratefully acknowledge D. Naydenova for her work on the current sense circuit as well as F. Wagner for his work on the microcontroller and graphical user interface.

References

Lombardo, S., Stathis, J. H., Lindner, B. P., et al.: Dielectric Breakdown Mechanisms in Gate Oxides, J. Appl. Phys., 98, 121301, 1–36, 2005.

Depas M., Verneire, B., Mertens, P. W., et al.: Determination of Tunnelling Parameters in Ultra-Thin Oxide Layer Poly-Si/SiO2/Si Structures, Solid-State Electronics, 38, 8, 1465–1471, 1995.

Avellan Hampe, A.: Charakterisierung von MOS-Transistoren vor und nach dem Gateoxiddurchbruch, ISBN 3-18-337109-X, VDI Verlag, 29 pp. , 2004.

20

On the synthesis and optimization of cascaded continuous-time Sigma-Delta modulators

M. Keller, A. Buhmann, M. Kuderer, and Y. Manoli

Chair of Microelectronics, Department of Microsystems Engineering (IMTEK), University of Freiburg, Georges-Koehler-Allee 102, D-79110 Freiburg, Germany

Abstract. Up to now, there exist two completely different approaches for the synthesis of cascaded CT Sigma-Delta modulators. While the first method is based on a DT prototype and thus on the application of a DT-to-CT transformation, the second one is entirely performed in the CT domain. In this contribution, the method of lifting will be applied to overcome the disadvantages afflicted with the first method (e.g. less ideal anti-aliasing filter performance, increased circuit complexity) and to establish a time efficient DT simulation model for the second method. Thereby, optimal modulator coefficients as well as optimal digital cancellation filters for an arbitrary cascaded CT modulator can be simulated in an efficient and rapid manner. For illustrative purposes, the complete synthesis procedure is demonstrated by the example of a 2-1-1 cascaded CT modulator.

1 Introduction

Owning to their inherent robustness to most circuit non-idealities afflicted with deep-submicron CMOS technologies, the concept of Sigma-Delta A/D conversion has proven itself as a very efficient technique for the implementation of A/D converters in such technologies over the last decades. While their classical field of application is found in low to medium bandwidth with medium to high resolutions such as telephone or audio, nowadays, they have to deal with ever-growing bandwidth standards altogether with the demand for higher or even adjustable resolution as it may be imposed by video or broadband data communication applications.

For several reasons, especially the principle of cascaded continuous-time (CT) Sigma-Delta modulators seems to be a most promising concept for such applications. Compared to a single stage realization of the same overall modulator order N ($N \geq 3$), they achieve a higher resolution (Marques et al., 1998). This benefit stems from the a priori stability of such modulators when only first and second order modulators are used in the cascade, which enables a noise-shaping filter behavior more closely to the one of an ideal high-pass filter of the same order. Moreover, the resolution of cascaded architectures can simply be adjusted by switching on or off several stages in the cascade which makes them very attractive for multi-standard adaptive circuits (Rodríguez-Vázquez, 2005). Compared to DT realizations, CT modulators show the advantage of a lower power consumption due to their lower bandwidth requirements for their operational amplifiers. Finally, any CT modulator offers an intrinsic anti-aliasing filter, thus giving it the additional advantages of reduced die area and further power savings compared to DT realizations.

After a short introduction to the state-of-the-art techniques for the synthesis of cascaded CT modulators and their afflicted drawbacks in the next Section, the method of lifting will be introduced using the example of a CT 2-1-1 modulator to overcome these drawbacks.

2 State-of-the-art techniques

2.1 DT-to-CT transformation

The first method (Ortmanns et al., 2001) benefits from the extensive work and its ongoing results performed on cascaded DT modulators for which a broad spectrum of literature is available, among others (Marques et al., 1998; Medeiro et al., 1999). By means of the DT-to-CT transformation, which was already well known from single stage modulators (Cherry et al., 2000), it was not only possible to translate any DT coefficients into CT ones, but also to re-use the very simple digital cancellation filters developed for cascaded DT modulators in CT architectures. Unfortunately, the resulting cascaded CT architecture as shown for the example of the 2-1-1 modulator in Fig. 1 differs from its DT equivalent in that addi-

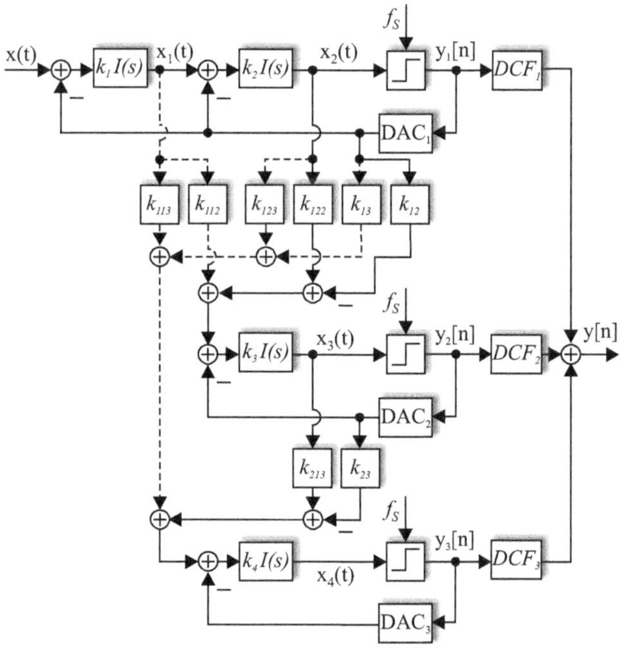

Fig. 1. Continuous-time 2-1-1 modulator based on the classical DT architecture (solid) and as obtained from a DT-to-CT transformation (including dashed signal paths).

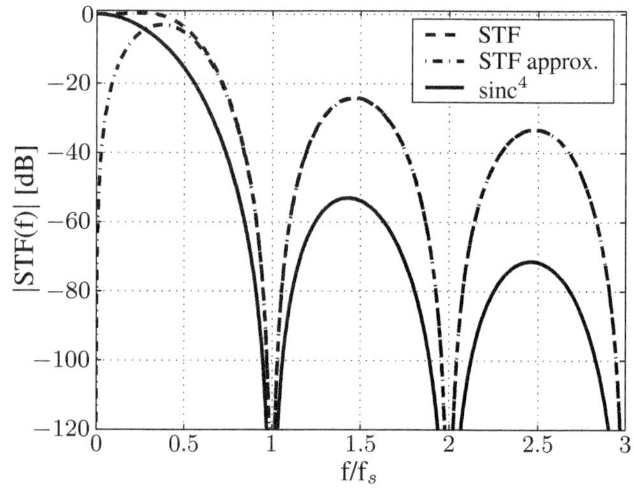

Fig. 2. Signal transfer function of the CT 2-1-1 modulator shown in Fig.1 with and without additional signal paths (dashed line vs. solid line) and approximation of the former by Eq. (1) (dashdot line).

tional signal paths (Fig. 1, dashed lines) are needed in order to enable the DT-to-CT transformation. Other than the mentioned drawback of increased circuit complexity (Tortosa et al., 2005), it has been shown that due to the direct connection from the first stage to the third stage via k_{113}, the high frequency behavior of the signal transfer function $STF(s)$ and therewith the anti-aliasing feature is given by Keller et al. (to be published):

$$STF(s) \approx k_1 k_{113} k_4 \left(\frac{f_s}{s}\right)^2 \frac{DCF_3[z]}{LF_3[z]}\bigg|_{z=e^{s/f_s}} \tag{1}$$

where f_s/s represents the integrator transfer function $I(s)$, $DCF_3[z]$ the digital cancellation filter and $LF_3[z]$ the loop filter of the last stage, respectively. The magnitude of the STF together with its approximation for high frequency input signals according to Eq. (1) is shown in Fig. 2 (dashed line vs. dashdot line). All CT coefficients have been derived via a DT-to-CT transformation of the DT coefficients given in Marques et al. (1998).

2.2 Direct synthesis

To overcome the drawback of increased circuit complexity afflicted with the DT-to-CT transformation, it has been proposed to skip the additional signal paths, which are solely needed to make the DT-to-CT transformation possible, and rather to re-adjust the digital cancellation filters (Tortosa et al., 2005). The first step toward that end is to transform the given modulator into its mathematical representation as shown in Fig. 3. A detailed description to this approach is given in Keller et al. (to be published). Replacing the strongly nonlinear quantizers by their linear white noise models and transforming every conglomerate of DAC, CT loop-filter LF or CT connecting loop-filter CLF and sampler to its DT equivalent allows for each noise transfer function, i.e. the transfer function from one quantization error to the output $y[n]$, to be entirely calculated in the z-domain. Therewith, the digital cancellation filter can be adjusted in the traditional manner, namely to cancel the quantization errors of all stages but the last. Furthermore, the overall noise transfer function related to the quantization error of the last stage should result in an ideal high-pass filter whose order is equal to the overall modulator order, i.e. the sum of the orders of all single stage modulators in the cascade, while maintaining a DC gain of one for the signal transfer function.

The application of this method overcomes the drawback of increased circuit complexity. Moreover, the resulting signal transfer function is equal to an ideal fourth order sinc-filter and therewith offers a superior anti-aliasing feature compared to the approach of the DT-to-CT transformation (Fig. 2, solid line vs. dashed line). However, it is detrimental in that one still has to resort to time consuming CT simulations. This becomes even more apparent whenever optimal coefficients as well as optimal digital cancellation filters for any cascaded CT architecture have to be found or to be verified by simulation.

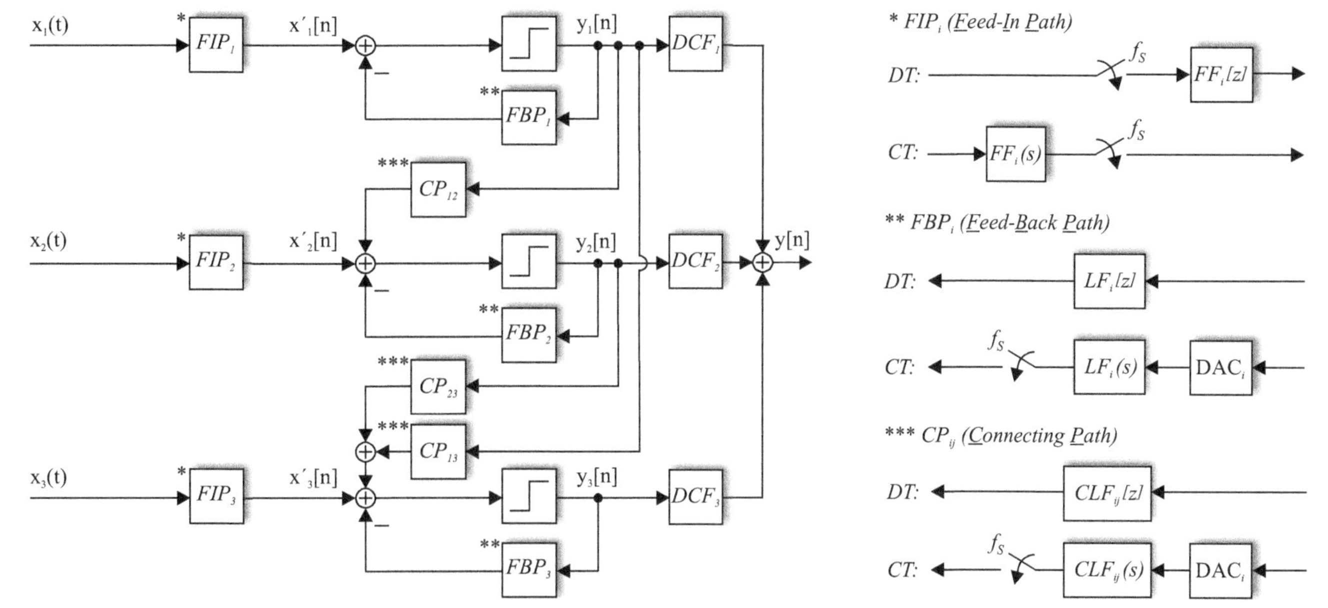

Fig. 3. System theoretical approach for a DT or CT modulator realized as a cascade of three single stage modulators possessing arbitrary orders. All $x_i(t)$ are equal to the input signal $x(t)$ - generally applied to the first stage of the cascade only - and may therefore be shorted.

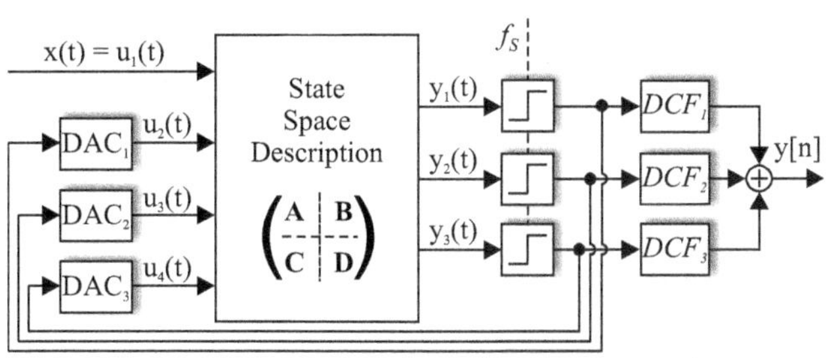

Fig. 4. State space representation of a cascaded CT modulator composed of three single stage modulators possessing arbitrary orders.

3 Lifting

To obtain a fast but yet precise DT simulation model for any CT cascaded modulator, the method of lifting is introduced using the example of the CT 2-1-1 modulator without additional signal paths in Fig. 1. Toward that end, the state space description of this modulator has to be found. Assuming a sample frequency equal to one and following the notations in Figs. 1 and 4, the state space description of this multiple input and multiple output system results in:

$$\begin{aligned} \dot{\boldsymbol{X}}(t) &= \boldsymbol{A}\,\boldsymbol{X}(t) + \boldsymbol{B}\,\boldsymbol{U}(t) \\ \boldsymbol{Y}(t) &= \boldsymbol{C}\,\boldsymbol{X}(t) + \boldsymbol{D}\,\boldsymbol{U}(t) \end{aligned} \tag{2}$$

with

$$\boldsymbol{A} = \begin{pmatrix} 0 & 0 & 0 & 0 \\ k_2 & 0 & 0 & 0 \\ 0 & k_{122}k_3 & 0 & 0 \\ 0 & 0 & k_{213}k_4 & 0 \end{pmatrix} \quad \boldsymbol{C} = \begin{pmatrix} 0\,1\,0\,0 \\ 0\,0\,1\,0 \\ 0\,0\,0\,1 \end{pmatrix}$$

$$\boldsymbol{B} = \begin{pmatrix} k_1 & -k_1 & 0 & 0 \\ 0 & -k_2 & 0 & 0 \\ 0 & -k_{12}k_3 & -k_3 & 0 \\ 0 & 0 & -k_{23}k_4 & -k_4 \end{pmatrix} \quad \boldsymbol{D} = \boldsymbol{0} \tag{3}$$

The DT state space description is then easily obtained as (Yamamoto, 1999):

$$\begin{aligned} \boldsymbol{X}[n+1] &= e^{\boldsymbol{A}}\boldsymbol{X}[n] + \int_0^1 e^{\boldsymbol{A}(1-\tau)}\boldsymbol{B}\boldsymbol{U}[n](\tau)d\tau \\ \boldsymbol{Y}[n] &= \boldsymbol{C}\boldsymbol{X}[n] \end{aligned} \tag{4}$$

$\boldsymbol{U}[n](\tau)$ is a vector containing the lifted input signals to the state space system block, each defined as a sequence of functions:

$$\begin{aligned} &\mathcal{L} : u \mapsto \{u[n](\tau)\}_{n=0}^{\infty}, \\ &u[n](\tau) = u(n+\tau), \\ &0 \le \tau \le 1 \end{aligned} \tag{5}$$

$$DCF_1[z] = \frac{(k_1 + 12b_1k_1 + 4)z^3 + (11k_1 + 12 - 12b_1k_1)z^2 + (11k_1 - 12 - 12b_1k_1)z + k_1 + 12b_1k_1 - 4}{24k_1z^4}$$

$$DCF_2[z] = \frac{(2b_2 + 1)z^3 - (6b_2 + 1)z^2 + (6b_2 - 1)z + 1 - 2b_2}{2c_1z^4}$$

$$DCF_3[z] = \frac{(1 - z^{-1})^3}{c_1c_2} \tag{6}$$

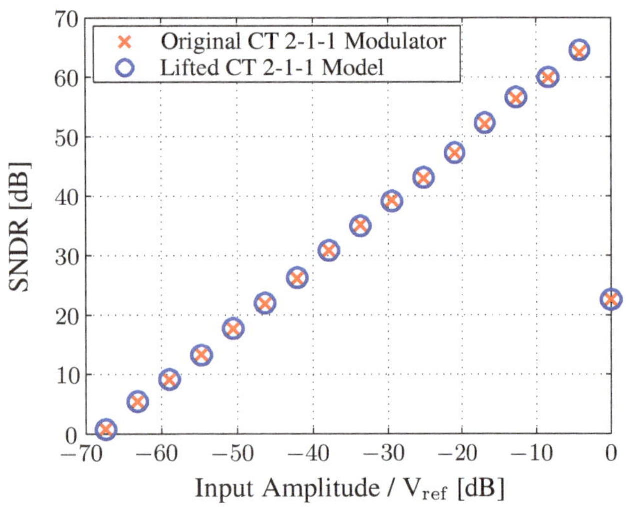

Fig. 5. SNDR of the CT 2-1-1 modulator, simulated on the original modulator (x-mark) and its lifted model (o-mark).

Unlike the classical DT state space transformation, which assumes the input signals $u_i(t)$ to be constant over one sampling interval, this approach allows for arbitrary waveforms of both the input signal as well as the feedback signals. With regard to the mathematical modelling of cascaded CT Sigma-Delta modulators, the former has its justification only in case of a DC input signal in combination with non-return-to-zero (NRZ) DAC output signals while it is erroneous for every other type of input or DAC output signal.

The only condition to be fulfilled for an efficient implementation of this model in a simulator like MATLAB is imposed by the integral in Eq. (4) to be solvable in closed form. Since usually criteria like stability, dynamic range or achievable signal-to-noise-ratio are determined by applying a sine wave input signal to the modulator (Schreier et al., 2005), troubles in solving the integrals may only be encountered depending on the choice of the DAC signal waveforms.

The application of lifting to the CT modulator thus results in an exact DT state space representation of the CT modulator. This in turn allows for fast and yet precise DT simulations to be performed without paying the penalty of loosing the inter-sampling behavior of the CT system and therewith loosing the information of the exact states of the state space variables. In this way optimal coefficients for a cascaded CT modulator based on the traditional DT architecture, i.e. feeding the quantization error of one stage to the next stage only, may be found by resorting to fast simulations. Moreover, scaling of the state space variables can easily be performed.

In order to validate the correctness of the application of lifting, exemplary simulations have been performed on the lifted model of the 2-1-1 modulator in Fig. 1 to find optimal values for the connecting coefficients k_{112}, k_{12}, k_{213} and k_{23}. More precisely, these connecting coefficients have been substituted by $c_1/(k_1k_2)$, c_1b_1, c_2/k_3 and c_2b_2 and the new coefficients b_i and c_i have been optimized. Although beyond the scope of this paper, this allows for a direct comparison with the optimal DT connecting coefficients as presented in Marques et al. (1998). Note that already a parametric sweep over these four coefficients between 0.1 and 2 with a step width of 0.1 results in 160000 simulations. In this context, a set of coefficients is defined as optimal in case that the highest signal-to-noise-and-distortion ratio (SNDR) within a signal bandwidth f_b of 1/32 (OSR 16, f_s 1) is achieved using a sine wave input signal with an amplitude of 0.7 times the reference voltage of the DAC and a frequency of $f_b/4$. The remaining coefficients k_1 to k_4 (2/3, 3/8, 1/2, 1/2) have been chosen according to a DT-to-CT transformation of the DT coefficients given in Marques et al. (1998). For the reason of simplicity, only NRZ digital-to-analog converters have been used. The digital cancellation filters given in Eq. (6) are obtained by calculations using effective quantizer gains of $1/(k_1k_2)$ in the first stage, $1/k_3$ in the second stage and $1/k_4$ in the third stage, respectively. With the obtained coefficients b_1, b_2, c_1 and c_2 (9/10, 1/5, 1/5, 11/10), simulations concerning the dynamic range of the modulator have been performed on both the original modulator and its lifted model. Obviously, the results shown in Fig. 5 are almost identical and therewith do confirm the correctness of the lifting approach.

4 Conclusions

In this contribution, it has been shown that cascaded CT modulators based on the classical DT architecture, i.e. feeding the quantization error to the next stage only, have two important advantages over those derived by DT prototypes: reduced circuit complexity and superior anti-aliasing filter behavior. The drawback of long simulation runtime afflicted with behavioral models of cascaded CT modulators was overcome by the application of lifting to the CT modulator. Time efficient simulations can then be performed on the equivalent DT model with almost identical results. However, lifting is thus far only applied to the simulation of ideal modulators. The embedding of non-idealities such as finite opamp gain or even jitter in these simulations will further improve both the quality and the profitableness of this method.

Acknowledgements. This work was supported by Deutsche Forschungsgemeinschaft (German Research Foundation) under grant MA 2193/6-1.

References

Marques, A., Peluso, V., Steyaert, M. S., and Sansen, W. M.: Optimal Parameters for $\Delta\Sigma$ Modulator Topologies, IEEE Trans. Circuits Syst. II, 45, 1232–1241, 1998.

Rodríguez-Vázquez, A.: Using $\Sigma\Delta$ Converters for Communications and their Application to Multi-Standard Receivers part of Multi-Standard Adaptive Circuits and Systems for Wireless Communications, Half Day Tutorial, Int. Sym. on Circuits and Systems, Kobe, Japan, 2005.

Ortmanns, M., Gerfers, F., and Manoli, Y.: On the Synthesis of Cascaded Continuous-Time $\Sigma\Delta$ Modulators, in Proc. IEEE Int. Symp. on Circuits and Systems, 419–422, 2001.

Medeiro, F., Pérez-Verdú, B., and Rodríguez-Vázquez, A.: Top-Down Design Of High-Performance Sigma-Delta Modulators, Dordrecht, The Netherlands: Kluwer Academic Publishers, 1999.

Cherry, J. A. and Snelgrove, W. M.: Continuous-Time Delta-Sigma Modulators for High-Speed A/D Conversion, Norwell, MA: Kluwer Academic Publishers, 2000.

Tortosa, R., de la Rosa, J. M., Rodríguez-Vázquez, A., and Fernández, F. V.: A Direct Synthesis Method of Cascaded Continuous-Time Sigma-Delta Modulators, in Proc. IEEE Int. Symp. on Circuits and Systems, 5585–5588, 2005.

Keller, M., Buhmmann, A., Ortmanns, M., Gerfers, F., and Manoli, Y.: On the Implicit Anti-Aliasing Feature of Continuous-Time Multistage Noise-Shaping Sigma-Delta Modulators, submitted to IEEE Trans. Circuits Syst. I.

Yamamoto, Y.: Digital Control, Wiley Encyclopedia of Electrical and Electronics Engineering, 5, 445–457, 1999.

Schreier, R. and Temes, G. C.: Understanding Delta-Sigma Data Converters, Piscataway, NJ: IEEE Press, 2005.

21

Influence of gate tunneling currents on switched capacitor integrators

W. Kraus and D. Schmitt-Landsiedel

Institute for Technical Electronics, Technische Universität München, Germany

Abstract. In order to achieve a higher level of integration in modern VLSI systems, not only the lateral geometrical dimensions have to be scaled. Lowering the supply voltage also requires scaling down the oxide thickness of the transistors. While the oxide thickness is scaled down proportionally with the supply voltage, the gate tunneling currents grow exponentially, which results in special issues concerning deviations in charge based analog and mixed signal circuitry. The influence of gate tunneling currents on this kind of circuits will be demonstrated at a fully differential switched capacitor integrator. The used process data is derived from the International Technology Roadmap for Semiconductors (ITRS Roadmap, 2006). The Parameter sets for the simulations are based on the Predictive Technology Model of the Arizona State University Modelling Group for the 65 nm Technology node (Predictive Technology Model, 2008).

1 Introduction

Physical constraints are the burden of modern VLSI circuits. One of the emerging problems is the gate tunneling current. The ITRS (ITRS Roadmap, 2006) forecasts a straight increase in the next few years, like it can be seen in Fig. 1. In comparison to the source drain leakage like gate induced drain leakage or subthreshold currents, not only the source and drain nodes are incorporated. The tunneling currents also run over the gate node. Figure 2 shows the different parts of the gate tunneling currents. For the operating points of our circuit the gate to bulk current I_{GB} can be neglected. The gate to channel current can be divided in one part to the source node I_{GCS} and one to the drain node I_{GCD}. These two are only relevant for an existing channel. The currents I_{GS} and I_{GD} in the gate to drain and gate to source overlap regions have always to be considered. Figure 3 shows the simulation of the different gate tunneling currents for the source node at V_{SS} and the drain node at V_{DD}. The voltages from gate to source and gate to drain have different signs. The corresponding gate tunneling currents therefore result in a zero crossing of the overlap component $I_{GS}+I_{GD}$ for $V_{GS}=V_{DD}/2$. For gate voltages above the threshold voltage the additional gate to channel currents have to be added to the overlap parts and result in a curve with a zero crossing near but not exactly at $V_{GS}=V_{DD}/2$. The incorporation of the additional node to the leakage path requires a more complex analysis of the circuit. The following considerations will start at simple building blocks and will transfer this knowledge to the fully differential integrator.

2 The transfer gate

The used transfer gate consists of one p-MOSFET and one n-MOSFET of the same size. This is necessary for conduction over the full voltage range and to eliminate the channel charge injection due to switching it on or off.

There are two different states of a transfer gate that have to be discussed. The conducting state allows the gate tunneling currents to flow across the channel. Both endpoints of the transfer gate will be influenced in the same way by the tunneling currents flowing from the gate to each endpoint. When the transfer gate blocks, both endpoints are separated and do not influence each other anymore. First we have a look at the blocking state. Figure 4a shows a transfer gate with a potential V_1 at the top node above $V_{DD}/2$. This causes a high voltage drop at the overlap region of the n-MOSFET and a low voltage drop at the overlap region of the p-MOSFET. Considering the exponential dependency of the gate tunneling currents to the voltage drop it is clear that the influence of the n-MOSFET is dominant in this case. When the voltage

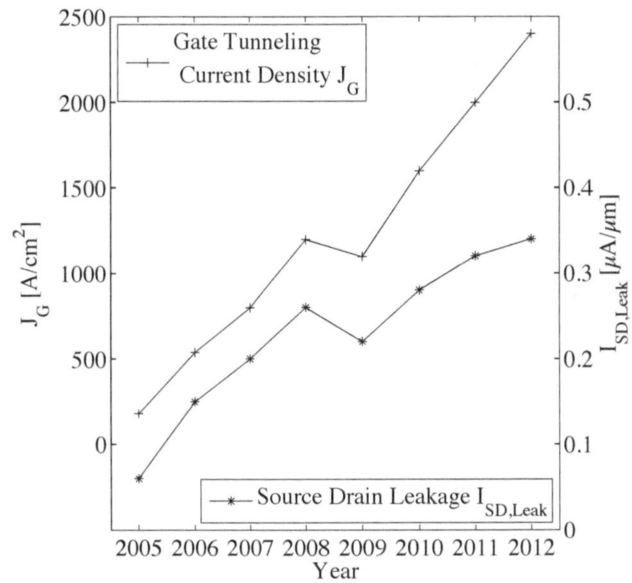

Fig. 1. The evolution of gate tunneling current density as predicted by the ITRS (ITRS Roadmap, 2006).

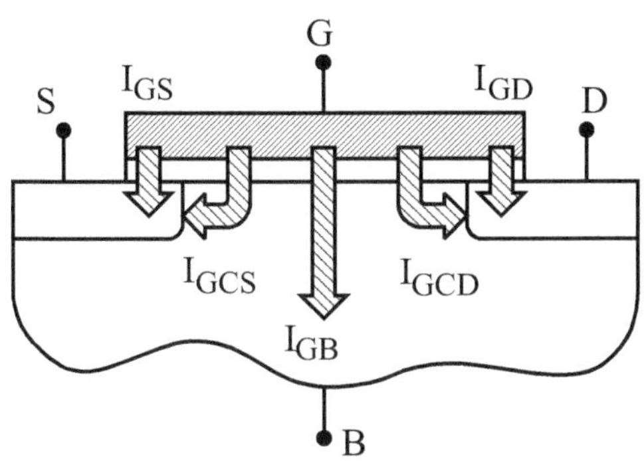

Fig. 2. The different parts of the gate tunneling currents in a MOSFET.

V_1 reaches a value near $V_{DD}/2$ the gate tunneling currents to the p- and the n-MOSFET are of the same size but of the opposite direction. In this case the gate tunneling currents eliminate each other. Due to the different tunneling characteristics of p- and n-MOSFETs this zero crossing will not be exactly at $V_{DD}/2$ (see Fig. 5).

In the conducting state both transistors establish a channel and add their gate to channel tunneling currents to the overlap part. But not only the absolute values of the gate tunneling currents change, also the zero crossing changes. The different threshold voltages of the n- and the p-MOSFET cause different tunneling characteristics, which lead to a shifted zero crossing. Figure 5 shows the comparison of both states depending on the input voltage V_1.

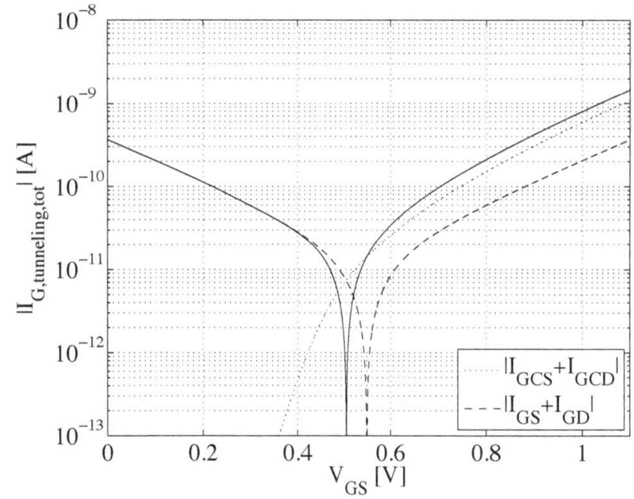

Fig. 3. Simulation results of the gate tunneling currents of a single n-MOSFET with minimal dimensions.

3 The switched capacitor integrator

3.1 Single ended integrator

Switched capacitor circuits handle and process signals by manipulating the charge stored on the involved capacitors. To see the influence of gate tunneling currents on this signals, it has to be shown in which part of the circuit and in which switching state this influence is really relevant.

At the switched capacitor integrator there are two phases, the charging phase and the integration phase. Figure 6 shows the charging phase. The coupling capacitor C_R is charged to the input voltage V_{in}. Both nodes of C_R are connected to a definite potential with a low resistance. Therefore the gate tunneling currents flow to these sinks and do not affect the amount of charges stored on C_R. One node of the integration capacitor C_I is connected to the low resistance output of the operational amplifier. The other is connected to the high resistance input of the operational amplifier and a blocking transfer gate, see Fig. 6. This node is affected by the gate tunneling currents of the transfer gate and of the operational amplifier. The influence of the operational amplifier will be discussed later. The high resistance node of C_I, which is directly connected to this transfer gate has an input voltage of $V_{DD}/2$ and will cause a very low drift of the stored charge due to a very low gate tunneling current (compare to Fig. 5).

In the integration phase the charge from C_R will be transferred to C_I. When the circuit has settled, the high resistance node at the input of the operational amplifier will have the potential of $V_{DD}/2$. Then again gate tunneling currents do not influence this node. But during the charge transfer the voltage at this node can be significantly different from $V_{DD}/2$. Figure 5 shows that the gate tunneling current can thus be about two to three decades higher than in the settled state. By integrating the gate current over this short time the

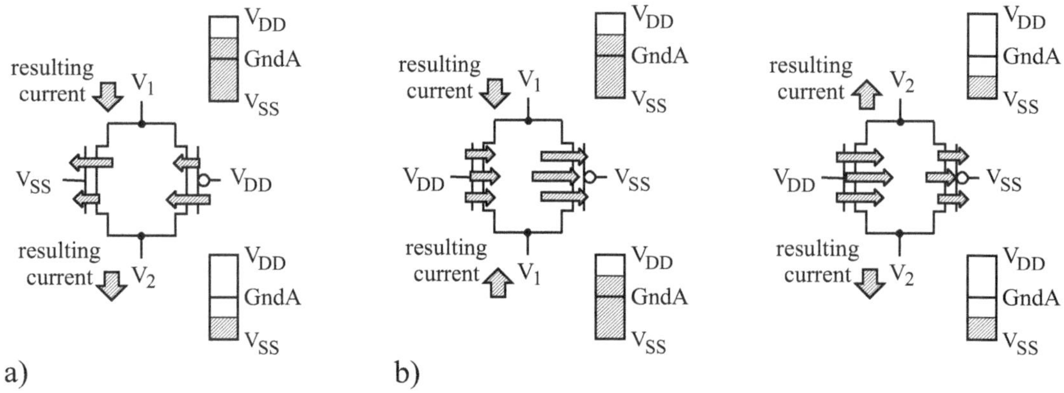

Fig. 4. Gate tunneling currents of a transfer gate in **(a)** the blocking and **(b)** the conducting state.

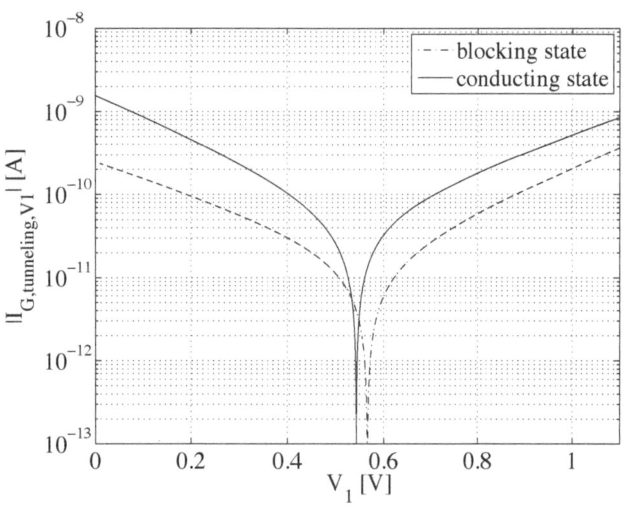

Fig. 5. Simulation results of the gate tunneling currents for the blocking and the conducting state of a transfer gate.

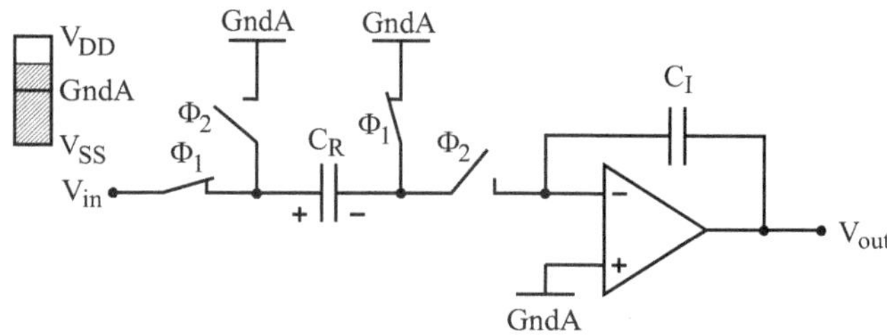

Fig. 6. Gate tunneling currents in the charging phase of a single ended switched capacitor integrator.

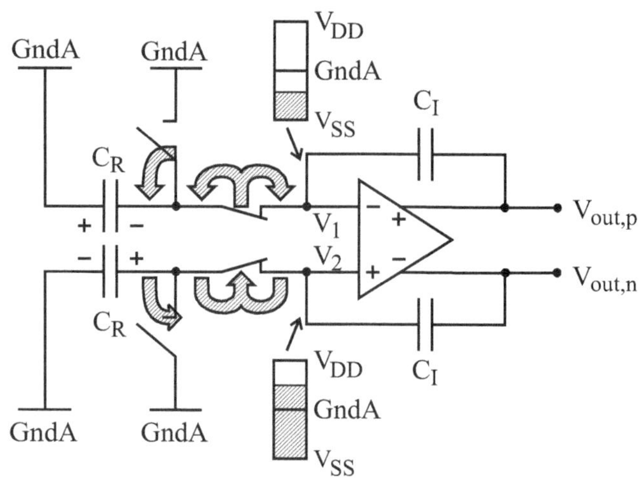

Fig. 7. Gate tunneling currents in the integration phase of a fully differential switched capacitor integrator.

charge deviation can be calculated. It is proportional to the signal deviation caused by the gate tunneling currents.

3.2 Fully differential integrator

Extending the circuit with an inverted path results in the fully differential architecture. Common mode effects can be eliminated by using this technology. During the charging phase the high resistance nodes are again at $V_{DD}/2$ and the gate tunneling current in both paths is of the same value and direction. It results in a common mode effect which can be eliminated by the fully differential architecture. Local mismatch of the relevant transistors causes a residual error, but this is small in the regarded technology. This can be neglected. The same applies for the settled state of the integration phase.

During the settling phase the high resistance nodes at the input transistors of the operational amplifier have inverted potentials referenced to $V_{DD}/2$ (see Fig. 7). The resulting high gate tunneling currents are also opposite in sign and result in a differential mode signal. This can not be eliminated with the fully differential architecture.

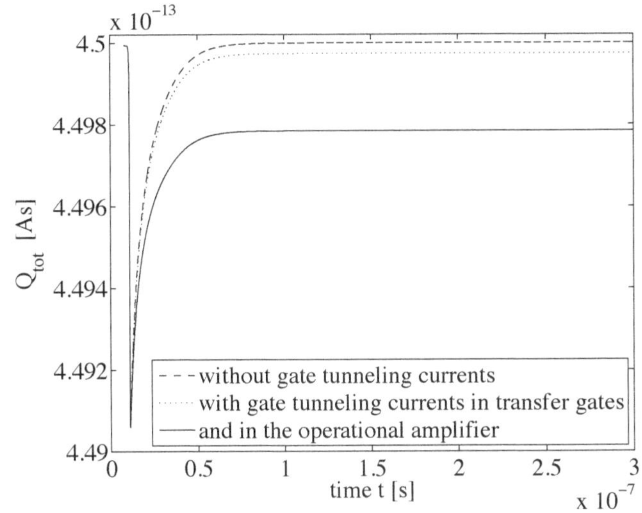

Fig. 8. Time dependency of the stored charges for the three simulations without tunneling currents, with gate tunneling currents in all transistors but the input transistors of the operational amplifier and in all transistors. At the beginning of this simulation C_R had been charged to the highest input value.

3.3 Gate tunneling currents in the input transistors of the operational amplifier

An operational amplifier has a symmetrical input stage. Both input transistors show the same gate tunneling current for equal potentials at both nodes. Also the direction of the gate currents is the same. When local mismatch is negligible, gate tunneling currents can be eliminated by the fully differential architecture in this operation mode. That's why the operational amplifier does not affect the stored signal during the charging phase or in the settled state of the integration phase.

During the settling phase the potentials at both input nodes are significantly different from each other. This results in gate tunneling currents in the same way as in the transfer gates, but the currents are much higher. Because of matching constraints and speed requirements, the input transistors have to be designed with a larger area compared to the transfer gates. The gate tunneling currents are proportional to this area. During the settling state of the integration phase the influence of the input transistors of the operational amplifier can not be eliminated by the fully differential architecture.

4 Simulation results

The considerations in the last sections showed that the influence of the gate tunneling currents is relevant just in the settling state of the integration phase. Right in this moment, when the transfer gates are also switched, some other effects overlay the gate tunneling currents. In the two transistors of the transfer gates the gate voltage switches from V_{DD} to V_{SS} and vice versa. This voltage ramp at both transistors does not

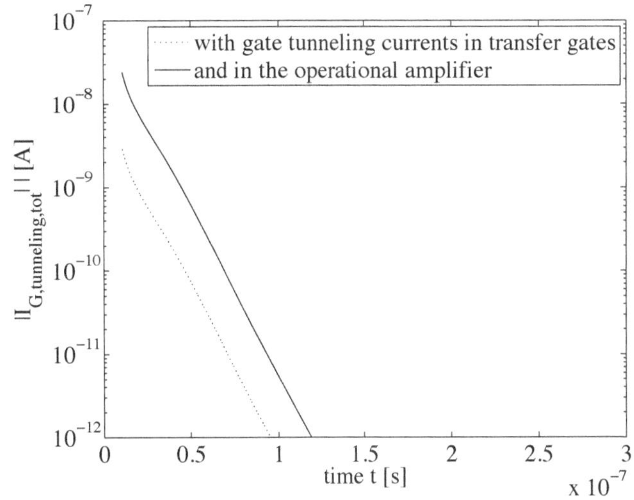

Fig. 9. Time dependency of the gate tunneling currents for the simulation with gate tunneling currents in all transistors but the input transistors of the operational amplifier and in all transistors.

switch them on or off at the same time, because the threshold voltages of the n- and the p-MOSFET are different. So the charge flowing from the transistor channels to source and drain will be stored temporarily on the integration capacitor C_I. During this settling mechanism the currents do overlay each other and cannot be separated properly.

The simulation tools offer us the possibility to switch the gate tunneling currents on or off for selected transistors by changing the transistor parameters. Three simulations have been made. The first was completely without gate tunneling currents. The second simulation took the gate tunneling currents of all but the input transistors of the operational amplifier into account. With the third simulation the input transistors of the operational amplifier were also added.

To observe the influence of the gate tunneling currents, the charge was calculated, which is stored in the complete integrator. After the charging phase has finished, the stored charges only will be transferred from one to the other capacitor. During the settling state there will be a slight deviation of the summed charges, but when it is settled, the temporary charge injection has vanished again. Figure 8 shows this behavior for the simulation without gate tunneling currents.

Taking the tunneling currents of the transfer gates into account, a small amount of the stored charge will be removed by these currents. Figure 8 also shows the reduced charge stored in the capacitors. The difference of the stored charge with and without gate tunneling currents at any time gives the deviation of charge up to this moment. By differentiation of this charge deviation the gate tunneling current at this time can be calculated and is shown in Fig. 9.

The input transistors of the operational amplifier have to be designed wider than the ones used in the transfer gates because of matching and speed constraints. The larger area causes larger gate tunneling currents. In addition to that the operation mode of these input transistors normally is in inversion, which always causes tunneling currents to the channel. Figure 9 shows about one decade larger gate tunneling currents compared to the transfer gates only.

A closer look at the time dependency in Fig. 9 shows a nearly exponential behavior. This can be explained considering the settling process in the integration phase for a high input signal. The charge from Capacitor C_R will be transferred to C_I driven by the output current of the operational amplifier, which is in the overdrive mode. In this mode the current is constant and therefore the voltage at the high resistance node will change linearily. Gate tunneling currents depend on the voltage in an exponential way, which results in the exponential time dependency.

The absolute values of the simulation with all gate tunneling currents show a deviation of the stored charges of $2.2{\cdot}10^{-16}$As, which is equivalent to about 0.05% of the signal. Taking into consideration that the integrator was designed for 10 bit resolution, the dissipated charge is equivalent to a half bit. But the gate tunneling currents depend exponentially on the voltage. So it is clear that this deviation only can be seen at high input levels.

5 Conclusions

Gate tunneling currents make the leakage problem more complex. But not every transistor contributing gate tunneling currents affects the signal path. Only the high resistance nodes at the signal storing capacitors are sensitive to them. In comparison of all relevant transistors, the input transistors of the operational amplifier had the biggest influence. They caused about one decade higher deviations of the stored signal than the rest of the relevant transistors.

References

ITRS Roadmap: Semiconductor Industry Association, The International Technology Roadmap for Semiconductors, http://www.itrs.net/, 2006.

Predictive Technology Model: Nanoscale Integration and Modelling Group, Arizona State University, Predictive Technology Model, PTM, http://www.eas.asu.edu/~ptm/, 2008.

Impact of negative and positive bias temperature stress on 6T-SRAM cells

S. Drapatz[1,2], G. Georgakos[2], and D. Schmitt-Landsiedel[1]

[1]Lehrstuhl für Technische Elektronik, Technische Universität München, Germany
[2]Infineon Technologies AG, München-Neubiberg, Germany

Abstract. With introduction of high-k gate oxide materials, the degradation effect *Positive Bias Temperature Instability* (PBTI) is starting to play an important role. Together with the still effective *Negative Bias Temperature Instability* (NBTI) it has significant influence on the 6T SRAM memory cell. We present simulations of both effects, first isolated, then combined in SRAM operation. During long hold of one data, both effects add up to a worst case impact. This leads to an asymmetric cell, which, in a directly following read cycle, combined with the generally unavoidable production variations, maximizes the risk of destructive reading. In future SRAM designs, it will be important to consider this combination of effects to avoid an undesired write event.

1 Introduction

The ongoing miniaturization in semiconductor industry has already led to decananometer transistors (ITRS roadmap). As the voltage cannot be scaled according to geometry and gate oxides are getting thinner, electric fields and tunneling currents in the devices are increasing. This was exacerbating *Negative Bias Temperature Instability* (NBTI) and leakage in the past years. In order to reduce these harmful effects (especially leakage), thicker gate oxide materials are needed. To keep the ability to store charges and control the MOSFET channel, the dielectric constant must then be higher than in conventional SiO_2. This is done with the introduction of high-k gate oxide materials.

One disadvantage of these new materials is an upcoming effect similar to NBTI: *Positive Bias Temperature Instability* (PBTI), which occurs predominantely on nMOS transistors, while NBTI is effective mainly on pMOS transistors. In the past, NBTI alone fortunately did not have huge impact on the SRAM memory cell performance (Fischer et al., 2007). But does the additional effect PBTI together with the still effective NBTI have a significant influence on SRAM core cell performance?

The following section gives a brief overview on the degradation mechanisms NBTI and PBTI. To quantitatively measure the influence of these effects on the SRAM 6T core cell, a benchmark is required. This is done with some metrics to describe the quality of the memory cell, introduced in Sect. 3. Using these metrics, the impact of NBTI and PBTI on SRAM is simulated in Sect. 4. Next, it is examined to what extent normal usage of the memory cell, i.e. hold, read and write, leads to BTI stress conditions of the single transistors. The worst use case will then be isolated and considered in the last section separately together with variation and yield considerations.

2 NBTI and PBTI

In this section, both degradation effects *Negative Bias Temperature Instability* and *Positive Bias Temperature Instability* are introduced.

2.1 NBTI degradation

NBTI is an effect only influencing pMOS transistors. It is caused by the following scenario: negative potential at the gate relative to higher potentials on source and drain. For standard CMOS circuits this means gate connected to V_{SS}, source and drain connected to V_{DD}. This leads to a conducting transistor without current in the channel (Fig. 1).

The failure mechanism NBTI is still under discussion, but it seems to be clear that the stress condition breaks silicon-hydrogen bonds, and the hydrogen ions diffuse away. These vacant positions, acting as traps, can bond holes. These additional holes in the gate oxide cause a decreasing threshold

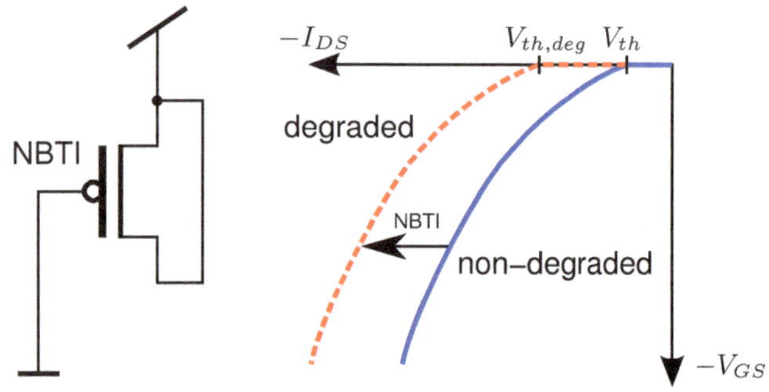

Fig. 1. Stress conditions and impact of NBTI to pMOS transistors

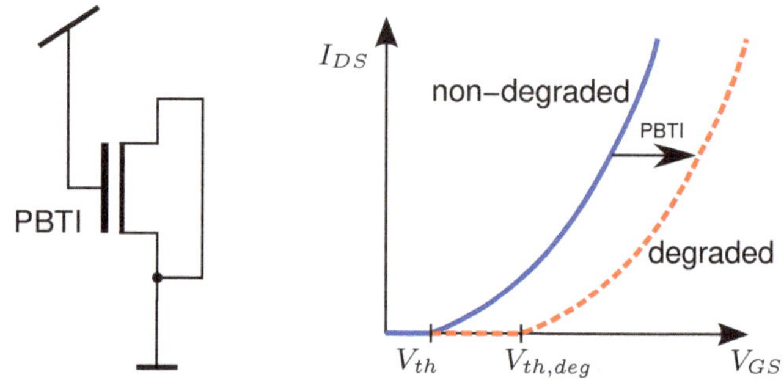

Fig. 2. Stress conditions and impact of PBTI to nMOS transistors.

voltage, e.g. from −0.5 V it goes down to −0.55 V. This makes the transistor weaker over time, and a circuit built of this transistor is getting slower (Fig. 1). Typical threshold voltage drift values in 65 nm technology: after 8 years of stress at $V_{DD,\mathrm{nom}}$ and 125°C there will be a mean drift value of approx. 60 mV.

2.1.1 NBTI distribution

NBTI is a statistical process. There is a distribution of single drift values, but the simulation tools so far can only handle and predict the mean value of this distribution. Analyses have shown that a stress of 10 000 s at 125°C lets the mean value drift for approx. 45 mV, but single devices have shown a drift of up to 150 mV (Fischer et al., 2007). Especially for SRAM with a huge number of cells in an array, this means that single memory cells will experience big drifts.

2.1.2 NBTI annealing

Additionally, NBTI shows a recovery effect: directly after end of stress, the threshold voltage drift is worst, but is annealing with a short time constant, depending on the duration of the stress (Heinrigs et al., 2007). This means that for a circuit it would be worst to work with this high amount of threshold voltage drift during or very shortly after stress. If the circuit only experiences the annealed drift values, the impact would be much less. It is not analysed so far how strong relaxation effects influence the SRAM circuits.

2.2 Expected PBTI degradation

Contrary to NBTI that is relatively well-known in the meantime, PBTI was not much researched in the past. Many papers are available for NBTI, but PBTI is just coming up. The reason is clear: up to now, it did not occur on semiconductor chips without high-k gate oxide.

Only with the introduction of these new materials, the PBTI effect appears. This is because NBTI is based on traps for holes, which only happens at the interface states. PBTI, however, is based on electron trapping, which mainly happens in high-k bulk as well as in high-k SiON interfacial layer. PBTI only works on nMOS transistors, and according to that the stress condition is contrary to NBTI: positive potential on the gate, zero potential on source and drain. This lets the threshold voltage drift to higher values, so both NBTI and PBTI weaken the devices over time (Fig. 2).

In summary, PBTI is not yet as good analyzed as NBTI, but the behavior seems to be very similar (Zafar et al., 2006). Within the scope of this work, we therefore assume the PBTI threshold voltage drift to behave like NBTI in terms of absolute value, distribution and annealing.

3 Modeling and SRAM metrics

For modeling, we use the following approach: ΔV_{th} together with Sensitivity$_{\mathrm{SRAM}}$ results in Degradation$_{\mathrm{performance}}$.

The transistor parameter drift ΔV_{th} is the direct result of the NBTI and PBTI effects, as described in the previous section. Together with the sensitivity of the SRAM circuit Sensitivity$_{\mathrm{SRAM}}$, it results in a degraded performance of the SRAM cell Degradation$_{\mathrm{performance}}$. The schematic of the 6T-SRAM cell that is used for this work is shown in Fig. 3.

While “performance” at e.g. logic gates would refer to delay, in memories it refers to some metrics that describe the “quality” of the memory cell. Three standard metrics are described in the next sections.

3.1 “Static Noise Margin (SNM)” or “Is it safe to read?”

When a data is read from memory, it is important that the cell does not flip because of this read cycle. The metric determining how safe is the read operation is called SNM(read). It is measured in Volts and determines how much additional noise voltage is necessary to flip the cell. The smaller this metric, the less stable the cell. SNM is also available for hold case and determines how much additional noise voltage is necessary to flip the cell under hold conditions. Under hold conditions, the cell is much more stable than under read conditions, which would make this SNM(hold) metric dispensable. The reason to keep this metric is that under hold conditions, it is common to reduce V_{DD} in order to minimize

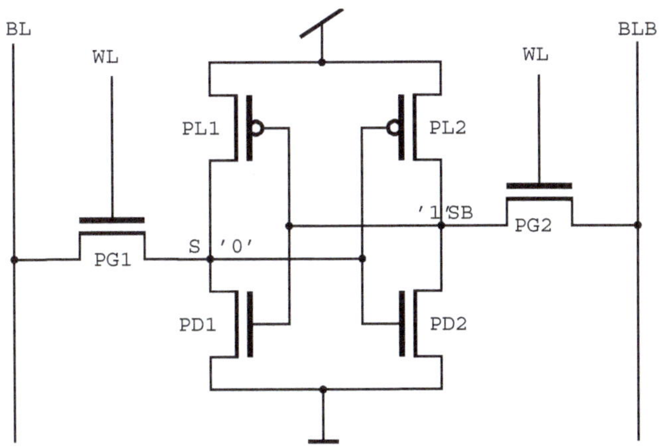

Fig. 3. Schematic of 6T SRAM circuit with naming conventions and assumed memory state: "0"on left side.

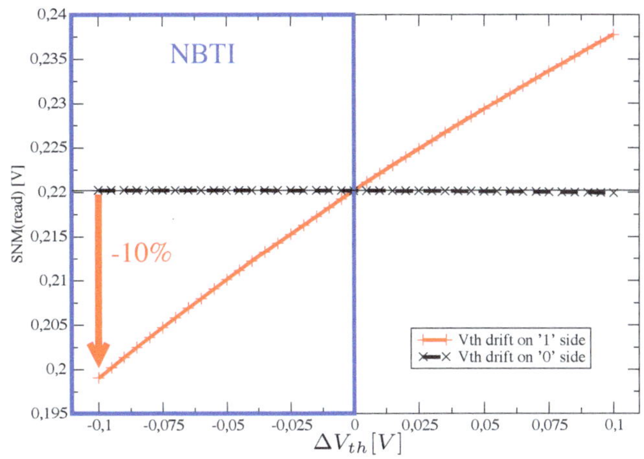

Fig. 4. Impact of threshold voltage drift of pullup transistor on read stability: 10% loss at -100 mV Vth drift

static power dissipation, which on its part reduces read stability as well. In this work, the focus is on SNM(read), because it is the most critical memory cell parameter.

3.2 "Write Margin" or "Is writing possible?"

Besides reading, it is also necessary for a memory cell that safe writing is possible. When writing conditions are applied to the memory cell, it must be sure that the cell will flip accordingly. This metrics called "Write Margin" determines how easy is writing, i.e. at which voltage applied to the according bitline the cell will flip. The higher the flip-voltage, the easier a zero can be written into the cell from the "low" BL.

3.3 "I(read)" or "How fast is reading?"

Speed, or in other words access time, is mainly determined by the current flowing in the channel of the access transistors. If it allows big current during the read cycle, faster discharging of the bitline capacity is possible. The bigger this metric "I(read)", the faster a memory cell can be read.

3.4 Safe reading or fast and writable?

All the mentioned SRAM metrics have opposite requirements. Either a cell can be optimized to safe reading or fast and easy writing, both optimizations are not possible together. For example, a cell which is optimized to safe reading will be characterized by a small access transistor and a big pulldown transistor to keep the output of the voltage divider (formed by the access and pulldown transistor) low. But this automatially slows down the cell, because fast access would require a wide access transistor. Additionally, it is getting harder to write, because this would require a small voltage drop across the access transistor.

In summary, the critical point in SRAM design besides minimizing the cell area is adjustment of the bias point as a tradeoff between readable, writeable and speed. Ongoing downscaling, especially in SRAM design, lets the margin get smaller. A back door for this increasing problem would be alternative SRAM circuit design, e.g. with more than 6 transistors, which of course needs more area per cell.

4 Sensitivity of SRAM circuit to BTI degradation

In this section, the sensitivity of the 6T SRAM circuit (Fig. 3) to BTI degradation of single transistors is examined. Since BTI degradation mainly causes V_{th} shifts, the impact of drifted threshold voltages is examined. The following analysis is based on hypothetical V_{th} drifts.

4.1 Sensitivity to NBTI-degraded pullup transistor

We present the consequence of ± 100 mV V_{th} drift of one of the pMOS pullup transistors to SNM(read). This, as well as all following simulations, is done with a Spice-based simulation tool. DC simulations were performed with varying threshold voltage of PL1 or PL2. We see that -100 mV V_{th} drift result in approx. 10% SNM loss (Fig. 4). Two curves exist in the graph because of two states of the memory cell: if the threshold voltage drift is applied to the transistor on side where the "0" is stored (PL1), the dashed line results: no influence on the read stability. On the other hand, if the threshold voltage drift is applied to the side where the "1" is stored (PL2), the solid line results, which means reduction of the read stability. Of course, one cannot choose the better memory state, because the user operation is deciding. As a consequence, the worse result has to be chosen in these graphs.

Altogether, NBTI in the pullup transistor makes an SRAM cell more vulnerable to destructive reads, but better writeable

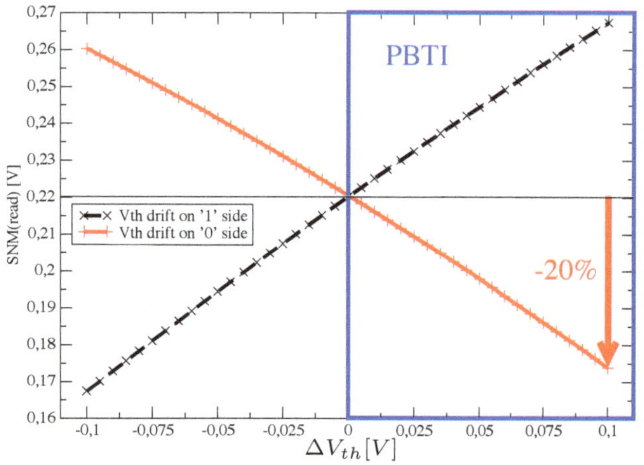

Fig. 5. Impact of threshold voltage drift of pulldown transistor on read stability: 20% loss at +100 mV V_{th} drift.

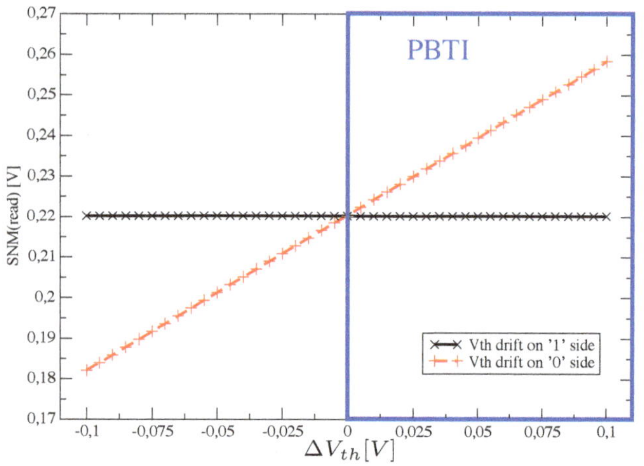

Fig. 6. Impact of threshold voltage drift of access transistor on read stability: no change for worst case.

in one memory state. It has no effect on access time since it does not change the read current.

4.2 Sensitivity to PBTI-degraded pulldown transistor

Again, V_{th} drifts up to ±100 mV are applied. This time, one pulldown transistor (PD1 or PD2) gets varied, and +100 mV V_{th} drift result in 20% SNM loss. So the influence of V_{th} on read stability is approx. double for pulldown compared to pullup, i.e. if PBTI will be in the same range of V_{th} drift, the effect on read stability will be twice as high (Fig. 5).

4.3 Sensitivity to PBTI-degraded access transistor

Analysis of threshold voltage drift of the access transistor showed improvement of read stability for one memory state, but no worst case change (Fig. 6).

However, a weak access transistor decreases both the write level as well as the read current.

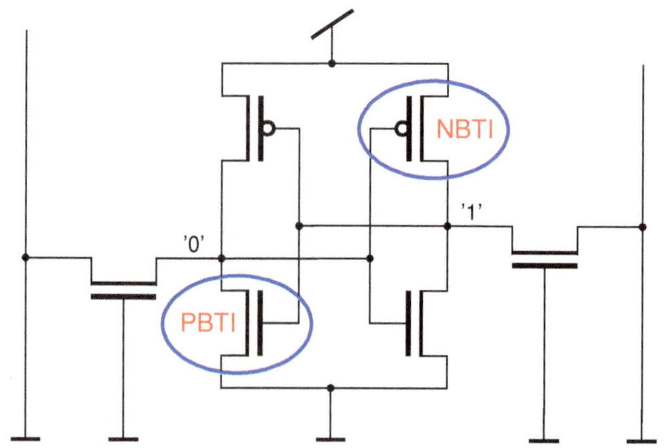

Fig. 7. NBTI stress on pullup and PBTI stress on pulldown during hold results in an asymmetrical cell.

5 BTI stress during normal SRAM operations

Now it is examined what BTI stress effects arise from normal SRAM operations that are namely hold, read and write.

5.1 Hold

During hold, data is kept in the feedbacked inverter-pair. During this period, PL2 experiences NBTI stress conditions (compare Figs. 1 and 3). Furthermore, the pulldown transistor on the opposite side (PD1) experiences PBTI stress conditions (compare Figs. 2 and 3). The access transistors do not see PBTI stress conditions, since they are in non-conducting state. Summing up, this means that for the complete time of data hold, one pullup and its opposite pulldown are affected by BTI (Fig. 7), and this state can last for years!

5.2 Read

In a read cycle, data is extracted from the memory cell and transferred to the bitlines via the access transistors. During this cycle, the situation is almost the same as during hold. Additionally, there might be a time when the access transistor on the "0" side experiences PBTI stress conditions, after the bitline got discharged, and the wordline is still open. But this is not regarded to be critical for the access transistor.

5.3 Write

During a write cycle, two cases must be distinguished: either the same data is written to the cell that was in memory before. Then the access transistor on the "0" side of the cell experiences PBTI stress for the cycle time, additionally to the hold stress for the pullup and pulldown. Or the cell flips because of the write, then the access transistor on the newly written "0" side only sees PBTI stress for the time after the cell has flipped. The pullup and pulldown stress are changing

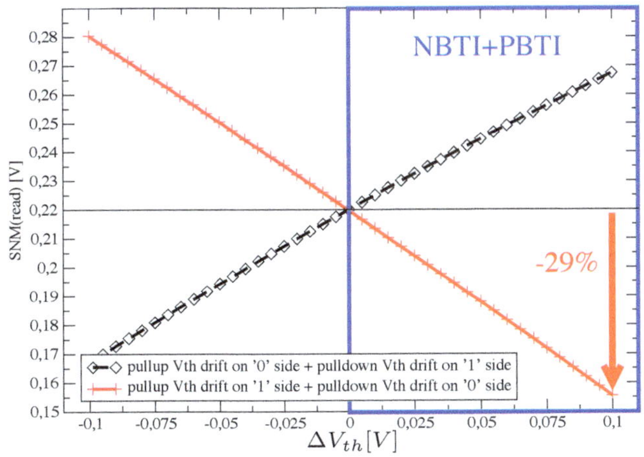

Fig. 8. Impact of threshold voltage drift of pullup- and opposite pulldown-transistor on read stability: 30% loss at ±100 mV drift.

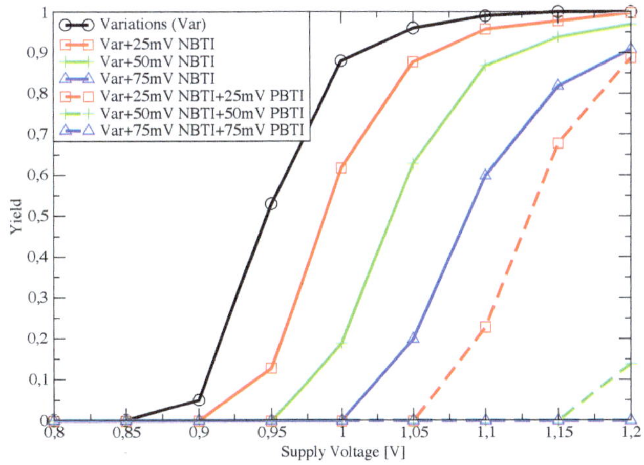

Fig. 9. Yield including global and local variations after long hold of one data with NBTI and PBTI degradation for 256 kBit cell array.

to the other pullup/pulldown pair, respectively. In summary, the write cycle also is not considered to be critical.

6 Long hold followed by direct read: the worst case BTI scenario for SRAM

The last section was pointing out: the hold state is the most critical one for BTI impact on 6T SRAM, because during the complete data hold phase, one pullup and one pulldown both experience full BTI stress conditions. It is now additionally taken into account that threshold voltage drift during or directly after stress is maximal, as there is no time for BTI annealing. The conclusion is that long hold of one data, followed by direct read, is the most critical BTI scenario: during this read, there is the biggest threat of destructive reading.

6.1 NBTI and PBTI effects are adding

Simulations have shown 10% and 20% loss of read stability for pullup/pulldown drifts, respectively. Now what happens during "hold" state, when both pullup and pulldown experience BTI stress? In Fig. 8 one can see that both impacts are adding! In other words: −100 mV pullup drift together with +100 mV pulldown drift result in almost 30% loss of read stability. This addition of the two effects can be explained with the small threshold voltage drifts relative to V_{DD}, which makes linearizing and therefore addition possible. If it is now additionally taken into account that

1. single devices can have more than 100mV V_{th} drift due to the wide BTI distribution and that
2. all devices can have more than 100mV V_{th} drift due to a direct read access without time for BTI annealing,

then it gets clear that these both effects together might result in a yield deterioration of SRAM over time.

6.2 Variations in manufacturing

One additional item was completely neglected so far: the unavoidable variations during manufacturing, which can be divided into global and local variations. While the degradation effects occur over long time of usage, the variations exist right after the production process. Global variations occur because the wafers cannot be treated exactly identically: they have slightly different doping, the mask will be slightly misaligned in the stepper etc. But modern processing technologies allow reducing these irregularities. In the scope of this work we regard a global variation according to the width of 2σ.

Local variations, however, are more severe for SRAM because of the huge number of memory cells in one array. For 1 MBit (2^{20}=1 048 576 cells), the variation corresponds to approx. 5,2σ. This is why in the scope of this work, we regard a local variation with a maximum value according to the width of 5σ.

6.3 Yield considerations

The global variations of 2σ and the local variations of 5σ described in Sect. 6.2 are now additionally applied to the 6T core cell. In some devices, the NBTI and PBTI drifts will be reduced because of the variations, i.e. the threshold voltage drifts due to variations and degradation effects will eliminate each other. But in some cells, the threshold voltage drifts will be boosted. These cells are best candidates for destructive reads, because their SNM(read) value will drop dramatically. In this yield consideration, cells with SNM(read)<5% V_{DD} are considered as failing cells.

These calculations were done with the "Worst Case Distance" algorithm (Antreich et al., 1994), which gives the same result as Monte Carlo simulations, but with less computing effort. Figure 9 shows the yield in terms of "all

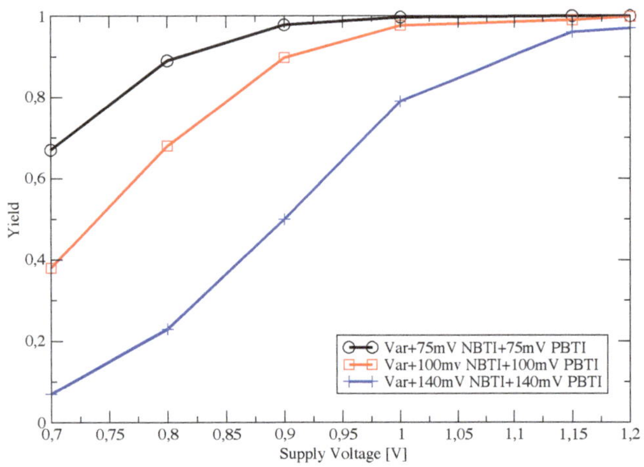

Fig. 10. Yield including global and local variations after long hold of one data with NBTI and PBTI degradation for a single Bit cell.

memory cells working in a 256 kBit array". If at least one cell in this complete array does not meet the specification, the whole array is considered as failing. One can see that lowering V_{DD} generally decreases the yield, because SNM(read) is reduced. But the main information is that variations and NBTI together do not have huge impact on SRAM. The combination of variations and NBTI plus PBTI, however, is much worse: even at $V_{DD,\mathrm{nom}}$, the yield for 25 mV BTI stress each is already below 90%. And the line for each 75 mV BTI stress does not even appear in the graph, because it is always zero! The yield in a 256 kBit array for Variations+NBTI+PBTI is so low that in Fig. 10, it is drawn for a single cell. Then the term "yield" is not the probability of all cells working in the array anymore, but the probability that this single cell fails. This can then be used to design appropriate cell circuits and redundancy schemes, so of course in realistic SRAM designs sufficient yield will still be obtained.

7 Conclusions

The impact of global and local variations and NBTI on 6T SRAM core cell are not negligible, but still do not dramatically impact the SRAM design. This is because the two pMOS pullup transistors are the only ones to be affected by NBTI. The sensitivity of the SRAM circuit to small threshold voltage drifts of these transistors is not excessive. The situation including PBTI is much worse: this effect works on the nMOS pulldown as well as the access transistors, and the sensitivity especially of the pulldown transistor is much higher: with the same threshold voltage drift, the impact on read stability is twice as high for pulldown compared to pullup devices.

Moreover, during hold state of the memory, the two effects are adding, so that the cell weakens drastically in terms of read stability. The cell loses approx. 30% for 100 mV NBTI and PBTI induced threshold voltage drift on pullup and pulldown transistor. Here is the biggest threat of undesired write during a read cycle, which is equal to data loss. BTI behavior like distribution and annealing also plays an important role, especially if a long hold is directly followed by a read cycle, because then annealing is not possible.

In future technologies including high-k materials, it is very important to consider these effects during SRAM memory design in order to be able to produce long-life memory cells.

Acknowledgements. This work has been supported by the German Ministry of Education and Research (BMBF) within the project "Honey" (Project ID 01M3184A). The content is the sole responsibility of the authors.

References

Antreich, K. J., Graeb, H. E., and Wieser, C. U.: Circuit analysis and optimization driven by worst-case distances, IEEE T. Comput. Aid. D., 13(1), 57–71, 1994.

Fischer, Th., Olbrich, A., Georgakos, G., Lemaitre, B., and Schmitt-Landsiedel, D.: Impact of Process Variations and long-term degradation on 6T-SRAM cells, Adv. Radio Sci., 2007.

Fischer, Th., Amirante, E., Hofmann, K., Ostermayr, M., Huber, P., and Schmitt-Landsiedel, D.: A 65nm teststructure for the analysis of NBTI induced statistical variations in SRAM transistors, Proceedings of the European Solid State device research conference, 51–54, 2008.

Heinrigs, W., Reisinger, H., Gustin, W., and Schluender, C.: Consideration of revocery effects during NBTI measurements for accurate lifetime predictions of state-of-the-art pMOSFETs Reliability physics symposium, 288–292, 2007.

Seevinck, E., List, F., and Lohstroh, J.: Static Noise Margin Analysis of MOS SRAM Cells, IEEE J. Solid-St. Circ., 22, 748–754, 1987.

Zafar, S., Kim, Y. H., Narayanan, V., et al.: A comparative study of NBTI and PBTI (Charge Trapping) in SiO2/HfO2 Stacks with FUSI, TiN, Re Gates, Proceeding of the Symposium on VLSI Technology, 23–25, 2006.

23

Efficiency modeling for MHz DCDC converters at 40 V input voltage range

J. Wittmann, A. Seidel, and B. Wicht

Robert Bosch Center for Power Electronics, Reutlingen University, Reutlingen, Germany

Correspondence to: J. Wittmann (juergen.wittmann@reutlingen-university.de)

Abstract. Size and cost of a switched mode power supply can be reduced by increasing the switching frequency. This leads especially at a high input voltage to a decreasing efficiency caused by switching losses. Conventional calculations are not suitable to predict the efficiency as parasitic capacitances have a significant loss contribution. This paper presents an analytical efficiency model which considers parasitic capacitances separately and calculates the power loss contribution of each capacitance to any resistive element. The proposed model is utilized for efficiency optimization of converters with switching frequencies > 10 MHz and input voltages up to 40 V. For experimental evaluation a DCDC converter was manufactured in a 180 nm HV BiCMOS technology. The model matches a transistor level simulation and measurement results with an accuracy better than 3.5 %. The accuracy of the parasitic capacitances of the high voltage transistor determines the overall accuracy of the efficiency model. Experimental capacitor measurements can be fed into the model. Based on the model, different architectures have been studied.

1 Introduction

Switched mode power supplies can achieve high efficiency in power management systems but require passive filter elements like inductors, transformers and capacitors which mainly dominate the system size and thus the costs. Especially at low power, high volume applications up to 10 W a high degree of integration is important. This can be achieved by increasing the switching frequency as this scales down the passive components. In many cases, this also has a positive effect on reliability as it relaxes the requirements on assembly and interconnect technology. Today, power supplies with a switching frequency up to 2 MHz are common, but the passive components remain dominant. With a further increase of the frequency, the switching losses significantly reduce the efficiency of the regulator. For low input voltages of less than 10 V, the efficiency still remains acceptable. However, many applications require an efficient power supply with an input voltage range up to 100 V. Higher input voltages significantly impacts the switching losses. Such high-voltage applications include growth areas like energy conversion from renewables. Moreover, in cars the battery voltage can go up to >40 V. In e-mobility systems, the medium board net voltage is defined to be 48 V.

To analyze the efficiency of converter architectures for high switching frequencies and high input voltage range, standard efficiency calculations turn out to be insufficient because the influence of parasitic capacitances becomes dominant, but it is modeled poorly and it is not possible to extract the root causes of the losses. As proposed by Wang et al. (2010), Wang and Huang (2011), the losses caused by parasitic capacitances in all current paths need to be considered. This approach is extended in this paper. This way, both critical loss contributions (root causes) and loss elements (locations) in an architecture can be obtained. The losses are calculated for each switching phase separately. There are two main benefits of the proposed model: (1) design parameters, e.g. the on-state resistance, can be optimized for a given architecture, (2) various converter architectures can be studied and compared in terms of efficiency by comparing the particular circuit elements. The focus in this paper is on a supply up to 40 V, 10 MHz and $V_{out} = 5$ V. The proposed methodology is suitable also for circuits operating at higher input voltages and higher frequencies as long as parameters, such as the parasitic capacitances, can be provided precisely enough.

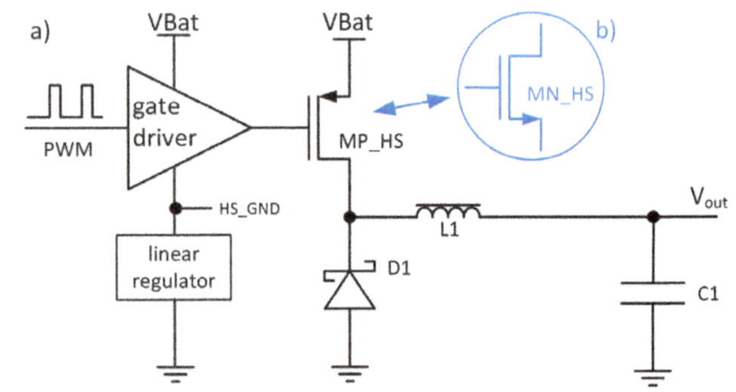

Figure 1. Overview of a buck converter in various architectures. **(a)** with PMOS high side switch; **(b)** with NMOS high side switch.

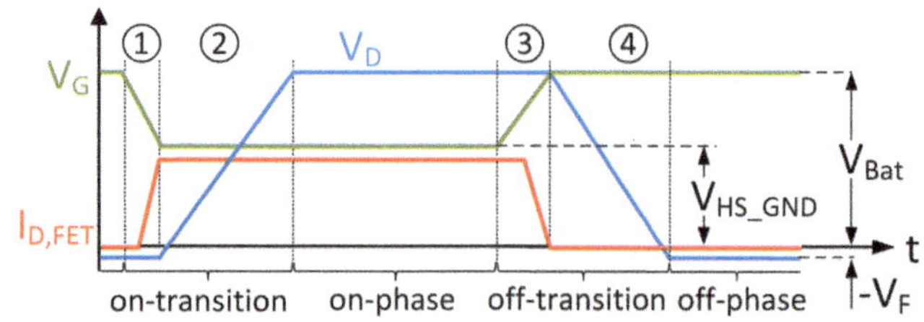

Figure 2. Separation of the switching transients into four phases.

The model has been developed in first place for an asynchronous buck converter architecture with a PMOS high side power FET as shown in Fig. 1a. The regulation circuit of the converter creates a pulse width modulated signal (PWM) which controls the input to output voltage ratio by regulating the duty cycle Wittmann and Wicht (in press). To fully open the PMOS FET MP_HS during the on-phase, the gate driver is connected between the supply voltage V_{Bat} and the HS_GND rail. A linear regulator regulates the HS_GND to $(V_{\mathrm{Bat}} - 5\,\mathrm{V})$.

The PMOS high side switch can be replaced by an NMOS high side switch (Fig. 1b) which requires a boot strapping circuit to generate the gate overdrive. The HS_GND for the NMOS switch is typically connected to the source of the NMOS switch.

The efficiency model for the buck converter is derived in Sect. 2. A discussion of the efficiency results and the loss elements including an architecture comparison is given in Sect. 3.

2 Efficiency model

To allow a separation of the power losses for each element of the converter, all current paths during the switching transients have to be analyzed. The charge or discharge of a parasitic capacitance in a switching phase contributes to the losses in each resistive element along the current path. Figure 2 shows the idealized drain current $I_{\mathrm{D,FET}}$, the gate voltage V_{G} and the drain voltage V_{D} of the PMOS power FET over one switching period. In order to obtain a single charge or discharge event for each capacitance per phase, the switching transitions are separated into four phases. A very fast gate driver is assumed and thus the rise at the drain voltage until the gate is fully charged and discharged, respecively, is not significant Wittmann and Wicht (in press)Wittmann et al. (2012).

Figure 3 shows the current paths which have been derived for each switching phase. In phase 1 and phase 3, the gate is discharged and charged, respectively, to turn on and off the power FET. The capacitances C_{SG} and C_{GD} are charged and discharged by the gate voltage change ΔV_{G}. In phases 2 and 4, the drain voltage sees a swing of approximately V_{Bat}. In these phases, the parasitic capacitances in the power FET C_{GD} and C_{DB} (C_{DS} included in = DB for simplification) and the junction capacitance of the Schottky diode are charged and discharged. The gate driver is designed strong enough such that the miller-plateaus in phases 2 and 4 are not significant.

The modeling approach will be demonstrated below for the discharge of C_{GD} by ΔV_{G} during phase 1 (Fig. 3a). The charge ΔQ_{CGD} on C_{GD} also flows through the gate driver's low side and through the linear regulator to ground on one side and through the Schottky diode (I_{DIO}) on the other terminal of C_{GD}. During discharge, the voltage at the gate driver output falls from 5 V to 0 V while the high side ground voltage is constant at $V_{\mathrm{HS\_GND}} = V_{\mathrm{Bat}} - 5\,\mathrm{V}$. The charge $\Delta Q_{\mathrm{CGD}} = \Delta V_{\mathrm{G}} \cdot C_{\mathrm{GD}}$ thus creates the energy loss $w_{\mathrm{lr}} = \Delta Q_{\mathrm{CGD}} \cdot V_{\mathrm{HS\_GND}}$ in the linear regulator when the charge is discharged to ground. The energy loss in the gate driver is $w_{\mathrm{driver}} = 0.5 \cdot \Delta Q_{\mathrm{CGD}} \cdot \Delta V_{\mathrm{G}}$ and in the Schottky diode $w_{\mathrm{d}} = \Delta Q_{\mathrm{CGD}} \cdot V_{\mathrm{F}}$ with forward voltage V_{F}.

These calculations can be done for all capacitances in all phases to obtain the loss contributions of all parasitic capacitances for each circuit element. This is summarized in Table 1, which shows the root causes and locations of the losses in all phases during a switching period. By summing up the losses per line, the loss contributions to the particular resistive elements (i.e. loss location) can be determined. Respectively, the columns yield the losses caused by each capacitance per phase. If a parasitic capacitance is discharged through the inductor $L1$ (e.g. C_{J}, C_{GD}, C_{DB} in phase 4) to the output, no losses occur, since the energy is stored in the inductor or can be used as output current. The losses caused by the parasitic capacitances are time invariant and occur once in a switching period. The according power dissipation thus increases linearly with the switching frequency.

In the turn-on transition phase (Fig. 2), the transition losses are calculated in the conventional way by $P_{\mathrm{SWon}} = 0.5 \cdot I_{T\mathrm{begin}} \cdot (V_{\mathrm{Bat}} + V_{\mathrm{F}}) \cdot t_{\mathrm{SWon}} \cdot f_{\mathrm{sw}}$ with the assumption that the drain voltage decreases linearly. t_{SWon} is the duration of the transition phase (see phase 2 in Fig. 2) and $I_{T\mathrm{begin}}$ the source-drain current through the transistor at the beginning of phase 2. The analytical calculation of the transition time t_{SWon} is usually not suitable, as it depends on various

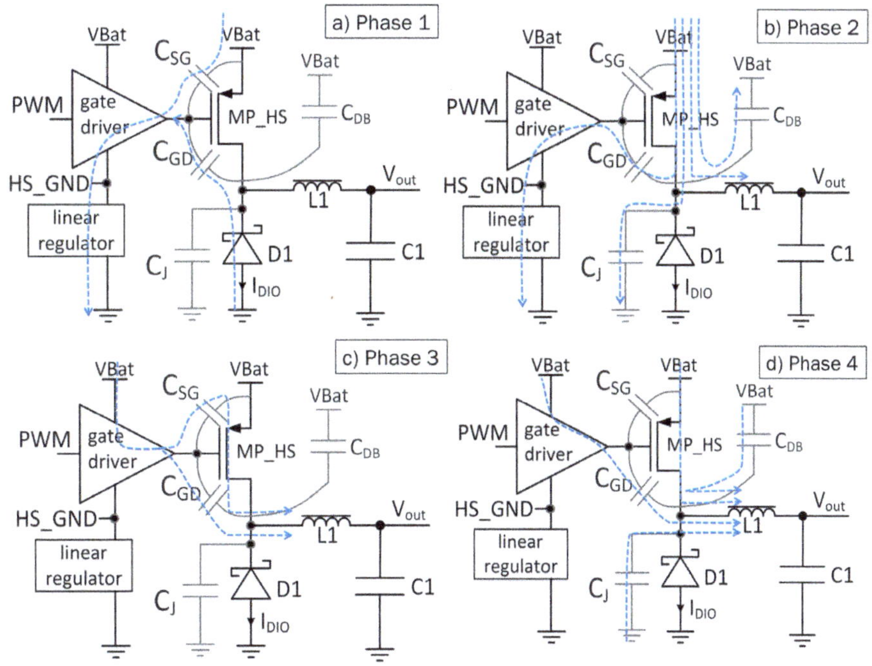

Figure 3. Current paths in a converter with PMOS switch.

Table 1. Location and root causes of losses due to parasitic capacitances in the FET and the Schottky Diode.

Loss location	Loss cause			
	Phase 1	Phase 2	Phase 3	Phase 4
Gate driver pull-down path	C_{GD}, C_{SG}			
Gate driver pull-up path			C_{DG}, C_{SG}	
Linear regulator	C_{GD}, C_{SG}	C_{GD}		
Schottky diode	C_{GD}			
FET R_{DSon}		C_{GD}, C_{DB}, C_J Transition losses, static losses		

parameters including the R_{DSon} transition, the level of the miller plateau and the stability of the HS_GND. Therefore it is appropriate to take t_{SWon} from transistor level simulation or experimental measurement. For the architecture comparison, t_{SWon} was taken from the transistor level simulation. Measurements show that t_{SWon} is roughly independent of the load current I_{out}. Additional losses due to reverse recovery are extracted in a transistor level simulation and scaled with the output current in the efficiency model.

During the turn-off transition phase (see phase 4 in Fig. 2), no losses occur as the load current is fully provided by the discharge current of the parasitic capacitances. The losses caused by the output current in the FET during the charging of the gate in phases 1 and 3 are negligible due to the fast discharge time with a strong gate driver. The static losses in the on-phase and off-phase occur when the power FET is fully turned on during the time t_{on} and fully turned off during the time t_{off}. During the on-phase, the power loss is determined by the average output current I_{out} and the on-state resistance R_{DSon} and thus is $P_{on} = R_{DSon} \cdot I_{out}^2 \cdot t_{on} \cdot f_{sw}$. During t_{off} the current flows through the forward conducting Schottky diode and the power loss is $P_{on} = V_F \cdot I_{out} \cdot t_{off} \cdot f_{sw}$. The losses due to reverse recovery in the Schottky Diode can be neglected.

3 Results and circuit architecture comparison

To verify the efficiency model, an asynchronous buck converter with a DEMOS p-channel HS FET was implemented in a 180 nm high-voltage BiCMOS technology. Figure 4 shows a comparison of the simulated total efficiency at $V_{Bat} = 18\,V$ (traces a) and c) in Fig. 4) and $V_{Bat} = 40\,V$ traces (b) and d) in Fig. 4) over the switching frequency. The traces a) to d) show a very good matching between the transistor level simulation and the efficiency model. Trace e) shows the efficiency from a real measurements at the same operation point as the simulation in a). The deviation between

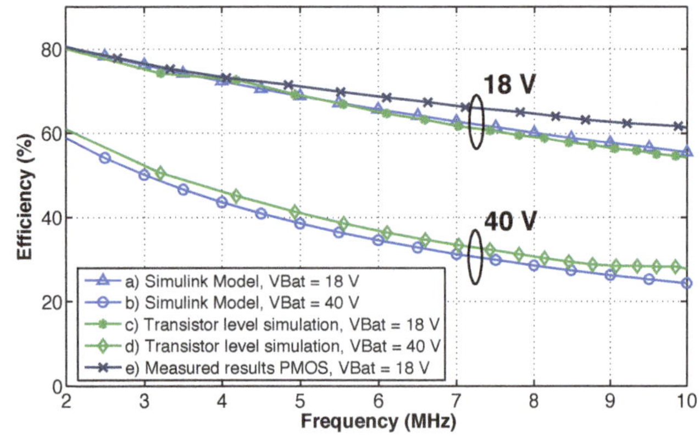

Figure 4. Efficiency comparison of circuit simulation, model and measured results over frequency at $V_{\text{Bat}} = 18\,\text{V}$, $40\,\text{V}$ and $I_{\text{out}} = 200\,\text{mA}$ (right).

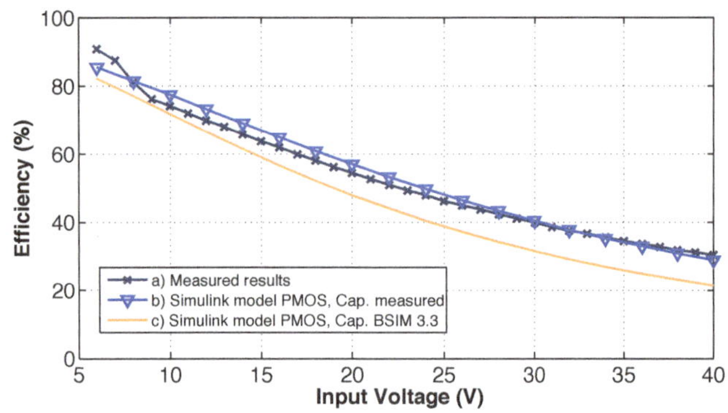

Figure 5. Efficiency comparison of model and measured results with parasitic capacitance extraction from the BSIM 3.3 model and measured capacitance extraction over input voltage at $f = 10\,\text{MHz}$ and $I_{\text{out}} = 200\,\text{mA}$.

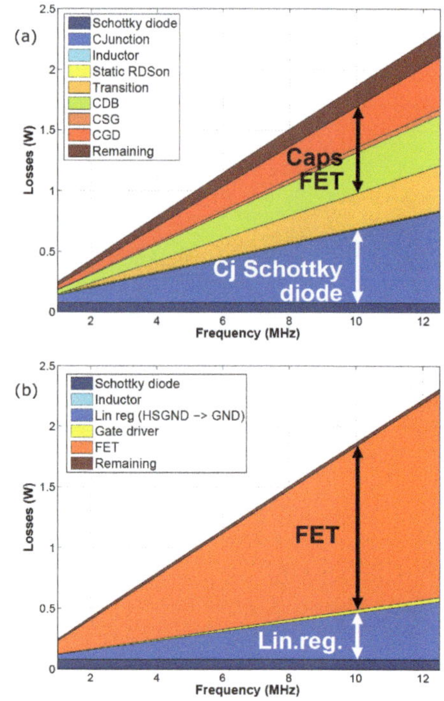

Figure 6. **(a)** Loss causes and **(b)** loss locations of an asynchronous buck converter with PMOS high side switch ($V_{\text{Bat}} = 40\,\text{V}$, $I_{\text{out}} = 200\,\text{mA}$, $R_{\text{DSon}} = 1.4\,\Omega$).

the curves appears mostly due to the known limited accuracy in the modeling of parasitic capacitances of high voltage transistors in the BSIM 3.3 model. This will be addressed by more advanced models like HiSIM-HV in future Mattausch et al. (2010). To provide more accurate capacitance values to the efficiency model, the values of the parasitic capacitances were measured in the according operation regions of the FET (saturation, triode and cutoff) with a gate charge and output charge measurement. Figure 5 shows that the measured capacitances improve the accuracy of the efficiency simulation compared to the measured efficiency.

The results of a frequency simulation of the efficiency model of the asynchronous buck converter with a high side PMOS FET are shown in Fig. 6 for operating conditions of $V_{\text{Bat}} = 40\,\text{V}$, $I_{\text{out}} = 200\,\text{mA}$. The power FET was optimized by an efficiency simulation over the on-state resistance. The size of the FET resulted to be optimum at $R_{DSon} = 1.4\,\Omega$. All relevant loss causes of the particular elements are plotted over the frequency in Fig. 6a while the loss locations, i.e. the resistive elements in which the losses are dissipated, are shown in Fig. 6b. The loss causes labeled "Remaining" in Fig. 3 mainly include the losses caused by the first five stages of the six stage gate driver and its current flowing through the HS_GND regulator. The capacitances of the PMOS FET cause losses, which are in the range of the losses caused by the capacitance of the Schottky diode. At higher frequencies, the losses caused by the capacitances C_{GD} and C_{DB} and the transition losses in the FET are becoming dominant, while in the Schottky diode, the losses caused by its junction capacitance become superior compared to the forward losses. Analyzing the loss locations in Fig. 6b, it can be seen that most of the losses caused by parasitic capacitances are occurring in the resistive channel of the FET. These results lead to the assumption that the efficiency can be significantly increased by using an NMOS power FET, which has either a smaller on-resistance compared to a PMOS FET with the same size, or the parasitic capacitance can be reduced with a smaller transistor size keeping the on-resistance equal.

An architecture comparison is done by replacing the PMOS FET by an NMOS FET of half the size (see Fig. 1b) with an on-state resistance of $R_{\text{DSon}} = 0.8\,\Omega$. For a good model accuracy, the R_{DSon} and the parasitic capacitances of an implemented LDMOS NMOS FET were measured and fed to the model, as described in Sect. 2. The simulation of the loss causes with NMOS FET at $f = 10\,\text{MHz}$ over V_{in} is shown in Fig. 7, in this case as a function of the

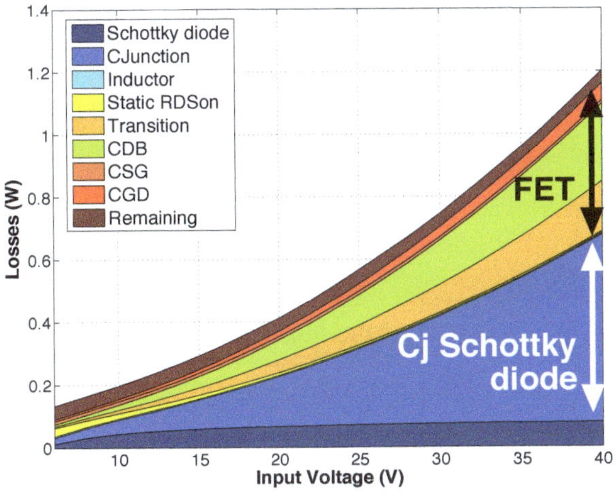

Figure 7. Loss causes of an asynchronous buck converter with NMOS high side switch over the input voltage V_{Bat} ($f_{\mathrm{sw}} = 10\,\mathrm{MHz}$, $I_{\mathrm{out}} = 200\,\mathrm{mA}$, $R_{DSon} = 0.8\Omega$).

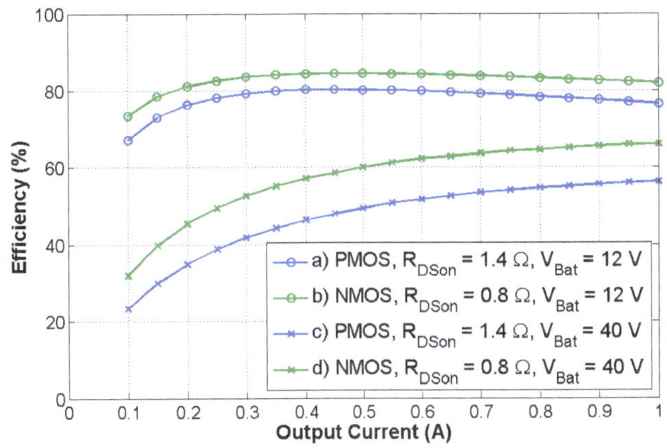

Figure 8. Comparison of the efficiency with PMOS and NMOS power FET over the output current I_{out} ($V_{\mathrm{Bat}} = 12\,\mathrm{V}, 40\,\mathrm{V}$, $f = 10\,\mathrm{MHz}$)

input supply voltage. The losses in the FET are significantly reduced. The current through the HS_GND contributes to the output current and, hence, the corresponding losses through the HS_GND regulator are eliminated. The transition losses are decreased due to lower R_{DSon}. The losses caused by the Schottky diode remain equal. The total losses are reduced by more than 30 % at $V_{\mathrm{Bat}} = 40\,\mathrm{V}$.

Figure 8 shows a comparison of the overall efficiency of the converter with a PMOS to an NMOS FET over the output current at a switching frequency of 10 MHz, $V_{\mathrm{Bat}} = 12\,\mathrm{V}$ and $V_{\mathrm{Bat}} = 40\,\mathrm{V}$. In all architectures, the efficiency first increases, since the output power increases more than the losses. With higher output current, the losses in the R_{DSon} get dominant at low input voltage. Due to the smaller ratio of R_{DSon} to the parasitic capacitances, the NMOS high side FET leads in average to 6% higher efficiency (curves a) vs. b)) at $V_{\mathrm{in}} = 12\,\mathrm{V}$ and to a 10 % higher efficiency (curves c) vs. d)) at $V_{\mathrm{in}} = 40\,\mathrm{V}$ compared to a PMOS FET.

4 Conclusions

By considering all parasitic capacitances and analyzing the current paths in the separate switching phases, the model is suitable to perform different efficiency analysis in a voltage range up to 40 V especially at high switching frequency. The possibility to extract all loss causes and locations separately allows to determine the critical elements. In particular, the model approach allows to (1) optimize design parameters for a given architecture, (2) to study and compare various converter architectures. To verify the model, an asynchronous buck converter with PMOS high side FET was manufactured. A model accuracy of around 3.5 % was achieved up to 10 MHz compared to measurements. It was demonstrated that the limited modeling accuracy of parasitic capacitances of high voltage transistors in the BSIM 3.3 can be circumvented by feeding measured capacitance values into the model.

A comparison based on the efficiency model was presented for an asynchronous buck converter switch resulting in a higher efficiency by using an NMOS high side switch. As a general conclusion, with higher frequency smaller FETs achieve a higher efficiency.

Edited by: D. Killat

References

Mattausch, H. J., Sadachika, N., Yokomichi, M., Miyake, M., Kajiwara, T., Oritsuki, Y., Sakuda, T., Kikuchihara, H., Feldmann, U., and Miura-Mattausch, M.: Power/HVMOS Devices Compact Modeling, chap. 2: HiSIM-HV: A Scalable, Surface-Potential-Based Compact Model for High-Voltage MOSFETs, Springer, 2010.

Wang, X. and Huang, A.: Capacitor energy variation based designer-side switching losses analysis for integrated synchronous Buck converters in CMOS technology, in: Applied Power Electron. Conf. and Exposition (APEC), 2011 Twenty-Sixth Annual IEEE, 1130–1137, doi:10.1109/APEC.2011.5744736, 2011.

Wang, X., Park, J., Van Brunt, E., and Huang, A.: Switching losses analysis in MHz integrated synchronous Buck converter to support optimal power stage width segmentation in CMOS technology, in: Energy Conversion Congress and Expos. (ECCE), 2010 IEEE, 2718–2724, doi:10.1109/ECCE.2010.5618055, 2010.

Wittmann, J. and Wicht, B.: MHz-Converter Design for High Conversion Ratio, in: 2013 IEEE 25rd International Symposium on Power Semiconductor Devices and ICs (ISPSD), 2014.

Wittmann, J., Funk, T., and Wicht, B.: Gate Driver Design for Fast Switching DC-DC Converters, in: 48th MPC Workshop, Aalen, 17–24, 2012.

24

Investigation of carbon nanotube antennas using thin wire integral equations

N. Fichtner, X. Zhou, and P. Russer

Institute for High Frequency Engineering, Technische Universität München, Arcisstr. 21, 80333 München, Germany

Abstract. In this paper the characteristics of small carbon nanotube (CNT) dipole antennas are investigated on the basis of the thin wire Hallén integral equation (IE). A surface impedance model for the CNT is adopted to account for the specific material properties resulting in a modified kernel function for the integral equation. A numerical solution for the IE gives the current distribution along the CNT. From the current distribution the antenna driving point impedance and the antenna efficiency are computed. The presented numerical examples demonstrate the strong dependence of the antenna characteristics on the used material and show the limitations of nanoscale antennas.

1 Introduction

The electromagnetic behavior of nanoscale circuits, transmission lines and antenna elements is of special interest for future highly integrated electrical circuits operating at extremely high clock rates. Especially the wireless transmission of signals between nanoscale circuit components and the microelectronic periphery is a key issue in present-day nanoelectronics. Carbon nanotubes (CNT) are interesting components for further miniaturization of electrical circuits and can be used as signal transmission lines as well as for antenna elements. Their small dimensions however demand a special treatment. Usually in antenna theory the metallic conductors in the investigated structures are treated as perfect electric conductors (Balanis, 1997). This assumption does not hold at nanoscale dimensions. Here the inherent material losses are considerable and impose limitations to the used materials like copper or aluminum for components at the nanoscale. Carbon nanotubes (CNT) exhibit ballistic transport over length of several μm and have a relaxation time which is approximately 50 times greater than in copper (Jishi et al., 1993). This offers the applicability of CNTs for the transmission and reception of electromagnetic signals.

2 Theory

2.1 Integral equation method

An incident electromagnetic wave induces an electric current $I_z(z')$ flowing in the thin wire dipole element (transverse currents are neglected because of the small diameter). From the current distribution and the excitation the antenna driving point impedance and the antenna efficiency can be computed. The two classical integral equations for linear antennas are the Pocklington's and Hallén's integral equations (Balanis, 1997). As Hallén's IE exhibit better stability properties a numerical solution thereof is less error-prone and will be used in the following (Sarkar, 1983). Hallén's integral equation is given by

$$\int_{-l/2}^{+l/2} K(z,z')I_z(z')dz' = \frac{-j}{Z_{F0}}\left[c_1\cos(kz) + c_2\sin(k|z|)\right] \tag{1}$$

with k as the wavenumber. The integral kernel $K(z,z')$ will be discussed later in Sect. 2.2. For a delta gap excitation in the dipole center the constant c_2 is given by $c_2{=}V_0/2$ with V_0 as the driving voltage and c_1 is obtained from the condition $I_z(\pm l/2){=}0$. For a numerical solution of Eq. (1) a Method of Moments (MoM) approach is applied (Balanis, 1997). There the unknown current density $I_z(z')$ is replaced by a series of known expansion functions. As expansion functions pulse-functions

$$u_i(z) = \begin{cases} 1 & \text{for } z_i - \frac{\Delta z}{2} < z < z_i + \frac{\Delta z}{2} \\ 0 & \text{elsewhere} \end{cases} \tag{2}$$

are used with $\Delta z{=}l/N$. The coefficients of the expansion of $I_z(z')$ are chosen such that at specific sampling points

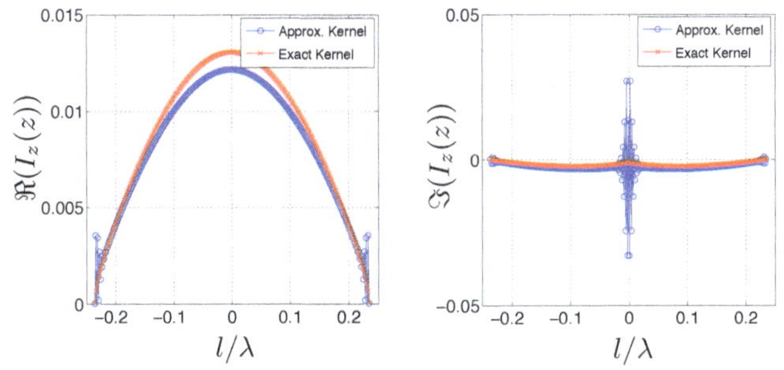

Fig. 1. Current distribution along the dipole for approx. kernel (blue line) and exact kernel (red line).

$z_i=(2i-1)\Delta z/2$ the IE in Eq. (1) is fulfilled (point matching). This results in a linear system of equations which can be solved by matrix inversion.

2.2 Modified kernel function

In CNTs the electric current is confined to a thin cylindrical shell and can be computed from the surface conductance σ (Slepyan et al., 1999). Hence the total current flowing in the CNT dipole with radius a is given by

$$\frac{I_z(z)}{2\pi a} = \sigma\left(E_z^s(z) + E_z^i(z)\right) \tag{3}$$

where E_z^s is the scattered electric field and E_z^i is the incident electric field. Combination of Eq. (3) with the Hallén integral equation yields

$$\left(k^2+\partial_z^2\right)\int_{-l/2}^{+l/2} K(z,z')I_z(z')dz' = j\omega\epsilon\left(Z_i' I_z(z) - E_z^i(z)\right) \tag{4}$$

where Z_i' is the impedance per unit length given by

$$Z_i' = \frac{1}{2\pi a\sigma} \tag{5}$$

In order to solve Eq. (4) in the same way then Eq. (1) the unknown current $I_z(z)$ on the right side of Eq. (4) must be eliminated. With the help of the Green's function for the differential operator $\left(k^2+\partial_z^2\right)$ the current term $I_z(z)$ can be written as an integral over the dipole length. Thus the new IE comprising the surface conductance is given by

$$\int_{-l/2}^{+l/2} K'(z,z')I_z(z')dz' = c_1\cos(kz) + c_2\sin(k|z|) + \frac{\omega\varepsilon}{2k}e^{-jk|z|}V_0 \tag{6}$$

with the modified integral kernel

$$K'(z,z') = K(z,z') + \frac{e^{-jk|z-z'|}}{2Z_{F0}Z_i'} \tag{7}$$

Unlike in Eq. (4) in Eq. (6) the current $I_z(z')$ is solely given under the integral with the modified integral kernel $K'(z,z')$ and can be solved in the same way than Eq. (1). Now both constants c_1 and c_2 must be chosen such that $I_z(\pm l/2)=0$ is fulfilled.

2.3 Kernel selection and stability

In computations with a large discretization number N, i.e. a small discretization length Δz, oscillations in the current distribution are observed near the dipole center and the dipole ends. Fikioris and Wu (2001) have shown that the current oscillations arise from the approximate kernel function $K_{\mathrm{app}}(z)$ if the number of basis functions N satisfies $N \gg l/a$. The exact and approximate kernels are

$$K_{\mathrm{ex}}(z,z') = \frac{1}{2\pi}\int_{-\pi}^{\pi}\frac{e^{-jk\sqrt{(z-z')^2+4a^2\sin^2\theta/2}}}{\sqrt{(z-z')^2+4a^2\sin^2\theta/2}}d\theta \tag{8}$$

and

$$K_{\mathrm{app}}(z,z') = \frac{e^{-jk\sqrt{(z-z')^2+a^2}}}{\sqrt{z^2+a^2}} \tag{9}$$

In Fig. 1 the current distribution along a 0.47λ long dipole with radius $a/\lambda=0.007$ and $N=1.5\ l/a$ is shown for the approximate and exact kernel. The driving point impedance for the particular cases are $Z_{\mathrm{in,app}}=37.3-j41\ \Omega$ and $Z_{\mathrm{in,ex}}=76.6-j42\ \Omega$. Thus the choice of the Kernel and the number of expansion functions N strongly affects the input impedance. Here the exact kernel is used at the cost of a additional integration which increases the computational effort.

3 Complex surface conductance

The complex surface conductance for a CNT was derived by Slepyan et al. (1999). Slepyan solved the single electron Boltzmann transport equation under consideration of the band structure and the impulse quantization in transverse direction. For small radii the CNT conductance can be approximated by

$$\sigma_{cnt} \cong -j\frac{2e_0^2 v_F}{\pi^2\hbar(\omega - j/\tau)} \tag{10}$$

where e_0 is the elementary charge and $v_F \approx 9.71\cdot10^5$ m/s is the Fermi velocity of graphene. The relaxation time τ at room temperature is about $\tau=1.4\cdot10^{-12}$ s. In order to compare the characteristics of a CNT dipole with a copper dipole the surface conductance for copper is needed. From Hanson (2005) we get

$$\sigma_{cu} = -j\frac{e_0^2 N_e^{2D}}{m_e(\omega - j/\tau_{cu})} \tag{11}$$

with N_e^{2D} is the number of electrons per m^2.

Table 1. Input impedance and efficiency η for perfect conducting (pc), CNT and copper dipole with length l=0.47 λ. The operating frequency is 100 GHz and N=51 for the half dipole ($N_e^{2D} \approx 1.93 \cdot 10^{19}$ 1/m²).

Radius a	Input impedance/Ω	Efficiency η/%
3 μm	$Z_{pc} = 67 - j19$	100
	$Z_{cnt} = 140 - j2575$	$1.6 \cdot 10^{-3}$
	$Z_{cu} = 75 - j1727$	$2.9 \cdot 10^{-2}$
30 nm	$Z_{pc} = 63 - j82$	100
	$Z_{cnt} = 950 - j7934$	$1.1 \cdot 10^{-3}$
	$Z_{cu} = 666 - j4733$	$2.6 \cdot 10^{-3}$
3 nm	$Z_{pc} = 62 - j109$	100
	$Z_{cnt} = 1819 - j10730$	$6.8 \cdot 10^{-4}$
	$Z_{cu} = 1978 - j7913$	$7.6 \cdot 10^{-4}$

4 Results

Firstly the input impedances and the achievable efficiencies are computed for dipoles at nanometer radii with a length of l=0.47 λ. Thus the surface conductances defined in Eqs. (10) and (11) are combined with the IE (6) and solved for the unknown current distribution $I_z(z)$. The input impedance is obtained from $Z_{in}=V_0/I_z(0)$ and the radiation efficiency is given by

$$\eta = \frac{R_{in}}{R_{in}+R_{hf}} \quad \text{with} \quad R_{hf} = \Re\left\{\frac{l}{2\pi a}\sqrt{\frac{\omega\mu_0}{2\sigma}}\right\} \qquad (12)$$

and $R_{in}=\Re\{Z_{in}\}$. The results are summarized in Table 1. From Table 1 follows that for smaller radii the impedance values as well as the efficiency of CNT and copper dipoles are comparable. In the present discussion the efficiency of copper dipoles may even decrease at nanometer radii if we account for the additional surface and grain scattering (Hanson, 2005) which is not included in Eq. (11). The conductivity of CNTs however is not affected by surface or grain scattering because of their regular structure and can even be increased up to a factor of 30 if the CNT is doped (Dekker, 1999).

Besides the efficiency the frequency dependence of the input impedance of a CNT dipole is investigated. In Fig. 2 the dipole driving impedance from 0 to 1000 GHz is shown. Remarkable are the positions of the resonances. Usually the first resonance would be at $f=c_0/2l$=7.5 THz whereas here the first resonance is at f=160 GHz. Consequently the wavelength λ on the CNT dipole must be reduced by a factor of $\lambda_0/\lambda \approx 50$ which is denoted as the slow wave factor (Hanson, 2005). This behavior can be explained by the presence of retarded plasmon modes in CNTs (Fichtner and Russer, 2006).

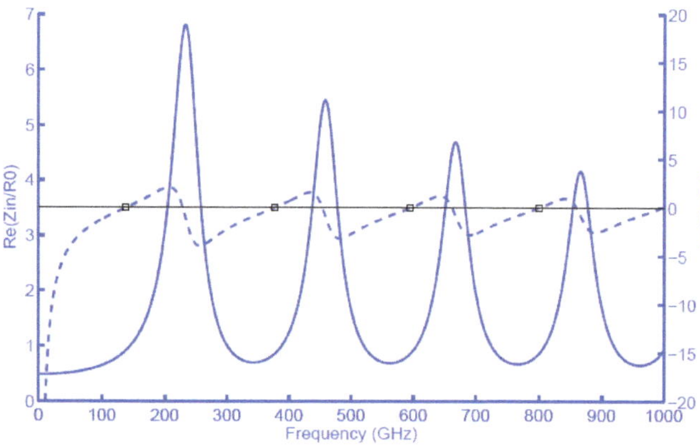

Fig. 2. Normalized input impedance for a CNT nanoantenna for a=5 nm and l=20 μm (R_0=12.9 kΩ).

5 Conclusions

Nanoantennas either of copper or of carbon exhibit inherently a very high resistance per unit length. A lower resistance may be achievable due to doping of the CNTs. The low efficiency of nanoantennas restrict the signal transmission to very short distances as present in e.g. chip-to-chip or on-chip communication systems (Yordanov et al., 2008).

References

Balanis, C.: Antenna Theory – Analysis and Design, John Wiley & Sons, 2 edn., 1997.

Dekker, C.: Carbon Nanotubes as molecular quantum wires, Physics Today, 52, 22–28, 1999.

Fichtner, N. and Russer, P.: On the Possibility of Nanowire Antennas, in: Microwave Conference, 36th European, 870–873, 2006.

Fikioris, G. and Wu, T.: On the Application of Numerical Methods to Hallén's Equation, IEEE Trans. Antennas Propagat., 49, 383–392, 2001.

Hanson, G.: Fundamental Transmitting Properties of Carbon Nanotube Antennas, IEEE Trans. Antennas Propagat., 53, 3426–3435, 2005.

Jishi, R., Dresselhaus, M., and Dresselhaus, G.: Electron-phonon coupling and the electrical conductivity of fullerene nanotubules, Phys. Rev. B, 48, 11 385–11 389, 1993.

Sarkar, T.: A study of the various methods for computing electromagnetic field utilizing thin wire integral equations, Radio Sci., 18, 29–38, 1983.

Slepyan, G., Maksimenko, S., Lakhtakia, A., Yevtushenko, O., and Gusakov, A.: Electrodynamics of carbon nanotubes: Dynamic conductivity, impedance boundary conditions, and surface wave propagation, Phys. Rev. B, 60, 17 136–17 149, 1999.

Yordanov, H., Ivrlac, M. T., Mezghani, A., Nossek, J., and Russer, P.: Computation of the Impulse Response and Coding Gain of a Digital Interconnection Bus, in: Proceedings of the 23rd Annual Review of Progress in Applied Computational Electromagnetics ACES, Niagara Falls, Canada, accepted, 2008.

25

Circuit design with Independent Double Gate Transistors

M. Weis, R. Emling, and D. Schmitt-Landsiedel
Technische Universität München, Lehrstuhl für Technische Elektronik

Abstract. Circuits with transistors using independently controlled gates have been proposed to reduce the number of transistors and to increase the logic density per area. This paper introduces a novel Vertical Slit Field Effect Transistor with unique independent double gate properties to demonstrate the possible advantages for independent double gate circuits. A new adder circuit is proposed, where the power could be reduced by one fifth and the area by on third compared to a tied gate configuration.

1 Introduction

Transistors with two gates have been investigated from the early days of semiconductor devices – namely the "Unipolar 'Field Effect' Transistor" proposed by Shockley in 1952 (Shockley, 1952). However for the last decades the single gate MOSFET in planar integrated CMOS technology dominated the semiconductor industry. Enormous progress has been made to scale transistors to ever smaller dimensions to obtain faster transistors, as well as to lower the effective costs per transistor in terms of transistors per area. With scaling device dimensions and increasing short channel effects, multiple gate transistors have been investigated to obtain an improved gate control. However multiple gate transistors can do more than just overcome the problems of increased short channel effects. If the multiple gates can be controlled independently, the logic functionality per transistor and logic density per area can be increased too. The basic concept of independent double gate transistors for logic circuits is discussed in Sect. 2. A novel transistor with excellent independent double gate properties is introduced in Sect. 3. The impact on circuit design using independent double gate transistors is outlined in Sect. 4. Basic NAND and NOR circuits are shown in Sect. 5 and adder circuits in Sect. 6.

2 Independent Double Gate Transistors

Standard CMOS transistors within logic circuits can be seen as a switch where the gate turns the transistor "ON" and "OFF". Using more than one gate, the behaviour of the switch depends on the number of controlling gates and additional logic can be implemented into one transistor. Transistors using independently controlled gates are not limited to two gates, but for reasons of the transistor geometry and the connectivity of the terminals it is convenient to use only two gates. Independent double gate transistors can be used to implement a logic "OR" or "AND" functionality within one transistor. For the "OR" type transistor this means that if gate 1 is "low" and gate 2 is "high" the transistor is turned "ON" and vice versa for gate 1 is "high" and gate 2 is "low" (where "low" and "high" are the logic voltage levels). Similar as for the "AND" type transistor and the gates biased in the same way as just described the transistor is turned "OFF". For the rest of the paper this is referred to the "independent gate configuration" (IGC). For the case that both gates are "high", the transistor is turned "ON" for a n-type transistor and "OFF" for a p-type transistor and vice versa for both gates are "low". For the rest of the paper this is referred to the "tied gate configuration" (TGC), where the transistor can be treated as standard transistor with one gate only using the standard transistor symbols.

For CMOS multigate (or FinFET) transistors the "OR" can easily be implemented, as for the double gate transistor there are two parallel inversion channels. However implementing the "AND" functionality is rather difficult for the IGC, as the transistor must turn "OFF" though one gate is still "low" or "high", which leads in general to a significant leakage current. Circuits proposed in Chiang et al. (2006)

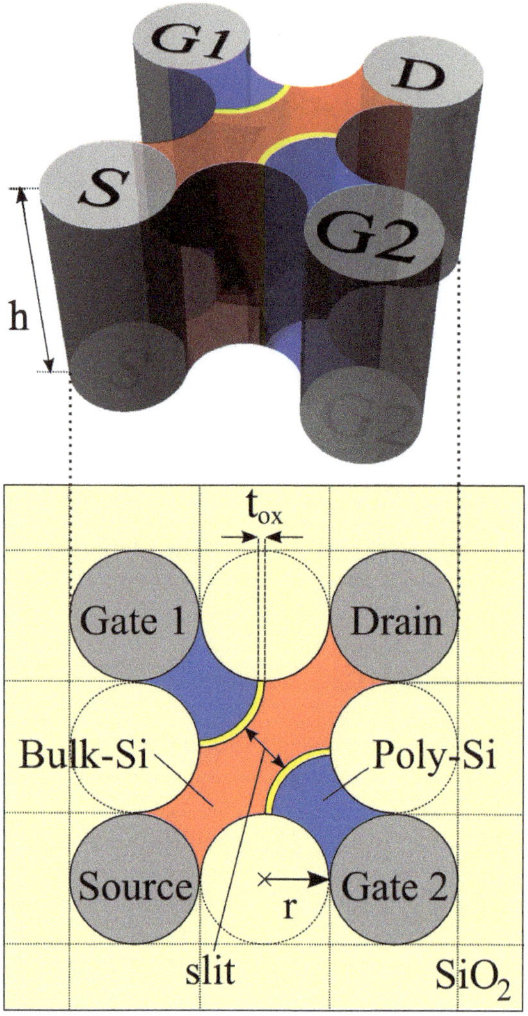

Fig. 1. Vertial Slit Field Effect Transistor.

and Mukhopadhyay et al. (2005) using transistors with the "AND" functionality suffer from these high leakage currents. Recently published Vertical Slit Field Effect Transistors (VeSFET) can offer "OR" and "AND" functionality without suffering high leakage for the "AND" type device for the IGC (Maly and Pfitzner, 2008; Weis et al., 2008; Lin et al., 2008).

3 Vertical Slit Field Effect Transistor

The VeSFET was developed in order to achieve an extreme regularity in layout of integrated circuits (Maly, 2007; Maly and Pfitzner, 2008). The implementation shown in Fig. 1 provides a square shaped unit device built on a silicon on insulator (SOI) substrate and all transistors are isolated from each other. The geometry in Fig. 1 is based on a standard feature size represented by the radius r. The four terminals are implemented using vertical metal pillars. The channel volume is defined by the distance between the source and drain terminal as well as the height h and the width of the vertical "slit" between the two gates.

3.1 Working principle of VeSFET

The VeSFET is a pn-junction-less transistor with a gate controlled bulk current using either a p- or n-type substrate for the complementary transistor types. It can be regarded as a hybrid of a Junction Field Effect Transistor (JFET) and a MOSFET. The operation is JFET-like because it is based on transport of majority carriers in a bulk channel, whose effective width is controlled by depleted regions induced by two gates on both sides of this channel. The gates, however, are separated from the channel by an insulating layer, like in a MOSFET. The difference to the MOSFET is that the current is not bound to an inversion channel but flows through the whole bulk volume between source and drain terminal. In the off-state the depletion regions induced by the two gates fill the whole channel and the device is non-conducting. For a normally "OFF" transistor the gate material has to be chosen such that for a n-type transistor and the gates biased "low", the depletion widths are close to their maximum expansion, and vice versa for a p-type transistor and the gates biased "high". Applying appropriate voltages to the gates the depletion regions withdraw and a conducting path from drain to source is established.

3.2 Independent Double Gate Operation of VeSFET

Modulation of the depleted regions induced by the two independently controlled gates can implement "AND" and "OR" functionality. This can be achieved via low and high substrate doping, as the depletion region is a function of the substrate doping for a given set of geometry and applied voltages at the terminals. The independent double gate feature of the VeSFET is illustrated in Fig. 2.

For the OR-type the depletion regions from gate 1 and gate 2 are overlapping in the middle of the channel for the case that both gates are "low" and the transistor is turned "OFF", as shown in Fig. 2a. The overlap region must be at least in the range of the debye length to ensure the transistor is "OFF". If an appropriate voltage is applied to gate 1 the corresponding depletion width will withdraw. A conducting path between source and drain will be established as shown in Fig. 2b and the transistor is "ON". The complementary case for gate 2 is shown in Fig. 2c. Applying a voltage to both of the gates both depletion widths will withdraw and the maximum opening of the channel is established, see Fig. 2d.

For the AND-type the depletion width of the gates reaches to the opposite gate, see Fig. 2e. Hence the overlap region extends over the whole channel and the transistor is "OFF". If now a voltage is applied to gate 1 the corresponding depletion region will withdraw. However the depletion with resulting

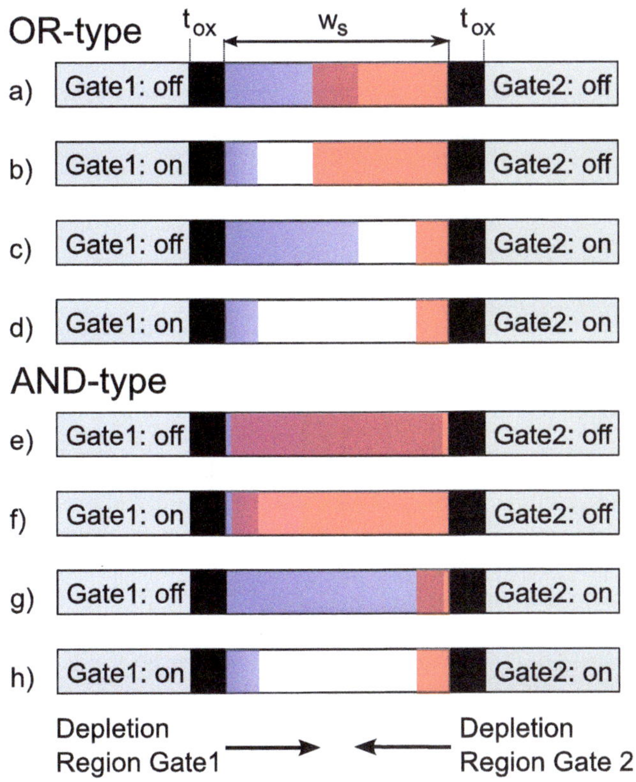

Fig. 2. Independent Double Gate for VeSFET OR- and AND-type.

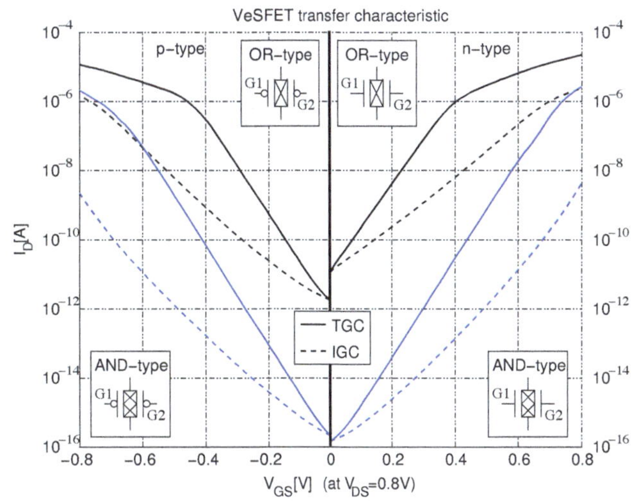

Fig. 3. Independent Double Gate VeSFET transfer characteristics.

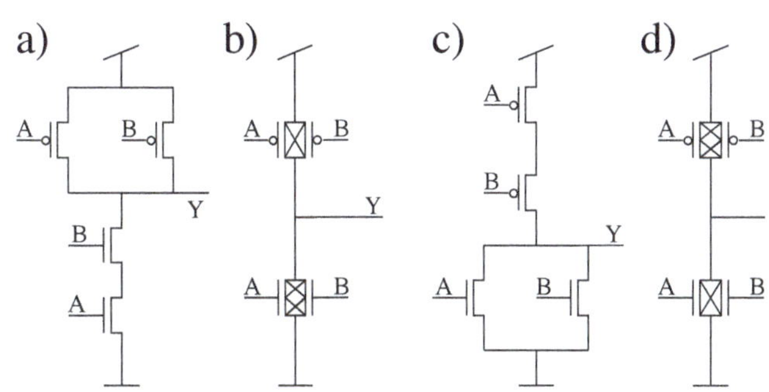

Fig. 4. 2-input NAND and NOR with TGC **(a)**, **(c)** and IGC **(b)**, **(d)**.

from gate 2 prevents establishment of a low-ohmic connecting path between source and drain as shown in Fig. 2f. Consequently the transistor is "OFF". The complementary case for gate 2 is shown in Fig. 2g. Only if an appropriate voltage is applied to both gates a low-ohmic connecting path between source and drain can be established as shown in Fig. 2h.

From device simulation low power optimized transistors were derived, where Fig. 3 shows the transfer characteristic for p- and n-type for IGC and TGC and introduces the transistor symbols (Weis et al., 2008). The transistor dimensions used for the circuits in this paper are t_{ox}=4 nm, r=50 nm and h=320 nm with substrate dopings of $1.5\times10^{17}\,\text{cm}^{-3}$ for the AND-type and $5\times10^{17}\,\text{cm}^{-3}$ for OR-type using polysilicon as gate material. For the AND-type IGC good turn-off is achieved with "OFF" currents around 4 nA and 2 nA for n- and p-type. TGC "ON" currents are 2.2 μA and 2.8 μA, respectively. For the OR-type TGC the "ON" currents are 22 μA and 12 μA for n- and p-type and for the IGC the "ON" currents are 2 μA and 3 μA respectively.

4 Impact on circuit design

From circuit design perspective the use of independently controlled transistors can reduce the number of transistors, as two transistors connected in parallel or series can be merged into one transistor, which is exemplary shown in Figs. 4 and 5. This technique reduces area, transistor stack height and power consumption. However all independent double gate transistors have in common that for the IGC the transistors "ON" and "OFF" currents are degraded compared to the TGC. For "OR" type devices this leads to higher gate delay and for the "AND" type devices this leads to higher leakage for the IGC. For this reason during times when the circuit is inactive input patterns should be applied in a way that both gates are at the same potential to reduce the leakage to a minimum. Multiple devices can be used in critical paths to compensate for the lower drive current in IGC. However there are cases where the TGC is a better choice instead of using multiple IGC. For this reason a trade-off between area reduction, gate delay and leakage has to be chosen when using transistors with independently controlled gates. The use of IGC is best for gates with small fan out and short capacitive wire loads.

In the following circuits using VeSFET with independently controlled gates are investigated by simulation. The proposed circuits are not limited to VeSFET and could be implemented with any type of transistor with independently controlled gates offering the "AND" and "OR" functionality.

Table 1. 2-input NAND and NOR for VeSFET and CMOS PTM.

confi-guration	power [nW]	leakage current min/max [nA]	rise/fall time [ps]	rise/fall delay [ps]
4a VeS	7.12	0.17/3.50	32.6/83.8	37.1/24.5
4b VeS	6.23	0.09/29.5	34.3/58.7	36.4/23.9
4a PTM	11.75	1.88/5.14	38.9/74.5	16.9/21.6
4c VeS	10.12	0.06/1.74	143/53.6	72.7/26.7
4d VeS	3.75	0.01 / 2.31	177/96.4	89.5/59.4
4c PTM	13.74	0.07/5.25	140/27.2	70.8/9.6

5 2-input NAND and NOR

Two input NAND and NOR gates in static CMOS are built in general with four transistors. Using independent double gate transistors the number of transistors can be reduced from four to two, merging transistors in parallel and in series as shown in Fig. 4. As mentioned in the previous section this will impact the power, delay and leakage currents. This structure was shown for the first time in Chiang et al. (2006) for double gate MOSFET, but with extremely high leakage in the independent gate configuration. A two transistor NAND, NOR using independent double gate VeSFET is compared to a four transistor NAND, NOR using OR-type VeSFET and to 65 nm CMOS based on the predictive technology model (PTM) (NIMO, 2008). The input is a piecewise linear voltage source with 50 ps rise and fall time. Each gate under test drives four identical copies of itself as load. The power for VeSFET compared to CMOS is reduced up to 40% due to lower device capacitances of VeSFET. Comparing IGC to TGC VeSFET the power can be further reduced by more than 10%. The leakage current for the IGC VeSFET gate is strongly dependent on the input pattern. For the case that both gates are on the same potential the leakage can be significantly reduced. However if the gates are on different potential the leakage can be increased by a factor of 8. Therefore clever input patterns should be applied during times when the circuit in inactive. The delay and the fall and rise times for all configurations for NAND and NOR are in same order of magnitude. The results are shown in Table 1 where the configuration refers to the circuits in Fig. 4. For the case that the NAND or NOR gate drives a fixed load capacitance of 1 fF the TGC VeSFET implementation is up to a factor 1.8 slower than CMOS 65 nm PTM and the VeSFET IGC is up to 3 times slower than the TGC VeSFET. However the power is reduced comparing CMOS and TGC VeSFET and can be further reduced using IGC VeSFET. Independently controlled double gate transistors can reduce the power, and the transistor count, respectively area reduction up to 50%, but it comes at the cost of an increased delay.

Table 2. One bit full adder.

confi-guration	average active power	minimum standby power	maximum $\overline{\text{Sum}}$ delay	maximum $\overline{\text{Cout}}$ delay
5a, OR	243.09 nW	16.091 pW	340.45 ps	198.69 ps
5a, AND	172.90 nW	2.5957 fW	2.9069 ns	1.3101 ns
5b	207.58 nW	85.89 pW	1.3998 ns	1.1076 ns
5c	203.14 nW	8.863 nW	1.2928 ns	590.84 ps
5a, PTM	220.81 nW	65.41 pW	327.58 ps	245.79 ps

6 Adder circuits

A new one bit full adder circuit using independent double gate VeSFET is proposed, which will be used to build a 16 bit ripple carry adder. Primary goal was to reduce the area and the transistor count. To find a metric for comparison all transistors are minimum sized. The area of the CMOS layout is assumed equal to that for TGC VeSFET. The independent gate VeSFET circuits are compared to the tied gate circuits using VeSFET as well as to 65 nm CMOS based on the predictive technology model (PTM) (NIMO, 2008).

6.1 One bit full adder

For the one bit full adder the standard mirror adder proposed by Weste is used (Weste and Harri, 2005). The tied gate configuration has 24 transistors without output inverters for the sum and the carry signal as shown in Fig. 5a. With independently controlled gates the transistor count can be reduced by one third to only 16 transistors. Two different circuits are shown in Fig. 5b and 5c. For the circuit shown in Fig. 5b "OR" and "AND" type transistors are used, whereas for the circuit shown in Fig. 5c only "AND" type transistors are used. The advantage for the latter circuit is that the maximum number of stacked transistors for the carry signal is reduced from two to one and the delay is decreased. For the simulation setup the inverted carry out signal is connected to another one bit full adder as capacitive load and the inverted sum signal is connected to an inverter with a fan out of two. All capacitive wire loads from the area optimized layout are included. All 56 possible transitions for the input pattern are applied using an ideal input driver with a rise and fall time of 50 ps. The results for the average power per pattern, the minimum standby power and the maximum delay for the inverted sum and carry out signals are given in Table 2.

For a supply voltage of 0.8 V one can see from Table 2 that for the circuit 3a using "OR" type TGC VeSFET the maximum inverted carry out delay ($\overline{\text{Cout}}$) is close to standard CMOS using the PTM 65nm model. Using 'AND' type TGC VeSFET leads to a dramatic increase of delay for the $\overline{\text{Cout}}$ and the $\overline{\text{Sum}}$, but also to a dramatic decrease of standby power (5a, AND). Comparing the IGC to the TGC the delay is increased by 2.4 and the leakage is higher for

Table 3. 16 bit ripple carry adder.

configuration	average active power	average standby power	maximum $\overline{\text{Cout}}$ delay	transistor count
5a, OR	0.32171 uW	11.273 nW	4.19 ns	432
5b	0.24891 uW	24.064 nW	24.49 ns	304
5c	0.25768 uW	11.676 nW	20.69 ns	304
5a, PTM	0.39701 uW	67.165 nW	2.77 ns	432

the configuration 3c. For the configuration 5b, the leakage is almost the same but the delay is five times higher. However the active power for both IGC circuits is reduced compared to the TGC (5a, OR and 5a, PTM) due to the lower capacitances and, most important, the area is reduced by one third, where the IGC area is $2.56\,\mu m^2$ compared to $3.24\,\mu m^2$ for the TGC.

6.2 16 bit ripple carry adder

The 16 bit ripple carry adder is built from the one bit full adder block discussed in the previous section. To reduce the number of transistors one can alternate stages using inverted and non-inverted signals for the input vectors. For a supply voltage of 0.8 V the results are shown in Table 3 for average active and standby power for random patterns, as well as the maximum $\overline{\text{Cout}}$ delay for the last stage and the transistor count. For the area calculation the inverters for the inverted input signals as well as the inverters for the sum signals are included, where the area for the TGC is $92.16\,\mu m^2$ and for the IGC $61.44\,\mu m^2$. For the input pattern an ideal input driver with a rise and fall time of 50 ps is used and all capacitive wire loads from the layout are included.

For TGC "OR" type VeSFET implementation (configuration 5a, OR) the delay is increased by 1.5 compared to the standard CMOS using PTM 65 nm (5a, PTM), however the active power can be reduced by 20%. As expected, due to lower drive current for the IGC VeSFET, the delay increases for the IGC implementations (5b, 5c) by factors 6 and 5 compared to the TGC VeSFET. The active power consumption for both IGC implementations is reduced to one fifth compared to CMOS. The area reduction for the IGC compared to the TGC is again one third.

7 Conclusions

In this paper the circuit design using independently controlled gates using a novel Vertical Slit Field Effect Transistor was demonstrated and the impact on figures of merit for power consumption, delay, leakage behaviour and area was presented. It was shown that the transistor count and respectively the area can be significantly reduced for logic gates, hence the logic density per area increases. For low power circuits standby patterns are favorable where both transistor gates are on the same potential offering even less leakage current. However due to the lower drive current in the IGC mode of operation the delay will be increased. For this reason independent double gate transistors are not the first choice for high speed circuits but fit perfectly in the low power application domain.

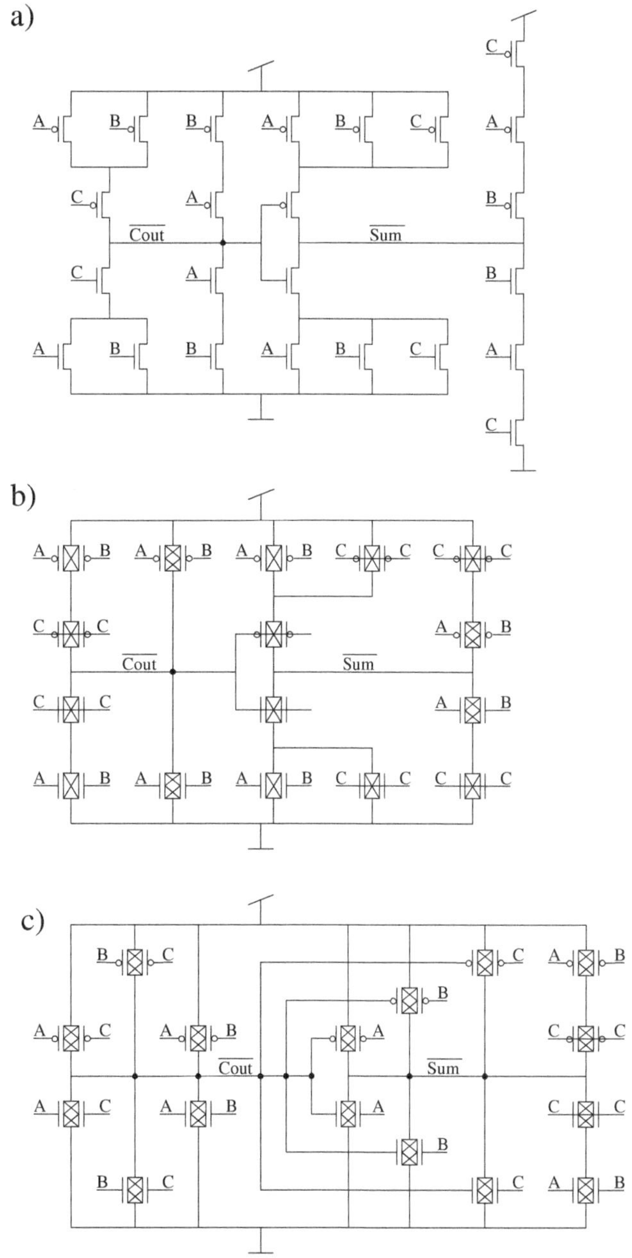

Fig. 5. One bit full adder: **(a)** Tied Gate **(b)** and **(c)** Independent Gate.

Acknowledgements. The authors express their thanks to Andrzej Pfitzner and Dominik Kasprowicz from Warsaw University of Technology for intensive discussions and for providing the VeSFET transistor models. They thank Wojciech Maly for introducing them to the concept of VeSFET.

References

Shockley, W.: A Unipolar "Field-Effect" Transistor, Proceedings of the IRE, 40(11), 1365–1376, 1952.

Chiang, M.-H., Kim, K., Chuang, C.-T., and Tretz, C.: High-Density Reduced-Stack Logic Circuit Techniques Using Independent-Gate Controlled Double-Gate Devices, IEEE T. Electron. Dev., 53(9), 2370–2377, 2006.

Mukhopadhyay, S., Mahmoodi, H., and Roy, K.: Design of High Performance Sense Amplifier Using Independent Gate Control in sub-50 nm Double-Gate MOSFET, Quality of Electronic Design, 2005, ISQED, 2005.

Maly W. and Pfitzner A.: Complementary Vertical Slit Field Effect Transistors, Carnegie Mellon University, CSSI Techreport No. CSSI 08-02, 01/2008, 2008.

Weis, M., Pfitzner, A., Kasprowicz, D., Lin, Y.-W., Fischer, Th., Emling, R., Marek-Sadowska, M., Schmitt-Landsiedel, D., and Maly, W.: Low Power SRAM Cell Using Vertical Slit Field Effect Transistor (VeSFET), European Solid-State Circuits Conference, ESSCIRC Fringe, 2008.

Lin, Y.-W., Marek-Sadowska, M., Maly, W., Pfitzner, A., and Kasprowicz D.: Is There Always Performance Overhead for Regular Fabric?, in: Proc. IEEE International Conference on Computer Design ICCD 2008, 557–562, 2008.

Maly, W.: Integrated Circuit Fabrication and Associated Methods, Devices and Systems, U.S. Non-Provisional Patent Application Serial Number CMU Docket 06-091; DMC Docket 06-001PCTCMU, 11/2007, 2007.

Nanoscale Integration and Arizona State University Modeling Group, Predictive Technology Model (PTM), http://www.eas.asu.edu/~ptm/, 2008.

Weste, N. H. E. and Harris, D.: CMOS VLSI DESIGN, A Circuits and Systems Perspective, Pearson Addison Wesley, Third Edition, 2005.

26

A novel technique for CAD-optimization of analog circuits with bipolar transistors

B. Dimov, V. Boos, T. Reich, Ch. Lang, E. Hennig, and R. Sommer
Institut für Mikroelektronik- und Mechatronik-Systeme gGmbH, Ilmenau, Germany

Abstract. In this paper, a novel approach for robust automatic optimization of analog circuits with bipolar transistors is presented. It includes additional formal parameters into the device model cards, which sweep the model parameters smoothly between the different device types. In this way, not only the sizing, but also the choice of the device type is committed to the optimization tool, thus improving the efficiency of the design process significantly.

1 Introduction

Very important step in the design of analog integrated circuits is the choice of the device types. The latter is a typical problem especially by the design of BiCMOS circuits. There, the bipolar transistors should be selected among several device types offered by the fabrication foundry, which have only few (or even no) adjustable parameters (X-FAB AG, 2008). In this way, the circuit performance depends strongly on the initial choice of the device types, which is done by the designer and is based mainly on his experience and basic engineering considerations. Once this choice has been made, a subsequent CAD-based circuit optimization is strongly restricted by the insufficient number of remaining adjustable device parameters.

Up to now, few approaches for optimization of the device types exist and are mostly based on discrete optimization methods (e.g. Cadence Design Systems, 2008). Here, we present a novel approach for automatic selection of device types using a gradient-based optimizer. Its general idea is described in Sect. 2 of this paper, with some important remarks about its applicability pointed out in Sect. 3. In Sect. 4, the optimization process of an integrated BiCMOS buffer amplifier with moderate complexity is described, thus demonstrating the efficiency of the proposed approach. Section 5 concludes the paper.

2 General description of the new approach

A typical analog circuit optimization tool can be represented with the 3-level structure shown in Fig. 1. The optimization goal is to find the global minimum of the objective function OBJ($\boldsymbol{Q}$), with $\boldsymbol{Q}=(q_1,\ldots,q_n)$ – the vector of the circuit performances under optimization. At each iteration step, a set of design parameters $\boldsymbol{P}=(p_1,\ \ldots,\ p_m)$ is generated by the optimizer. The corresponding equivalent electrical circuit is composed and simulated with a SPICE-like electrical simulator, and the resulting performance vector $\boldsymbol{Q}$ is obtained. For this, electrical device models are used with model parameters $\boldsymbol{R}=(r_1,\ldots,r_k)$ specified by the fabrication foundry, i.e. $\boldsymbol{Q}=\boldsymbol{Q}(\boldsymbol{P},\boldsymbol{R})$.

The efficiency of the design process can be improved significantly, if not only the sizing, but also the choice of the type of the devices is performed automatically by the optimization tool. For this, additional design parameters $\boldsymbol{S}=(s_1,\ldots,s_l)$ should be incorporated, which switch between the different device model cards. The latter is schematically shown in Fig. 2. A drawback of this approach is the abrupt change of the corresponding circuit performances $\boldsymbol{Q}$, which results from the discrete change of the device type. The latter is hardly compatible with any gradient-based optimization techniques, thus prohibiting the utilization of their great fastness.

In order to facilitate the application of gradient-based algorithms, a smooth transition of the circuit performances $\boldsymbol{Q}$ across the whole range of available device types should be ensured. We do this by incorporating the additional design parameters $\boldsymbol{S}$ directly into the device model cards. They are used to map the model cards of the available device

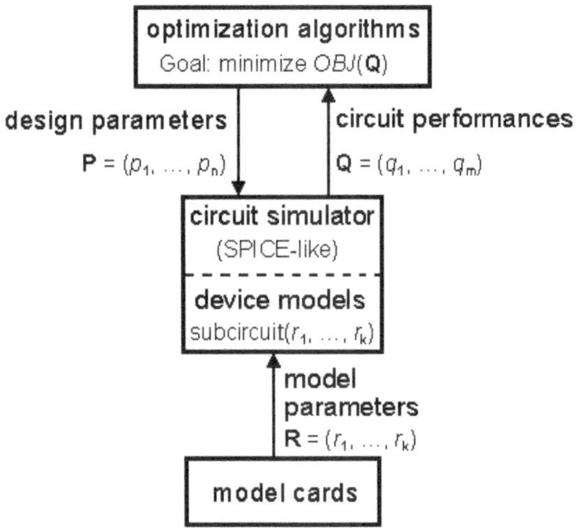

Fig. 1. Typical structure of analog circuit optimization tool.

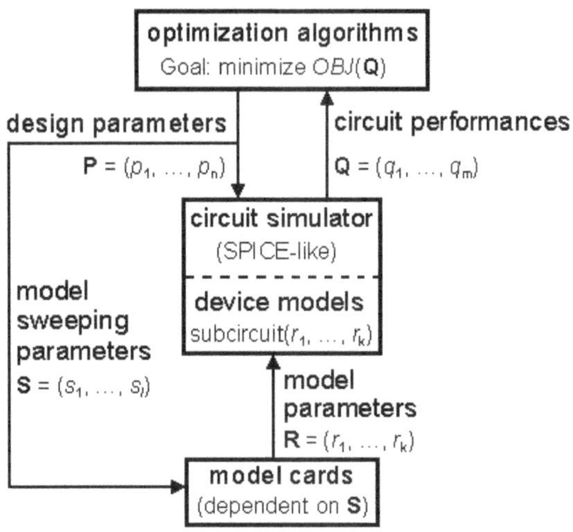

Fig. 2. Modified structure of analog circuit optimization tool permitting automatic selection of the device types.

types into a single model card with redefined model parameters $\boldsymbol{R}$:$\boldsymbol{R}(\boldsymbol{S})=(r_1(\boldsymbol{S}), \ldots, r_k(\boldsymbol{S}))$. This is schematically shown in Fig. 3, where the model cards of two PNP bipolar transistor types *pnp1* and *pnp2* are mapped into the single model card *pnp12*. In this example, the devices are modeled within the *S*PICE *G*ummel-*P*oon (SGP) model (Gummel and Poon, 1970; Berkner, 2002). Only k=3 model card parameters are considered: $\boldsymbol{R}$:={BF, VAF, IS}, with BF, VAF, and IS being the forward current gain, the forward Early voltage, and the saturation current model parameters. One additional parameters s is used to map the *pnp1* and *pnp2* model cards: $\boldsymbol{S}$:={s}. As also shown in Fig. 3, this mapping is assumed to be: $pnp12|_{s=0} \leftrightarrow pnp1$; $pnp12|_{s=1} \leftrightarrow pnp2$, i.e. $r_1(0)$=`bf1`, $r_2(0)$=`vaf1`, $r_3(0)$=`is1`, $r_1(1)$=`bf2`, $r_2(1)$=`vaf2`, $r_3(1)$=`is2`.

```
model  pnp1  bjt type=pnp struct=lateral
+ BF=bf1     VAF=vaf1     IS=is1

model  pnp2  bjt type=pnp struct=lateral
+ BF=bf2     VAF=vaf2     IS=is2

model  pnp12 bjt type=pnp struct=lateral
+ BF=r1(s)   VAF=r2(s)    IS=r3(s)
```

Fig. 3. Bipolar transistor types `pnp1` and `pnp2` merged into the single model card `pnp12`.

If the functions $r_1(\boldsymbol{S}), \ldots, r_k(\boldsymbol{S})$ are continuous, they sweep the circuit performances also continuously through the model cards of all available device types:

$$\boldsymbol{Q} = \boldsymbol{Q}(\boldsymbol{P}, \boldsymbol{R}(\boldsymbol{S})). \tag{1}$$

In this way, the automatic choice of the device type can be supported also by deterministic optimization algorithms, thus increasing significantly the efficiency of the optimization process. However, the following requirements should also be fulfilled:

- the sequence, in which the device types are swept, should be chosen with respect to minimize the number of the local extrema of the circuit performances, i.e. $\boldsymbol{S}$ should sweep the circuit performances $q_1, \ldots, q_n$ as monotonic as possible.
- because the sweeping functions $r_1(\boldsymbol{S}), \ldots, r_k(\boldsymbol{S})$ are continuous, nonphysical interim values of the model parameters $\boldsymbol{R}(\boldsymbol{S})$ (and thus also of the circuit performances $\boldsymbol{Q}(\boldsymbol{P},\boldsymbol{R}(\boldsymbol{S}))$) cannot be avoided, however their evaluation as optimum ones should be suppressed somehow.
- the number l of the model sweeping parameters $s_1, \ldots, s_l$ should be kept as low as possible, because it directly increases the total order of the optimization problem.

Below, a possible strategy to fulfill these requirements is described.

3 Implementation of the new approach

3.1 Optimal arrangement of the device types

Although there is no full equivalence between the total circuit performance and the performances of the circuit's devices alone, they are usually strongly correlated. Utilizing this correlation, the devices are sorted with respect to their figures-of-merit, which have to be swept with $\boldsymbol{S}$ as monotonic as possible. Ideally, this sorting requirement should be fulfilled about *all* device's figures-of-merit, but this is possible only in case of device models with few model parameters and figures-of-merit (e.g. resistors and capacitors). The optimal choice of the type of such devices is usually a trivial task; the real challenge is met with the complex device models (e.g. transistors) having many and often weak correlated figures-of-merit. In this case, a simultaneous monotonic sweeping of *all* figures-of-merit is not possible, but the designer should ensure it for those figures-of-merit *dominating* the device performance within the particular application. As a typical example for this, the sorting of amplifier's transistors can be considered. The matching of the transistors in case of differential amplification is essential for the circuit functionality. Therefore, the designer should include it within the dominating sorting criteria, contrary to the case of nondifferential amplification.

This arrangement can be performed dynamically also about devices with equal models within one and the same analog circuit. Considering again amplifier's transistors, the dominant figures-of-merit for input-stage ones are usually the generated noise and the transit frequency. However, they are less important about output-stage transistors, whose high-current performance is more dominant. Therefore, the device types of the input and output transistors can be sorted (and correspondingly swept with S) in different ways, although the devices have identical models and thus share one and the same set of possible device types.

It should be pointed out, that the device type arrangement described above is performed before the circuit's optimization, i.e. at an early design stage, when even the *dc* operating point may be unknown. Therefore, the devices' figures-of-merit should be calculated out of the context of the particular analog circuit. Being dynamically dependent on both the circuit functionality in general and the device's specific function within the circuit, the device types arrangement should be usually performed multiple times during the circuit's design phase. Therefore, an efficient approach for automatic evaluation of the devices' figures-of-merit should be implemented in the optimization strategy.

One very promising candidate for this is the direct analytic approach described in Dimov et al. (2008). It is based on explicit analytic formulas calculating the device's figures-of-merit directly from its model card. Such estimation requires no electrical simulations or similar numerical calculation techniques, thus being extremely robust and fast. The formulas proposed in Dimov et al. (2008) can be easily implemented in self-written hand calculators or conventional computation tools such as Wolfram Research (2008). In this way, the device's figures-of-merit are evaluated immediately, and the device type arrangement is performed with negligible computation efforts.

3.2 Choice of the sweeping functions $R(S)$

As pointed out in Sect. 2 of the paper, the main advantage of the proposed novel approach is its compatibility with gradient-based optimization algorithms. The gradient calculations are done numerically using finite differences (Graeb, 2007). About the optimization problem considered in this paper, the latter can be expressed using Eq. (1):

$$\operatorname{grad}\left(\mathrm{OBJ}\left(\boldsymbol{Q}\left(\boldsymbol{P}^n,\boldsymbol{R}\left(\boldsymbol{S}^n\right)\right)\right)\right) \overset{\sim}{=} \frac{\mathrm{OBJ}\left(\boldsymbol{Q}\left(\boldsymbol{P}^n+\Delta\boldsymbol{P}^n,\boldsymbol{R}\left(\boldsymbol{S}^n+\Delta\boldsymbol{S}^n\right)\right)\right)-\mathrm{OBJ}\left(\boldsymbol{Q}\left(\boldsymbol{P}^n,\boldsymbol{R}\left(\boldsymbol{S}^n\right)\right)\right)}{\left|\left(\Delta\boldsymbol{P}^n,\Delta\boldsymbol{S}^n\right)\right|}$$

with superscript n representing the nth optimization iteration. Despite of the particular optimization strategies implemented into the gradient-based algorithm, the decision about the next design parameter set ($\boldsymbol{P}^{n+1}$, $\boldsymbol{S}^{n+1}$) as well as the next numeric step ($\Delta\boldsymbol{P}^{n+1}$,$\Delta\boldsymbol{S}^{n+1}$) is derived always from the gradient(s) of the objective function calculated at the current (and optionally: previous) design parameter set(s) (Graeb, 2007). Therefore, the sweeping functions $\boldsymbol{R}(\boldsymbol{S})$ should be not only continuous and thus numerically differentiable, but their gradients calculated at nonphysical interim values of the device model parameters should always push the optimization algorithm towards feasible model parameter values.

This requirement is successfully met, if piece-wise linear sweeping functions are used. About the simple example described with Fig. 3, the sweeping functions have to be defined as:

$$r_1(s) = (\texttt{bf2} - \texttt{bf1})s + \texttt{bf1}$$
$$r_2(s) = (\texttt{vaf2} - \texttt{vaf1})s + \texttt{vaf1}$$
$$r_3(s) = (\texttt{is2} - -\texttt{is1})s + \texttt{is1}$$

3.3 Minimizing the number l of the sweeping parameters

Being incorporated into the device model cards, the model sweeping parameters $s_1, \ldots, s_l$ appear as additional instance parameters of the devices under optimization. Let assume, that $i \leq l$ sweeping parameters are used to sweep the model card parameters of j uncorrelated devices from one and the same type. The latter results in $i \times j$ additional uncorrelated optimization parameters, i.e. the total order of the optimization problem increases rapidly with i. Therefore, the number of the sweeping parameters should be kept as low as possible.

A powerful way for this is the multiple reusing of already existing instance parameters. This idea will be illustrated about two types of bipolar transistors – QNA and QNB, which have to be merged in a common model card QNAB without increasing the number of optimized parameters. Let QNA be a scalable transistor with instance parameter EA=1...10 corresponding to emitter areas 1...10 μm^2, respectively. Then, its model card parameters are functions of EA, which will be designated as $\boldsymbol{R}^{\text{QNA}}(EA)$. Let QNB be a transistor with fixed layout and emitter area of 1 μm^2, i.e. having no instance parameters. Its model card parameters will be designated as $\boldsymbol{R}^{\text{QNB}}$. A possible definition of the model card parameters $\boldsymbol{R}^{\text{QNAB}}(s)$ is:

$$0.5 \leq s < 1: \boldsymbol{R}^{\text{QNAB}}(s) := 2(\boldsymbol{R}^{\text{QNA}}(1) - \boldsymbol{R}^{\text{QNB}})s + (2\boldsymbol{R}^{\text{QNB}} - \boldsymbol{R}^{\text{QNA}}(1))$$
$$1 \leq s \leq 10: \boldsymbol{R}^{\text{QNAB}}(s) := \boldsymbol{R}^{\text{QNA}}(EA \rightarrow s),$$

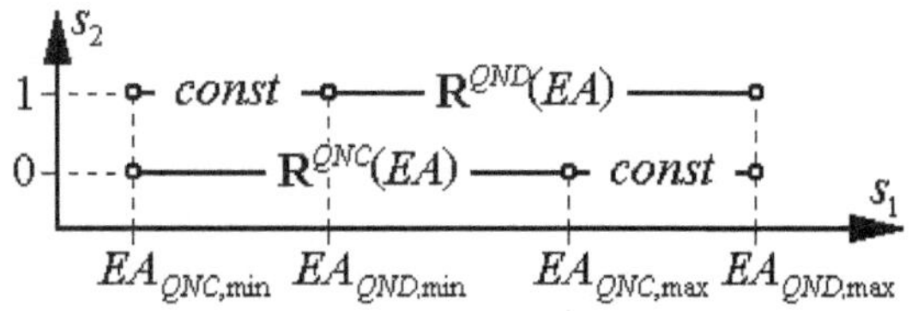

Fig. 4. Arrangement in case of different feasible ranges of the instance parameters.

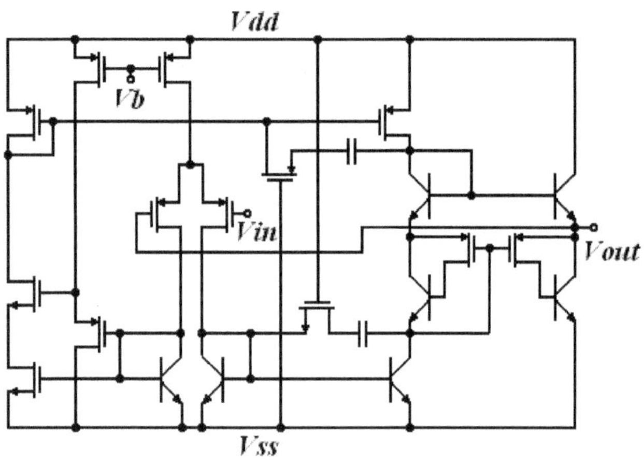

Fig. 5. BiCMOS integrated buffer amplifier under optimization (topology from Furth and Andreou, 1993).

i.e. the single sweeping parameter s incorporated into the model card of QNAB is used both to sweep linearly between QNB and QNA and to model the emitter area of QNA, thus replacing the instance parameter *EA*.

Note, that the proposed arrangement and the resulting sweeping functions $\boldsymbol{R}^{\text{QNAB}}(s)$ should fulfill all sorting preconditions described in Sects. 3.1 and 3.2. Otherwise, local minima of the objective function OBJ($\boldsymbol{Q}$) can appear for interim values of s between 0.5 and 1 pointing to nonexistent type of the QNAB device.

Another typical problem is the arrangement of devices with identical instance parameters, but different feasible ranges of the latter. Considering again two bipolar transistor types QNC and QND, let both have variable emitter area *EA*, i.e. model card parameters being functions of *EA*: $\boldsymbol{R}^{\text{QNC}}(EA)$ and $\boldsymbol{R}^{\text{QND}}(EA)$, respectively. Let $EA_{\text{QNC,min}} \leq EA \leq EA_{\text{QNC,max}}$ and $EA_{\text{QND,min}} \leq EA \leq EA_{\text{QND,max}}$ are the feasible ranges of the emitter area of QNC and QND, respectively. Without loss of generality, let us assume: $EA_{\text{QNC,min}} \leq EA_{\text{QND,min}} < EA_{\text{QNC,max}} \leq EA_{\text{QND,max}}$. The aim is again to merge the both device types in a common model card QNCD with minimum number of additional sweeping parameters.

Table 1. Performances of the buffer amplifier: specification, before, and after the optimization.

Performance	Specification	Initial	Optimized
Bandwidth, MHz	maximized	110.1	150.1
Phase margin, degree	>60	70.6	60.0
Supply current, mA	<2	1.896	1.900
SlewRateUp, MV/s	>500	609.8	519.3
SlewRateDown, MV/s	>500	787.2	691.6
Offset, mV	<1	0.6136	0.043
PSRR, dB	<−70	−74.15	−73.88
output rmsNoise, μV	<300	277.5	192.7

For this, the sweeping parameter s_1 is introduced and the model cards of QNC and QND are extended as:

$$EA_{\text{QNC,min}} \leq s_1 \leq EA_{\text{QNC,max}} : \boldsymbol{R}^{\text{QNC1}}(s_1) := \boldsymbol{R}^{\text{QNC}}(EA \to s_1)$$
$$EA_{\text{QNC,max}} < s_1 \leq EA_{\text{QND,max}} : \boldsymbol{R}^{\text{QNC1}}(s_1) := \boldsymbol{R}^{\text{QNC}}(EA_{\text{QNC,max}})$$
$$EA_{\text{QNC,min}} \leq s_1 < EA_{\text{QND,min}} : \boldsymbol{R}^{\text{QND1}}(s_1) := \boldsymbol{R}^{\text{QND}}(EA_{\text{QND,min}})$$
$$EA_{\text{QND,min}} \leq s_1 \leq EA_{\text{QND,max}} : \boldsymbol{R}^{\text{QND1}}(s_1) := \boldsymbol{R}^{\text{QND}}(EA \to s_1)$$

Note, that the feasible ranges of s_1 are identical about the new model cards QNC1 and QND1. The QNCD model card is obtained by linear sweeping with a second parameter s_2, assuming e.g. $\text{QNCD}|_{s2=0} \leftrightarrow \text{QNC1}$ and $\text{QNCD}|_{s2=1} \leftrightarrow \text{QND1}$:

$$\boldsymbol{R}^{\text{QNCD}}(s_1, s_2) = (\boldsymbol{R}^{\text{QND1}}(s_1) - \boldsymbol{R}^{\text{QNC1}}(s_1))s_2 + \boldsymbol{R}^{\text{QNC1}}(s_1).$$

The latter is schematically shown in Fig. 4. It should be pointed out, that again one of the sweeping parameters (i.e. s_1) is used also to replace the instance parameter *EA*, thus reducing the order of the optimization problem. Note, that with such device type arrangement, nonphysical parameter values (e.g. s_2=0, $EA_{\text{QNC,max}} < s_1 \leq EA_{\text{QND,max}}$) may be evaluated as final optimum, but they can be rounded to the closest feasible parameter set (e.g. s_2=0, $s_1 = EA_{\text{QNC,max}}$) without loosing the optimum circuit performance $\boldsymbol{Q}$.

4 Application of the new approach

The described novel approach for automatic optimization of the device types has been tested optimizing the BiCMOS integrated buffer amplifier with moderate complexity proposed in Furth and Andreou (1993). Its electrical scheme is shown in Fig. 5. The circuit contains seven bipolar transistors and has to be realized with a 0.6 μm BiCMOS fabrication technology providing 13 different types of bipolar transistors with fixed as well as continuously and discrete scalable emitter lengths.

The goal of the optimization has been the increasing of the bandwidth of the amplifier, keeping the rest of its performances within the specification constraints (see Table 1). Therefore, the transit frequency has been chosen to be the dominant figure-of-merit for the arrangement of the whole variety of bipolar transistors. The arrangement has been done with two sweeping parameters, one of which has replaced the emitter length instance parameters, as described in Sect. 3.3. The smooth sweeping between the different transistor types has been done using piece-wise linear functions. The amplifier has been optimized with the commercial analog circuit optimization tool WiCkeD (MunEDA GmbH, 2008). By this, only gradient-based optimization strategies have been used. During this test, only the bipolar transistors' parameters (type and sizing) have been optimized; all other circuit devices have been kept unchanged.

The results are given in Table 1: the bandwidth has been improved significantly at the cost of reasonable phase margin degradation, keeping all other performances nearly unchanged. This optimum has been achieved with only 5 optimization steps, thus demonstrating not only the efficiency of the proposed novel approach, but also its excellent compatibility with the modern CAD-tools for analog circuit design.

5 Conclusions

A novel approach for analog circuit optimization involving automatic selection of the device types has been presented. Including additional formal parameters to the model cards of the devices, it permits a smooth transition of the model parameters across the range of available device types. In this way, the optimal device type can be evaluated using gradient-based optimization strategies. At the same time, the proposed approach is fully compatible with stochastic optimization algorithms. Thus, the universality of the latter can be utilized together with the efficiency of the deterministic optimization techniques.

Acknowledgements. This work was supported by the BMBF research project HONEY under grant No. 01 M 3184.

References

Berkner, J.: Kompaktmodelle fuer Bipolartransistoren, expert verlag, 2002.

Cadence Design Systems: Virtuoso NeoCircuit Circuit Sizing and Optimization, http://www.cadence.com/products/custom_ic/neocircuit/index.aspx, 2008.

Dimov, B., Hennig, E., Lang, Ch., and Sommer, R.: Direct Performance Evaluation of Bipolar Transistor Devices for Analog Circuit Design, Proc. Xth Int. Workshop on Symbolic and Numeric Methods, Modeling and Applications to Circuit Design SMACD'08, 49–55, 2008.

Furth, P. M. and Andreou, A. G.: A High-Drive Low-Power BiCMOS Buffer Using Compound PMOS / NPN Transistors, Proc. 36th Midwest Symposium on Circuits and Systems, 2, 1369–1372, 1993.

Graeb, H.: Analog Design Centering and Sizing, Springer, 2007.

Gummel, H. K. and Poon H. C.: An Integral Charge Control Model of Bipolar Transistors, AT&T Tech. J., 49, 827–852, 1970.

MunEDA GmbH: WiCkeD Tool Suite, http://www.muneda.com, 2008.

Wolfram Research: Mathematica 6, http://www.wolfram.com/products/mathematica/index.html, 2008.

X-FAB Semiconductor Foundries AG: Design Kit Documentation, http://www.xfab.com/fileadmin/X-FAB/Download_Center/Technology/BiCMOS/XB06_Data_sheet.pdf, 2008.

27

Energy modelling in sensor networks

D. Schmidt[1], M. Krämer[2], T. Kuhn[2], and N. Wehn[1]

[1]Microelectronic System Design Research Group, University of Kaiserslautern, Germany
[2]Networked Systems Group, University of Kaiserslautern, Germany

Abstract. Wireless sensor networks are one of the key enabling technologies for the vision of ambient intelligence. Energy resources for sensor nodes are very scarce. A key challenge is the design of energy efficient communication protocols. Models of the energy consumption are needed to accurately simulate the efficiency of a protocol or application design, and can also be used for automatic energy optimizations in a model driven design process. We propose a novel methodology to create models for sensor nodes based on few simple measurements. In a case study the methodology was used to create models for MICAz nodes. The models were integrated in a simulation environment as well as in a SDL runtime framework of a model driven design process. Measurements on a test application that was created automatically from an SDL specification showed an 80% reduction in energy consumption compared to an implementation without power saving strategies.

1 Introduction

In the vision of a world of ambient intelligence in which people are constantly interacting with an adaptive, proactive, and user friendly environment, wireless sensor networks are a vital component, because of their inherent redundancy, their robustness, and their transparency to the user. As the nodes in AmI sensor networks have very limited energy resources and computing power, a key challenge is the development of energy efficient application specific communication protocols. In this paper, we present a component based modelling methodology for energy modelling of sensor nodes for these kinds of networks. The proposed approach is applicable for a wide range of wireless sensor nodes and applications. The models serve two purposes:

- At design time the models are used to estimate or simulate the energy consumption of the designed application and communication protocols.
- As the models are very lightweight they can be used to estimate the energy consumption at runtime. This way the node can estimate the remaining lifetime of its energy source and adjust its quality of service accordingly.

We show how the models can be integrated into a simulation environment that is capable of simulating complete networks of nodes accurately regarding their behavior and energy consumption. This allows for optimization of network behavior and network lifetime rather than single node lifetime.

As a side-effect of the modelling methodology, the model's semantics can be adapted to reflect basic energy-optimal strategies, without having to add energy optimization strategies explicitly to the model. This is done in the model runtime environment as well as in the code patterns that are used for transforming the model into code. Using this approach, the code for the hardware nodes can be generated automatically and validated using the integrated simulation framework.

The remaining parts of this paper is structured as following: Sect. 2 provides an overview over different energy modelling approaches and introduces our proposed energy modelling methodology. The simulation environment and the underlying integration methodology is presented in Sect. 3. An examplary energy model for the MICAz (MICAz, 2006), a state-of-the-art sensor node, that was obtained using the proposed energy modelling methodology is shown in Sect. 4 before Sect. 5 shows in a case study how these models can be used in a MDD process for energy aware communication protocols. Section 6 summarizes our conclusions and points out areas of future research.

2 Energy modelling methodology

To model the energy consumption of microelectronic devices several different approaches can be taken. Highly accurate estimates can be obtained by simulation of a microelectronic circuit on transistor level or on even lower abstraction levels, for example with SPICE (SPICE, 2006). While these simulations cover all effects including leakage, switching energy,

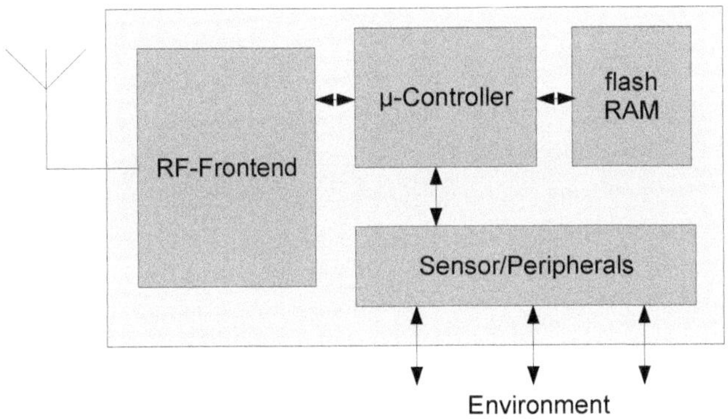

Fig. 1. Architecture of a typical sensor node.

etc., it is very tedious to create the models, as a detailed in-depth knowledge of the hardware is mandatory. Also, simulation on circuit level is very time consuming and is not suitable to simulate whole networks with a large number of individual nodes.

Hence, for many microprocessors energy models that simulate the hardware on the instruction level have been created. This is usually done by measurement of synthetic benchmarks (Senger et al., 2005). These contain loops that execute only one kind of instruction, so that the energy consumption of every single instruction can be calculated from the measurements. Even more measurements have to be done to model inter-instruction dependencies or the impact of operands on energy consumption. These models still allow for relatively accurate energy simulations but require a lot less knowledge of the hardware circuit. Another benefit over circuit simulation is the improved simulation runtime. However, the cost of model creation is relatively high and the resulting model is still too complex to be used at runtime, as sensor nodes typically have very limited computational power.

To overcome the mentioned problems we propose a component based high level modelling approach. A typical sensor node's architecture is depicted in Fig. 1. It is made up of several components like a microcontroller, a transceiver chip, sensor electronics, and various other devices like LEDs, flash RAMs, etc. Each of these components can be operated in different states, e.g. power down, idle, transmitting, and receiving in case of a transceiver chip. Based on the assumption that instruction level effects are negligible for the components used on today's sensor nodes we consider the energy consumption in each state to be constant. Thus, we can model the energy consumption of each individual component using a finite state machine (FSM) in the following way:

- Every operational state of a component is modeled as one state in the FSM.
- Every possible change from one operational state to another is modeled as a state transition in the FSM.

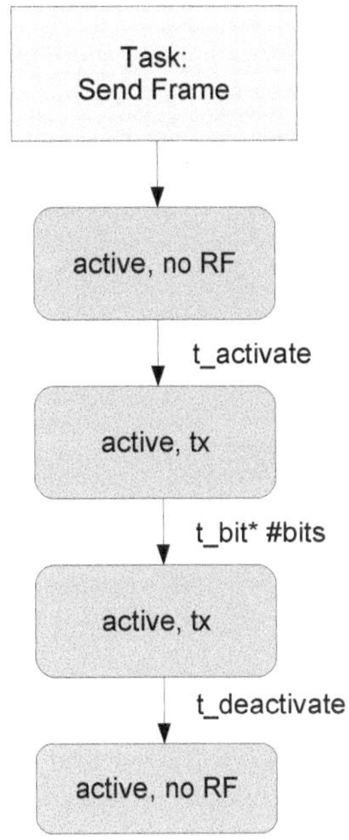

Fig. 2. Parametric flowchart for frame transmission.

- Every state in the FSM is associated with an energy consumed per time.
- Every transition in the FSM is associated with a time needed to switch between two operational states.
- The FSM has a well-defined initial state that corresponds to the stable state a component reaches after power up.

All the times and energy consumptions associated with states and state transitions can be obtained by relatively simple measurements or from data sheets.

Once every component of a sensor node has been modeled as a FSM a similar model for the whole sensor node can be constructed. Basically it is the product automaton of all component models. As usually only the microcontroller can trigger state changes the product automaton can be greatly reduced, because only in states with an active microcontroller other components can change their states.

The resulting FSM only represents the possible states the system can be in, but does not reflect constraints on state changes that appear in normal operation. Wireless communication is a good example as after the transceiver has begun frame transmission it can not be interrupted. This is not represented in the FSM model. Hence, in a second step dynamic behavior of the sensor node in typical scenarios is analyzed to model these constraints. A node's dynamic behavior is defined by the SDL model, the runtime environment implementing the SDL semantics, and the code transformation patterns. Power saving strategies can also be integrated anywhere, ranging from explicit power saving strategies that are part of the application model to implicit strategies that

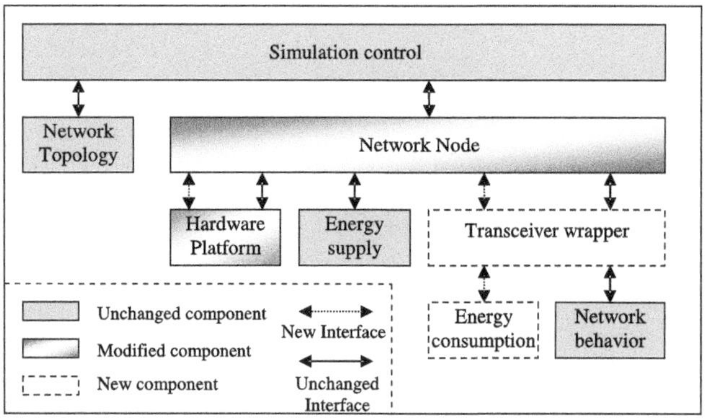

Fig. 3. Structure of the simulator integration framework.

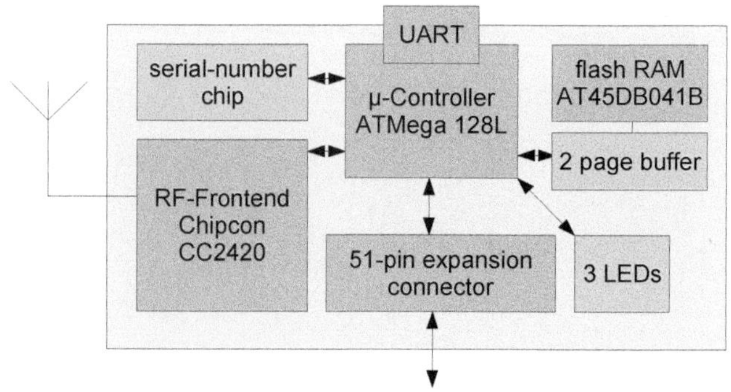

Fig. 4. Architecture of MICAz.

are part of the runtime environment. The complete dynamic behavior of a node can be formalized as a set of communicating state machines. This holds for the SDL model, which is based on state machines, as well as the runtime environment that executes the SDL model. These state machines have common actions with the state machine describing the nodes hardware. An example for a part of a nodes dynamic behavior is given in Fig. 2. It represents the task of sending one frame of a given number of data via the wireless interface to a distant node. It starts in the state where the microcontroller is the only active component of the sensor node. It then triggers the transceiver to transmit one frame. The number of data bytes to be transmitted is the parameter in this case and determines the amount of time spent in transmit mode. After transmission the transceiver is deactivated again to save energy and the initial state is reached again.

From these flowcharts it is very easy to calculate the energy consumed by the specified task, conforming to Eq. (1). Here P_{state} is the power consumed in the state "state", t_{state} is the time spent in this state. P_{trans} and t_{trans} are the power and time for the transitions between to states.

$$E = \sum_{\text{state}} P_{\text{state}} \cdot t_{\text{state}} + \sum_{\text{trans}} P_{\text{trans}} \cdot t_{\text{trans}} \qquad (1)$$

This analytic model can be integrated in a simulator, offering the possibility to perform simulations that predict the energy consumption of a system very accurately.

3 Simulator integration methodology

To simulate the energy consumption of an array of sensor nodes accurately, it is necessary to capture all relevant simulation artifacts that affect the behavior of the sensor node array, i.e. platform resources, network resources, and energy resources. This is due to the fact that one simulated artifact can significantly affect all of the other simulated artifacts, and therefore change the outcome of the simulation. For example, the energy consumption of a sensor node participating in a large sensor network is significantly affected by its wireless transceiver chip. The transceiver chip is used for communicating with other network nodes. Due to network collisions and bandwidth limitations, it is possible that a large number of transmission errors occur, changing the communication pattern of every network node, and thus affecting the energy consumption. Furthermore, platform limitations like inaccurate timers can make this effect worse by introducing jitter into time-synchronized networks. To capture all of these effects, a number of highly specialized simulators are necessary (SAM, 2006).

In (SAM, 2006), it has been shown that simulators for simulating network behavior are already available, that these specialized simulators can be turned into simulation components, and that they can be interconnected using a message based interface to form a system simulator. To integrate the simulation of energy consumption, two steps must be performed: The state machines describing a nodes energy consumption must be integrated, and the message based interfaces for coupling simulation of energy consumption with the remaining parts must be integrated into the system simulator. Therefore, two methodologies are supported:

– Integration of energy consumption into already existing simulators is feasible in the case, that a simulation of the energy consumption of a hardware component is already included in a simulation component. In this case, only an interface must be implemented into the simulation component. The Avrora simulator (Avrora, 2005) includes already the simulation of energy consumption.

– Integration of energy consumption as new simulation components is used when it is not feasible to add another interface to a simulator component. Using this approach, the simulated artifacts are clearly separated, i.e. one simulation component simulates the behavior of a hardware component, and another component simulates its energy consumption by implementing the energy model.

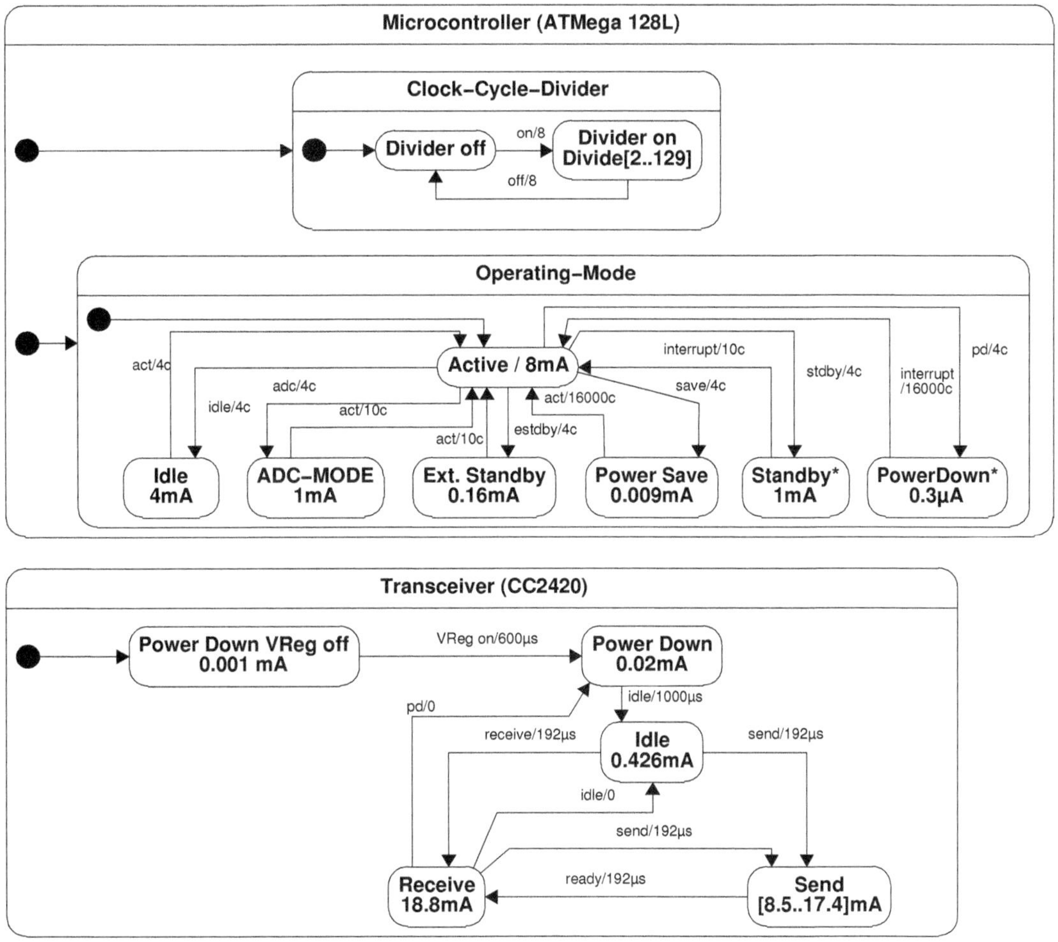

Fig. 5. Energy model of MICAz (only microcontroller and transceiver).

Both methodologies can easily be integrated into the simulator framework. Figure 3 shows the integration of both methodologies for the simulated hardware platform, and for the simulated network. It is shown, that the central component "Network Node", which is part of the simulator framework needs to be changed to integrate the energy consumption into the simulation. This is necessary, because this component controls all simulators that simulate one node. For the simulated hardware, it was feasible to integrate the energy consumption into the already available simulator. For the simulation of the wireless network, a new component implementing the energy model was created, and the simulation component for the network behavior was replaced with a wrapper, distributing simulator messages among the original simulation component for network behavior and the added component that tracks the energy consumption.

4 Energy model for MICAz

The MICAz from Crossbow Technology is a typical wireless sensor node. It features an 8-bit Atmel microcontroller with RISC architecture, clocked at 7.3728 MHz, with 4 kB of internal SRAM and 4 kB data EEPROM as well as 128 kB of internal flash memory. The ZigBee-compliant wireless transceiver chip operates at data rates up to 250 kbit/s. A 512 kB flash memory can be accessed via two SRAM page buffers of 264 bytes each. Three LEDs can be used to show the operational status of the device and each node is equipped with a serial-number chip that gives a node its unique ID. The MICAz has a 51-pin expansion connector as an interface to arbitrary sensors. Figure 4 shows the overall architecture.

For our energy model we only considered the microcontroller, the transceiver chip, and the flash memory. The LEDs can be turned off to reduce energy consumption and the energy consumption of the serial-number chip is negligible. Two component models for the transceiver and the microcontroller of the MICAz are shown in Fig. 5. Here, transition times are given in seconds for the transceiver and in clock cycles for the microcontroller. The microcontroller can be in different power reduced states that differ in the number of timers running and interrupt sources active.

As the MICAz operates at a constant input voltage of 3 V, the energy consumption for every state is given in mA. For the transceiver send state, however, there is not a single value given, but a range of energy consumption. The actual consumption in this state depends on the chosen output power of the transceiver chip. It is remarkable that even at the highest output power the energy consumption is lower than in receive mode.

5 Energy-optimization using energy-models

Energy optimization can be performed on several levels. It can be optimal for the node, for a few nodes or for a whole network. With the simulators described above it is possible to analyze and optimize the consumed energy on each level. For the optimization of a network, special protocols like MacZ (MacZ, 2006), which is currently under development, are needed. The first attempt is the reduction of consumed energy for a single node. For this task accurate models for the used hardware are needed. With our models of the MICAz we have extended the scheduler of the specification language SDL to optimize energy. SDL is a reactive language with additional timers. This enables the language to react on external events, like interrupts or even after a specified time period. SDL expects a continuous timebase and therefore the microcontroller has to wake up in regular intervals to increase the timebase. If no timer is expired and no external signal has arrived the microcontroller can enter a lower energy state. The entered energy state is chosen from the model and is based on the active components. In some cases the operating mode with the lowest consumption cannot be entered, e.g. due to an active transfer via the serial (UART) interface or because the wake up-time is longer than to the next scheduled timer.

To verify the results of the simulator and test the scheduler, a small example application was specified. The application gets temperature measurements every 10 s via serial communication link and send an average value every minute to another station of the same kind.

The measurement of the consumed energy for a time of 10 min shows that the consumption using the scheduler is reduced from 16.2 J to 3.09 J. Using the scheduler for this example application saves 80% energy. This consumption can be reduced in a future version if the precision of the timers is not so important. In the current implementation of the scheduler the precision of the timebase is set to $\frac{1}{16384}$ s, which is the same as without the new scheduler. Since the microcontroller has to wake up at least every 7 ms, it is planned to reduce this precision, if energy is more important.

6 Conclusions and future work

In this paper we presented a simple yet effective methodology to create energy models for wireless sensor nodes. These models consist of two parts: a FSM modelling energy consumption in the different states a node can be in, and flowcharts that represent power saving best-practice-strategies for the dynamic behavior of the node in typical scenarios. We have shown how the energy models can be integrated in platform simulators to simulate a complete network of sensor nodes taking platform resources, network resources, and energy resources into account.

In a case study, we have used the proposed methodology to obtain an energy model for a MICAz sensor node and have used the modified SDL scheduler for an automated generation process for energy optimized code. This way we could show that a MDD process can generate energy optimized code directly from an SDL specification. Compared to an implementation that does not facilitate power saving strategies we measured an improvement in energy consumption of 80%.

Future work will focus on the modelling of energy sources enabling the nodes to estimate their remaining run time. Quality of service degradation or changing roles in a network can benefit from this.

Acknowledgements. This work was financed in context of the BelAmI-Project (BelAmI, 2006).

References

Avrora: Avrora Simulator, http://compilers.cs.ucla.edu/avrora, 2005.

BelAmI: Bilateral German-Hungarian Collaboration Project on Ambient Intelligence Systems, http://www.belami-project.org, 2006.

International Telecommunications Union: Specification and DescriptionLanguage (SDL), ITU-T Recommendation Z.100, August 2002.

Kuhn, T.: MacZ – a QoS MAC Layer for Ambient Intelligence Systems, in: Advances in Pervasive Computing 2006, edited by: Pfeifer, T., Schmidt, A., Woo, W., et. al., Adjunct Proceedings of the 4th International Conference on Pervasive Computing, books@ocg.at, Vol. 207, Austrian Computer Society (OCG): Vienna, 244 pp., ISBN 3-85403-207-2, 2006.

MICAz: MICAz ZigBee Series (MPR2400), http://www.xbow.com/Products/productsdetails.aspx?sid=101, 2006.

SAM06: Kuhn, T. and Becker, P.: A Simulator Interconnection Framework for the Accurate Performance Simulation of SDL Models, Springer LNCS 4320, ISBN 3-540-68371-2, 2006.

Senger, R., Marsman, E., and Brown, R.: Methodology for Instruction Level Power Estimation in Pipelined Microsystems, https://domino.research.ibm.com/acas/w3www_acas.nsf/images/conf05/$FILE/senger.pdf, 2005.

SPICE: Simulation Program with Integrated Circuit Emphasis, http://bwrc.eecs.berkeley.edu/Classes/IcBook/SPICE/, 2006.

28

Modeling of temperature scenarios in a multicore processor system

E. Glocker and D. Schmitt-Landsiedel

Lehrstuhl für Technische Elektronik, Technische Universität München, Theresienstraße 90, 80333 München, Germany

Correspondence to: E. Glocker (elisabeth.glocker@tum.de)

Abstract. In modern CMOS integrated Systems-on-Chip global temperature variations arise as well as local fluctuations in regions of high activity, resulting in the arise of local hot spots. This in turn can greatly affect reliability and lifetime of a chip. Economically affordable processor packaging cannot be provided for the worst case hot spot scenario. In a multicore system a reciprocal influence between the temperatures of neighbouring cores occur leading to increasing core temperature compared to a single core. This results in the need to monitor and regulate the operating temperature during runtime in order to keep it at tolerable values. This can be done in an easy way in an invasive architecture. In this paper the temperature distributions of cores in a multicore system are simulated for various scenarios. Different task allocation techniques and application characteristics as well as different physical conditions such as package types, material parameters and cooling all result in different system power scenarios. The impact of different scenarios which affect the system temperature scenario is investigated. The results are analysed and compared to determine the worst case scenario. With regard to simulation results and practicability the best temperature levelling measures are chosen.

1 Motivation

Integrated circuits today and even more in the future are subject to significant variations: Between different components (resulting from fluctuations in manufacturing process), over space (e.g., "hot spots") and time (short-term: resulting from fluctuations in operating conditions, long-term: resulting from degradation effects due to ageing). This results in significant differences in processing capabilities and in susceptibility to degradation because of varying processing loads for different processing elements even on the same chip. Especially as we can foresee system-on-chip (SoC) architectures with 1000 or more processors on a single chip, static, central management concepts for execution control may meet their limits. So ways must be found to deal with these imperfections resulting in more and more unreliable components.

Invasive architectures for multicore systems provide the required self-organising behaviour: the main idea of invasive computing is a resource-aware programming support, where parallel programs get the ability to explore the system and make decisions for execution (e.g. number of processors to execute on) based on the current state – including physical hardware properties – of the hardware platform (Teich et al., 2011).

To realise invasive architectures, a closed-loop between applications and the underlying hardware is necessary with the need of physical hardware for monitoring and regulating the physical conditions of the processor system, including temperature.

In modern CMOS technologies core temperature in a multicore is increasing and distributed non-uniformly: power densities in modern CMOS technologies increase, because of decreasing feature size and continuing demand for higher performance. Dependent on the thermal conductivity of chip and package, this will influence chip temperature, leading to global variations and local fluctuations in regions of high activity. Therefore, in case of non-uniform workload of different circuit blocks, local temperature hot spots will arise. This together in turn can greatly affect reliability and life-time of a chip (Semenov et al., 2006; Brooks et al., 2007). Additionally, with the power density also the costs for packaging and cooling increase. As a result, economical, reasonably priced processor packaging cannot be provided for the worst case hot spot scenario anymore (Gunther et al., 2001).

In a multicore system, the temperature of a processor block not only depends on its own power density, but also on the power density of neighbour blocks. This leads to a reciprocal influence between the core temperature of neighbouring cores, and with that an even higher increasing core temperature can arise. By monitoring the operating temperature

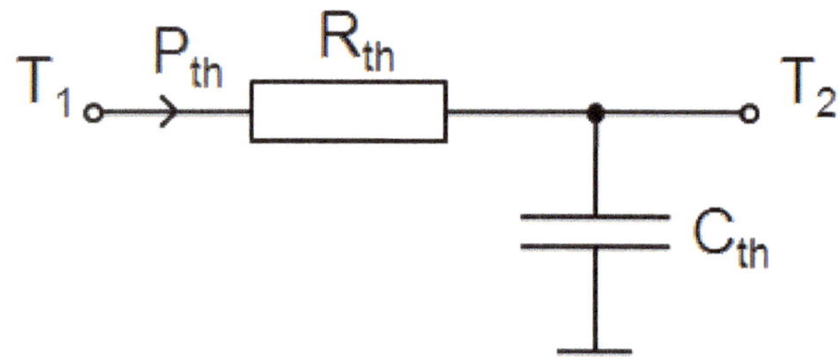

Fig. 1. Due to the duality between heat transfer and electrical phenomenas a RC circuit based thermal model is possible.

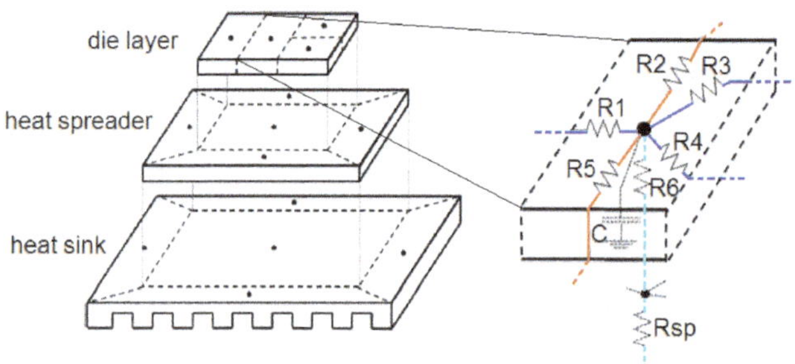

Fig. 2. Modelling of a whole microarchitecture with equivalent RC circuits.

during runtime, it is possible to keep it at tolerable values by load balancing.

After describing the used temperature model in Sect. 2, the simulation environment is presented (Sect. 3). The temperature distribution for various scenarios in a multi-processor system is presented in Sect. 4 and the best temperature limiting measures are presented in Sect. 5.

2 Temperature model

Due to the duality between heat transfer and corresponding electrical parameters – current and heat flow are described by the same potential difference differential equations – a thermal model based on thermal resistances and capacitances can be made: as in an electrical RC circuit, the thermal Rs and Cs lead to exponential rise and fall times characterized by thermal RC time constants. Figure 1 shows the basic components of the thermal model: the temperature difference ΔT can be considered as analogue to the voltage drop and the input power P_{th} as analogue to the current flow. So heat flow can be described as current passing through a thermal resistance R_{th} leading to a temperature difference ΔT. The capacitance C_{th} determines the transient behaviour, i.e. the delay before a power at the input results in reaching the steady state temperature (Boehm, 200; Skadron et al., 2004).

For modelling a whole microarchitecture with the thermal RC model, heat sink, heat spreader and die layer are broken up into several blocks as shown in Fig. 2 (left side). For the die layer the blocks directly correspond to the microarchitecture units. Heat convection is modelled at the package-air interface by a thermal resistance. Inside the package and the silicon only heat conduction has to be modelled, since this is the dominant mechanism there. For each of the blocks, the equivalent RC circuit consisting of lateral and vertical circuit elements is constructed as shown on the right side of Fig. 2.

In the shown example the resistances $R6$ and Rsp model the vertical heat flow. $R1$ to $R5$ model the lateral heat flow from the block center to the center of each edge: $R1$, $R3$ and $R4$ model the heat exchange with neighbour blocks and $R2$ and $R5$ model the heat flow to the next layer (heat spreader/sink in this case). In combination with the capacitance C the transient temperature changes can be modelled. The block node models the power source P as each circuit block is modelled as a heat (power) dissipator (Skadron et al., 2004).

3 Simulation environment

The temperature simulations are done with HotSpot, a thermal model tool suitable for architectural studies (Huang et al., 2004). This compact simulator shows good agreement with finite-element simulations: For steady-state temperature simulations the errors in HotSpot (with respect to ambient temperature of 45 °C) compared to the finite-element simulator Floworks were found to be always smaller than 5.8 % (usually smaller than 3 %) (Skadron et al., 2004). HotSpot was validated against experimental results obtained from a test chip and showed good agreement, with largest errors for steady-state temperature smaller than 5 % (Huang et al., 2004). As input for the temperature simulation, a floorplan and power values per block and time step are necessary. The power values are generated via power simulations with PTScalar, a cycle-accurate microarchitecture-level performance and power simulator for SuperScalar architectures (Liao et al., 2005).

The simulations are done at a base processor frequency of 3.47 GHz and a supply voltage of 1.5 V. Ambient temperature is 35C and chip thickness is 0.5 mm. Convection capacitance is 140.4 J K^{-1}, convection resistance is 0.1 K W^{-1}, silicon thermal conductivity is 100 W $\mathrm{m}^{-1}\,\mathrm{K}^{-1}$ and silicon specific heat is 1.75 MJ $\mathrm{m}^{-3}\,\mathrm{K}^{-1}$ (Skadron et al., 2004). A sampling time step of 2.8818 μs, corresponding to 10 k clock cycles at 3.47 GHz, is chosen. This means the power values are simulated in each time-step, averaged over the last 10 k cycles. The temperature error is less than 0.1 C when using a time step every 10k clock cycles or lower (Skadron et al., 2004). The used processor floorplan is based on a 0.13 μm DEC ALPHA 21364 processor, where the entire periphery is treated as L2 cache (Skadron et al., 2004), as shown in Fig. 3. For the multicore floorplan 16 single DEC ALPHA's (4 × 4) were used. As benchmark example for the power simulations an anagram program was used. Based on the power

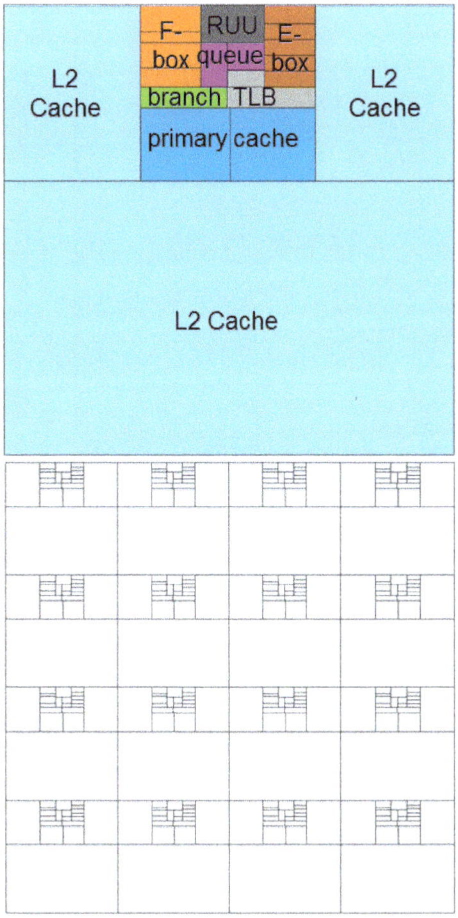

Fig. 3. Single core floorplan based on DEC ALPHA 21364 processor, where entire periphery is treated as L2 cache and resulting 4 × 4 multicore floorplan.

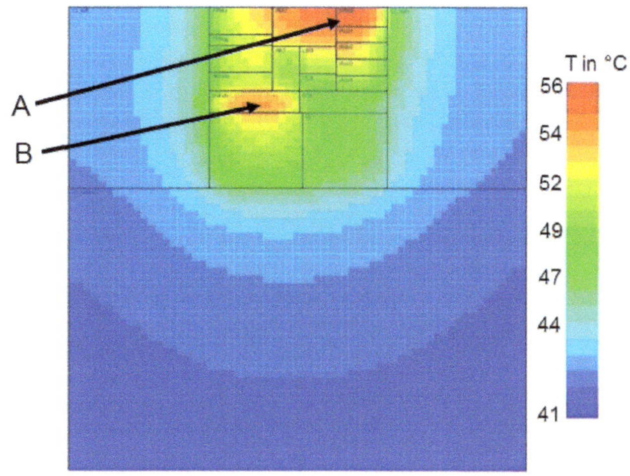

Fig. 4. Temperature distribution for single core. The hotspots are hotspot "A" in RUU/IntReg region (56 °C) and hotspot "B" in branch region (54 °C).

simulation results with the benchmark, a power scenario with a time step that results in an average power per clock cycle of 39 W (corresponding to the average power of the benchmark) is chosen.

4 Evaluation of temperature scenarios

In the following simulations the steady-state temperature is simulated. In general the steady-state temperature is based on average power dissipation and is a good estimation for the temperature present when the average task execution time is large enough. The rate of heat dissipation is affected by the duty cycles of the power sources where different thermal time constants are involved: for package/substrate and die, thermal time constants are non-critical since they are within seconds or milliseconds, respectively. The individual device temperature peak is within 100s of microseconds, so an instantaneous peak junction temperature much higher than the steady-state temperature is possible due to accumulated heat (Chandra, 2006). This is why at first the results of steady-state and transient temperature simulation for a single core (anagram program, runtime 0.28 ms) is evaluated: for calculating steady-state temperature an averaged power value over the whole runtime per core unit is considered. The transient temperature is calculated for every time step during the runtime by using the power values per core unit for the desired time step. The transient temperature per core unit over the whole runtime can be considered as constant, since it varies in all cases by a value smaller than 0.1 C. The steady-state temperature for all the core units normally differs by around 1 °C from its transient temperature. In the most extreme case the steady-state temperature for the IntReg core unit was found to be 3.67 °C lower than its transient temperature.

4.1 Single vs. multicore scenario

The temperature distribution for a single core is shown in Fig. 4. Temperatures range from 41 °C to 56 °C, where the lowest temperature is reached in the L2 cache. There are two hotspots: hotspot "A" in the RUU and IntReg region (56 °C) and hotspot "B" in the branch region (54 °C). IntReg region contains the integer register files and RUU is the register update unit that acts as a combination of register stages and reorder buffer. Branch is the branch prediction unit that predicts which way of a branch (e.g. if-then-else construct) is taken. For the single core scenario the hotspot with highest temperature is hotspot "A".

For the multicore system hotspot "A" is not as critical as hotspot "B": for single core hotspot "A" is located at the edge of the die. The die is sealed inside its package, so there is low airflow along the edge of the die. This results in limited heat removal via convection. Also the surface area for convective transfer is minimal since chip thickness is low. As a consequence temperature in hotspot "A" is higher than temperature in hotspot "B". In a multicore system hotspot "A" is usually

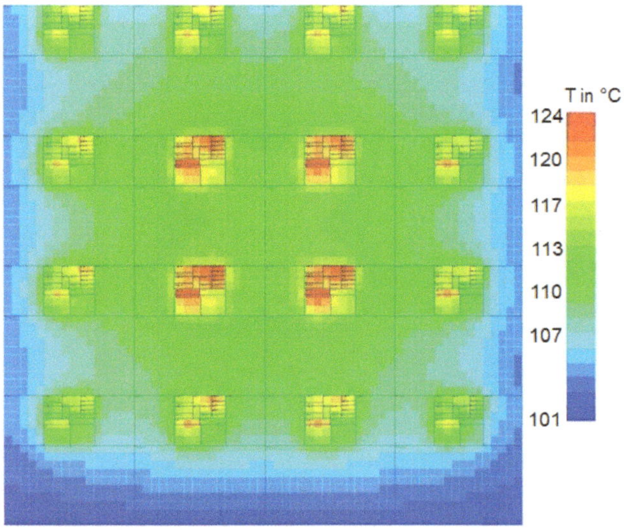

Fig. 5. Temperature distribution for a 4 × 4 multicore system. Hotspot "B" temperature of the middle cores increases by 124 % and hotspot "B" temperature of the corner cores increases by 112 % compared to single core.

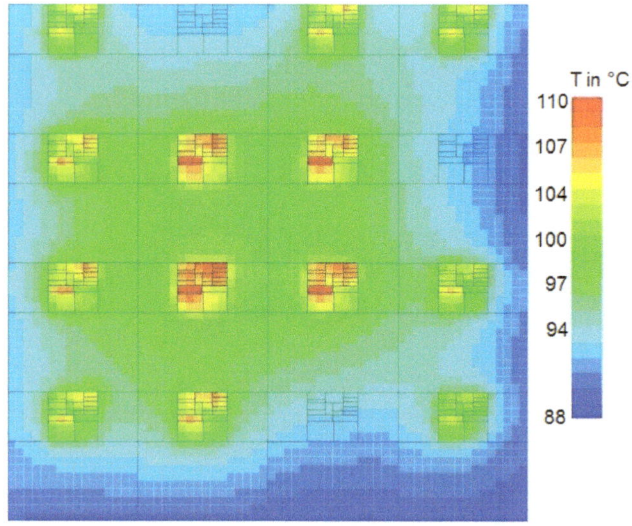

Fig. 6. For 80 % usage scenario, maximum and hotspot "B" temperature of active cores decrease by 11 %.

not located at the edge of the die (only for the four cores in the north edge of the die). In contrast to the fact that for a single core hotspot "A" has a higher temperature, hotspot "B" is the critical one in the multicore scenario (see Fig. 5). Temperature distribution for a 4 × 4 multicore system is shown in Fig. 5. Temperatures range from 101 °C to 124 °C. Since the temperatures of individual cores influence each other, the highest temperature is reached in the middle of the multicore system. The overall maximum temperature of the multicore scenario increases by 120 % (to 124 °C) compared to the single core scenario. The hotspot "B" temperature of the middle cores increases by 124 % (to 121 °C) and the hotspot "B" temperature of the corner cores increases by 112 % (to 114 °C).

4.2 Other usage scenario

In a realistic multicore scenario only about 80 % of the processors (13 cores in a 4 × 4 multicore system) are used. Figure 6 shows the temperature distribution for 4 × 4 multicore system with 13 used cores: the temperature distribution has changed depending on whether neighbouring cores are active or non-active. The overall temperature of active cores decrease by about 11 % (about 13 °C). The change between active and non-active mode of a processor depends on the application and/or the operating system. It is a typical procedure to use power gating in a multicore system to turn off not used cores (Lee et al., 2009). Since additional energy is needed to turn them on again, from an energy point of view it is best to not use cores that are turned off, when there is the possibility to use a not used – but still on – processor instead. This may lead to an unbalanced use of the cores, which is disadvantageous from a reliability, life-time and temperature point of view.

4.3 Other usage scenario and intelligent task allocation

During times of lower usage rates, it is possible to choose the placement of active and non-active cores. For a 50 % usage scenario (8 cores, 4 × 4 multicore system) the resulting temperature distribution for square configuration and checkerboard configuration is shown in Fig. 7: in square configuration the bottom 8 cores are active, the upper 8 cores are inactive: Maximum and hotspot "B" temperature of active cores decrease by about 29 % (about 34 °C) compared to full usage. In a lower usage scenario an intelligent task allocation and scheduling is possible, resulting in suitable combinations of application needs/characteristics and individual core health status to reduce overall temperature values. In our simulation every active core performs the same application. In this scenario intelligent task allocation is possible by placing active cores as far away from each other as possible to lower the overall temperature: In checkerboard configuration maximum and hotspot "B" temperature of active cores decrease by about 31 % (about 37 °C), the lowest temperature decrease by about 33 % (about 33 °C) compared to a full-usage scenario. Maximum and hotspot "B" temperature of active cores decrease by about 3 % (about 3 °C) compared to square configuration.

For a 25 % usage scenario (4 cores, 4 × 4 multicore system) the resulting temperature distribution for square configuration and no-middle configuration is shown in Fig. 8: in square configuration the left bottom 4 cores are active, all others are inactive: maximum and hotspot "B" temperature of active cores decrease by about 44 % (about 54 °C) compared to full usage. In no-middle configuration all middle

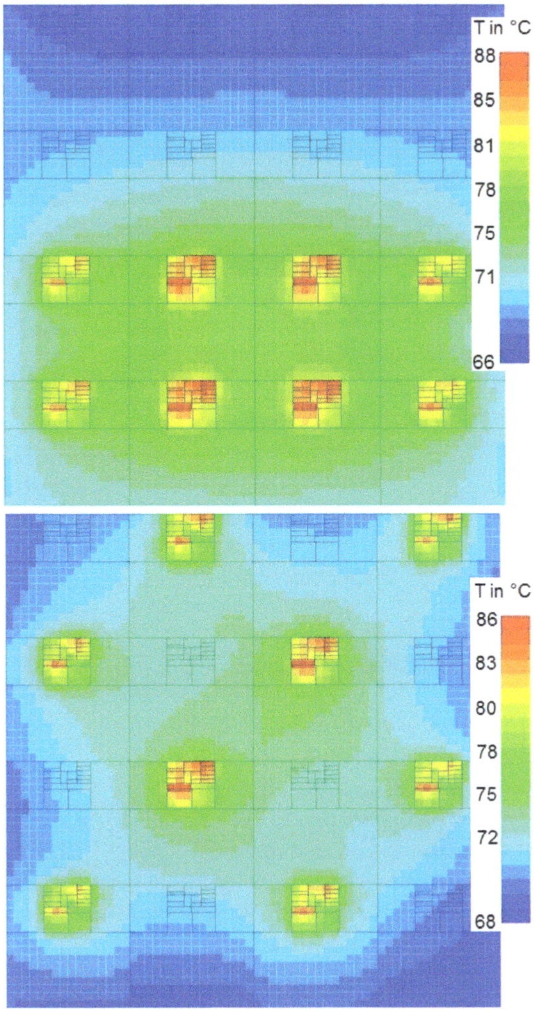

Fig. 7. 50 % usage scenario: in square configuration (left side) maximum and hotspot "B" temperature (active cores) decrease by 29 %. In checkerboard configuration (right side) maximum and hotspot "B" temperature (active cores) decrease by 31 % compared to full-usage scenario and by 3 % compared to square configuration.

cores are inactive. Maximum and hotspot "B" temperature of active cores decrease by about 47 % (about 57 °C), the lowest temperature by about 50 % (about 50 °C) compared to full-usage scenario. Maximum and hotspot "B" temperature of active cores decrease by about 5 % (about 3 °C) compared to bottom square configuration.

4.4 Better cooling

When reducing the ambient temperature from 35C to 30C, the overall temperature values decrease by about 4 % (about 5 °C). When choosing a better packaging by doubling the thickness of the heat sink, the maximum and hotspot "B" (middle cores) temperature decrease by about 4 % (about 5 °C). Hotspot "B" (corner cores) and lowest temperature stay nearly the same. However, both techniques are cost in-

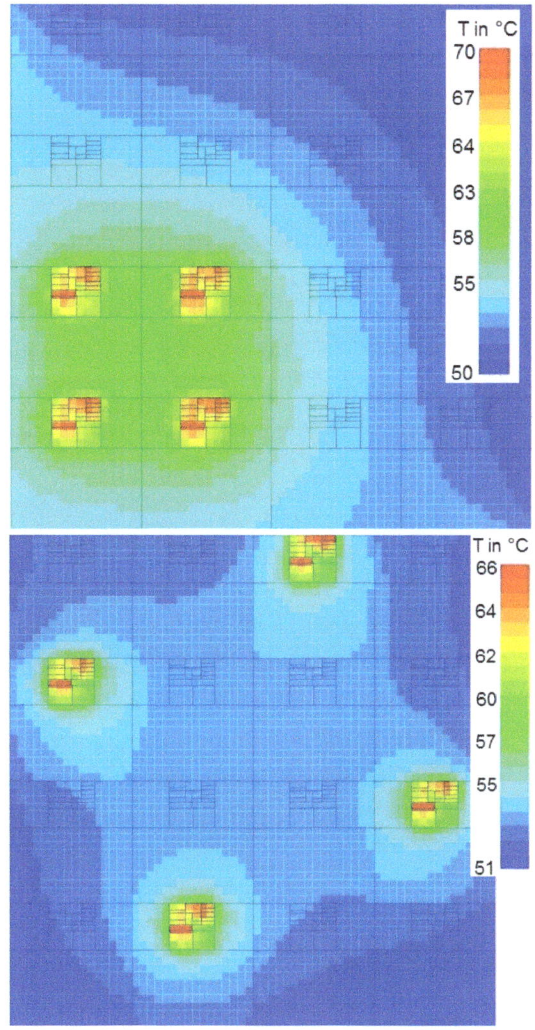

Fig. 8. 25 % usage scenario: in square configuration (left side) maximum and hotspot "B" temperature (active cores) decrease by 44 %. In no-middle configuration (right side) maximum and hotspot "B" temperature (active cores) decrease by 47 % compared to full-usage scenario and by 5 % compared to square configuration.

tensive: as power increases there is no longer a linear relationship between cooling capabilities and cost of solution (Gunther et al., 2001).

4.5 Other power scenario

Temperature is directly depending on input power. By using lower input power for the individual cores, temperatures are decreased. A second power scenario with a time step that results in an average power per clock cycle of 32 W (corresponding to the start-up phase of the benchmark) is evaluated. Compared to the standard power scenario with average power per clock cycle of 39 W temperature distribution stays the same, but overall temperature values decrease by about 13 % (about 15 °C).

Table 1. Comparison of temperature limiting measures.

Measures	Temperature improvement	Implementation
other power scenario: 1) average power per cycle 32 W 2) voltage scaling ($V_{DD} = 1.3V$) ⇒average power per cycle 29 W	overall temperature decrease by about 13 % (15 °C) by about 18 % (22 °C)	e.g. improving task fragmentation & scheduling to decrease individual input powers implementing of (adaptive) voltage /frequency scaling
80 % usage	overall temperature decrease by about 11 % (13 °C)	may lead to unbalanced core use if always the same cores stay (non) active; performance loss; choose of non-active cores
50 % usage, intelligent task allocation & scheduling	overall temperature decrease by about 31 % (37 °C) (compared to full-usage) 3 % (3 °C) (compared to square conf.)	see implementation of 80 % usage; also: implementation: intelligent task allocation & scheduling, choose suitable application-core pairs

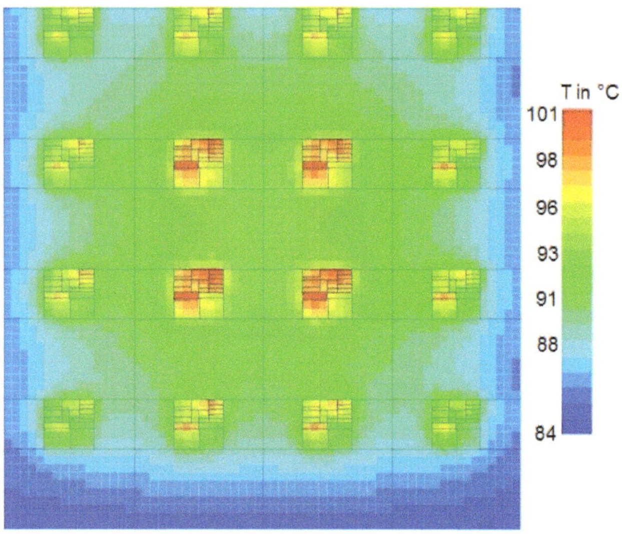

Fig. 9. For voltage scaling with reduced supply voltage of 1.3 V, maximum and hotspot "B" temperature (middle cores) decrease by 18 %.

There are various possibilities to change the power scenario e.g. using lower supply voltage: Fig. 9 shows the resulting temperature for reduced supply voltage of 1.3 V (instead of 1.5 V). This results in the use of a time step that results in a lowered average power per clock cycle of 29 W (instead of 39 W). The overall temperature distribution stays the same, the overall temperature values decrease by about 18 % (about 22 °C). Lowering supply voltage is realizable by e.g. using adaptive voltage (or frequency) scaling. In the approach presented in Wirnshofer (2011), the supply voltage is changed depending on the current system situation during runtime by detecting timing pre-errors. The supply voltage is regulated to a value where no pre-errors occur based on this information.

Improving the task execution can also realize lowered input powers per core: when parallel applications are executed on a multicore system, an improved task fragmentation and scheduling for an application lead to lower input powers for the individual cores. This is one of the basic principles of Invasive Computing (Teich et al., 2011).

5 Temperature limiting measures

Full usage of a multicore system is the worst case scenario. In a realistic usage scenario about 80 % of the cores are used, which makes it possible to e.g. choose active cores, choose suitable cores for an application to execute on and implement techniques that allow different treatment for cores in a multicore system like power gating, where not used cores are shutdown or adaptive voltage scaling. With regard to our simulation results and practicability the best temperature limiting measures are either changing the power scenario or changing the usage scenario combined with intelligent task allocation (see Table 1): with enforced lower usage scenarios, the temperature improvement is good. But performance may be lost, since not all cores are used. It may also lead to an unbalanced core use, if always the same cores stay active and inactive, respectively.

When adding intelligent task allocation to lower usage scenarios a further improvement of temperature is possible. Using another power scenario leads to good temperature improvement but some effort for implementation of e.g. intelligent task fragmentation and scheduling or (adaptive) voltage/frequency scaling is necessary. In case of using adaptive voltage/frequency scaling also supply voltage or frequency may be lowered. Since the temperature should be monitored and regulated during runtime, different techniques can and should be implemented in the same system. This makes it possible to choose the best temperature limiting measure in the current system state during runtime.

6 Summary and conclusions

Economically affordable processor packaging cannot cover the worst case hot spot scenario anymore. In this paper temperature distributions of different scenarios in a multicore system are analysed. With regard to simulation results and practicability the best temperature limiting measures were investigated and compared to each other: The best solution is either intelligent core choice combined with lower usage-rates or lowering of the input power e.g. by implementing supply voltage or frequency scaling or using parallel programs with intelligent task fragmentation techniques. To perform this, the temperature should be monitored and regulated during runtime, so we propose an implementation of different concepts and choosing a temperature limiting measure during runtime for individual situations. To conquer worst case hot spot scenarios in multicore systems, a collaboration of intelligent software, operating system and hardware solutions is necessary. Future work will include the simulation of bigger multicore areas, different applications on the same invasive multicore system and the investigation of the closed-loop interaction of software, operating system and hardware for invasive architectures.

Acknowledgements. This work was supported by the German Research Foundation (DFG) as part of the Transregional Collaborative Research Centre "Invasive Computing" (SFB/TR 89).

References

Boehm, R. F.: Conduction Heat Transfer, in: The CRC Handbook of Thermal Engineering, edited by: Keith, F., CRC Press LLC, 2000.

Brooks, D. et al.: Power, Thermal, and Reliability Modeling in Nanometer-Scale Microprocessors, In IEEE Micro, Vol. 27, No. 3, 49–62, 2007.

Chandra, R.: Transient Temperature Analysis, Robuspic Workshop on European Solid-State Device Research Conference (ESSDERC), 2006.

Gunther, S. H. et al.: Managing the Impact of Increasing Microprocessor Power Consumption, In Intel Technology Journal first quarter 2001, 2001.

Huang, W., et. al.: Compact Thermal Modeling for Temperature-Aware Design, in: Proceedings Design Automation Conference, 878–883, 2004.

Lee, J. et al.: Optimizing throughput of power- and thermal-constrained multicore processors using DVFS and per-core power-gating, in: ACM/IEEE Design Automation Conference, 47–50, 2009.

Liao, W., et al.: Temperature and supply Voltage aware performance and power modeling at microarchitecture level, in: IEEE Transactions on Computer-Aided Design of Integrated Circuits and Systems, 1042–1053, 2005.

Semenov, O et. al.: Impact of Self-Heating Effect on Long-Term Reliability and Performance Degradation in CMOS Circuits, in: IEEE Transactions on Device and Materials Reliability, Vol. 6, No. 1, 17–27, 2006.

Skadron, K., et al.: Temperature-Aware Microarchitecture: Modeling and Implementation, In ACM Transactions on Architecture and Code Optimization, Vol. 1, No. 1, 94–125, 2004.

Teich, J. et al.: Invasive Computing: An Overview, in: Multiprocessor System-on-Chip – Hardware Design and Tool Integration, edited by: Hübner, M. and Becker, J., 241–268, Springer, Berlin, Heidelberg, 2011.

Wirnshofer, M., et al.: A variation-aware adaptive voltage scaling technique based on in-situ delay monitoring, In IEEE International Symposium on Design and Diagnostics of Electronic Circuits & Systems (DDECS), 261–266, 2011.

Simulation of RF MEMS based matching networks and a single pole double throw switch for Multiband T/R Modules

S. A. Figur[1], F. van Raay[2], R. Quay[2], L. Vietzorreck[3], and V. Ziegler[1]

[1]EADS Deutschland GmbH, EADS Innovation Works Germany, 81663 München, Germany
[2]Fraunhofer Institute of Applied Solid-State Physics (IAF), Tullastr. 72, 79108 Freiburg, Germany
[3]TU München, Arcisstr. 21, 80333 München, Germany

Correspondence to: S. A. Figur (sascha.figur@eads.net), F. van Raay (riedbert.van.raay@iaf.fraunhofer.de), R. Quay (ruediger.quay@iaf.fraunhofer.de), L. Vietzorreck (vietzorreck@tum.de), and V. Ziegler (volker.ziegler@eads.net)

Abstract. This work presents concepts and designs for two essential components of a frequency agile transmit and receive module based on Radio Frequency Micro Electro Mechanical System (RF MEMS) switches for the frequency range from 3.5 GHz up to 8.5 GHz. The advantages of a variable power amplifier (PA) matching compared to a fixed broadband solution are examined and discussed in context with the designed components.

To demonstrate the principle functionality, an assembly concept is presented, which allows for the integration of a frequency agile 6 W power amplifier with surrounding components like phase shifters, switches and antennas. An RF MEMS switching element is introduced as a high isolation polarization switch, featuring low insertion loss as well as almost no DC power consumption.

1 Introduction

The general trend towards higher data throughput, frequency-agility as well as higher power together with higher efficiency, demands for more broadband and more flexible systems. Furthermore reconfigurability becomes an important factor to adapt to changing requirements during lifetime of a given system. In many systems, in special space borne systems, adjustments during lifetime can only be done, if these adjustments have already been foreseen and prepared during design stage and assembly.

Figure 1 depicts a frequency-agile transmit and receive module (T/R module), used in non-frequency-agile systems amended by reconfigurable components to deal with changing demands. The single components shall not be described in depth here, since in this paper the main focus lies on the power amplifier and high-power ultra-wide-band polarization switch.

With a fixed broadband matching network, compromises have to be accepted due to the Fano-Bode-criteria (see Sect. 2.2.1): the transistor cannot be optimally matched for all operating frequencies and conditions at a time. In contrast to the common method to match a transistor over a wide frequency range, trading bandwidth for good matching, this paper presents a multi-band concept to match the transistor optimally based on the current operating situation. Adaptive multi-band matching offers the possibility to operate the amplifier more flexibly, while choosing the situationaly best fitting state.

Figure 1.2 shows a schematic power amplifier based on a gallium nitride transistor (GaN), which is matched variably by radio frequency micro electro mechanical system (RF MEMS) switches. Similar systems based on different technologies were presented in (Malmqvist, 2009; Qiao et al., 2005; Lu et al., 2003). Due to the almost zero DC power consumption and low insertion losses (Siegel et al., 2006), RF MEMS are well suited for applications in efficient systems as well as in power amplifiers. The used RF MEMS technology is described in detail in Siegel et al. (2006).

The possibilities of an adaptive frequency agile matching network exceed the possibility of only matching the transistor for the current transmit or receive frequency, furthermore the operating mode of the transistor can be changed. Is the operating frequency kept constant, the transistor can be tuned to deliver maximum output power or to operate with maximum power added efficiency (PAE). While a broadband matching can be realized to certain extend with fixed matching networks, changing the operating mode of a transistor

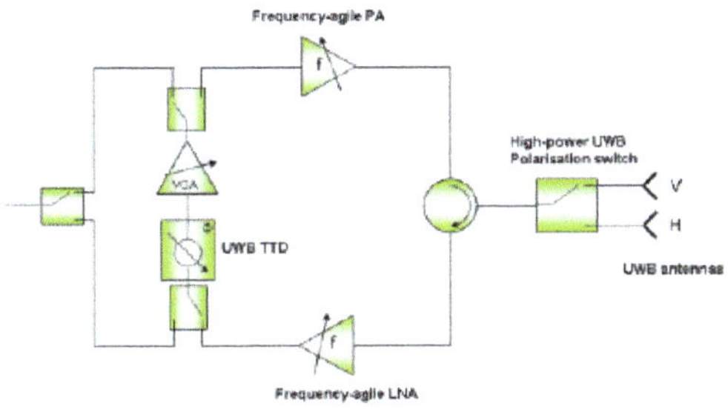

Figure 1.1 Single T/R module including antenna

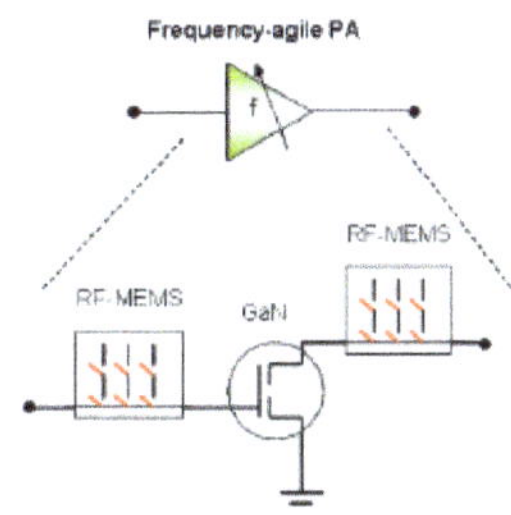

Figure 1.2 Power amplifier

Fig. 1. Frequency-agile T/R module. The figure depicts a single module, the complete system generally consists of an array of several single modules.

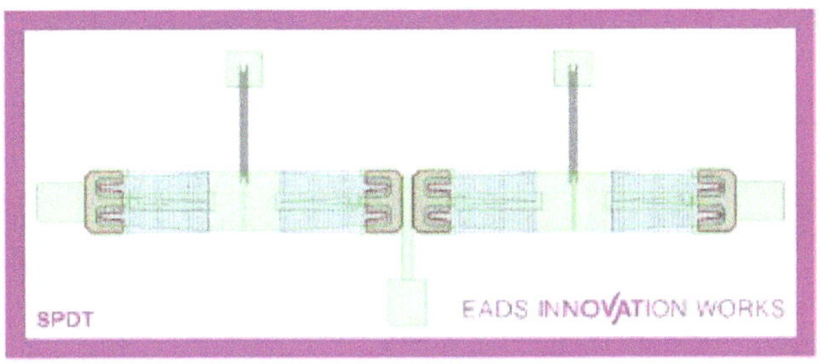

Fig. 2. Single pole double throw (SPDT) switch. Input at bottom side, outputs left and right hand side of the layout, control lines at top.

for a fixed frequency is only possible with adaptive matching networks.

Based on the frequency bands of common communication applications, the frequency range between 3.7 GHz and 8.4 GHz was chosen for demonstration purposes, while the range from 1.5 GHz up to 18.1 GHz is considered to be the target specification for further development.

2 Design of single components

2.1 High-power ultra-wide-band polarization switch

An elementary part of a frequency-agile and reconfigurable T/R module, such as shown in Fig. 1 is a switching element, which can switch between the vertically or horizontally polarized antennas. This element is also necessary to change between transmit and receive mode by changing the way the vertical signal path in Fig. 1.1 is connected with the rest of the module, adjusting magnitude and phase of the signal. Figure 2 depicts an RF MEMS based single pole double throw (SPDT) switch, which can be fabricated on silicon and integrated in a module.

The simulation results of this switch are given in Fig. 3. These simulations do not consider any external wiring such as bond wires or pads, since the external wiring has a strong influence on the performance and is itself strongly depending on the integration environment of the switch.

Insertion loss of less than 0.55 dB and a matching of lower than -10 dB over a frequency range from 4 GHz up to over 20 GHz show promising results. Within the complete target frequency range below 20 GHz isolation between the opened and the closed output is lower than -40 dB.

2.2 Frequency-agile power amplifier

2.2.1 General limitations of matching networks

A well-known limit on the performance of matching networks is given by the Fano-Bode-criteria. In Steer (2010) the Fano-Bode-criteria is simplified to

$$\frac{\mathrm{BW}}{\omega_0}\ln\left(\frac{1}{\Gamma_{\mathrm{avg}}}\right) \le \frac{\pi}{Q_{\mathrm{load}}}.$$

$\frac{\mathrm{BW}}{\omega_0}$: fractional bandwidth of the matching network in radians; Γ_{avg}: average in-band reflection coefficient; Q_{load}: quality factor of load impedance.

Among other general results Steer, 2010 concludes from this:

1. A higher Q_{load} results in a narrower bandwidth for a constant in-band average reflection coefficient.
2. Matching the load for all frequencies is only possible for $Q_{\mathrm{load}} = 0$.
3. With increasing Q_{load} specified bandwidths are increasingly difficult to achieve.

To cope with this restriction in multi-band applications, one can either:

- trade bandwidth for good matching in a broadband matching network
- use separate hardware for each band with optimum matching and switch between them

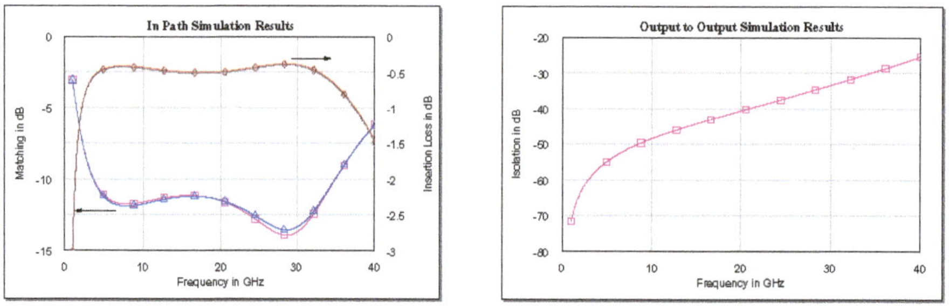

Figure 3.1 In path simulation results

Figure 3.2 Simulation of isolation between both outputs, when one of the outputs is routed to the input

Fig. 3. Simulation results of the SPDT from Fig. 2.

Table 1. Discrete design frequencies of matching networks for maximum output power.

Maximum Output Power	Frequencies in GHz				
Gate	4	5	6.5	7.5*	8.5
Drain	4	5.5	6.5	–	8.5

* Only in radials version.

– implement tunable narrowband matching networks.

While the first two solutions have drawbacks in terms of performance or hardware costs and area consumed, the latter one has the potential for very good performance with minimum hardware and area consumption. Therefore, the following tunable matching networks are designed to provide optimum matching conditions for each frequency-band/mode of operation for the multi-band application.

2.2.2 Chosen transistor

Due to the target of achieving a high output power with a single transistor, a GaN transistor was chosen. An example of a GaN based T/R module can be found in Schuh et al. (2008). The source and load impedances, which the transistor has to be connected to for frequencies between 3.5 GHz and 8.5 GHz, are shown in Fig. 4; data is given for maximum output power and maximum PAE.

Due to the position of the matching points, it can be expected, that the most broadband matching can be achieved for the output matching tuned for maximum output power. The output matching network for maximum efficiency should be slightly more broadband than both input matchings, but all three are expected to be rather narrowband.

2.2.3 Design principle of matching networks

The matching networks for gate and drain of the GaN transistor are realized as single stub networks on silicon. Therefore a stub line is connected in parallel at some distance to the load structure, which has to be matched to 50 Ω. The stub line can be realized with an open circuit at the end or can be shortcut to ground. Due to the available RF-MEMS technology, the stubs have to be realized open ended. The stub represents a certain impedance to ground, based on the electrical length and the configuration of the stub line – in this case open ended. The distance between stub and load structure is such chosen, that the parallel stub compensates for the imaginary part of the impedance seen at the stub position looking towards the load structure. Therefore a purely resistive impedance of 50 Ω is seen from the input of the matching network.

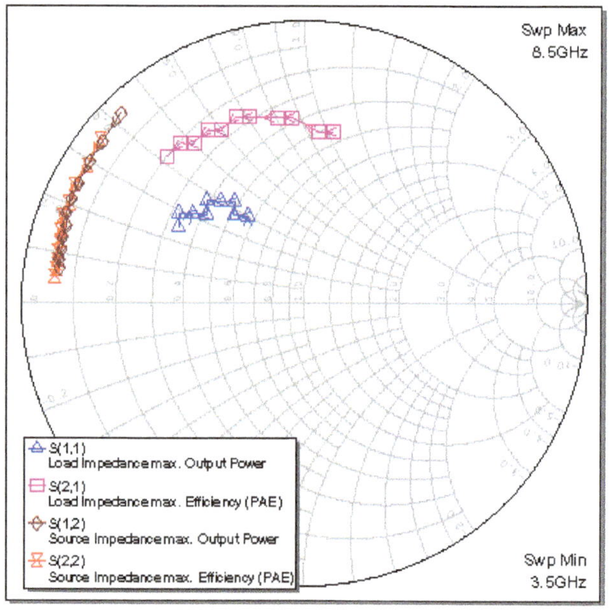

Fig. 4. Large signal matching points of the chosen GaN transistor. Data normalized to 50 Ω.

Figure 5 depicts a part of an RF-MEMS based matching network. Reconfigurability is achieved by several switches used to connect different stub lines at different positions to the through line. Based on the matching network's dimensions, the length of the through line is fixed. Using different

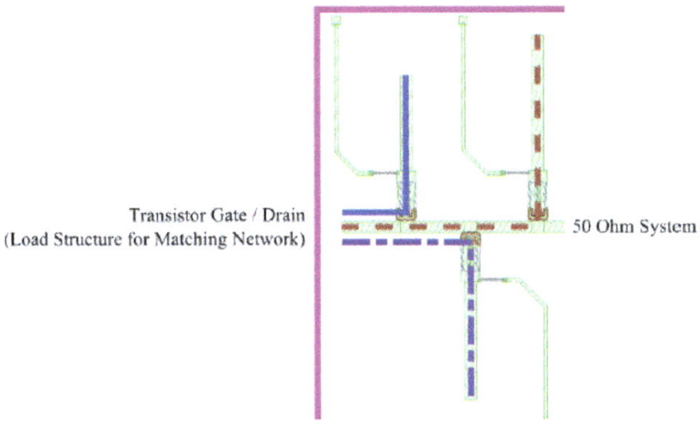

Fig. 5. Principle of a single stub matching network based on RF-MEMS switches. Different colors and linetypes indicate different states of the matching network chosen by throwing the according switch.

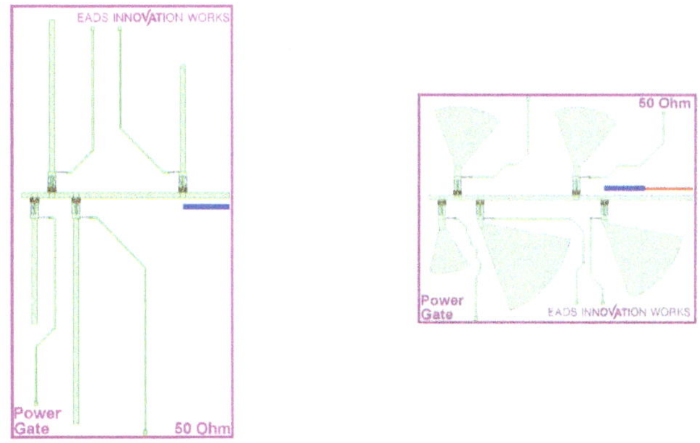

Figure 6.1Stub based

Figure 6.2 Radials based

Fig. 6. Gate matching network for maximum output power. The blue bar indicates the right hand side line length of the stub based network, the red, thinner bar shows the additional line length necessary for the radials based network.

stubs for matching results in varying residual line lengths between currently used stub and the end of the through line towards the 50 Ω system. Since the through line is designed with an impedance of 50 Ω, the varying residual line length does not affect the matching.

Stub based matching networks, as depicted in Fig. 5, are simple to design and well suited to match the load for a few, discrete frequency points and operating modes. The three depicted examples also illustrate the limitations of the approach. To avoid resonances and coupling between the stub lines, a minimum distance between two adjacent stubs has to be kept. As a result, the number of stub lines per length unit is limited.

A simple solution for this limitation is to put some stubs $\lambda/2$ or a multiple farther away from the load structure than calculated. This increases the possible number of stubs in the design, but introduces higher losses for all possible states, due to the longer through line.

2.2.4 Frequency-agile matching networks – maximum output power

To demonstrate the capabilities of this approach, frequency agile matching networks for gate and drain have been designed to tune the chosen transistor for maximum output power. Table 1 gives an overview of the matched discrete frequencies. Due to geometrical restrictions, the matching networks could not be designed to match gate and drain of the transistor both at 5 GHz or 5.5 GHz. Therefore gate and drain can only be matched at two slightly different frequencies around 5 GHz.

Figure 6 and Fig. 7 show the designed matching networks for maximum output power. As stated before, due to the available technology, the stubs have to be realized as open circuits. Therefore, a possibly shorter shortcut stub to ground is not applicable and some stubs are rather long in Fig. 6.1.

Another possibility to reduce the size of the matching networks incorporates radial stubs as shown in Fig. 6.2 and Fig. 7.2. The higher capacity to ground of the radial stubs decreases the length and can reduce space consumption. A major drawback of this solution is the increased width of the stubs, further decreasing the number of possible stubs per length unit. In Fig. 6.2 on the right hand side of the matching network, a piece of line (indicated by thin, red bar) had to be added to cope with the wide radial stubs. However, the necessity of adding additional line lengths is strongly dependent on the design. In Fig. 7.2 no adjustments were necessary.

Simulation results of the gate and drain matching networks are given in Fig. 8 and Fig. 9, respectively. The long stub lines of the stub based gate network result in higher insertion losses and limit the area of the smith chart, which can be matched well. In consequence matching is partially low, which could be significantly improved by using radial stubs.

Since the gate matching is rather narrowband, the frequency range around 7.5 GHz cannot be matched with one of the four states in the stub version. Consequently another stub has been added in the radial stub version to match the gate around 7.5 GHz. At the five design frequencies matching is below −10 dB and insertion loss is still high, but acceptable, since this can be compensated by a driver amplifier.

The simulation results of the drain matching networks in Fig. 9 show very good performance for the stub version as well as the radial stub version. For almost all frequencies between 3.5 GHz and 8.5 GHz it is possible to find one network state with insertion loss lower than 0.5 dB and matching lower than −15 dB. The radial stub version offers comparable performance and is reduced in size.

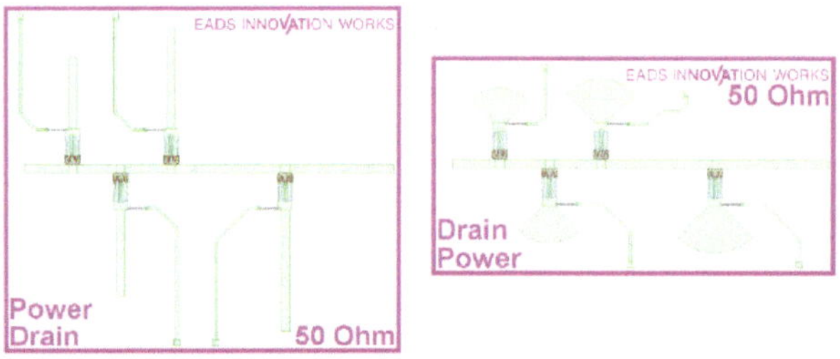

Figure 7.1 Stub based **Figure 7.2 Radials based**

Fig. 7. Drain matching network for maximum output power.

2.2.5 Frequency-agile matching networks – maximum PAE

Additional to the networks shown in Sect. 2.2.4, matching networks for maximum PAE have been designed. Since the design principle is the same, only the smaller radial versions are shown here. The layout is depicted in Fig. 10.

The matching is, as expected to be, more narrowband than shown for the maximum output power networks. Figure 11 gives the simulation results of these structures. Due to the low space consumption of the radial stubs in Fig. 10.2, additional stubs could be added to achieve a better average matching over the whole frequency range. The gate matching network has not been extended to keep the losses as low as possible. The corresponding design frequencies for both matching networks are given in Table 2.

The drain matching works quite well over a wide frequency range, exhibiting losses lower than 1 dB. However, the performance of the gate matching network is comparable to the one shown in Fig. 8 for maximum output power mode.

2.2.6 Combined mode frequency-agile matching networks

The formerly shown networks are designed to demonstrate the frequency-agility of this concept. As stated before, a major advantage of adjustable matching networks is the possibility to change the operating mode of the transistor during operation. Therefore a combined matching network, based on the single matching networks for maximum output power and maximum PAE, has been designed to show this functionality.

In Table 3 design frequencies for maximum output power and maximum PAE are summarized. The necessary source impedances for maximum output power and maximum PAE are not very different for the gate matching. Therefore the combined matching network is only designed for drain matching.

The according designs as stub based and partially radials based are depicted in Fig. 12. The two remaining line stubs in Fig. 12.2 could not be replaced with radials, since the radials would have been so big, that these would have been overlapping in any possible configuration. Therefore the two stubs have been kept. Furthermore the additional design frequencies of the drain matching network for maximum PAE had to be discarded, due to the high number of stubs, which had to be arranged.

Table 2. Discrete design frequencies of radials based matching networks for maximum PAE.

Maximum PAE	Frequencies in GHz					
Gate	–	4	5	–	6.5	8.5
Drain	3.5	4	5	5.5	6.5	8.5

Table 3. Discrete design frequencies of combined matching network.

Drain	Frequencies in GHz			
Maximum Efficiency (PAE)	4	5	6.5	8.5
Maximum Output Power	4	5.5	6.5	8.5

Figure 15.1 and Fig. 15.2 depict simulation results of the combined matching network for stub based and radials based realizations, respectively. Both versions are quite similar and show very good performance.

The maximum output power states are close to the results shown before, even though insertion loss is a bit higher, due to the additional stubs and therefore longer through line. The optimum load impedance for maximum PAE is varying more over frequency than the one for maximum output power and exhibiting a higher quality factor. As a direct result, the matching of the four discrete states is narrower, than the one for maximum output power. However, the matching is good for the design frequencies and insertion loss is acceptable.

3 Integration approach

In a later design step the power amplifier shall be integrated in a T/R module as depicted in Fig. 1. Figure 13 gives a top view of the integration approach, restricted to the close surrounding of the power amplifier.

In agreement with the manufacturer of the GaN transistor, the integration approach in Fig. 14 has been worked out. The basis of this approach is a conductive carrier board, providing a common ground plane. To keep the bond wires as short as possible, it is accepted, that the ground plane is not flat but contains a small step around the transistor. This is necessary, due to the different chip thicknesses of silicon matching networks and GaN transistor.

The circuit is assembled beginning from the GaN transistor in the center of Fig. 14. After soldering the transistor, the additional parts are conductively glued on the carrier board as close to the adjacent structures as possible to minimize the bond wire lengths.

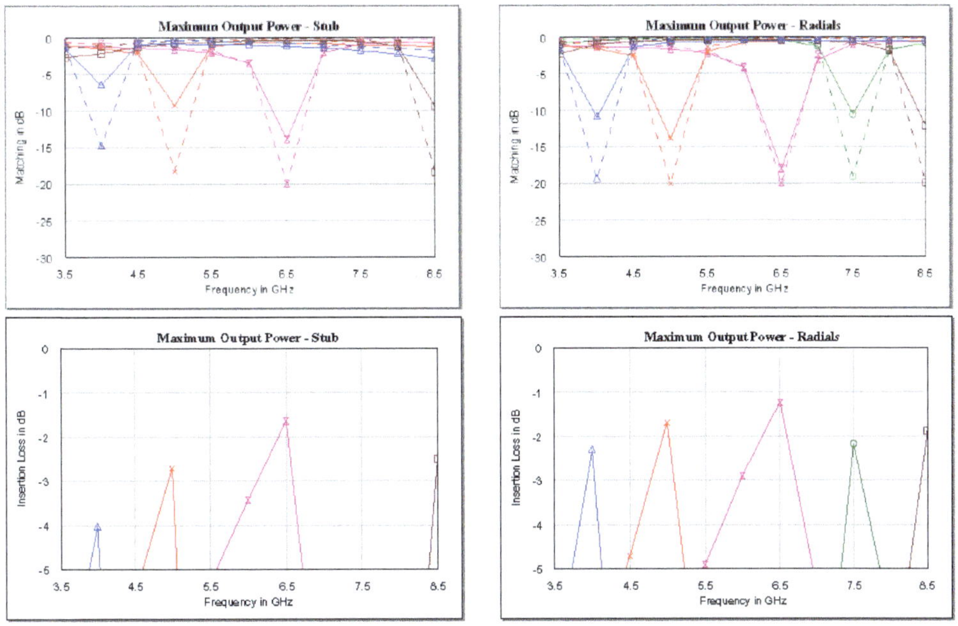

Fig. 8. Simulation results of gate matching networks for maximum output power – stub and radials based versions. Different colors and symbols denote different states of the networks. Matching graphs include S_{11} (dashed line) as well as S_{22} (solid line). Insertion Loss showsS_{21} and S_{12}, which are identical.

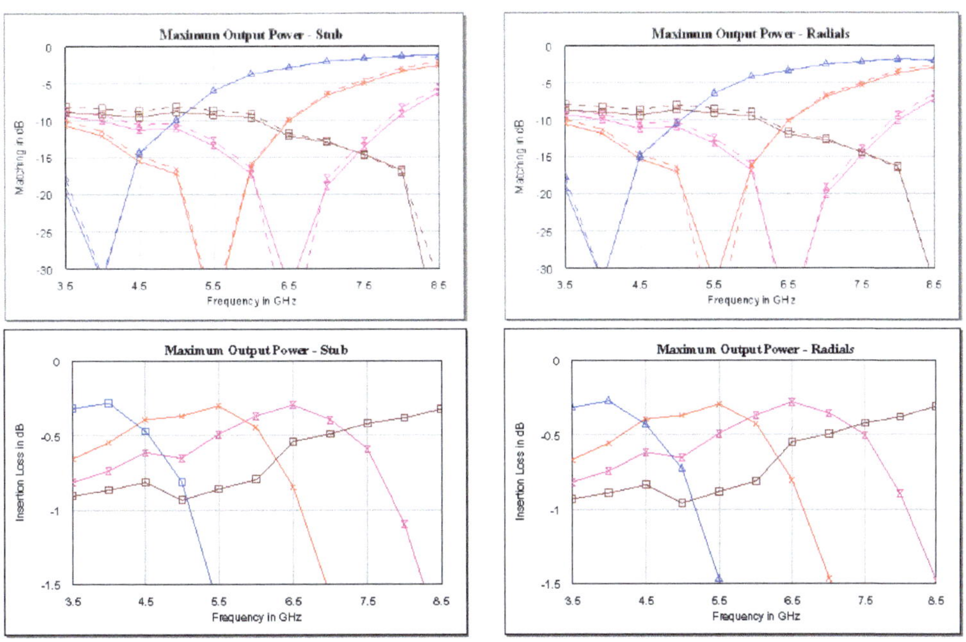

Fig. 9. Simulation results of drain matching networks for maximum output power – stub and radials based versions. Different colors and symbols denote different states of the networks. Matching graphs include S_{11} (dashed line) as well as S_{22} (solid line). Insertion Loss shows S_{21} and S_{12}, which are identical.

Table 4. Comparison with presented measurements of other groups.

Source	Remarks	Frequency range in GHz	States	S11/S22	S21
Malmqvist, 2009	Values given	23.9–26.4	4	≤ −15 dB	−1.5 dB to −2 dB
Lu, 2003	Input matching, only graphs shown	6 and 8	2	−10 dB to −15 dB	−2 dB to −5dB
	Output matching, only graphs shown	6 and 8	2	−10 dB to −15 dB	−1 dB
This work, simulated (Radials based)	Gate, Max output power	3.5–8.5	5	≤ −10 dB	−1 dB to −2.5 dB
	Drain, Max output power	3.5–8.5	4	≤ −15 dB	≤ 0.5 dB
	Gate, Max efficiency	3.5–8.5	4	−10 dB	−1 dB to −2 dB
	Drain, Max efficiency	3.5–8.5	6	≤ −15 dB	≤ −0.9 dB
	Drain, Combi, Max output power	3.5–8.5	4	≤ −15 dB	−0.5 dB to −0.6 dB
	Drain, Combi, Max efficiency	3.5–8.5	4	≤ −15 dB	−0.7 dB to 1.2 dB

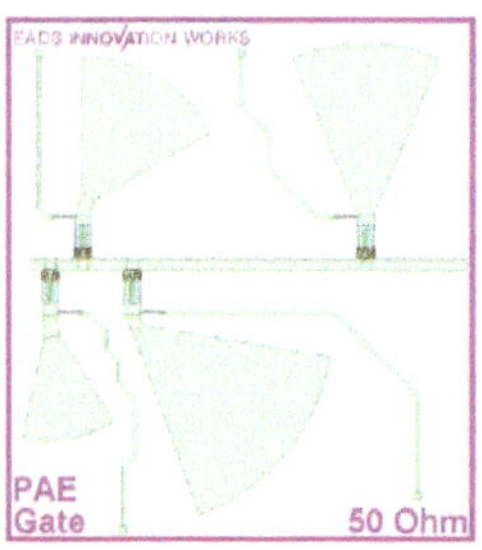

Figure 10.1 Gate matching network

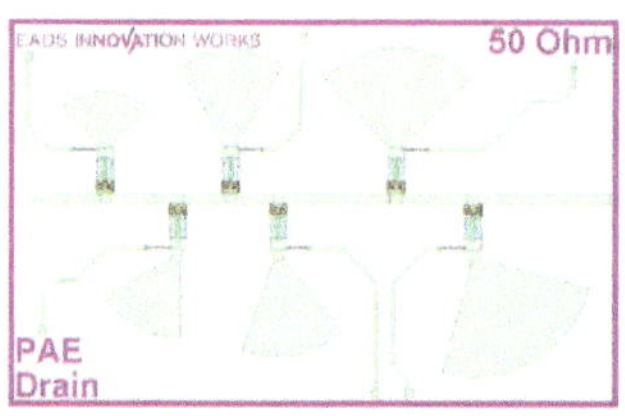

Figure 10.2 Drain matching network

Fig. 10. Radials based matching networks for gate and drain, optimized for maximum efficiency (PAE).

The Rogers RT/Duroid 5880 PCBs are fabricated in a thickness similar to the silicon chips. Thus the top sides of the silicon based matching networks and the PCBs are leveled without the necessity to mill steps into the conductive carrier. Milling tolerances for the conductive carrier could cause additional spacing between neighboring structures, resulting in longer bondwires. Additional surrounding components can be integrated in the same way or on top of the Rogers RT/Duroid 5880 PCBs.

4 Comparison with state of the art

A comparison with other works is difficult, since this kind of matching network is specifically designed for one transistor. Therefore generally the system's overall performance is shown and the performance of the matching network itself is not presented. Two similar works have been found, which give results on the matching networks. The main results given are summarized in Table 4 and compared with this work.

5 Conclusions

This paper shows the design of an RF-MEMS SPDT switch for frequency-agile systems as well as adaptive frequency-agile matching networks for adaptively matching power amplifiers to the current operating conditions. The presented simulation results promise a valuable benefit of adaptive matching networks compared to fixed matching networks in multi-band applications.

Apart from overcoming compromises necessary for broadband matching networks, the adaptive frequency-agile matching networks proposed in this paper, can tune the transistor between different operational modes, while keeping the operating frequency fixed. Therefore an amplifier can be tuned from maximum output power to maximum PAE, depending on current conditions. This feature is interesting for efficient systems, which are in non-constant environments. If propagation attenuation increases, the transistor can be tuned from maximum PAE mode to maximum output power to compensate for the higher losses. When the attenuation decreases, the mode can be changed back to maximum PAE to save energy.

The functionality of the components could be verified based on simulations in the frequency range between 3.5 GHz and 8.5 GHz.

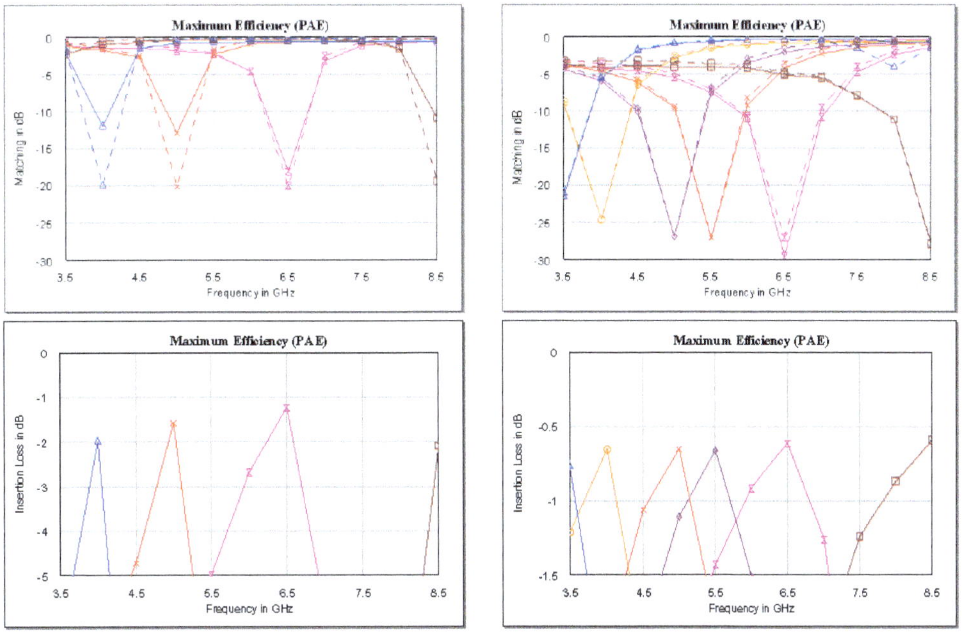

Fig. 11. Simulation results of gate and drain matching networks for maximum efficiency (PAE) – radials based. Different colors and symbols denote different states of the networks. Matching graphs include S_{11} (dashed line) as well as S_{22} (solid line). Insertion Loss shows S_{21} and S_{12}, which are identical.

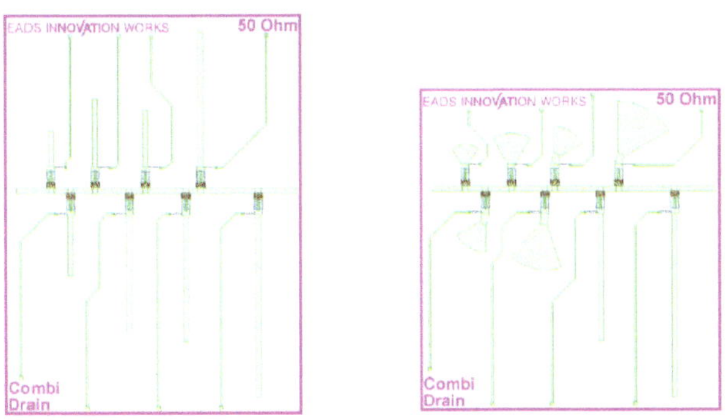

Figure 12.1 Stub based

Figure 12.2 Radials based

Fig. 12. Combined drain matching network. Transistor can be tuned for maximum output power or maximum efficiency (PAE).

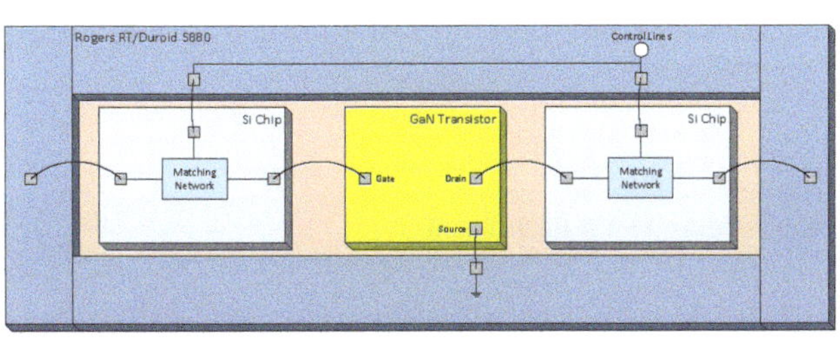

Fig. 13. Schematic integration approach in top view.

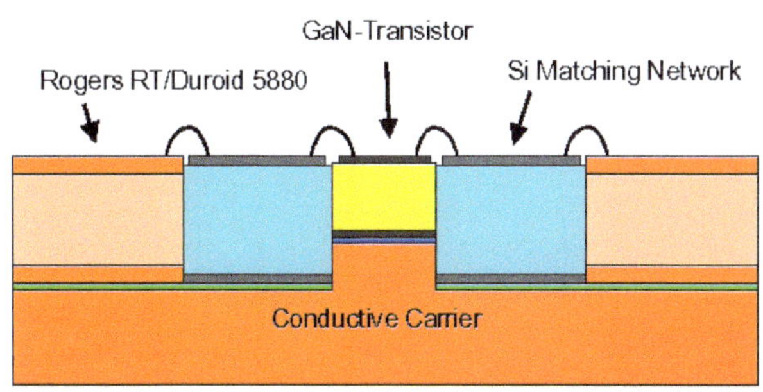

Fig. 14. Schematic integration approach in cross section.

6 Outlook

The designed structures are fabricated, so that they can be integrated and characterized. The components will be integrated in a module after a complete characterization and analysis of the results. Finally a redesign can improve the performance and the frequency range.

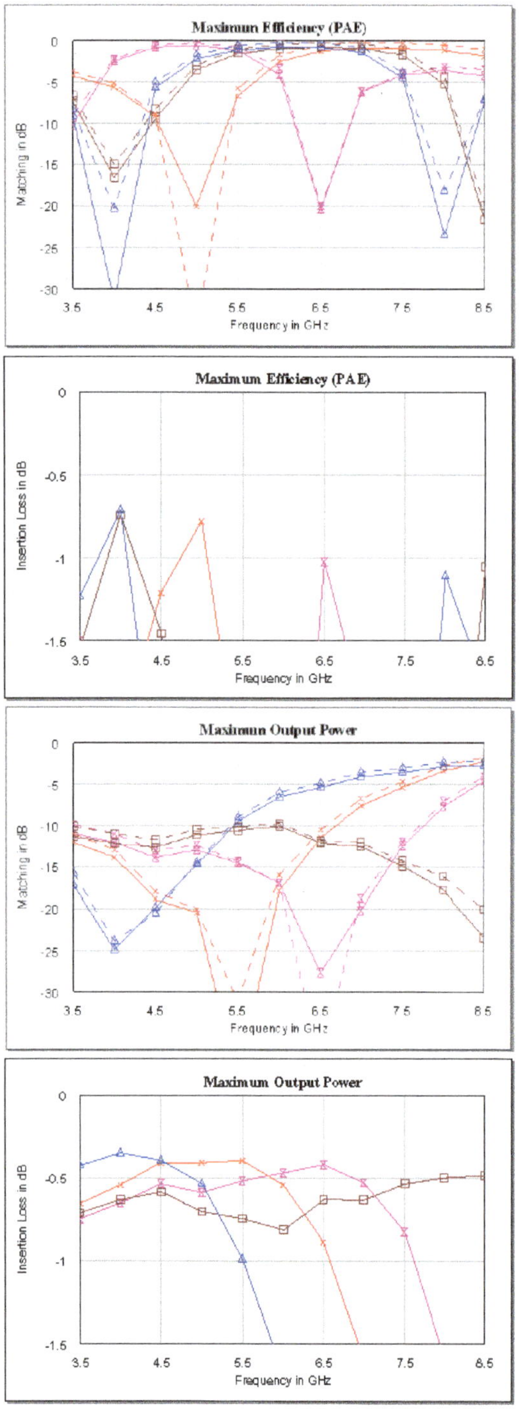

Figure 15.1 Stub based

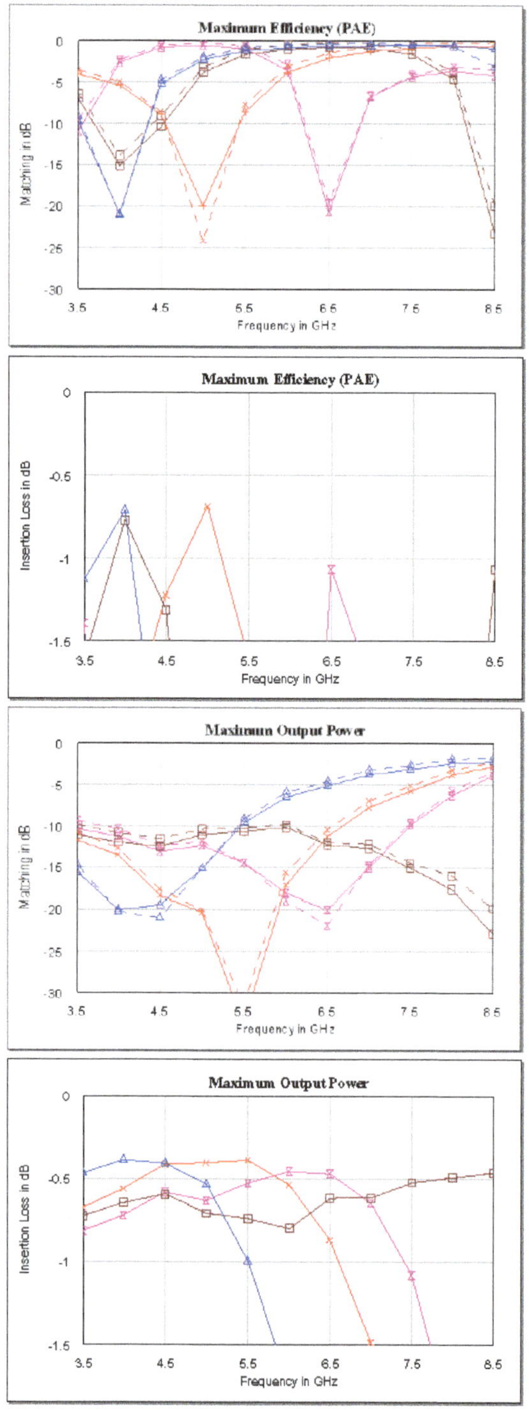

Figure 15.2 Radials based

Fig. 15. Simulation results of combined matching network for maximum output power and maximum efficiency (PAE) – stub and radial based versions for drain. Different colors and symbols denote different states of the networks. Matching graphs include S_{11} (dashed line) as well as S_{22} (solid line). Insertion Loss shows S_{21} and S_{12}, which are identical.

Acknowledgements. This work has been financially supported by FP7 – Space, Project Saturne, Grant agreement no. 242458.

References

Lu, Y., Peroulis, D., Mohammadi, S., and Katehi, L. P. B.: A MEMS Reconfigurable Matching Network for a Class AB Amplifier, IEEE Microwave and Wireless Components Letters, 13, 437–439, 2003.

Malmqvist, R., Rantakari, P., Samuelsson, C., Lahti, M., Cheng, S., Saijets, J., Vaha-Heikkila, T., Rydberg, A., and Varis, J.: RF MEMS based impedance matching networks for tunable multiband microwave low noise amplifiers, Semiconductor Conference, 2009, CAS 2009, International, vol. 1, October 2009.

Qiao, D., Molfino, R., Lardizabal, S. M., Pillans, B., Asbeck, P. M., and Jerinic, G.: An Intelligently Controlled RF Power Amplifier With a Reconfigurable MEMS-Varactor Tuner, Transactions on Microwave Theory and Techniques, 53, 1089–1095, 2005.

Schuh, P., Sledzik, H., Reber, R., Fleckenstein, A., Leberer, R., Oppermann, M., Quay, R., van Raay, F., Seelmann-Eggebert, M., Kiefer, R., and Mikulla, M.: GaN MMIC based T/R-Module Front-End for X-Band Applications, Microwave Integrated Circuit Conference, 2008, EuMIC 2008, European, 27–28 October 2008.

Siegel, C., Ziegler, V., Schonlinner, B., Prechtel, U., and Schumacher, H.: Simplified RF-MEMS switches using implanted conductors and thermal oxide, 36th European Microwave Conference, September 2006.

Steer, M.: Microwave and RF Design, Published by SciTech Publishing, Inc. 2010.

Comparison of in-situ delay monitors for use in Adaptive Voltage Scaling

N. Pour Aryan[1], **L. Heiß**[1], **D. Schmitt-Landsiedel**[1], **G. Georgakos**[2], **and M. Wirnshofer**[1]

[1]Lehrstuhl für Technische Elektronik, Technische Universität München, Germany
[2]Infineon Technologies AG, München-Neubiberg, Germany

Correspondence to: N. Pour Aryan (n.aryan@tum.de)

Abstract. In Adaptive Voltage Scaling (AVS) the supply voltage of digital circuits is tuned according to the circuit's actual operating condition, which enables dynamic compensation to PVTA variations. By exploiting the excessive safety margins added in state-of-the-art worst-case designs considerable power saving is achieved. In our approach, the operating condition of the circuit is monitored by in-situ delay monitors. This paper presents different designs to implement the in-situ delay monitors capable of detecting late but still non-erroneous transitions, called Pre-Errors. The developed Pre-Error monitors are integrated in a 16 bit multiplier test circuit and the resulting Pre-Error AVS system is modeled by a Markov chain in order to determine the power saving potential of each Pre-Error detection approach.

1 Introduction

In today's advanced integrated circuits, with ever increasing performance demands, methods and schemes aiming to minimize power consumption in digital circuits are becoming more of a concern. The state-of-the-art worst-case guardbanding approach adds several safety margins considering Process, Voltage and Temperature variability and also Aging (PVTA) to the supply voltage required for correct operation under nominal condition. This fixed supply voltage approach results in unnecessary power dissipation in non-worst-case scenarios. Adaptive Voltage Scaling (AVS) controls the supply voltage according to the operating condition of the circuit by exploiting unused safety margins.

The operating condition of the circuit is commonly monitored by measuring the timing of the circuit. The global speed monitor approach in Drake et al. (2007) uses delay lines (replica paths), aiming to track the effect of temperature changes or supply voltage reduction on the critical paths. Delay lines are only replicas of the real circuit and thus local variations, process gradients or aging mechanisms are not covered by these global monitors.

In contrast to the global speed monitor, in-situ delay monitors measure the timing inside the real circuit and thus provide reliable timing information considering both local and global variations. The in-situ delay monitors are enhanced flip-flops providing information about the timing of the circuit. The AVS approach in Bowman et al. (2009) and Das et al. (2009) uses in-situ delay monitors capable of detecting timing errors. Additional complexity is introduced in these approaches for error recovery. Moreover, repetition of erroneous computation in case of an error occurrence excludes the application of error detection approaches to real time systems.

Therefore, Wirnshofer et al. (2011) use in-situ delay monitors that are able to detect critical transitions instead of error detectors. These in-situ delay monitors, also called Pre-Error flip-flops, are inserted at the end of critical paths in the circuit. The supply voltage is adjusted during normal operation of the circuit based on the timing information obtained by Pre-Error flip-flops. As the voltage adaptation is based on Pre-Errors instead of errors, no additional circuitry and clock cycles are needed for data recovery. Therefore, the Pre-Error AVS is applicable to real time systems.

In this paper, different designs to implement the Pre-Error flip-flops are developed and investigated. The rest of this paper is organized as follows. In Sect. 2, an overview of Pre-Error AVS is given. Section 3 discusses the design of the developed Pre-Error flip-flops. In Sect. 4, the Markov model representing the Pre-Error AVS to estimate the power saving potential is introduced. Comparisons between the different Pre-Error flip-flops are presented in Sect. 5 and the paper is concluded in Sect. 6.

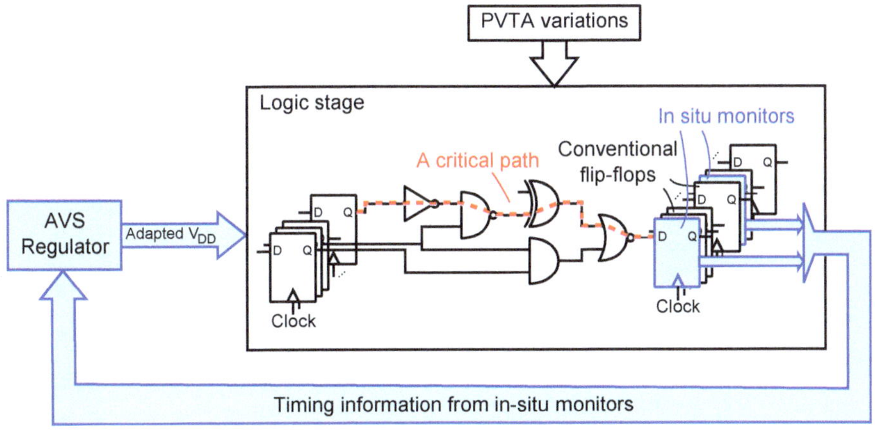

Fig. 1. Block diagram of the Pre-Error AVS system: timing information extracted by in-situ delay monitors is used to tune the supply voltage in a closed-loop configuration.

2 Overview of Pre-Error Adaptive Voltage Scaling

Pre-Error flip-flops are in-situ delay monitors with the ability to distinguish between relaxed and critical operation of the circuit. Timing information extracted by Pre-Error flip-flops inserted at the end of critical paths is used for the closed-loop voltage control, shown in Fig. 1. Data transitions during a certain time interval before clock rising edge, the Pre-Error detection window, result in a Pre-Error signal.

For different input patterns the delay of the digital circuit varies. Figure 2 shows the delay histogram of the 16 bit multiplier's critical output for random input patterns. The part of the distribution lying right of the dashed line denoting the detection window will result in Pre-Errors. Note that for the same input pattern, different operating conditions result in different delays for the digital circuit. Reducing the supply voltage increases the delay of the circuit and moves the histogram to larger delays, resulting in more Pre-Errors.

The supply voltage is adjusted based on the count of Pre-Errors during an observation interval, i.e. a defined number of clock cycles. To adapt the supply voltage, two decision limits for the Pre-Error count after an observation interval are defined. If n_{pre}, the count of Pre-Error occurrence in the previous observation interval, is under a lower threshold of $n_{\text{limit}\downarrow}$ the voltage is decreased. If n_{pre} is above an upper threshold of $n_{\text{limit}\uparrow}$ the voltage is increased. For Pre-Error counts between $n_{\text{limit}\downarrow}$ and $n_{\text{limit}\uparrow}$ the voltage is maintained. As the voltage adaptation is based on the statistics of Pre-error occurrence, an overcritical voltage reduction, i.e. the risk of timing errors, can never be completely ruled out. However, the error rate can be adjusted in a user-defined controlled manner by adjusting the number of clock cycles for the observation interval, the detection window length T_{Pre} or the decision limits $n_{\text{limit}\downarrow}$ and $n_{\text{limit}\uparrow}$.

With varying activity rates in digital circuits, regulating the voltage based on the number of occurred Pre-Error pulses during a *fixed* time interval would result in aggressive voltage reduction in phases with low activity. This would lead to unpredictable risk of timing errors. Therefore, the observation interval should only consist of active clock cycles. To distinguish between active and inactive clock cycles, we introduce a transition detector monitoring all data transitions, either relaxed or critical.

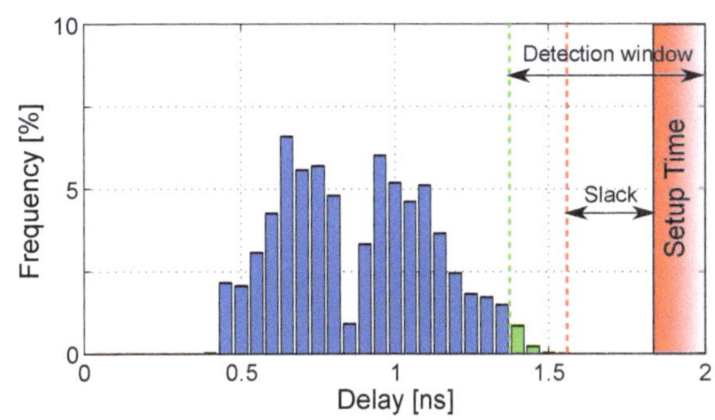

Fig. 2. Delay histogram of the multiplier's critical output for different input patterns at a fixed operating condition.

The AVS Regulator shown in Fig. 1 consists of an AVS control unit and a voltage regulator. The digital AVS control unit counts the Pre-Errors in an observation interval of N active clock cycles and decides whether to change the supply voltage. The communication to the voltage regulator is done via a binary control word representing the voltage level.

3 Implementation of Pre-Error flip-flops

A Pre-Error flip-flop is a conventional flip-flop with additional circuitry enabling it to detect Pre-Errors and data transitions. For realizing the Pre-Error detection window, either the duty-cycle of the clock signal or a delay element can be used. All Pre-Error flip-flops are designed in a 65 nm low power CMOS technology.

3.1 Duty-cycle based Pre-Error approach

In this approach the Pre-Error flip-flop is implemented in dynamic and static design style, respectively.

3.1.1 Dynamic Pre-Error flip flop

The Dynamic Pre-Error flip flop is designed in dynamic logic, shown in Fig. 3a. The low phase of the clock signal is exploited as the Pre-Error detection window. The transition pulse generator, which is an XOR gate with two inputs of data and delayed data, generates a pulse at its output X for every data transition. Data transitions occurring during the detection window are assigned as Pre-Errors by transistor M_2. All data transitions from one to the following clock rising edge are assigned as Transition pulses by transistor M_4.

For the AVS scheme it is important to define a detection window as accurate and robust as possible. Therefore, the variations of the detection window length are investigated in the following. Since the duty cycle of the clock signal is exploited as the detection window, variations caused by the Pre-Error flip-flop itself and the clock tree are considered. Thus, the clock tree as an H-tree with three levels of buffers is included in the design. The H-tree is the basic clock topology for many clock distribution systems (Tam et al., 2004).

Figure 3b illustrates the deviations from the ideal detection window length in the nominal and corner cases[1] over V_{DD}. The ideal detection window length is considered the length at the supply voltage of $V_{DD} = 1.0\,V$, nominal process and a temperature of $T = 27\,°C$. Moreover, Monte Carlo simulations are performed to determine the uncertainties of the detection window length due to local variations. For the supply voltage range of $V_{DD} = 1.2\,V$ down to $V_{DD} = 0.8\,V$, the 3σ interval due to local variations is derived from Monte Carlo simulations and depicted in Fig. 3b. In this design, global variations have a minor impact compared to local ones.

3.1.2 Static Pre-Error flip flop

For the Static Pre-Error flip-flop, illustrated in Fig. 4a, the falling edge of the clock signal is used as the starting point for the detection window.

The Static Pre-Error flip-flop is implemented in static design style with standard library elements, requiring less design effort than the Dynamic Pre-Error flip-flop. The Pre-Error detector is comprised of a flip-flop with inverted clock as its clock input. The data input and the output $Q2$ are compared by the XOR1 gate. The flip-flop with inverted clock

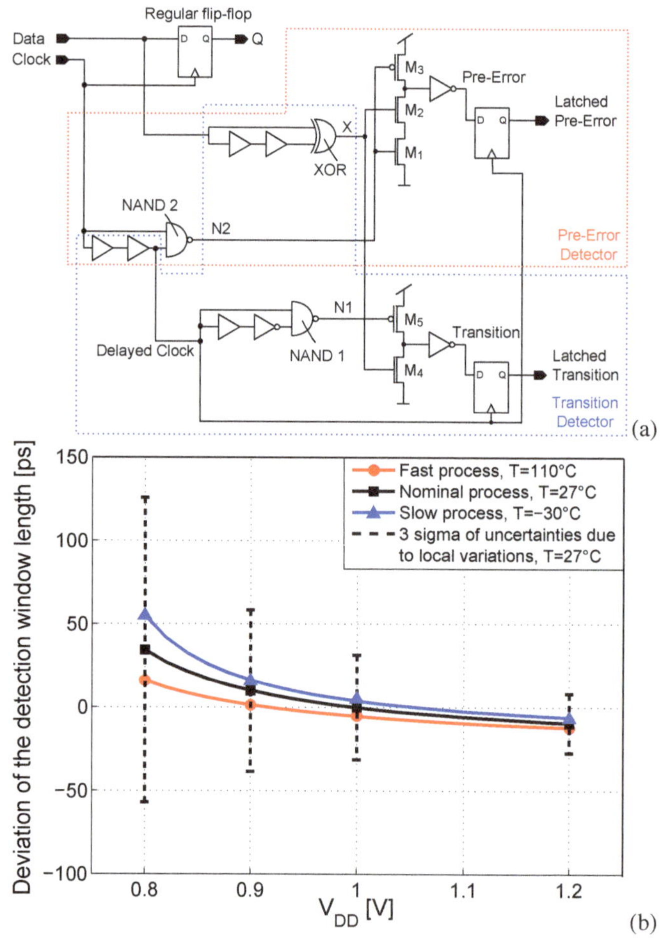

Fig. 3. Dynamic Pre-Error flip-flop, **(a)** Schematic, **(b)** Deviations of the detection window over V_{DD} under variations.

latches data at the beginning of the detection window. Therefore, if a data transition occurs during the detection window this flip-flop fails to detect valid data and a Pre-Error is generated by XOR1. For the Transition detector circuit, the inputs of XOR2 are the Data signal and the output Q of the regular flip-flop. In case of a data transition, the input signal Data will differ from its value in the previous clock cycle, stored as Q. Hence, a transition signal is generated by XOR2.

Deviations of the detection window over V_{DD} for corner cases and under local variations are shown in Fig. 4b. The Static Pre-Error flip-flop is more robust to global and local variations compared to the Dynamic Pre-Error flip-flop.

In duty-cycle based Pre-Error approaches, the deviations of the detection window over V_{DD} or due to variations are independent of the length of the nominal detection window.

3.2 Delay element based Pre-Error approach

In the Delay element based Pre-Error approach a shadow flip-flop with delayed data is added in parallel to the regular flip-flop (Eireiner et al., 2007). Figure 5a shows the schematic

[1] Note that typically the fast corner is at fast process and low temperature. In this example, however, the voltage is scaled to a point where the circuit is operated at temperature inversion. Here, the effect of decreasing threshold voltage with temperature exceeds the mobility degradation. Consequently, the circuit exhibits an inverted temperature characteristic, as it speeds up with increased temperature and vice versa.

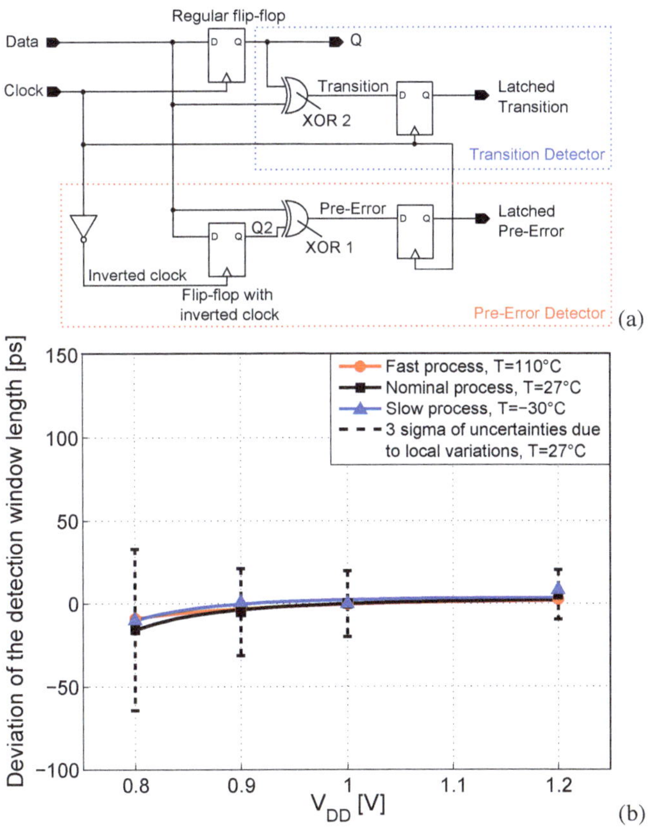

Fig. 4. Static Pre-Error flip-flop, **(a)** Schematic, **(b)** Deviations of the detection window over V_{DD} under variations.

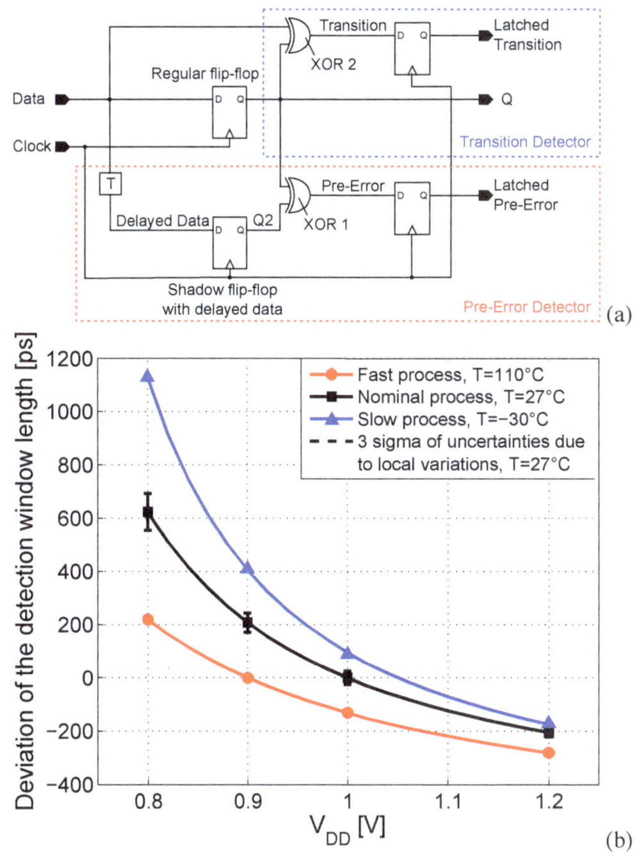

Fig. 5. Crystal-ball Pre-Error flip-flop, **(a)** Schematic, **(b)** Deviations of the detection window over V_{DD} under variations.

of this so-called Crystal-ball Pre-Error flip-flop. The delay element comprising an inverter chain specifies the length of the detection window. When a data transition occurs closer to the clock rising edge than the delay of the delay element, the shadow flip-flop will miss to latch the input data and the Pre-Error pulse is generated. Transition detection is implemented the same way as in the Static Pre-Error flip-flop.

For a delay element comprising 44 inverters designed to implement a nominal detection window of 650 ps at $V_{DD} = 1.0\,V$, deviations of the detection window over V_{DD} for corner cases and under local variations are shown in Fig. 5b. For the Crystal-ball Pre-Error approach, the detection window length is strongly dependent on the delay of the delay element, which has a large shift in corner cases. For an ideal detection window of $T_{Pre} = 650$ ps at a supply voltage of $V_{DD} = 1.0$ V, nominal process and temperature of $T = 27\,°C$, the shifts in the detection window length are more than 10 times larger than the shifts for the duty-cycle based Pre-Error flip-flops. Remember that an enlarged detection window increases the probability for a voltage increment and vice versa. Thus, the strong voltage dependence of the Crystal-ball Pre-error flip-flop might seem critical at first, but it stabilizes the adapted voltage around one level.

4 Analyzing the modeling of the Pre-Error AVS

As shown in Fig. 2, for different input patterns the output delays of the digital circuit vary due to changing signal propagation paths. The probability P_{pre} of a Pre-Error in one clock cycle is the number of critical delays (occurring during Pre-Error detection window) divided by the number of all delays. In the Pre-Error AVS, the supply voltage is regulated between a finite number of voltage levels, M, based on the number of Pre-Errors in the previous observation interval.

After each observation interval comprising N active clock cycles, the supply voltage is decreased if the count of Pre-Error occurrence, n_{pre}, is below $n_{limit\downarrow}$. Therefore the probability of decreasing the supply voltage, $P_{V_{DD}\downarrow}$, for each voltage level, V_{DD}, is calculated by

$$P_{V_{DD}\downarrow} = P\left[n_{pre} < n_{limit\downarrow}\right] = \sum_{n_{pre}=0}^{n_{limit\downarrow}-1} \binom{N}{n_{pre}} \cdot (P_{pre})^{n_{pre}} \cdot (1-P_{pre})^{N-n_{pre}} \quad (1)$$

Similarly the probability of increasing and maintaining the supply voltage, $P_{V_{DD}\uparrow}$ and $P_{V_{DD}\rightarrow}$ are calculated. The transition probabilities automatically adapt to the actual operating condition of the circuit. For a fixed operating condition,

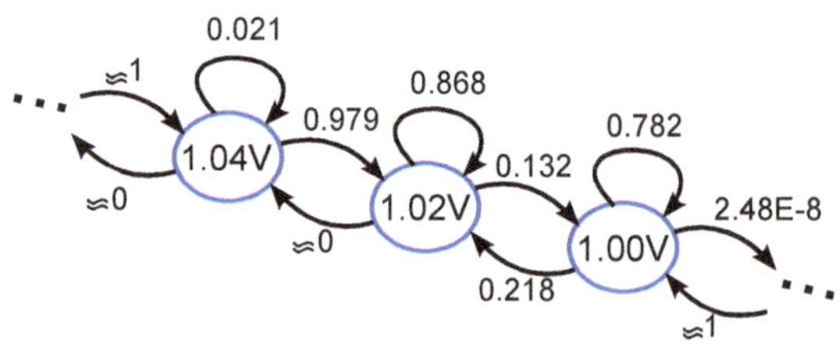

Fig. 6. Markov chain used to model the Pre-Error AVS. The values next to each arrow denote the transition probabilities.

the transition probabilities satisfy the Markov property, and thus we use a Markov chain to model the Pre-Error AVS. An example of the resulting Markov chain with a voltage granularity of 20 mV is shown in Fig. 6.

The corresponding Markov matrix is formed assuming a maximum voltage of $V_{DD} = 1.2$ V and a right Markov matrix (each row summing up to 1):

$$\mathbf{P} = \begin{pmatrix} P_{1.20\,V\rightarrow} & P_{1.20\,V\downarrow} & 0 & 0 & 0 & \cdots \\ P_{1.18\,V\uparrow} & P_{1.18\,V\rightarrow} & P_{1.18\,V\downarrow} & 0 & 0 & \cdots \\ 0 & P_{1.16\,V\uparrow} & P_{1.16\,V\rightarrow} & P_{1.16\,V\downarrow} & 0 & \cdots \\ \vdots & \vdots & \vdots & \vdots & \ddots & \end{pmatrix} \tag{2}$$

For a fixed operating condition, the voltage adaptation comes into the steady state. The corresponding steady state vector of probabilities π reads as

$$\pi = \begin{pmatrix} P_{1.20\,V} & P_{1.18\,V} & P_{1.16\,V} & \cdots \end{pmatrix} \tag{3}$$

where $P_{V_{DD}}$ is the final probability of being at the voltage level V_{DD}. The steady state vector of probabilities satisfies the equation

$$\pi \mathbf{P} = \pi \tag{4}$$

which defines π as the left eigenvector of the Markov matrix, with the corresponding eigenvalue 1. For changing operating condition, π automatically adapts.

4.1 Power saving potential of the Pre-Error AVS approach

The designed Pre-Error flip-flops are integrated into a 16 bit multiplier test circuit in a 65 nm CMOS technology. The clock frequency is chosen as 500 MHz ($T = 2$ ns). Pre-Error flip-flops are placed at the end of the three most critical paths (9.4 % of all 32 outputs). An OR tree combines the outputs of three Pre-Error flip-flops to generate the overall Pre-Error. Similarly, the Transition signals are ORed together. The power saving potential is evaluated using the Markov Model for each Pre-Error approach.

When analyzing the power saving of each Pre-Error approach, the overhead of the AVS circuitry has to be considered. Additional power consists of the extra circuitry for the integrated Pre-Error flip-flops, OR trees and the AVS control unit. For all Pre-Error approaches integrated in the 16bit multiplier test circuit, the power overhead of the AVS control unit is evaluated as 3.4 %. The power overhead corresponding to the OR trees is less than 1 %. Finally the power overhead due to the three Pre-Error flip flops is 9.8 %, 4.3 % and 10.1 %, resulting in the total power overhead of 13.3 %, 7.8 % and 13.7 % for the Dynamic, Static and Crystal-ball Pre-Error approaches ($T_{Pre} = 650$ ps), respectively. To evaluate the risk of timing errors, the error probability is determined for each supply voltage level with SPICE simulation. The overall error rate is determined using the simulated error rates at each voltage level, and the steady state vector of probabilities:

$$P_{err} = \sum_{V_{DD}} P_{V_{DD}} \cdot P_{err,V_{DD}} \tag{5}$$

where $P_{V_{DD}}$ is the final probability of being at a certain voltage level V_{DD} after voltage adaptation and $P_{err,V_{DD}}$ is the error probability of the corresponding supply voltage level of V_{DD}. Similarly, the total power consumption is evaluated as:

$$P_{dyn} = \sum_{V_{DD}} P_{V_{DD}} \cdot P_{dyn,V_{DD}} \tag{6}$$

in which $P_{dyn,V_{DD}}$ is the simulated dynamic power consumption of the supply voltage level of V_{DD}.

5 Results and comparison

Different variations can be included into the Markov model: the delay variations of the multiplier test circuit, the setup-time variations of all flip-flops and the Pre-Error detection window variations for each approach. To be able to compare between the proposed approaches, we excluded the local variations of the multiplier test circuit for the following results.

For the Static Pre-Error approach, Fig. 7 shows the resulting power saving potential considering detection windows with nominal lengths of 650 ps (lowest error rates), 550 ps, 400 ps and 250 ps (highest error rates). For all designs, the error rate deviation from the nominal values in fast and slow corners is small compared to the impact of local variations. To compare the robustness of the proposed Pre-Error designs, we inspected the uncertainty of an adjusted error rate. The chosen error rate is 10^{-9}, which requires a detection window of approximately 550 ps. The Crystal-ball Pre-Error flip-flop implementing the detection window of 550ps has a total power overhead of 12.9 %. The normalized error uncertainty in presence of local variations is obtained by dividing the error uncertainty shown in Fig. 7 by the nominal error rate. For an error rate of 10^{-9}, the resulting normalized error uncertainty is 2.3 and 4.8 for Static and Dynamic Pre-Error flip-flops. For the Crystal-ball Pre-Error flip-flop, it is only 1.6 as a result of the strong voltage dependence for the Pre-Error detection length, which reduces voltage alterations and provides even more stabilized voltage control. Table 1 shows the power saving with the nominal error rate of 10^{-9}.

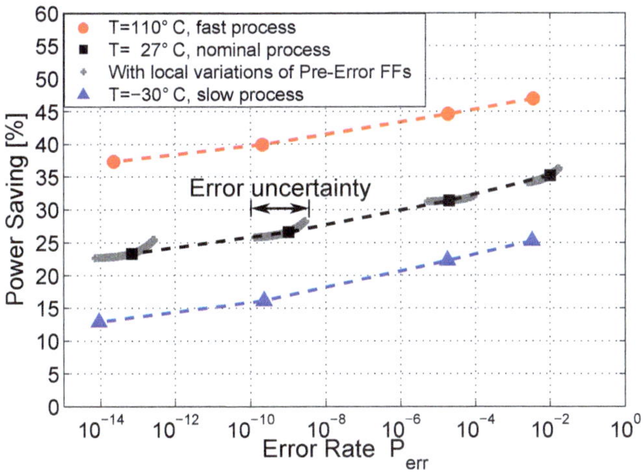

Fig. 7. Power saving potential in relation to the error rate for the Static Pre-Error approach.

As shown in Table 1, the power saving potential of the Static Pre-Error approach is larger than for the other two approaches as it has less power overhead. The power saving potential of the Static Pre-Error approach in the nominal case is 26.6 % compared to the conventional fixed voltage approach. For the fast corner ($T = 110\,°C$, fast process) lower voltages are more likely due to increased circuit speed, resulting in higher power savings. As the AVS also exploits the conservative timing margins produced by the synthesis tool, there is considerable power saving even in the slow corner.

6 Conclusions

Three Pre-Error detection approaches for use in Pre-Error AVS were designed and optimized. The Dynamic Pre-Error flip-flop requires a lot of design effort to achieve a robust and reliable design. In contrast, both Static and Crystal-ball Pre-Error flip-flops are designed in static design style using standard library elements, requiring less design effort. Monte-Carlo simulations and corner analysis were performed to evaluate the variations of the detection window length. The maximum 3σ of the detection window length deviation occurring at the lowest evaluated supply voltage ($V_{DD} = 0.8$ V) is 50 ps, 70 ps and 90 ps for Static, Crystal-ball and Dynamic Pre-Error designs, respectively. To determine the power saving and robustness, the Pre-Error AVS was modeled by a Markov chain. The variations of detection window length were included in the Markov model. For each Pre-Error design, the power saving potential and error rate uncertainties were determined for local and global variations. For a nominal error rate of 10^{-9} at a clock frequency of 500 MHz, the Static Pre-Error design has the largest power saving of 26.6 % and Crystal-ball and Dynamic Pre-error designs have power savings of 21.6 % and 21.1 %, respectively. The Crystal-ball Pre-Error design has the smallest normalized error uncertainty of 1.6 and Static and Dynamic Pre-Error designs have the normalized error uncertainties of 2.3 and 4.8.

Table 1. Power saving potential of Pre-Error AVS scheme applied to the 16 bit multiplier test cicruit (in %).

Pre-Error detection approach	Fast process $T = 110\,°C$	Nominal process $T = 27\,°C$	Slow process $T = -30\,°C$
Static	40.1	26.6	16.9
Dynamic	34.6	21.1	10.7
Crystal-ball	34.9	21.6	11.1

Acknowledgements. This work was supported by the German Research Foundation (DFG) within Grant SCHM 1478/8-1.

References

Bowman, K. A., Tschanz, J. W., Nam Sung Kim, Lee, J. C., Wilkerson, C. B., Lu, S.-L. L., Karnik, T., and De, V. K.: Energy-efficient, metastability-immune resilient circuits for dynamic variation tolerance, IEEE J. Solid-State Circuits, 44, 49–63, 2009.

Das, S., Tokunaga, C., Pant, S., Ma, W.-H., Kalaiselvan, S., Lai, K., Bull, D. M., and Blaauw, D. T.: RazorII: In situ error detection and correction for PVT and SER Tolerance, IEEE J. Solid-State Circuits, 44, 32–48, 2009.

Drake, A., Senger, R., Deogun, H., Carpenter, G., Ghiasi, S., Nguyen, T., James, N., Floyd, M., and Pokala, V.: A Distributed critical-path timing monitor for a 65nm high-performance microprocessor, IEEE International Solid-State Circuits Conference (ISSCC), 11–15 February 2007, 398–399, 2007.

Eireiner, M., Henzler, S., Georgakos, G., Berthold, J., and Schmitt-Landsiedel, D. : In-Situ delay characterization and local supply voltage adjustment for compensation of local parametric variations, IEEE J. Solid-State Circuits, 42, 1583–1592, 2007.

Tam, S., Limaye, R. D., and Desai, U. N. : Clock generation and distribution for the 130-nm Itanium^{reg} 2 processor with 6-MB on-die L3 cache, IEEE J. Solid-State Circuits, 39, 636–642, 2004.

Wirnshofer, M., Heiß, L., Georgakos, G., and Schmitt-Landsiedel, D.: A variation-aware Adaptive Voltage Scaling technique based on in-situ delay monitoring, IEEE 14th International Symposium on Design and Diagnostics of Electronic Circuits Systems (DDECS), 261–266, 2011.

Envelope tracking of a radio frequency amplifier for Long Term Evolution using a three-level class-G modulator

A. Bräckle, L. Rathgeber, F. Siegert, S. Heck, and M. Berroth

University of Stuttgart, Institute of Electrical and Optical Communications Engineering, Stuttgart, Germany

Correspondence to: A. Bräckle (alexander.braeckle@int.uni-stuttgart.de)

Abstract. In this contribution, efficiency enhancement of a radio frequency (RF) power amplifier (PA) for Long Term Evolution (LTE) base stations using envelope tracking is investigated. The dynamically adapted drain bias voltage is provided by a three-level class-G modulator. The potential of this bias modulator is demonstrated with a class AB PA built in gallium nitride (GaN) technology with an operation frequency of 2.75 GHz, a maximum output power of 41 dBm and a peak power-added efficiency (PAE) of 60 %. When driven by an LTE signal with a bandwidth of 20 MHz, overall system efficiency can be increased from 17 % with a constant supply voltage up to 29 % with envelope tracking. This is accompanied by a reduction of linearity. Adjacent channel leakage ratio (ACLR) decreases by 10 dB in the first and by 18 dB in the second adjacent channel, respectively. Error vector magnitude (EVM) increases from 9 % up to 29 %.

1 Introduction

In an RF transmitter, by far most of the power is consumed by the PA. Typically, more than 50 % of the base station input power is dissipated in this component (Blume et al., 2010). Therefore, efficiency enhancement of RF PAs is in the focus of many research activities. In this contribution, PAE is considered to quantify efficiency. It is defined as

$$\mathrm{PAE} = \frac{P_{\mathrm{out}} - P_{\mathrm{in}}}{P_{\mathrm{DC}}}, \tag{1}$$

where P_{out} is the RF output power of the PA, P_{in} denotes the RF input power, and P_{DC} is the DC power consumed by the PA. PAs can be operated very efficiently when driven in the compression region, but PAE rapidly drops for low output powers (Cripps, 1999).

This poses severe challenges especially for modern mobile communications systems such as LTE. Due to the constantly increasing bandwidth requirements, these systems utilize higher-order modulation schemes resulting in a high peak-to-average power ratio (PAPR) of 10 dB and more. Apart from efficiency, linearity is a second characteristic to be considered in order to avoid distortion of the transmitted signal. The mobile communications standards pose obligatory linearity requirements on all components in transmitters (ETSI, 2011). Therefore, nowadays mainly PAs of classes A, AB, and B are in use due to their high linearity. However, average efficiency is in the order of only about 20 %.

Several concepts for efficiency enhancement without reducing amplifier linearity are currently investigated, such as the Doherty amplifier (Doherty, 1936), the Chireix amplifier (Chireix, 1935), Class-S amplifiers (Raab, 1973), and Envelope Tracking (ET). The latter one is studied in more detail in this paper.

2 Envelope tracking

The concept of ET was first proposed by Saleh and Cox (1983). In order to understand the principle of this idea, the drain-source voltage u_{DS} of the transistor in a PA with an amplitude-modulated input signal is considered. As it can be seen in Fig. 1, u_{DS} oscillates around the drain bias voltage U_{DC}. For high output powers the available voltage swing between 0 V and U_{DC} is fully used. However, for lower output powers, the amplitude of u_{DS} decreases, resulting in higher losses. Those can be represented by the area below the envelope of the drain-source voltage.

The idea of ET is to dynamically adapt the drain bias voltage to the instantaneous output power of the PA. The corresponding drain-source voltage is depicted in Fig. 2. The area below the envelope of u_{DS} is minimised, resulting in an increased efficiency of the circuit.

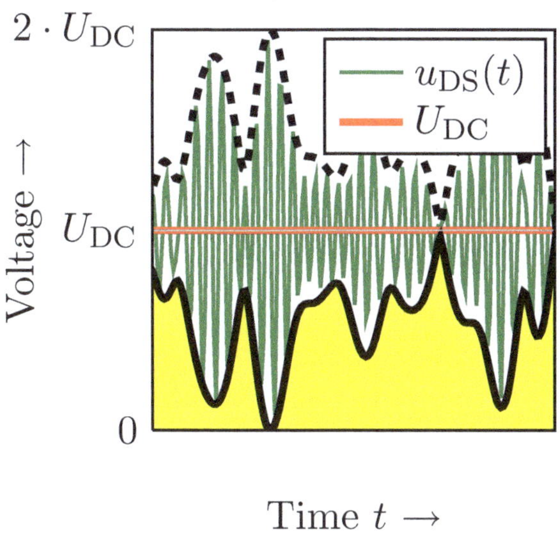

Fig. 1. Drain-source voltage of a PA with fixed drain bias.

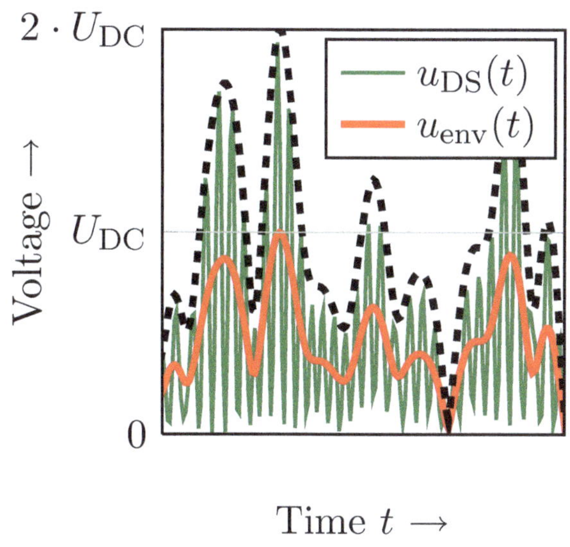

Fig. 2. Drain-source voltage of a PA in an ET system.

3 Class-G amplifier architecture

There are several architectures for voltage modulators. Due to the high efficiency requirements, they are all based on switch-mode topologies.

Buck modulators are pure switch-mode converters. In order to generate a continuous-valued drain bias voltage, the desired waveform is generated by pulse-width modulation (PWM) and subsequent low-pass filtering. As PWM is based on oversampling, the switching frequency needs to be significantly higher than the bandwidth of the output signal. For broadband signals such as LTE, this results in switching frequencies above 100 MHz, leading to high switching losses for large power transistors. Therefore, buck converters with output powers above 2 W are either used in narrow-band applications (Raab, 1998; Høyerby and Andersen, 2007, 2009) or implemented as integrated circuits in RF power technologies such as LDMOS (Pinon et al., 2008) or GaAs (Busking et al., 2012).

In switch-mode assisted linear modulators, a linear stage and a switching stage are operated in parallel. When properly designed, the main portion of the modulator output current is provided by the switching stage. The linear stage works as an active filter and compensates for the ripple generated by the switching stage. Therefore, the switching frequency can be chosen much lower and switching losses are reduced. Switch-mode assisted linear modulators are widely used (Kimball et al., 2006; Jeong et al., 2009; Hoversten and Popovic, 2010), but suffer from two disadvantages. On the one hand, for very high output powers, the absolute power dissipated in the linear stage becomes very high even if it is small compared to the output power. This poses severe challenges on the design of a broadband linear stage. On the other hand, modulator efficiency drops for low output powers due to the inherent switching operation, which results in constant switching losses.

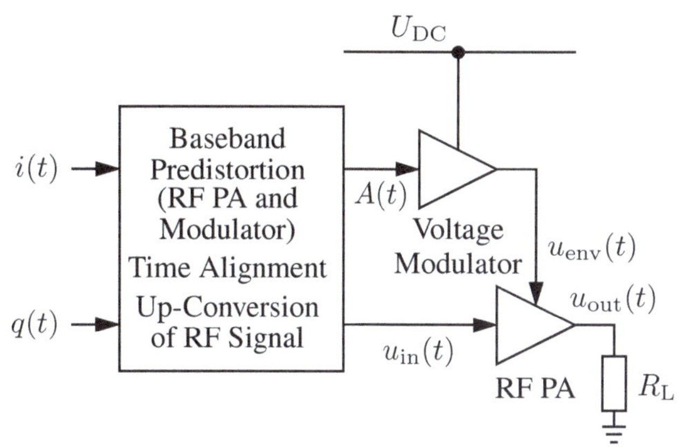

Fig. 3. Block diagram of an ET system.

One way to overcome these drawbacks is the class-G concept, which was first proposed by Sampei et al. (1978) for audio amplifiers and then implemented for RF applications by Raab (1986). In contrast to the modulators introduced above, the drain bias can only be set to discrete-valued voltage levels. The simplest configuration of a two-level class-G modulator is depicted in Fig. 4. If the envelope $A(t)$ is above a fixed reference voltage $U_{ref,1}$, the transistor T_1 is turned on and the modulator output voltage u_{env} equals the high supply voltage $U_{DC,M+}$. For low envelopes, the transistor is in off-state and the diode D_1 connects the low supply voltage $U_{DC,M-}$ to the output.

Although the RF PA does not work in its highest-efficiency operating point most of the time when using a class-G modulator, overall system efficiency can be significantly increased in the backoff region. Furthermore, the voltage modulator

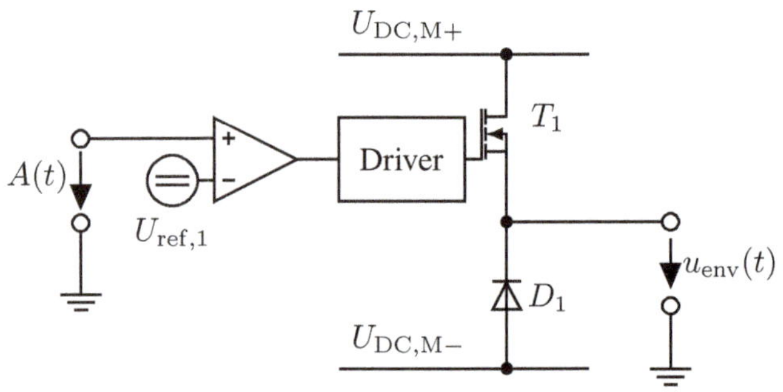

Fig. 4. Simplified circuit diagram of a two-level class-G modulator.

can be designed very efficiently due to its simplicity. Especially at low modulator output voltages, very high efficiencies can be achieved because no inherent switching occurs.

The efficiency can be improved further by inserting more than two discrete voltage levels. In this work, a three-level class-G modulator is presented.

4 Radio frequency power amplifier

The potential of multi-level class-G modulators is demonstrated with an RF PA based on a gallium nitride (GaN) high electron mobility transistor (HEMT). The device is the commercially available CGH40010 manufactured by Cree and has a nominal supply voltage of 28 V. With this transistor, a class AB PA operating at a center frequency of 2.75 GHz is designed.

Figure 5 shows the measured PAE of the PA with a fixed drain bias voltage of 28 V when driven by a continuous-wave (CW) input signal. The amplifier achieves a peak PAE of 60 % at an output power of 41 dBm. However, as expected, efficiency drops in the backoff region. As derived by Cripps (1999), PAE is halved for a reduction of output power by approximately 6 dB.

Furthermore, Fig. 5 also depicts the probability density function (PDF) of an LTE signal. It can be seen that the PA is operated at an output power of about 29 dBm most of the time under this excitation. At this output power, efficiency with a CW signal drops below 20 %. With a dynamic LTE input signal, average efficiency is 22.4 %, which is significantly less than the maximum PAE.

The potential of the class-G architecture becomes clear when performing CW measurements of the RF PA with reduced supply voltages. Figure 5 shows the PAE versus output power not only for the maximum supply voltage of 28 V, but also for $U_{DC} = 10$ V and $U_{DC} = 20$ V. For lower supply voltages, the point of peak PAE is shifted towards lower output powers. For instance, PAE in CW operation can be increased up to about 44 % at an output power of 29 dBm when reducing the drain bias to 10 V.

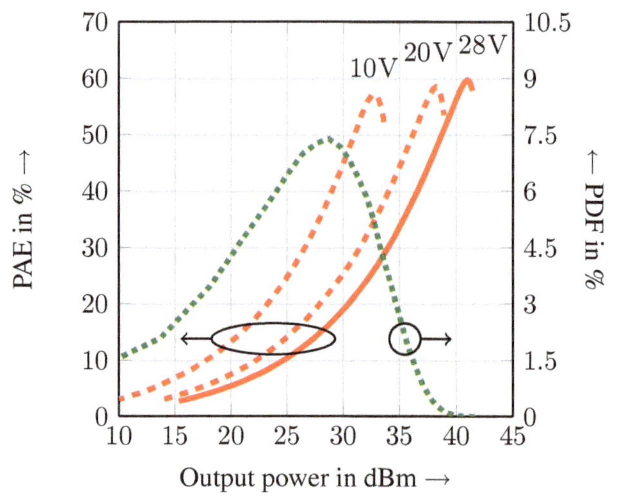

Fig. 5. Measured PAE of the PA under CW excitation for a fixed supply voltage of 28 V (solid), reduced supply voltages (dashed) and PDF of an LTE signal (dotted).

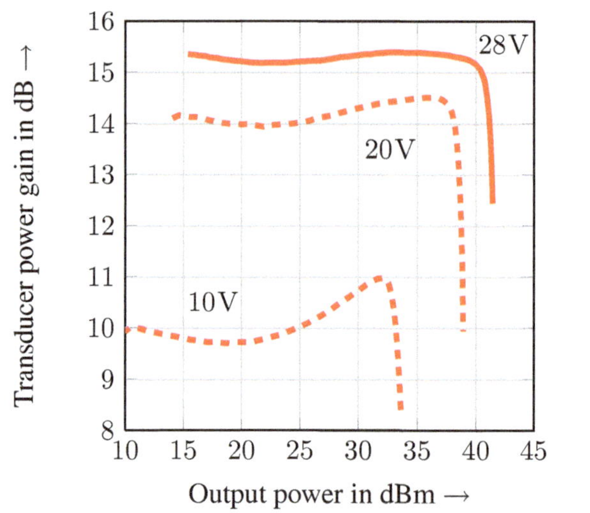

Fig. 6. Measured transducer power gain of the PA under CW excitation for several supply voltages.

However, the supply voltage of the PA cannot be arbitrarily reduced. As shown in Fig. 6, the transducer power gain of the PA, which is defined by

$$G_T = \frac{P_{out}}{P_{avs}}, \tag{2}$$

where P_{avs} denotes the available source power, drops for lower supply voltages. In this work, G_T shall not be lower than 10 dB. This leads to an allowable drain bias range from 10 V to 28 V.

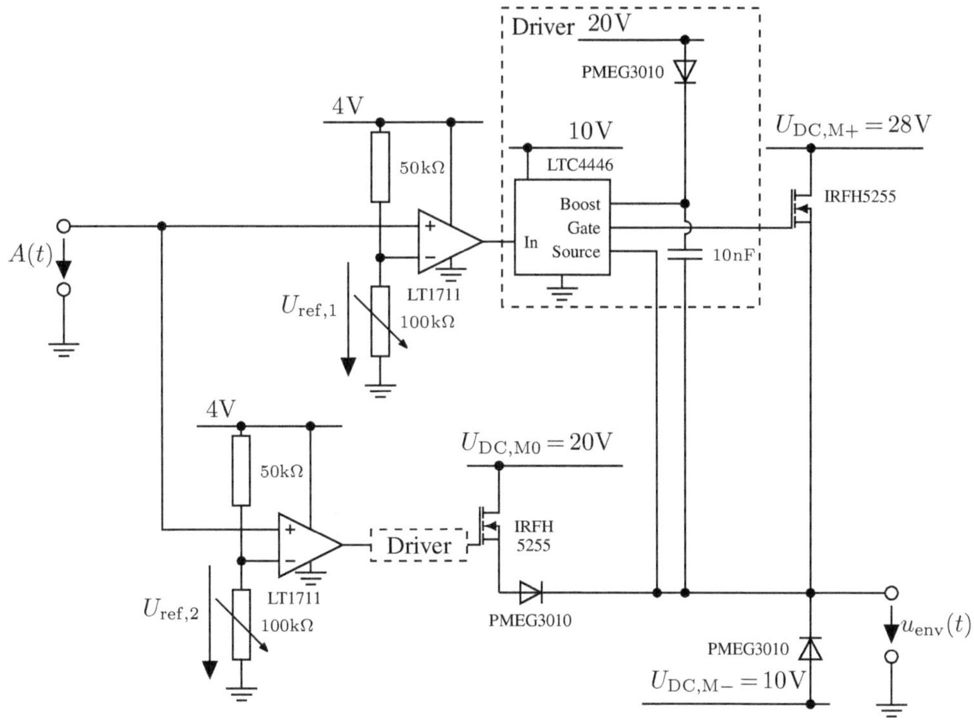

Fig. 7. Schematic of the designed three-level class-G modulator. For the sake of simplicity, shunt capacitors in the supply lines are omitted.

5 Class-G modulator design

Based on the analysis of the RF PA in the previous section, design requirements for the three-level class-G voltage modulator can be deduced. The highest drain bias voltage has to be equal the nominal value, that is $U_{DC,M+} = 28$ V. For the low supply voltage, $U_{DC,M-} = 10$ V are chosen, which makes sure that gain does not drop below 10 dB. The third voltage level $U_{DC,M0}$ has to be between those two. In this work, it is arbitrarily chosen to be $U_{DC,M0} = 20$ V. Furthermore, the modulator needs to be able to deliver a maximum output current of 1 A, which is the supply current required by the PA. A high switching speed is also desirable in order to follow the envelope of an LTE signal with a bandwidth of 20 MHz.

The schematic of the designed class-G modulator is depicted in Fig. 7. It consists of two identical paths to control the power transistors switching the discrete supply voltages. The envelope $A(t)$ of the RF signal, which serves as input signal of the circuit, is compared to constant reference voltages $U_{ref,1}$ and $U_{ref,2}$, respectively. If the envelope is above the reference level, a subsequent gate driver circuit turns on the associated power transistor and thus connects $U_{DC,M+}$ or $U_{DC,M0}$ to the modulator output node. If $A(t)$ is smaller than both reference voltages, none of the transistors is turned on, the diode attached to the output node conducts and u_{env} equals $U_{DC,M-}$.

For the power transistors, n-channel metal-oxide semiconductor field-effect transistors (MOSFETs) are used due to their higher switching speed compared to p-channel MOSFETs. As they are controlled by the gate-source voltage and due to the fact that the gate is connected to the output and thus has no fixed potential, a gate driver integrated circuit (IC) is used in a bootstrap configuration.

6 Measurement results

In this section, the measurement results achieved when operating the RF PA with a dynamic supply voltage generated by the three-level voltage modulator are presented. The generation of the modulator supply voltages is not included in the efficiency calculations. A photograph of the circuit setup can be seen in Fig. 8.

6.1 Static measurements

First, measurements under CW excitation are performed. In this case, one of the three possible supply voltages is constantly turned on by the modulator and the power of the RF PA input signal is increased.

The measured PAE of the entire system is depicted in Fig. 9. The efficiency for a constant supply voltage of 28 V is drawn as well for comparison. It can be clearly seen that efficiency in the backoff region is significantly increased by the class-G modulator. For output powers up to 33 dBm, the

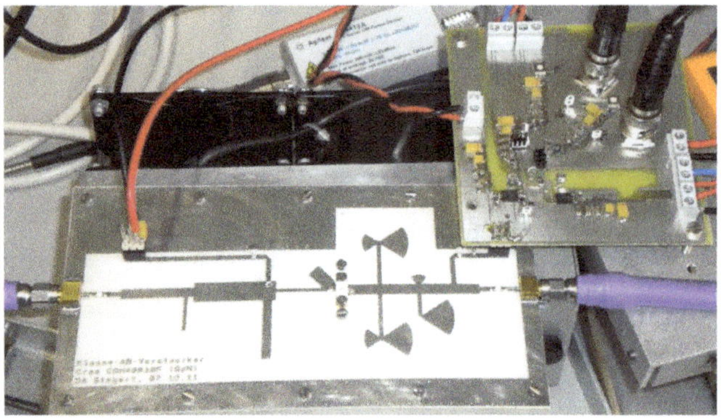

Fig. 8. Class-G modulator (top right) connected to the RF PA (bottom left).

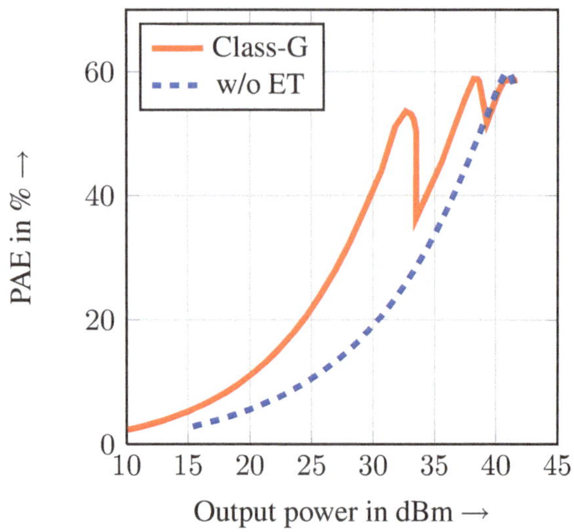

Fig. 9. Measured PAE of the entire class-G system under CW excitation.

PA is supplied by the low bias voltage $U_{DC,M-} = 10\,V$. For output powers between 33 dBm and 38 dBm, the drain bias is set to $U_{DC,M0} = 20\,V$. If the output power is even higher, u_{env} equals $U_{DC,M+} = 28\,V$. By this, PAE at an output power of 29 dBm can be increased from 17 % up to almost 38 %.

6.2 Dynamic measurements

The class-G system is also characterised with excitation by an LTE signal at the input. It has a bandwidth of 20 MHz and a PAPR of 11.6 dB. The class-G modulator is driven by the according envelope $A(t)$. The reference voltages are computed by

$$U_{ref,1} = \sqrt{\frac{\max\{P_{out}\}|_{u_{env}=U_{DC,M0}}}{\max\{P_{out}\}|_{u_{env}=U_{DC,M+}}}} \cdot A_{max} = 0.75 \cdot A_{max} \quad (3)$$

$$U_{ref,2} = \sqrt{\frac{\max\{P_{out}\}|_{u_{env}=U_{DC,M-}}}{\max\{P_{out}\}|_{u_{env}=U_{DC,M+}}}} \cdot A_{max} = 0.43 \cdot A_{max}, \quad (4)$$

Table 1. Measured efficiency and linearity of the class-G system under LTE excitation.

	PAE	ACLR1	ACLR2	EVM
Requirement		−44.2 dB	−44.2 dB	9.0 %
w/o ET	16.7 %	−27.2 dB	−43.9 dB	10.0 %
3-level class-G	29.2 %	−17.9 dB	−26.1 dB	29.3 %
2-level class-G	32.2 %	−18.9 dB	−24.8 dB	31.6 %

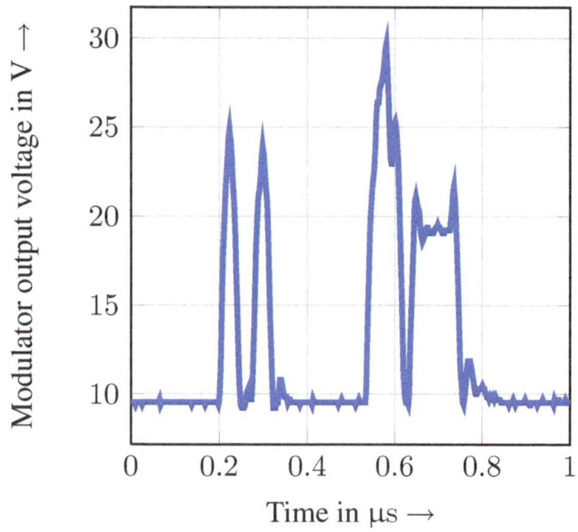

Fig. 10. Time-domain measurement of the modulator output voltage with LTE excitation.

where A_{max} denotes the maximum of the modulator input signal.

Figure 10 shows a time-domain measurement of the modulator output voltage. The results are summarized in Table 1. For comparison, measurement results of a two-level class-G modulator presented in (Bräckle et al., 2012) are included as well. Efficiency can be enhanced from 16.7 % up to 29.2 % by using the three-level modulator, which is a significant increase. However, in contrary to theory, PAE of the two-level configuration is even slightly higher. This is caused by the additional circuitry required to drive two switches. Furthermore, in the present three-level configuration, both switches are turned on at high output powers although only the MOSFET for $U_{DC,M+}$ is needed. This can be overcome by a digital control circuit. And finally, the medium supply voltage $U_{DC,M0}$ is arbitrarily chosen to be 20 V. A further increase in efficiency is expected after optimization of $U_{DC,M0}$ with respect to the statistical properties of an LTE signal.

In contrast, linearity decreases. ACLR in the first channel reduces by about 10 dB, in the second channel it is even 17 dB worse than for a constant supply voltage. EVM is 29.3 %, which is three times the value with a fixed drain bias. For the two-level configuration, linearity is even worse. This is mainly caused by the shift of gain for different supply voltages shown in Fig. 6. Every time u_{env} is changed, the gain changes as well, resulting in an additional amplitude modulation and distorting the signal.

7 Conclusions

This work has presented the potential as well as the challenges in designing multi-level class-G systems. By using a three-level modulator, overall system efficiency can be enhanced from 16.7 % up to 29.2 %. However, linearity is significantly degraded, which is mainly caused by the supply voltage dependency of the RF PA gain. Therefore, in order to benefit from the efficiency enhancement of class-G modulators, RF PAs need to be optimized for constant gain at several supply voltages.

Further optimizations are to be carried out on an intelligent driver system to avoid that two transistors are turned on at the same time. An optimized choice of the voltage levels will be investigated as well.

References

Blume, O., Zeller, D., and Barth, U.: Approaches to Energy Efficient Wireless Access Networks, in: 4th International Symposium on Communications, Control and Signal Processing, Limassol, Cyprus, 3–5 March 2010, 1–5, 2010.

Bräckle, A., Rathgeber, L., Siegert, F., Heck, S., and Berroth, M.: Power Supply Modulation for RF Applications, in: 15th International Power Electronics and Motion Control Conference, Novi Sad, Serbia, September 2012.

Busking, E., de Hek, P., and van Vliet, F.: 1 GHz GaAs Buck Converter for High Power Amplifier Modulation Applications, in: Proceedings of the 7th European Microwave Integrated Circuits Conference, 353–356, Amsterdam, October 2012.

Chireix, H.: High Power Outphasing Modulation, P. IRE, 23, 1370–1392, 1935.

Cripps, S.: RF Power Amplifiers for Wireless Communications, 1st edition, Arctech House, Boston, London, 1999.

Doherty, W.H.: A New High Efficiency Power Amplifier for Modulated Waves, P. IRE, 24, 1163–1182, 1936.

Evolved Universal Terrestrial Radio Access (E-UTRA); LTE: Base Station (BS) conformance testing (3GPP TS 36.141 version 9.6.0 Release 9), European Telecommunications Standards Institute, 2011.

Hoversten, J. and Popovic, Z.: System Considerations for Efficient and Linear Supply Modulated RF Transmitter, in: IEEE 12th Workshop on Control and Modeling for Power Electronics, Boulder, June 2010.

Høyerby, M. and Andersen, M.: Self-Oscillating Soft Switching Envelope Tracking Power Supply for Tetra2 Base Station, in: 29th International Telecommunications Energy Conference, 53–60, September 2007.

Høyerby, M. and Andersen, M.: TEDS Base-Station Power Amplifier Using Low-Noise Envelope Tracking Power Supply, IEEE T. Microw. Theory, 57, 1687–1693, 2009.

Jeong, J., Kimball, D., Kwak M., Draxler P., Hsia C., Steinbeiser, C., Landon, T., Krutko, O., Larson, L., and Asbeck, P.: High-Efficiency WCDMA Envelope Tracking Base-Station Amplifier Implemented With GaAs HVHBTs, IEEE J. Solid-St. Circ., 44, 2629–2639, 2009.

Kimball, D., Jeong, J., Hisa, C., Draxler, P., Lanfranco, S., Nagy, W., Linthicium, K., Larson, L., and Asbeck, P.: High-Efficiency Envelope-Tracking W-CDMA Base-Station Amplifier Using GaN HFETs, IEEE T. Microw. Theory, 54, 3848–3856, 2006.

Pinon, V, Hasbani, F., Giry, A., Pache, D., and Garnier, C.: A Single-Chip WCDMA Envelope Reconstruction LDMOS PA with 130 MHz Switched-Mode Power Supply, in: IEEE Int. Solid-St. Circ. Conf., 564–636, 2008.

Raab, F. H.: Radio Frequency Pulsewidth Modulation, IEEE T. Commun., 21, 958–966, 1973.

Raab, F.: Average Effciency of Class-G Power Amplifiers, IEEE T. Consum. Electr., 32, 145–150, 1986.

Raab, F., Sigmon, B., Myers, R., and Jackson, R.: L-Band Transmitter Using Kahn EER Technique, IEEE T. Microw. Theory, 46, 2220–2225, 1998.

Saleh, A. A. M. and Cox, D. C.: Improving the Power-Added Efficiency of FET Amplifiers Operating with Varying-Envelope Signals, IEEE T. Microw. Theory, 31, 51–56, 1983.

Sampei, T., Ohashi, S., Ohta, Y., and Inoue, S.: Highest Efficiency and Super Quality Audio Amplifier Using MOS Power FETs in class G Operation, in: IEEE T. Consum. Electr., 24, 300–307, 1978.

2.5 Gbps clock data recovery using 1/4th-rate quadricorrelator frequency detector and skew-calibrated multi-phase clock generator

S. Tontisirin and R. Tielert

Technische Universität Kaiserslautern, Germany

Abstract. A Gb/s clock and data recovery (CDR) circuit using 1/4th-rate digital quadricorrelator frequency detector and skew-calibrated multi-phase voltage-controlled oscillator is presented. With 1/4th-rate clock architecture, the coil-free oscillator can have lower operation frequency providing sufficient low-jitter operation. Moreover, it is an inherent 1-to-4 DEMUX. The skew calibration scheme is applied to reduce phase offset in multi-phase clock generator. The CDR with frequency detector can have small loop bandwidth, wide pull-in range and can operate without the need for a local reference clock. This 1/4th-rate CDR is implemented in standard 0.18 μm CMOS technology. It has an active area of 0.7 mm^2 and consumes 100 mW at 1.8 V supply. The CDR has low jitter operation in a wide frequency range from 1–2.25 Gb/s. Measurement of Bit-Error Rate is less than 10^{-12} for 2.25 Gb/s incoming data $2^7 - 1$ PRBS, jitter peak-to-peak of 0.7 unit interval (UI) modulation at 10 MHz.

1 Introduction

At the receiver side of optical communication or high speed serial communication, clock and data recovery circuit has to generate a synchronized clock with the incoming serial data for using it in data regeneration and demultiplexing. Regarding the factor of cost, the monolithic IC solution for clock and data recovery is attractive, especially, the CMOS IC without the on-chip coil, which occupies large area. At high frequency, the ring-based voltage control oscillator (VCO) is far inferior in jitter performance but at adequately low frequency it can provide sufficient low-jitter operation. With 1/4th-rate architecture, the required clock frequency can be reduced. Hence low-jitter operation is achievable. Moreover, power consumption can also be decreased.

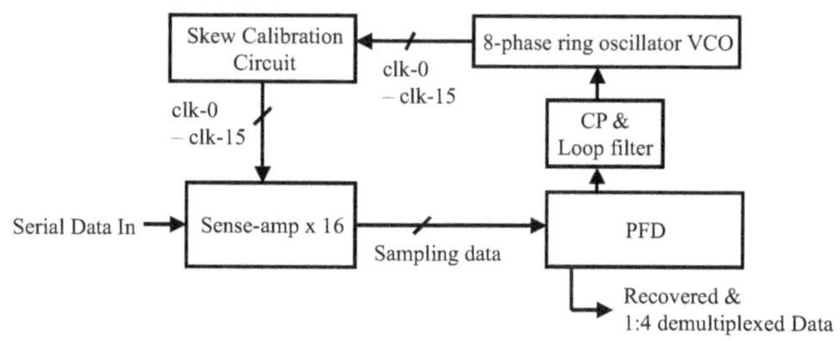

Fig. 1. Block diagram of 1/4th-rate CDR with frequency detector and skew calibration.

The jitter transfer function of CDR is normally small to improve noise performance. Conversely, pull-in range is reduced. CDRs with frequency initial loop (Cao, 2002) or with frequency synthesizer loop (Farjad-Rad, 2004) require a local reference clock. CDR with frequency acquisition loop using digital quadricorrelator frequency detector (DQFD) can have low jitter operation and wide pull-in range because there is no trade-off between loop bandwidth and pull-in range. It can operate without the need for a local reference clock. The full-rate clock architecture was presented in Pottbäcker (1992). The half-rate clock approach (Yang, 2004) can lower the power consumption and relax the required clock frequency.

In this work, a 1/4th-rate CDR using digital quadricorrelator frequency detector is described. The skew calibration scheme (Wu, 2001) is applied to reduce phase offsets in the required multi-phase clock generator.

2 Overall architecture

Block diagram of the proposed 1/4th-rate CDR using digital quadricorrelator frequency detector is depicted in Fig. 1. The CDR consists of a fully differential 8-phase ring-oscillator-

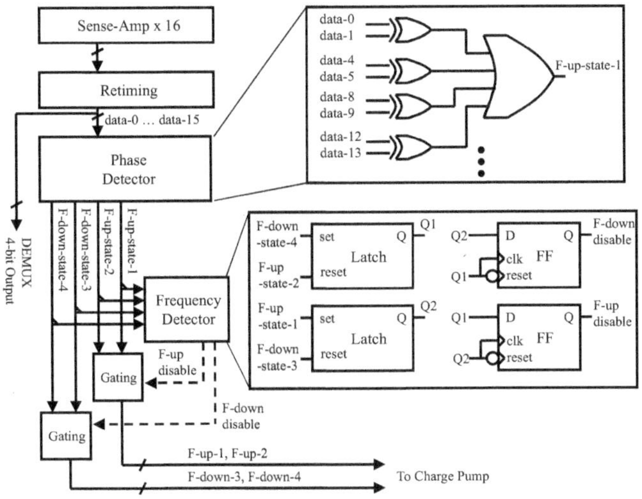

Fig. 2. Block diagram of phase frequency detector.

Table 1. Operation table of phase detector.

Serialdata transition	Meaning	Output
State-1	Clock too late	F-up-state-1 = '1'
State-2	Clock too late	F-up-state-2 = '1'
State-3	Clock too early	F-down-state-3 = '1'
State-4	Clock too early	F-down-state-4 = '1'

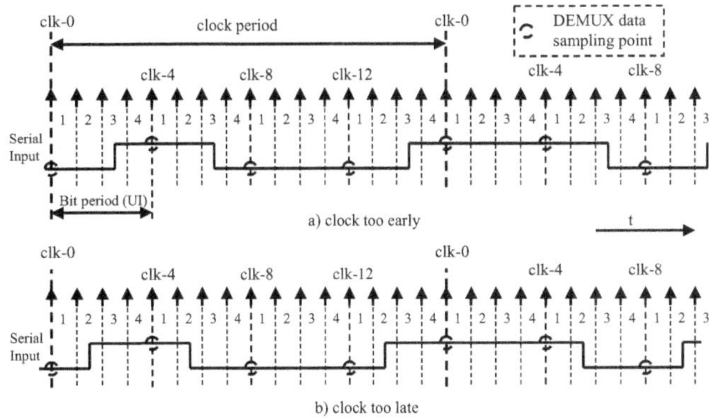

Fig. 3. Time diagram of phase detector.

Table 2. Operation table of frequency detector.

Signal	Set-condition	Reset-condition
Q1	F-down-state-4 = '1'	F-up-state-2 = '1'
Q2	F-up-state-1 = '1'	F-down-state-3 = '1'
F-down-disable	Rising edge of Q1 & Q2 = '1'	Falling edge of Q1
F-up-disable	Rising edge of Q2 & Q1 = '1'	Falling edge of Q2

based VCO, skew calibration circuit, 16 sense-amplifiers, phase frequency detector (PFD), charge pump and loop filter components. The skew-calibrated 16-phase clocks are provided by using both clock edges from VCO. They are applied in 16 sense-amplifiers for sampling the incoming serial data stream. The 16 sampling data, covering 4 serial data bit periods, are used in PFD to generate frequency up/down signals to control the frequency and phase of VCO synchronously to the incoming serial data.

2.1 Phase frequency detector

The block diagram of PFD is depicted in Fig. 2. PFD senses the phase difference between 1/4th-rate clock and input data. In order to have automatic data recovery, PFD have to sample the data by the clock instead of sampling the clock by the data, as described in Pottbäcker (1992) and Yang (2004). The data from sense-amplifiers are retimed to make all sampling data synchronous. When CDR locking is achieved, the sampling data of clk-0, clk-4, clk-8 and clk-12 represent the regenerated 4-bit output of the demultiplexer.

PFD operates at 1/4th-rate clock, therefore it can be implemented by CMOS logic which has lower power consumption compared to high speed current mode logic. The operation of PFD can be explained in the following. Each 1/4th of clock period, 1 unit interval (UI), is divided into 4 phase states, state-1, state-2, state-3 and state-4 , as shown in Fig. 3. The operation table of PD is depicted in Table 1. Phase detector (PD) steers clk-0, clk-4, clk-8 and clk-12 to achieve sampling at the middle of the data eye. If transitions of serial data occur in state-3 or state-4, as too early sampling points, PD generates F-down-state-3 signals or F-down-state-4 signals. If transition edges of serial data occur in state-1 or state-2, as too late sampling points, PD generates F-up-state-1 signals or F-up-state-2 signals. In the phase-locked condition, serial data transitions alternate between phase state-2 and state-3, hence, PD generates the same amount of F-up-state-2 and F-down-state-3 signals.

Frequency detector (FD) monitors the outputs from PD to detect a frequency difference. If VCO frequency is not equal to 1/4th of the data rate frequency, the serial data transition edges rotate around the phase states of VCO clock. The frequency difference defines this rotation frequency. As shown in Fig. 4, if VCO frequency is lower than 1/4th of data rate frequency, appearances of serial data transitions cyclically move in direction state-2, state-1, state-4, state-3, state-2 and state-1, respectively. It rotates in the opposite way, if VCO frequency is too high, represented in Fig. 5. PD generates the same amount of F-up and F-down signals because of the cyclical walking of serial data transitions, so that the loop could not be driven locked. Therefore, in this proposed CDR,

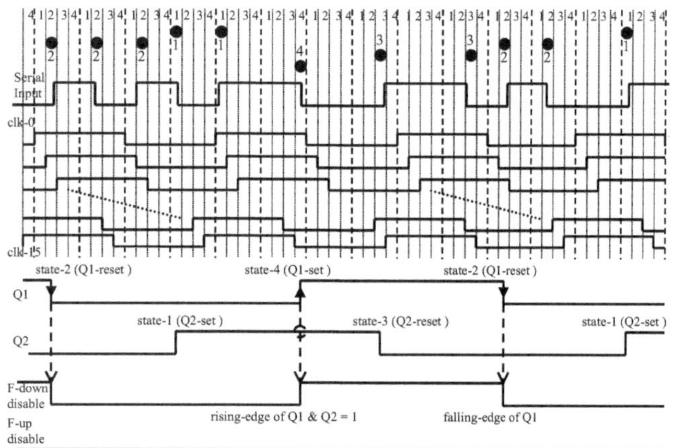

Fig. 4. Time diagram of frequency detector: frequency is too low.

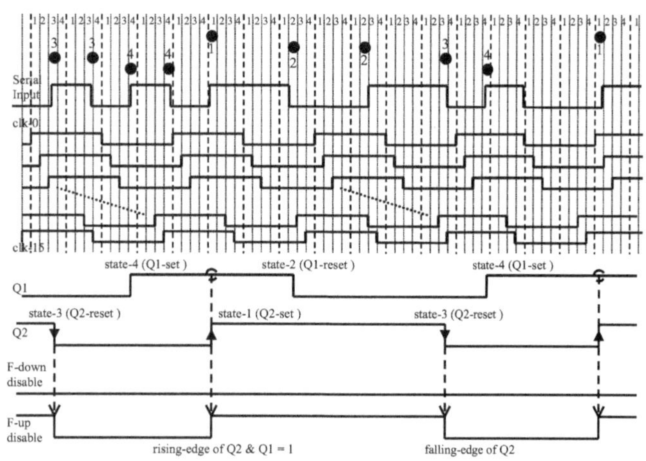

Fig. 5. Time diagram of frequency detector: frequency is too high.

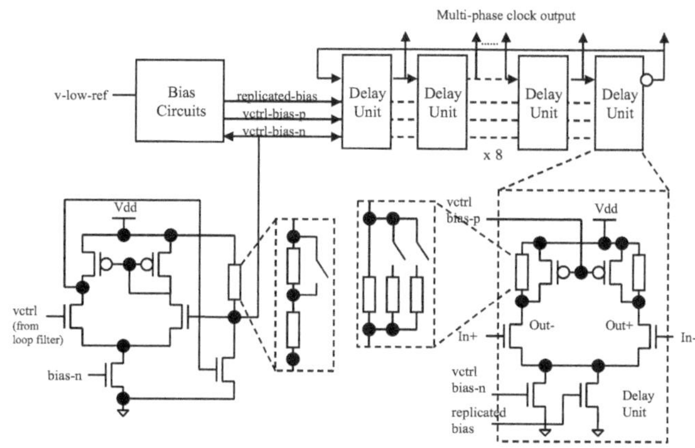

Fig. 6. Simplified VCO schematic.

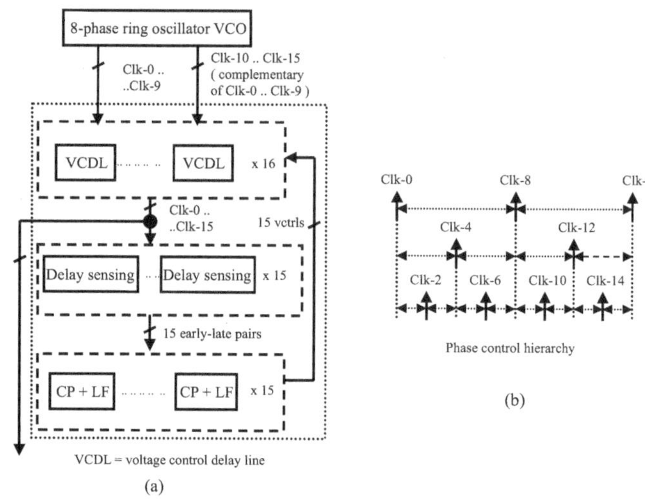

Fig. 7. Skew calibration scheme.

FD detects the cyclical walking of serial data transitions and generates signal F-up-disable and F-down-disable to produce the required unbalance of F-up and F-down signals to drive VCO frequency to 1/4th of the data rate frequency. The operation of FD is shown in Table 2. Q1 and Q2 are used to detect the cyclical walking of serial data transitions. If VCO frequency is too low Q2 leads Q1, shown in Fig. 4. F-down-disable is set by the rising edge of Q1 when Q2 is '1'. It is reset by the falling edge of Q1. F-up-disable is set by the rising edge of Q2 when Q1 is '1' and reset by the falling edge of Q2. If Q2 leads Q1, F-down-disable is generated but F-up-disable is always '0'. If VCO frequency is too high Q1 leads Q2, F-up-disable is generated while F-down-disable is '0', depicted in Fig. 5. In the lock condition, serial data transitions alternate between phase state-2 and state-3 hence Q1 and Q2 stop beating. Therefore, F-down-disable and F-up-disable automatically disappear. As a result, FD will not disturb the lock condition.

2.2 VCO design and skew calibration scheme

A simplified schematic of the VCO is shown in Fig. 6. It consists of 8 fully differential delay units. The 4 configurable frequency ranges can be selected by switching of resistive loads. Replicated bias is applied to control the voltage swing. A V-I converter circuit converts the control voltage to the bias current of the delay unit. The layout of VCO is designed for balancing delays between each unit. Common centroid topology and dummy cells are used to improve matching properties.

In order to reduce phase offsets further, a skew calibration, as described in Wu (2001), is utilized. The block diagram is shown in Fig. 7a. Each clock phase has its own delay-locked loop (DLL) to calibrate the phase positions. The phase control hierarchy is shown in Fig. 7b. Clk-8 is calibrated to be in the middle between 2 successive clk-0s. Clk-4 is adjusted to be in the middle between clk-0 and clk-8. Similarly, clk-12

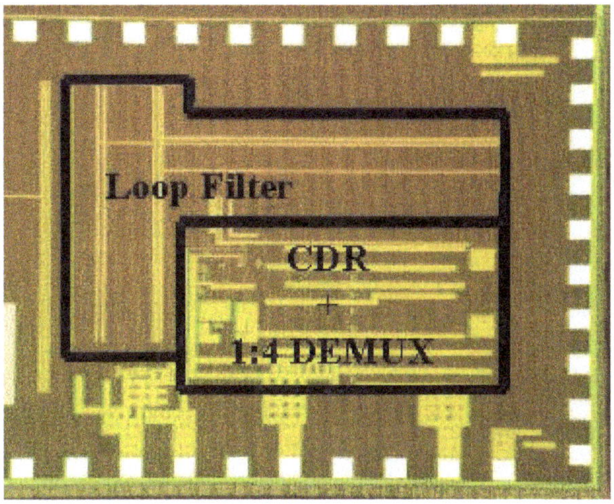

Fig. 8. Die photograph.

Table 3. Performance summary.

Technology	0.18 μm CMOS
Active area + loop filter	0.7 mm^2 + 0.65 mm^2
Power Consumption	100 mW at 1.8-Vdd
VCO Gain	80 MHz./V
VCO Frequency Ranges	240 – 570 MHz.
CDR Data Rate	1 – 2.25 Gb/s
Loop Bandwidth	1 MHz.
Pull-in Range	more than 100 MHz.
Bit-Error-Rate (BER)	less than 10^{-12} at 2.25 Gb/s, Jitter p-p 0.7 UI, modulation at 10 MHz.

a)

b)

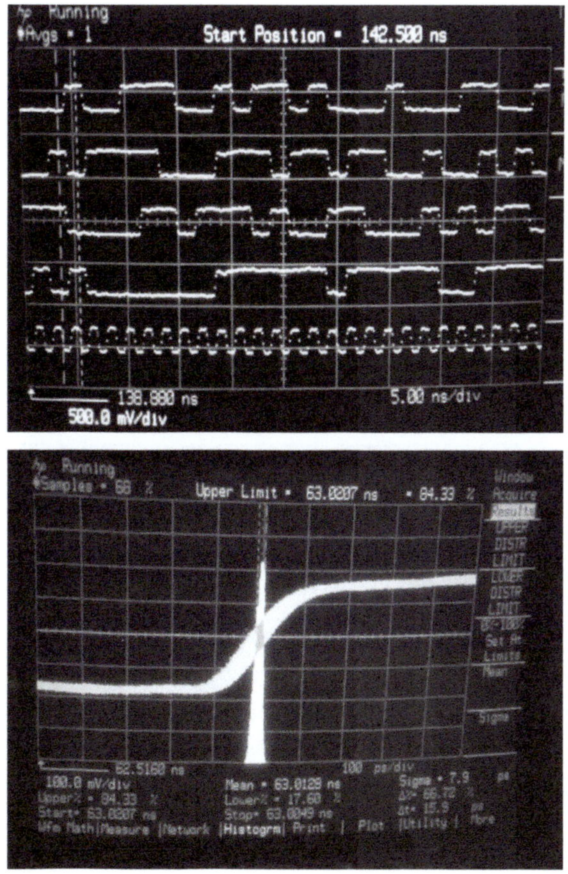

Fig. 9. **(a)** DEMUX output and recovered clock 562.25 MHz, (2.25 Gb/s) **(b)** Jitter histogram of corresponding recovered clock, jitter 5.9 ps,rms.

is controlled to be in the middle between clk-8 and the next clk-0. This method controls also the rest of the clocks.

2.3 Loop characteristic

Standard loop filter using on-chip passive capacitors and resistors are used. Linear model approximation for phase-locked loop (PLL) is still valid for the close-to-lock behavior of this CDR because the FD is not active when frequency acquisition is achieved. Loop behavior during frequency acquisition can be approximated as a control loop with one pole at origin, which is always stable.

3 Experimental results

The 1/4th-rate CDR is implemented in 0.18 μm CMOS technology. CDR excluding loop filter components occupies an area of 0.7 mm^2. The on-chip loop filter capacitor of 600 pF occupies an additional area of 0.65 mm^2. The die photograph is depicted in Fig. 8. The performance summary is shown in Table 3. The CDR can operate at data rate from 1 to 2.25 Gb/s and consumes 100 mW from 1.8 V supply. Its loop bandwidth is 1MHz and pull-in range is larger than 100MHz.

The CDR was measured by using NRZ data with PRBS of $2^7 - 1$. Fig. 9a shows the 4-bit output of DEMUX and recovered clock at 562.5 MHz., corresponding to the incoming data rate of 2.25 Gb/s. The recovered clock has 5.9 ps(rms) jitter. Its jitter histogram is depicted in Fig. 9b. The CDR was also tested using 2.25 Gb/s serial data with a peak-to-peak jitter of 312 ps, 0.7 UI, at 10 MHz modulation. The jitter histograms of incoming serial data and its corresponding recovered clock are shown in Fig. 10a and Fig. 10b, respectively. The recovered clock has a jitter of 15.9 ps (rms) and bit-error-rate (BER) was measured to be less than 10^{-12}.

4 Conclusions

The 1/4th-rate CDR using digital quadricorrelator frequency detector is described. The 1/4th-rate architecture allows a lower operation frequency, hence a coil-free oscillator can meet the jitter requirement. All logic units can be implea)

b)

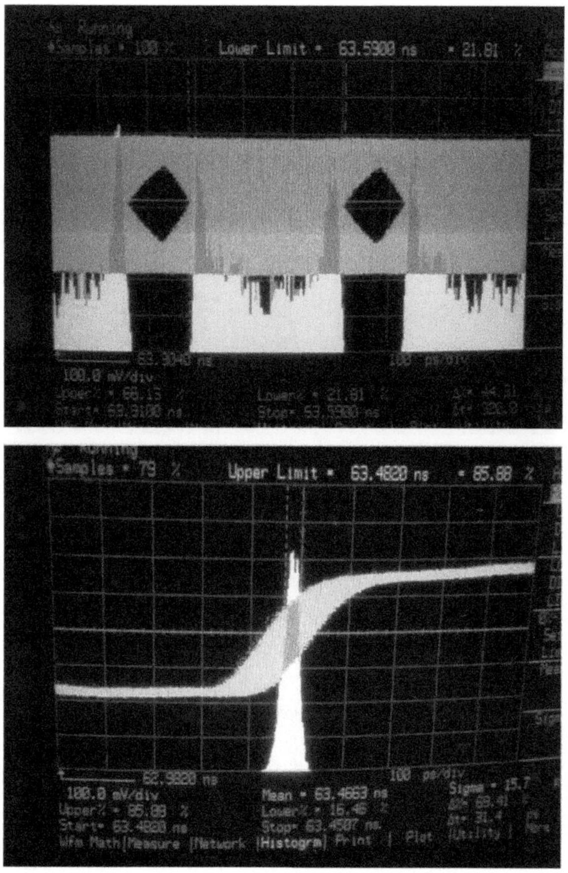

Fig. 10. (**a**) Jitter histogram of serial data 2.25 Gb/s with jitter, p-p 0.7 UI modulation at 10 MHz. (**b**) Jitter histogram of corresponding recovered clock, jitter 15.9 ps,rms.

mented in CMOS logic that leads to lower power consumption compared to high speed current mode logic. Furthermore, this 1/4th-rate CDR is an inherent 1-to-4 demultiplexer, therefore additional circuit is not required. The CDR can have small loop bandwidth and wide pull-in range. It operates without a local reference clock. The CDRs, implemented on 0.18 μm CMOS technology, showed low jitter operation at data rates from 1 to 2.25 Gb/s. Higher data rate will be achieved by parameter changes in the oscillator range.

Acknowledgement. The testchip is sponsored by the Gesellschaft für Schwerionenforschung (GSI) Darmstadt.

References

Cao, J., Green, M., Momtaz, A., Vakilian, K., Chung, D., Jen, K.-C., Caresosa, M., Wang, X., Tan, W.-G., Cai, Y., Fujimori, I., and Hairapetian, A.: OC-192 Transmitter and Receiver in Standard 0.18 μm CMOS, IEEE J. of Solid-State Circuits, 1768–1780, 2002.

Farjad-Rad, R., Nguyen, A., Tran, J. M., Greer, T., Poulton, J., Dally, W. J., Edmondson, J. H., Senthinathan, R., Rathi, R., Edward Lee, M.-J., and Ng, H.-T.: A 33 mW 8 Gb/s CMOS Clock Multiplier and CDR for Highly Integrated I/O, IEEE J. of Solid-State Ciruits, 1553–1561, 2004.

Pottbäcker, A., Langmann, U., and Schreiber, H.-U.: A Si Bipolar Phase and Frequency Detector IC for Clock Extraction up to 8Gb/s, IEEE J. of Solid-State Circuits, 1747–1751, 1992.

Wu, L. and Black Jr., C.: A low Jitter Skew-Calibrated Multi-Phase Clock Generator for Time Interleaved Application, ISSCC, 396–397, 2001.

Yang, R.-J., Chen, S.-P., and Liu, S.-I.: A 3.125 Gb/s Clock and Data Recovery Circuit for the 10-Gbase-LX4 Ethernet, IEEE J. of Solid-State Circuits, 1356–1360, 2004.

33

Power supply network design: a case study driven approach

M. Eireiner[1], **S. Henzler**[2], **T. Missal**[1], **J. Berthold**[2], **and D. Schmitt-Landsiedel**[1]

[1]Institute for Technical Electronics, Technical University Munich, Germany
[2]Infineon Technologies AG, Munich, Germany

Abstract. A study, based on product related scenarios, on power supply integrity issues is conducted. The effectiveness of specific design parameters depends strongly on the expected loading of the power distribution grid. Therefore, the commonly used approach to only use an even current distribution can lead to non-optimal power grid designs. For power grid optimization, a problem reduction from quadratic to linear order is presented. Simulations in a System-on-Chip (SoC) environment show, that power supply integrity mainly depends on the placing of the cores within the SoC die.

1 Introduction

With decreasing feature sizes, power supply integrity has become a serious concern in integrated circuit design. Lowered supply voltages, increasing current densities, increasing operating frequencies, and increasing sheet resistances have decreased the noise margins for every process technology. This trend is expected to continue with technology scaling (Nassif and Fakhouri, 2002, Larsson, 1999, Mezhiba and Friedman, 2004).

Up to now, the design of power distribution networks has been the only design method to combat power supply distortions. The design of a power distribution grids includes the choice of how many of the available metal layers are used for power distribution, as well as the sizing, in terms of widths and pitches, of the used metal layers. The thickness of a metal layer is given by technology and therefore is no design parameter.

The rest of the paper is organized as follows. In Sect. 2 and Sect. 3 the movitation and the simulation setup are outlined. Simulation results for single parameter sweeps are presented in Sect. 4, for combined parameter sweeps in Sect. 5. A SoC environment is analyzed in Sect. 6.

2 Motivation

Based on the International Technology Roadmap for Semiconductors (ITRS) (ITRS, 2005), for technology nodes from 130 nm to 45 nm, values for sheet resistance ($R_{\square}$) of local and intermediate signal wires, supply voltage (V_{DD}), power density (P/A), current density (I/A), and the resulting IR-Drop are displayed in Table 1. All values are normalized to their corresponding value in the 90 nm technology node. We see that from the 90nm to the 45nm technology, the sheet resistance will increase by a factor of 2.07 and 1.77 for local and intermediate signal wires, respectively. The current density will also increase by a factor of 1.46. This together results in an IR-Drop increase of about a factor of 2.6–3.0 depending on the usage of local and intermediate signal wires for supply voltage delivery.

Therefore, power supply integrity is a growing concern not only for high performance, but also for low power designs.

Another point which increases the challenge of power supply network design is that the power supply network has to be designed at an early stage in the design process. Therefore, only little is know about the actual power distribution within the chip (Benoit et al., 1998). Since after the routing the power grid hardly can be changed, it is a common approach to estimate the power consumption of a chip, multiply it by a factor of three to seven, depending on the circuit architecture and on how conservative the chip is designed, and distribute it evenly over the entire chip based on a DC simulation (Dharchoudhury et al., 1998). For these reasons, power distribution grids are always designed over conservative. Therefore, up to one third of the available metal resources in one layer might be used for power distribution.

A DC simulation is much faster compared to a transient simulation. However, this comes at the cost that transient effects are lost in the simulation setup. In (Saint-Laurent and Swaminathan, 2004) it was shown that the cycle average of the supply voltage can be taken as key metric for path delay degradation. To verify this, we did simulations on a critical path replica in a 90 nm technology. The nominal supply voltage is 1.2 V. In Fig. 1 the resulting path delay versus the cycle average of the variation of effective supply voltage

Table 1. Scaling properties of IR-Drop normalized to 90 nm technology node (ITRS, 2005).

	130 nm	90 nm	65 nm	45 nm
$R_{\square}$ (local)	0.7	1	1.54	2.07
(intermed.)	0.85	1	1.44	1.77
V_{DD}	1.13	1	1	0.9
P/A	0.87	1	1.14	1.31
I/A	0.77	1	1.14	1.46
IR-Drop	0.54–0.65	1	1.64–1.76	2.58–3.02

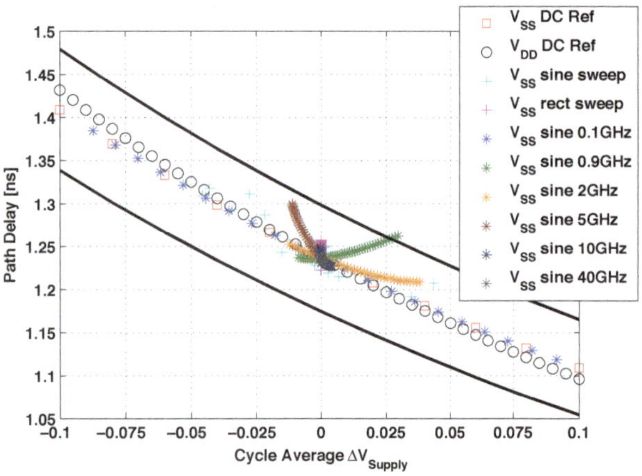

Fig. 1. Path delay versus cycle average of effective supply voltage V_{Supply} for varying V_{SS}/V_{DD} curves, with amplitudes +/–0.3 V. The solid black lines give a 5% boundary around the V_{SS} reference case.

ΔV_{Supply} is displayed. Here, the effective supply voltage V_{Supply} is defined as the difference between the cycle average of V_{DD} and V_{SS}, respectively. In turn ΔV_{Supply} is defined as the difference between the case of no noise, 1.2 V, and the noisy one: $\Delta V_{\text{Supply}}=1.2\,\text{V}-V_{\text{Supply noise}}$. The graphs with the circles and rectangular symbols are simulation sweeps, in which V_{DD} and V_{SS}, respectively, are kept constant during one simulation run. Both graphs act as reference. The solid black lines give a 5% boundary around the V_{SS} sweep. The graphs with crosses and stars show the results of simulations with transient varying V_{SS} curves. Sinodial and rectangular curves were taken with varying amplitude and/or frequency. The frequency range for the sinodial noise was swept from 100 MHz up to 40 GHz. However, most of the simulation results, which had almost perfect correlation with the reference simulation are not shown due to reasons of clarity. The peak to peak amplitude of the ground disturbance was up to 0.6 V. We see that all simulations, with one minor exception, show an error which is well below 5%. Simulations with transient varying V_{DD} curves showed similar results. Therefore, the

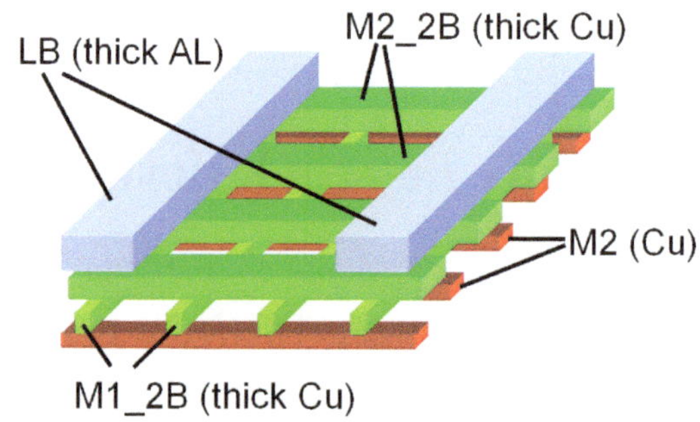

Fig. 2. Schematic of the used 4 stages power grid.

exact waveform of V_{DD} and V_{SS} are not of concern but only their cycle average has to be taken into consideration if the effect of power supply distortions on path delay variations is analyzed. Hence, fast DC simulations instead of expensive transient simulations were performed to allow for a fast evaluation of varying power grid sizings.

3 Simulation setup

We base our analysis on the example of an ARM926 core in a 90nm technology, as it is described in (Lueftner et al., 2006). As initial point of our analysis we take a power distribution grid, which consists of four layers. The top layer, LB, is a thick aluminum layer, followed by two copper metal layers with double height for intermediate signal wiring, M2_2B and M1_2B, and finally the local power rails in the M2 layer, which is a copper layer with single height. The layers are orthogonal to each other and are connected frequent with vias at the crossing points. The schematic is depicted in Fig. 2. For our analysis we take a symmetric 600 μm × 1200 μm section of the power grid. The end points of the layers at the section boundary are short-circuited with their starting points to supress artefacts.

From the power data given in (Lueftner et al., 2006) the average power consumption was taken and multiplied by a factor of three for temporal variations. For lateral variations of the grid loading, two different settings were analyzed. In the first loading scenario the power is distributed evenly across the whole chip. In the second scenario, a hotspot was created in the middle of the power grid. The hotspot is 150 μm × 300 μm and has a current density which is three times higher than in the case of the even loading, the schematic is depicted in Fig. 3. The current density in the rest of the chip was chosen such that for Even and Hotspot loading the same total current density resulted.

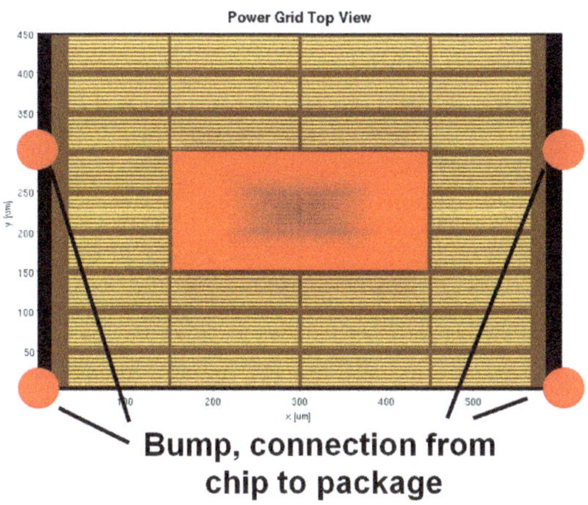

Fig. 3. Schematic of the hotspot loading. A hotspot, $150\,\mu m \times 300\,\mu m$, with a three times higher current density compared to the even case is placed in the middle of the power grid. The remaining power grid was loaded such that the overall current density matched for Hotspot and Even loading.

4 Single parameter sweeps

In the first analysis all design parameter of the power grid are varied one at the time, with the others at their initial value, and the resulting IR-Drop was simulated for both loading scenarios. The pitch of the lowest power rail is given through the height of a standard cell. The pitch of the LB layer is set through minimum bump pitch requirements. Therefore, six independent design parameters, four width and two pitch parameters, exist.

For every design parameter a sweep across a wide parameter range was conducted. The increase or decrease of a design parameter was translated into an equivalent relative increase or decrease in area. The resulting IR-Drop depending on the change in area is displayed in Fig. 4, here at the example of varying M2 width with Hotspot loading. The resulting graphs for the remaining five design parameters look similar to the one for the M2 width sweep in Fig. 4. Changing from Hotspot to Even loading results in a decrease of IR-Drop by preserving roughly the shape of the graph. However, it has to be mentioned that in our use case, in which the grid is connected by bumps to the package supply every $600\,\mu m$ see Fig. 2, virtually no voltage drop occurs in the LB layer. Therefore, changing the width of the LB layer has minimal impact on the resulting IR-Drop.

Based on the simulations described above, a cost performance analysis in terms of resulting change in IR-Drop and change in occupied area is conducted for both loading scenarios. Fig. 5 shows the effectiveness of a design change for the different design parameters. Effectiveness is defined as $\frac{\Delta \text{IR-Drop}}{\Delta \text{rel.area}}$. The effectiveness is derived from the criterion how much area is needed to achieve a given reduction in IR-

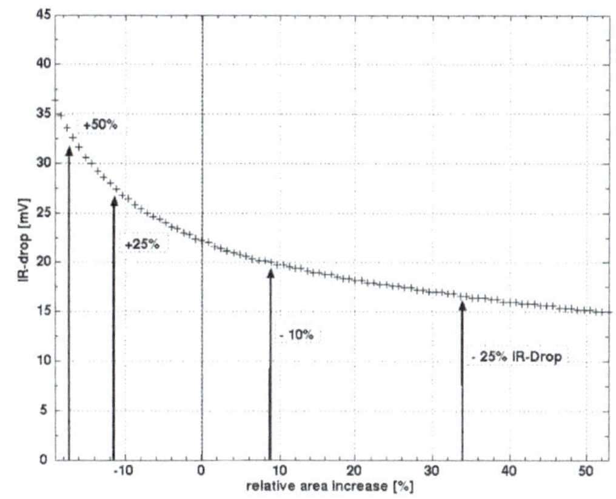

Fig. 4. Resulting IR-Drop for varying M2 width for the Hotspot loading. An increase/decrease of M2 width is translated into a relative increase/decrease of area.

Drop. The less area needed, the more efficient the design parameter is. The results in Fig. 5 are normalized to the maximum effectiveness in one loading scenario. The LB layer is omitted in this plot, since, as mentioned before, it does not contribute significantly to IR-Drop.

We see that for different loading scenarios of the power grid, the effectiveness of the design parameter drastically changes. For example for Even loading M1_2B pitch has the lowest effectiveness, whereas for a Hotspot loading the situation is vice versa, M1_2B pitch is the preferable parameter to reduce IR-Drop area efficient.

The loading of an actual design hardly will be an Even loading, and the critical case for IR-Drop is some hotspot on the chip. Therefore, we see that the common approach to use Even loading for initial power grid dimensioning gives wrong design incentives. This can lead to non-optimal power grid designs in terms of area efficiency. Therefore, we propose to use not only Even loading, but also some Hotspot loading for initial power grid design. With the additional loading scenario effects of lateral varying current densities on power grid dimensioning can be accounted for. Hence, right from the start a better power grid design is possible, which helps to achieve the area efficient optimum.

5 Combined parameter sweeps

Power grid design usually is not done by only varying one design parameter, but multiple at a time. The question we want to address in this section is, whether it is possible to estimate the resulting IR-Drop of varying two design paramters at the same time, by taking the simulation results of their single parameter sweeps. Therefore, it would be possible to achieve a problem reduction from a quadratic order, $O(N^2)$,

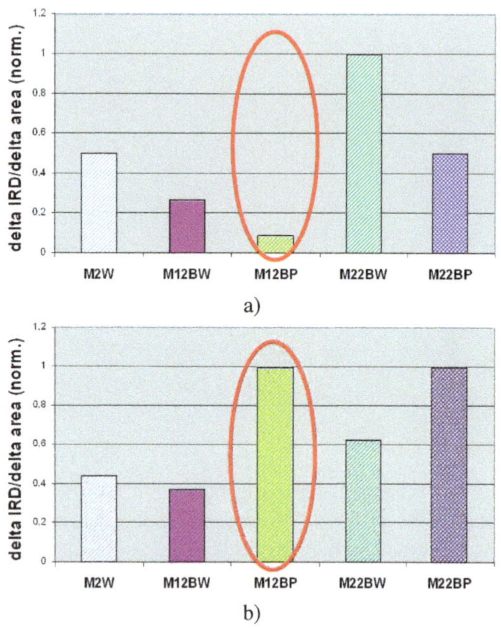

Fig. 5. Effectiveness in terms of IR-Drop over relative area change for given reduction in IR-Drop for varying design paramters. **(a)** Even loading **(b)** hotspot loading

to linear order, $O(N)$. This in turn decreases the simulation time signficantly.

First we tried to estimate the resulting IR-Drop with the additive equation:

$$\mathrm{IRD}_{\mathrm{est}}(v_1+\Delta v_1,\, v_2+\Delta v_2) = \mathrm{IRD}_{\mathrm{single}}(v_1+\Delta v_1) + \mathrm{IRD}_{\mathrm{single}}(v_2+\Delta v_2) - \mathrm{IRD}_{\mathrm{std}}$$

Where $\mathrm{IRD}_{\mathrm{est}}(v_1+\Delta v_1,\, v_2+\Delta v_2)$ is the estimated IR-Drop for changing design paramters v_1 and v_2, $\Delta\mathrm{IRD}_{\mathrm{single}}(v_1+\Delta v_1)$ and $\Delta\mathrm{IRD}_{\mathrm{single}}(v_2+\Delta v_2)$ are the changes in IR-Drop caused by single design paramter variation Δv_1 and Δv_2, respectively, IRD_{std} is the IR-Drop for the initial or standard dimensioning of the power grid. As reference all combinations of parameter sweeps were simulated. However, the additive IR-Drop estimation showed high deviation from the reference simulation and therefore can not be used for an acurate estimation.

As a second approach we tried to estimate the resulting IR-Drop with the multiplicative equation:

$$\mathrm{IRD}_{\mathrm{est}}(v_1+\Delta v_1,\, v_2+\Delta v_2) = \frac{\mathrm{IRD}_{\mathrm{single}}(v_1+\Delta v_1)\cdot \mathrm{IRD}_{\mathrm{single}}(v_2+\Delta v_2)}{\mathrm{IRD}_{\mathrm{std}}}$$

To visualize the normalized error, $\mathrm{Err}_{\mathrm{norm}}$, between the simulated and the estimated IR-Drop the error was calculated for all parameter cominations:

$$\mathrm{Err}_{\mathrm{norm}} = \frac{\mathrm{IRD}_{\mathrm{sim}} - \mathrm{IRD}_{\mathrm{est}}}{\mathrm{IRD}_{\mathrm{sim}}}$$

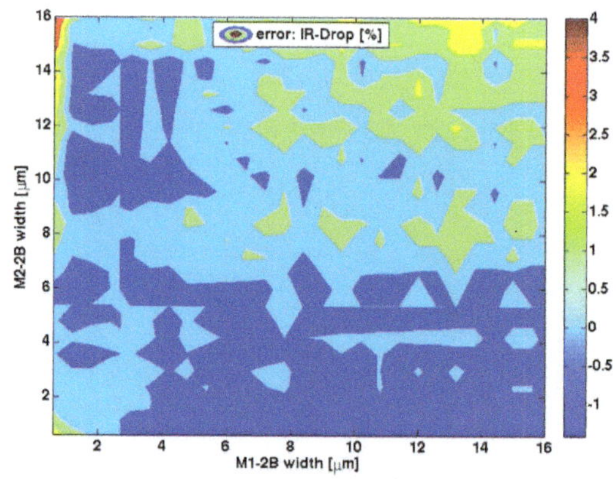

Fig. 6. Resulting error map for estimating IR-Drop multiplicatively, for varying M1_2B width and M2_2B width.

The resulting error map over the entire parameter range, here at the example for varying M1_2B width and M2_2B width, is displayed in Fig. 6. The color of the map reflects the resulting error in percentage. In Fig. 7 the histogram for the error map of Fig. 6 is displayed. The error in this case is between –2% and +4%. For all possible combinations of parameter variations over the entire relevant design space the error is in the range of $\pm 10\%$. Therefore, the resulting IR-Drop of changing two design paramters at the same time can efficiently and accurately be estimated by the proposed equation. Hence, a problem reduction from quadratic to linear order was achieved.

Up to now, no analytical derivation for the good fit of the multiplicative approach can be given. But it is expected, that parallel connections of power rails, e.g. LB || M1_2B and M2_2B || M2, are responsible for the malfunction of the additive as well as for the good fit of the multiplicative approach. However, more research has to be done to derive an analytical verification of the observed phenomenon.

6 SoC power grid example

As an other example, we consider an System on Chip (SoC) design, as it is presented in (Lueftner et al., 2006). We choose a chip size of 6 mm×6 mm. The schematic of an exemplary power distribution is given in Fig. 8. In the green region in the middle of the chip, regular bump connections are available. In the outer pink region no bump connections are available due to package constraints. In the red region at the edge of the chip, the I/O connections are located and therefore, no bump connections for power distribution and no power routing in the LB metal layer is possible.

As loading we again take an ARM926 core, with a die size of 1 mm×1 mm. The die is placed at varying positions within the SoC and the resulting IR-Drop is simulated. Due

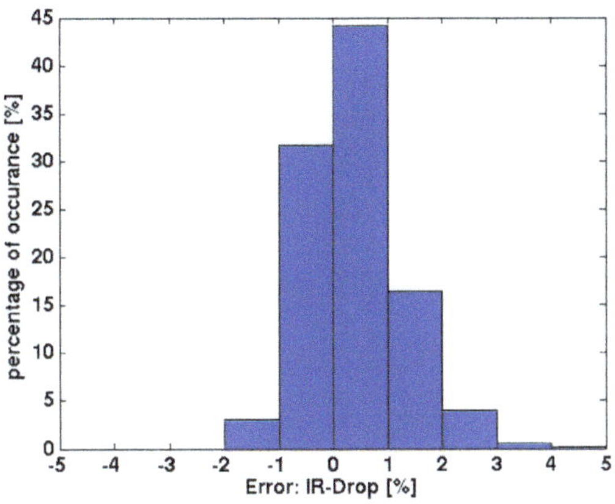

Fig. 7. Histogram for the error map of Fig. 6.

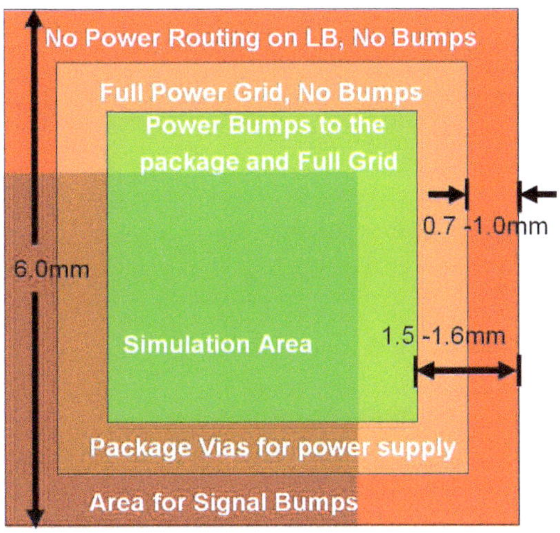

Fig. 8. Schematic of the power supply network of an example SoC with 6 mm×6 mm die size. ESD and I/O restiction prohibit the bump connections and LB usage in the outer regions of the chip.

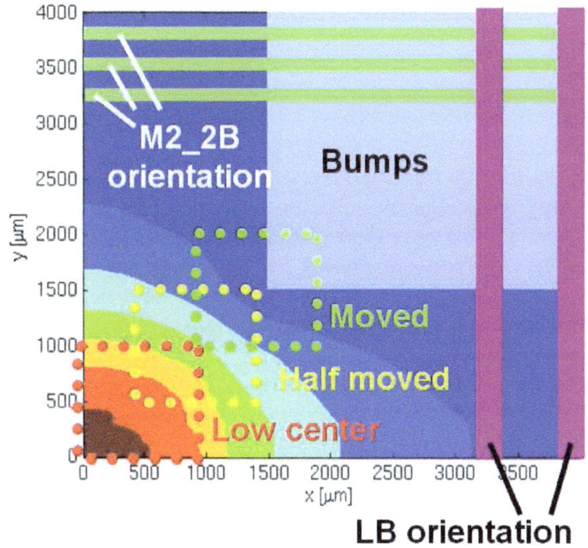

Fig. 9. Schematic of the power supply network of an example SoC with 6 mm×6 mm die size. ESD and I/O restiction prohibit the bump connections and LB usage in the outer regions of the chip.

Table 2. Power supply noise for varying core positions within SoC power grid.

	Low center	Half moved	Moved
Max	126.3 mV	42.4 mV	13.1 mV
Min	96.7 mV	30.1 mV	7.2 mV
Avg	35.9 mV	0.2 mV	0.2 mV

to reasons of symmertry and to reduce simulation time the grid was reduced to the shaded area, 4 mm×4 mm, in Fig. 8. Exemplarily, three positions of the ARM core within the SoC are depicted in Fig. 9. The resulting maximum, minimum, and average IR-Drop within the core area are tabulated in Table 2. For the placing in the lower left corner, almost the entire IR-Drop is on LB and M2_2B layer. M1_2B and M2 contribute only minor to the overall IR-Drop. We see from Table 2, that by moving the core only 1mm to the center, maximum IR-Drop is reduced be about a factor of 10, for minimum and average IR-Drop the factors are even higher. Similar simulations with different placings, e.g. starting in the left center, were also conducted and showed compareable results. Therefore, in a SoC environemnt the placing of the major power consuming cores is the most critical issue for power supply integrity. However, I/O and ESD restrictions might require a placing of such blocks to the edges and I/O pins of the SoC.

7 Conclusions

In this paper a case study driven approach to power supply integrity analysis was presented. Based on product related scenarios, it was shown that using only the assumption of Even loading for initial power grid design can lead to non-optimal power grids. For the optimization of power grids, it was shown that a problem reduction from quadratic to linear order is possible by using a multiplicative IR-Drop estimation. However, an analytical derivation of the used formula could not be given and is the goal of future research, as well as the extension from two to multiple design variables. In SoC environments in which I/O, packaging, and ESD requirements restrict the on-chip power routing, almost the entire IR-Drop is in the upper layers. Placing of cores within the SoC has the highest single impact on power supply integrity.

References

Benoit, M., Taylor, S., Overhauser, D., and Rochel, S.: Power Distribution in High-Performance Design, in: Proceedings of IEEE Symposium of Low-Power Electronics Design, 274–278, 1998.

Dharchoudhury, A., Panda, R., Blaauw, D., and Vaidyanathan, R.: Design and Analysis of Power Distribution Networks in PowerPC Mircoprocessors, in: Proceedings of Design Automation Converence, 738–743, 1998.

ITRS: International Technology Roadmap for Semiconductors, http://public.itrs.net/, 2005.

Larsson, P.: Power Supply Noise in Future IC's: A Crystal Ball Reading, in: IEEE Custom Integrated Circuits Conference, 467–474, 1999.

Lueftner, T., Berthold, J., Pacha, C., Georgakos, G., Sauzon, G., Hoemke, O., Beshenar, J., Mahrla, P., Just, K., Hober, P., Henzler, S., Schmitt-Landsiedel, D., Yakovleff, A., Klein, A., Knight, R., Acharya, P., Mabrouki, H., Juhoor, G., and Sauer, M.: A 90 nm CMOS Low-Power GSM/EDGE Multimedia-Enhanced Baseband Processor with 380 MHz ARM9 and Mixed-Signal Extensions, in: 2006 ISSCC Digest of Digital Papers, 252–253, 2006.

Mezhiba, A. V. and Friedman, E. G.: Scaling Trends of On-Chip Power Distribution Noise, IEEE Transactions on Very Large Scale Integration (VLSI) Systems, 12, 386–394, 2004.

Nassif, S. R. and Fakhouri, O.: Technology Trends in Power-Grid-Induced Noise, in: Proceedings of the 2002 international workshop on System-level interconnect prediction, 55–59, ACM Press, 2002.

Saint-Laurent, M. and Swaminathan, M.: Impact of Power-Supply Noise on Timing in High-Frequency Microprocessors, IEEE Transactions on Advanced Packaging, 27, 135–144, 2004.

50 Gbit/s real-time test environment for integrated photonic DQPSK receivers

T. Föhn, C. Fischer, and M. Berroth

University of Stuttgart, Institute for Electrical and Optical Communications Engineering, Stuttgart, Germany

Correspondence to: T. Föhn (thomas.foehn@int.uni-stuttgart.de)

Abstract. In this paper an FPGA-based test system for high-speed transmission experiments with integrated photonic receivers is presented. Pseudorandom binary sequences are generated inside the FPGA and encoded as either differential quadrature phase shift keying (DQPSK) or quadrature phase shift keying (QPSK) signals. The DQPSK encoder uses a 64-fold parallel-prefix-layers architecture for real-time operation which allows for a maximum internal encoder data rate of 64 Gbit/s. Two-fold parallel data streams of I and Q signals suitable for driving an optical IQ-modulator can be transmitted and received by four 12.5 Gbit/s transceivers. Integrated bit error testers are used to determine bit error rates in real-time.

1 Introduction

To response the ever increasing bandwidth requirements in optical transmission links, data rate per channel is continuously increased as well as the aggregated data rate in one fiber, e.g. by using many channels in wavelength division multiplexing systems. The modulation scheme in most installed commercial transmission lines is still on-off keying (OOK), as simple photodetectors can be used for the optoelectric conversion at the end of the optical link. However, future data links require 40 Gbit/s, 100 Gbit/s or even higher data rate per channel, which is an issue in OOK transmission systems as the pulses become more susceptible for dispersion effects. In particular polarization mode dispersion randomly disturbs the signal which makes it difficult for compensation. Furthermore, high power densities in the short pulses give rise to disturbing non-linear effects. Using high-order modulation formats like quadrature phase shift keying, data rates can be increased while the symbol rate is kept constant. However as the phase of the optical signal has to be recovered, the receiver architecture becomes more complicated. Due to the constant progress in semiconductor processing integrated silicon photonic devices can now be fabricated with sufficient accuracy. This platform offers the possibility to integrate complex interferometer-based phase detectors even in combination with polarization and wavelength division multiplexers in a CMOS process, perspectively integrated together with electronic signal processing circuitry (Knoll, 2014). In many cases it is sufficient to characterize single photonic devices with continuous-wave transmission measurements. Some devices need to be embedded into test benches or being cascaded for determining their characteristics with high accuracy. For more complex circuits, like in the case of integrated receivers, high-speed data transmission experiments are desired to verify the operation and to attain device properties from a system level of view. This kind of experiment is more complex as further components such as fast optical modulators and photodiodes are needed as well as high-speed electronic transceivers and modulator drivers. Many of today's field programmable gate arrays (FPGA) offer several Gbit/s transceivers together with a large number of programmable logic cells, making them a cost effective solution for Gbit/s transceivers in transmission experiments.

2 Integrated photonic DQPSK receiver

Differential quadrature phase shift keying is a variation of the standard QPSK where two data bits $b_0(i)$ and $b_1(i)$ are coded in the phase transitions between consecutive symbols instead of being coded in the absolute phase position of the signal. Figure 1 (left) shows a part of the DQPSK constellation diagram together with the phase transitions. The phase of the ith symbol can be described with the status bits $d_0(i)$ and $d_1(i)$,

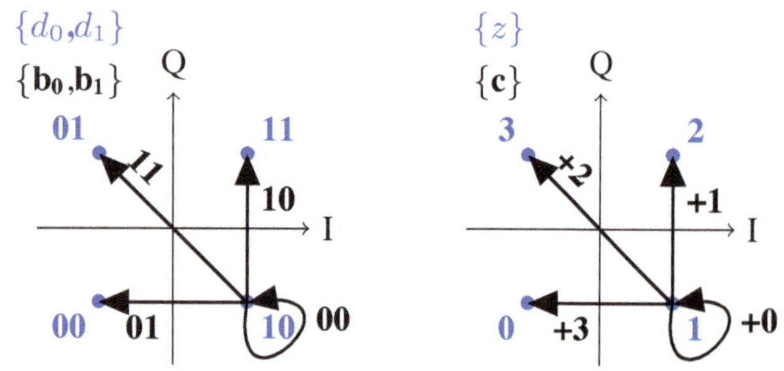

Figure 1. Left: Part of a DQPSK constellation diagram with initial phase status $\{d_0,d_1\} = 10$ and corresponding phase transitions for the transmission of the symbol $\{b_0,b_1\}$ to neighboring states. Right: Corresponding representation of phase states and transitions with arithmetic numbers for the modulo-4 based encoder.

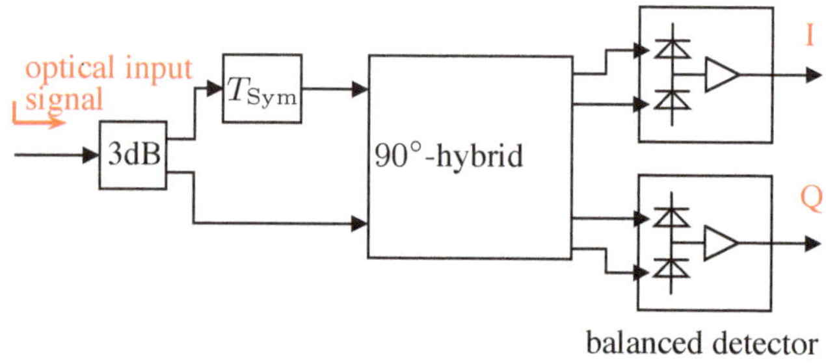

Figure 2. Block diagram of a direct receiver with delay line and 90°-hybrid for DQPSK signals. As the optical signal has to be delayed by one symbol duration T_{Sym}, the length of the delay line depends on the desired data rate.

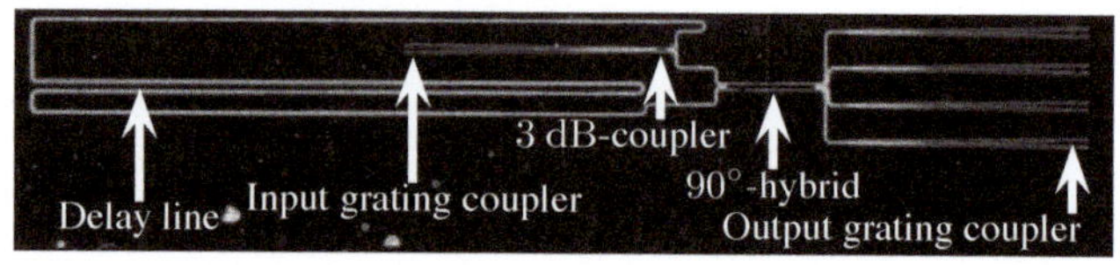

Figure 3. Photograph of an integrated DQPSK receiver in silicon-on-insulator without photodetectors. The footprint is about $250\mu m \times 2000\mu m$, mainly due to the delay line and the $500\mu m$ long tapers between the grating couplers and single-mode waveguides. Smaller footprints can be achieved with focusing grating couplers and meander-shaped delay lines.

that directly correspond to the in-phase (I) and quadrature (Q) signal for an optical IQ-Modulator.

While QPSK modulated signals require a coherent receiver, DQPSK coded symbols can be retrieved with a direct receiver. Possible implementations consists of either two delay line interferometer with 3-dB couplers or one delay line interferometer with a 90°-hybrid (Seimetz, 2009). A block diagram with the principal components of the latter receiver is depicted in Fig. 2. The passive part has been realized as integrated receiver at the Institute for Electrical and Optical Communications Engineering on a silicon-on-insulator (SOI) wafer.

Figure 3 shows a chip photo of a fabricated device. A grating coupler on the left hand side is used for the coupling between a standard single mode fiber and the planar photonic circuit. The 3-dB coupler is realized as symmetric multimode interference coupler (MMI) which distributes the incoming signal equally to two branches. A delay line is introduced in one of the branches, delaying the signal by one symbol period. Therefore the receiver has to be designed to work at a fixed data rate. The physical length of the delay line also depends on the waveguide geometry, i.e. the waveguide width and thickness. The simulated group index of a typical nanowaveguide with a cross-section of 400 nm × 230 nm is about $n_g = 4.40$ at a wavelength of $\lambda = 1550$ nm. For a target data rate of 25 Gbaud a length of about 2.7 mm is needed.

The actual detection of the phase difference between two consecutive bits is done in the 2×4 90°-hybrid, that is also realized as MMI. This device has two input and four output ports. The input signals are split equally and interfere at the output ports with a phase difference of $0, 0.5\pi, \pi$ and 1.5π plus a constant offset angle φ_{MMI} that depends on the device. The signals are coupled out again with grating couplers. The opto-electric conversion can be done with external photodiodes. However, future receivers will include integrated photodiodes. The signal and the delayed signal interfere at the four 90°-hybrid outputs with a phase difference of $\varphi + n\frac{\pi}{2}$, where n is an integer between 0 and 3. The constant phase $\varphi = \varphi_{MMI} + \varphi_{ext}$ is composed of the intrinsic MMI phase offset and an external phase offset defined by the exact length difference between the two interferometer arms. When balanced photodiodes are used and φ is set to $\frac{3}{4}\pi$, the receiver provides two output signals that correspond to the transmitted bits $\{b_0,b_1\}$ according to Fig. 1 without requiring further logic operations (Fischer, 2013). This fine adjustment can be achieved e.g. by using heaters on top of the delay line without high power dissipation due to the large waveguide length.

3 FPGA-based test environment

The principal component of the test environment for the integrated optical receiver is the electrical DQPSK transceiver, where the signals are created and the received signals are evaluated.

A block diagram of the realized transceiver is shown in Fig. 4. The principle functional blocks are the DQPSK encoder and the so-called GTX interfaces that include pseudorandom binary sequence (PRBS) generators, high-speed electrical interfaces and the bit error tester. Further components such as an overall control block for selecting the PRBS sequence or starting and stopping the transmission, a serial interface for the communication with a PC, memory and a reference decoder complete the transceiver design. A Xilinx Kintex-7 FPGA with 16 high data rate GTX interfaces, 50 950 logic slices and a total of about 2500 KiB

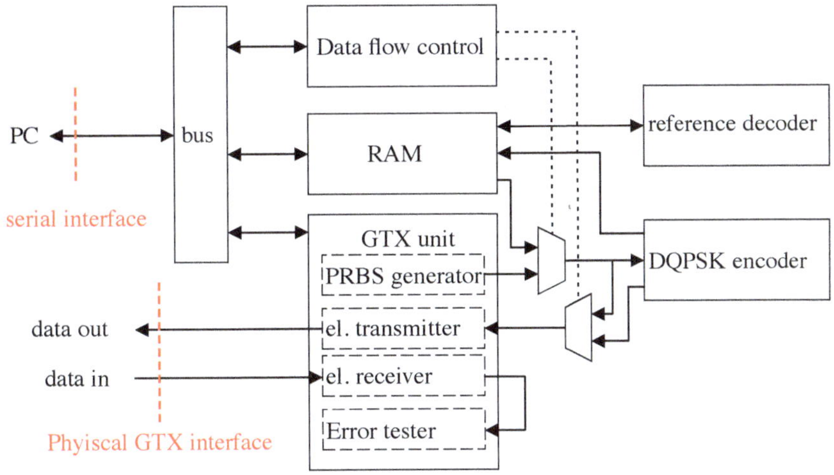

Figure 4. Basic structure of the DQPSK transceiver including the basic functional blocks of the FPGA design. The transceiver can be controlled externally over a serial interface. The input and output data is transmitted by the physical GTX interfaces.

random-access memory (RAM) is used. One of the GTX interfaces is available as electrical contact on the Xilinx evaluation board. Four more are accessible on a separate expansion board and are used to send and receive the data. For the data generation the GTX built-in PRBS generators are used. Different sequences like PRBS-7, PRBS-15, PRBS-23 and PRBS-31 can be generated and are shift out as 64 Bit vectors. In this implementation two pattern generators are used, one for the in-phase component and one for the quadrature component. Together they compose vectors with 64 two bit symbols. A multiplexer allows for bypassing the DQPSK encoder, so that a standard QPSK is transmitted.

4 DQPSK encoder

The coding of data bits $b_0(i)$ and $b_1(i)$ into phase transitions can be seen in the DQPSK constellation diagram in Fig. 1 (left). The transmission of the data 00, 10, 11 and 01 corresponds to a phase shift of 0, $\frac{\pi}{4}$, $\frac{\pi}{2}$, and $\frac{3\pi}{4}$. Starting with a reference phase the encoder therefore has to calculate the new phase position for each symbol, which is a function of the previous phase and the bits to be transmitted. There are different ways to implement this function. One approach with logic operations (Seimetz, 2009) requires very fast logic gates and flip-flops with a low propagation delay, as the status bits must be fed back within one symbol duration. Another way is the use of arithmetic modulo-4 additions. In this approach a different representation of the phase positions and transitions is chosen, as shown in Fig. 1 (right). The status bits d are coded to phase bits z that represent the phase as arithmetic numbers clockwise ascending from 0 to 3. The phase transitions are also coded as arithmetic numbers c, representing the phase transition directly as multiple of 90°. With initial phase bits $z(0)$, the phase bits of the nth symbol can then be calculated as

$$z(n) = z(0) + \sum_{i=1}^{n} c(i) \bmod 4. \qquad (1)$$

This calculation can be parallelized in parallel prefix network adders (PPN) (Harris, 2003), which moderates the timing requirements of the feedback path. The scheme of a PPN adder is shown in Fig. 5 (left). The scheme is limited to an 8-fold parallel structure for clarity. Modulo-4 adders are marked with squares. In the first stage, the sum of $c(i)$ and its successor $c(i+1)$ are calculated for each even index i. With these results, all possible sums of $c(i)$ and its successors $c(i+1)$, $c(i+2)$ and $c(i+3)$ are calculated. This can be continued further up to the last stage. A n-fold parallel prefix network adder therefore needs $\log_2(n)$ stages. At the end, the state $z(n)$ of the last transmitted symbol has to be added to $c(0)$ to maintain the phase difference to the first transmitted symbol. The feedback path is the critical path in this architecture. The propagation delays, on the Kintex 7 in the range of about 270 ps to 440 ps, depending on the cell type, then still limit the maximum achievable bit rate. To attain a minimum of 50 Gbit/s at least 256-fold parallelization is needed. The architecture of a parallel prefix layers adder (PPL), as presented in (Zhongxia, 2011), allows a more efficient use of the FPGA resources. The scheme of such an adder is shown in Fig. 5 (right). The main difference to a PPN adder are the flip-flops at each stage. Thus the state $z(n)$ has only to be added to all the sums of $c(0)$ to $c(i)$ at the end of the adder structure, which shortens the feedback path and allows for higher clock rates. Comparing both architectures, similar data rates can be attained at a lower degree of parallelization in the PPL architecture due to higher possible clock frequencies at the expense of a higher number of registers and adders. However, the extra effort for the additional cells is lower than that of a higher parallelized adder structure. Based on the

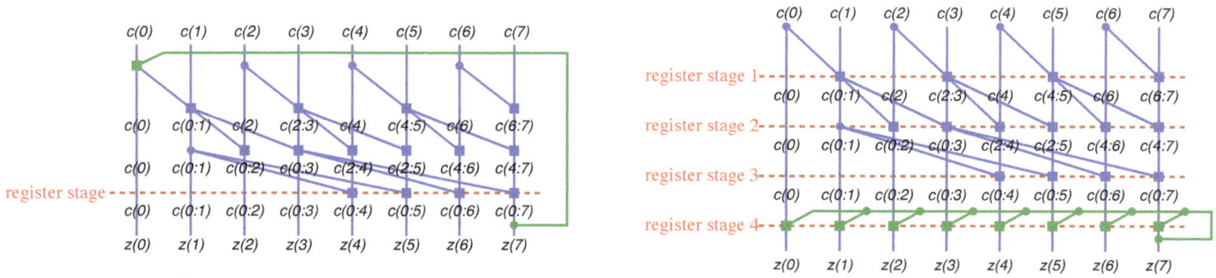

Figure 5. Example of an 8-fold parallel prefix network (PPN) adder (left) and parallel prefix layers (PPL) adder (right). 2-bit adders are represented by squares. The long feedback path in the PPN architecture over three adder stages is significantly shortened in the PPL architecture.

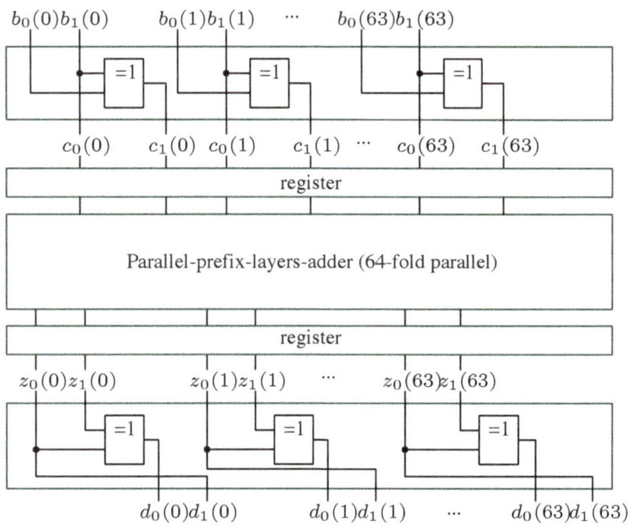

Figure 6. Complete architecture of the 64-fold parallel PPL DQPSK encoder. The conversion from binary data to arithmetic increments is done in the first stage. Phase differences are added up in the following PPL adder and are converted to I and Q signals in the last stage. Two register stages are inserted for buffering.

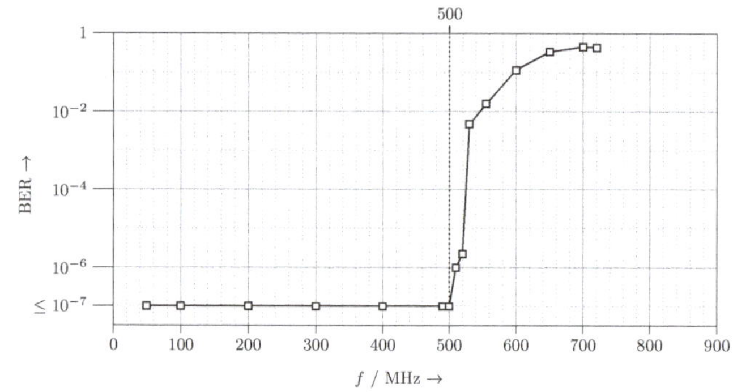

Figure 7. Measured bit error rate of the 64-fold parallel DQPSK encoder for different encoder clock frequencies. The BER increases suddenly at about 500 MHz, due to timing issues inside the design.

typical propagation delays a 64-fold parallel PPL adder is sufficient to achieve a bit rate higher than 100 Gbit/s, leaving sufficient headroom for additional line delays in the final configuration. The Kintex 7 logic cells are organized in slices, with one slice containing four look-up tables and eight flip-flops. Regarding the hardware resources needed, the encoder core with PPL architecture needs at least 289 slices out of about 50,000 slices available. Achieving similar performance with a PPN architecture would require a 512-fold parallelization, which would occupy about ten times more slices. In an implementation for 100 GBit/s with PPN architecture on a former upper-class FPGA a 1280-fold parallel encoder is needed due to the lower clock frequency, occupying about 20 times more resources in terms of look-up tables (Winzer, 2008).

The complete encoder is shown in Fig. 6. First the data bits are coded to the bits c by means of logic functions and buffered in registers. The next stage is the PPL adder that calculates the bits z. In the last stage the phase bits are translated to the driving signals for IQ-modulator, again using logic functions.

The synthesized encoder is tested in combination with a slow reference decoder, with alternating encoding and decoding runs. A BER measurement of the final encoder layout is shown in Fig. 7. Up to a system clock of 500 MHz no bit errors are detected in a sequence of 10^7 bits. As 64 two-bit words are encoded in each cycle a total data rate of 64 Gbit/s is achieved. The architecture can be expanded to higher degrees of parallelization, as the encoder only uses a small portion of FPGA resources. For real-time transmission experiments the GTX built-in bit error testers are used. No hardware decoder is needed, as the integrated photonic receiver already decodes the data inherently. The received data is checked against the selected PRBS sequence in blocks of 64 bit, i.e. a bit error counter does increment if one or more bits in one block are erroneous. For actual small bit error (BER) rates the measured BER is in the worst case by a factor of 64 higher, depending on whether random or burst errors occur.

The serial interfaces are limited to a maximum data rate of 12.5 Gbit/s, therefore two transmitters are used for the I and Q signal, respectively. The signals are interleaved with external multiplexers at the transmitter side and de-interleaved at

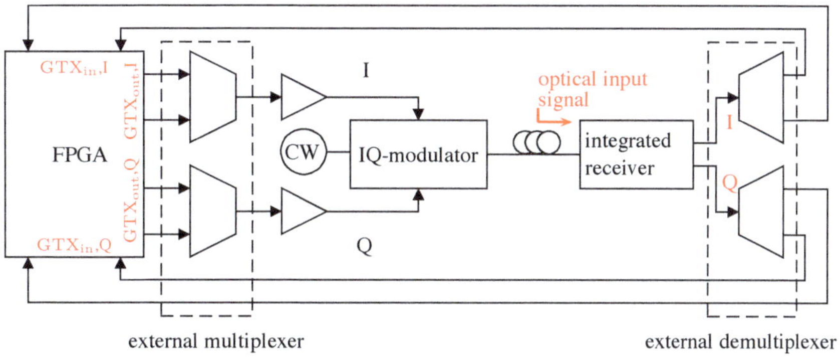

Figure 8. Block diagram of the transmission system. The external multiplexers are connected to the FPGA by the GTX interfaces and are followed by amplifiers to provide sufficient signal swing to drive the IQ-modulator. The optical signal is transmitted over a fiber link to the input of the integrated receiver. After the opto-electric conversion in the balanced photodetectors, the signal is de-interlaced by the demultiplexers and fed back to the transceiver by the GTX interfaces.

the receiver side with demultiplexers. For this purpose commercial Hittite ICs are used on printed circuit boards with Taconic high frequency substrate.

A common issue in FPGA-based transmitters with several parallel output channels is the synchronization. After each reset, the interfaces initialize and generally start to transmit the data at a random time, thus data streams are shifted against each other by a multiple of the bit duration. Furthermore, sub-bit delays can occur when the interfaces are clocked with different clock sources. In this design, one clock source is used for all the four GTX transceivers. However, different lengths in the clock paths result in sub-bit delays that have to be compensated externally with time delays. As a common clock source is used, the phase relations do not change even after a reset. Therefore this compensation has to be done only once, whereas the bit-wise synchronization has to be performed after each system reset.

The complete transmission system with its main components is shown in Fig. 8, with the FPGA test environment acting as transmitter and receiver. External multiplexers and demultiplexers allow for doubling the symbol rate. The transmitted signals have to be amplified to drive the IQ-modulator and are then sent over an optical fiber link to the integrated receiver. After the opto-electric conversion the data is fed back to the FPGA board where the BER is measured.

5 Conclusions

This work presents the possibility to use an FPGA as cost-effective test environment for high-speed optical transmission systems. A 64-fold parallel real-time DQPSK encoder has been implemented together with electrical interfaces and bit error testers. It works stable up to a clock frequency of about 500 MHz, allowing for 64 Gbit/s operation. The encoder PPL structure works more efficient than the more common PPN architecture and occupies ten times less resources for the encoder core. Even higher data rates can be attained with higher degree of parallelization, external multiplexers and an FPGA board with more electrical GTX interfaces.

Edited by: D. Killat

References

Fischer, C.: Design of a FPGA-based DQPSK transceiver for optical communication systems with data rates up to 50 Gbit/s, Diploma Thesis, University of Stuttgart, Institute for Electrical and Optical Communications Engineering, Germany, 2013.

Harris, D.: A taxonomy of parallel prefix networks, Conference Record of the Thirty-Seventh Asilomar Conference on Signals, Systems and Computers, Asilomar Grounds in Pacific Grove, CA, USA, 9–12 November, 2003.

Knoll, D., Lischke, S., Zimmermann, L., Heinemann, B., Micusik, D., Ostrovskyy, P., Winzer, G., Kroh, M., Barth, R., Grabolla, T., Schulz, K., Fraschke, M., Lisker, M., Drews, J., Trusch, A., Krüger, A., Marschmeyer, S., Richter, H., Fursenko, O.,Yamamoto, Y., Tillack, B., Wohlfeil, B., and Petermann, K.: Monolithically Integrated 25 Gbit/sec Receiver for 1.55 μm in Photonic BiCMOS Technology, Optical Fiber Communication Conference and Exposition (OFC), San Francisco, CA, USA, 9–13 March, 2014.

Seimetz, M.: High-Order Modulation for Optical Fiber Transmission, 1st edition Springer Series in Optical Sciences, edited by: W. T. Rhodes, Atlanta, Springer Verlag Berlin Heidelberg, 2009.

Winzer, P. J., Raybon, G., Song, H., Adamiecki, A., Gnauck, A. H., Fishman, D. A., Doerr, C. R., Chandrasekhar, S., Buhl, L. L., Xia, T. J., Wellbrock, G., Lee, W., Basch, B., Kawanishi, T., Higuma, K., and Painchaud, Y.: 100 – Gb/s DQPSK Transmission: From Laboratory Experiments to Field Trials, J. Lightwave Technol., 26, 3388–3402, 15 October, 2008.

Zhongxia, H., Wen, W., Jingjing, C., Yinggang, L., Stackenas, D., and Zirath, H.: FPGA-based 5 Gbit/s D-QPSK Modem for E-band Point-to-Point Radios, 41st European Microwave Conference (EuMC), Manchester, United Kingdom, 10–13 October, 2011.

35

Highly efficient integrated rectifier and voltage boosting circuits for energy harvesting applications

D. Maurath, C. Peters, T. Hehn, M. Ortmanns, and Y. Manoli

Chair of Microelectronics, Department of Microsystems Engineering (IMTEK), University of Freiburg, Georges-Koehler-Allee 102, 79110 Freiburg, Germany

Abstract. This paper presents novel circuit concepts for integrated rectifiers and voltage converting interfaces for energy harvesting micro-generators. In the context of energy harvesting, usually only small voltages are supplied by vibration-driven generators. Therefore, rectification with minimum voltage losses and low reverse currents is an important issue. This is realized by novel integrated rectifiers which were fabricated and are presented in this article. Additionally, there is a crucial need for dynamic load adaptation as well as voltage up-conversion. A circuit concept is presented, which is able to obtain both requirements. This generator interface adapts its input impedance for an optimal energy transfer efficiency. Furthermore, this generator interface provides implicit voltage up-conversion, whereas the generator output energy is stored on a buffer, which is connected to the output of the voltage converting interface. As simulations express, this fully integrated converter is able to boost ac-voltages greater than |0.35 V| to an output dc-voltage of 2.0 V–2.5 V. Thereby, high harvesting efficiencies above 80% are possible within the entire operational range.

1 Introduction

Especially electronic devices participate the development towards miniaturization and high performance with least power consumption only. Nowadays, very low power applications emerge increasingly in the field of stand-alone and wireless acting sensor/actuator-systems. A further rise of the applicability of such systems as fully embedded nodes suffers from the inconvenience with batteries. These electrochemical energy storages have limited lifespan, are very temperature sensitive and need to be replaced repetitively. In case of hardly accessible locations, there is little chance for having battery power supply based cost-efficient wireless embedded systems. Thus, the idea of power supplies, which are based on energy harvesting principles, is a helpful contribution in order to realize cost-efficient fully embedded sensor/actuator-systems.

A very space-efficient and robust kind of energy harvesting is possible with vibration-driven inductive micro-generators. However, such micro-generators supply only small open-circuit output voltage in the range of 2–3 V_{pp}. Beside this, a relatively high internal resistive impedance causes only a small load optimum, were the maximum possible output power is obtained. These circumstances enforce developments of novel kinds of rectifiers, since common full-wave rectifiers cause relatively high voltage losses.

In order to circumvent these crucial limitations, the second chapter shows enhanced rectifier circuitry with active diodes, which were designed, implemented and fabricated. These active diodes consist of an active circuit instead of passive devices only. Additionally, due to this low voltages and high internal resistance an adaptive generator-interface is required. This interface gives the opportunity of most appropriate loading of the generator. Since only a simple capacitive energy buffer connected to the output of the rectifier can not provide adequate load matching. Due to the realization of this interface with switched-capacitors, the functionality of the rectifier is improved with respect to minimum input voltages. Conclusively, the forth chapter presents simulated as well as measured results. So, the active diodes of the proposed rectifier without the generator interface can operate above 1.75 V generator output voltage amplitude. Together with the interface, generator output amplitudes of less than 1 V are sufficient for operation. Whereas the rectification, matching and conversion efficiencies can be kept above 80% over the entire operating range.

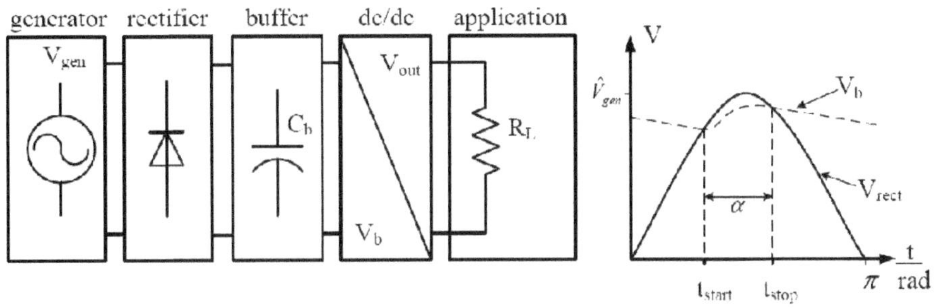

Fig. 1. Common single stage power conversion and generator interfacing approach **(a)**. The generator can only supply power within conduction degree α **(b)**.

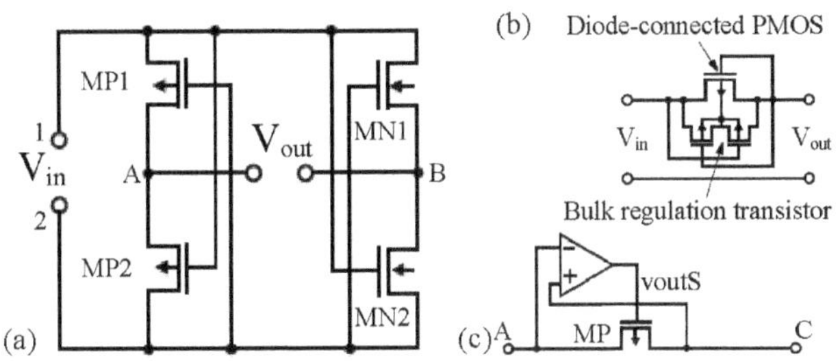

Fig. 2. Schematic of a negative voltage converter **(a)**, the bulk regulation configuration **(b)**, and the active diode principle **(c)**.

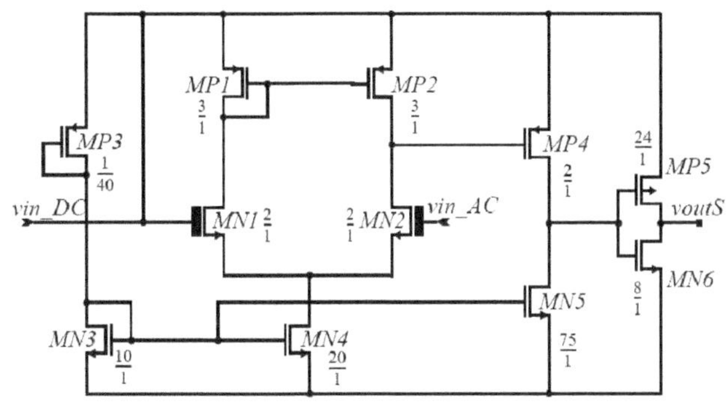

Fig. 3. Schematic of the comparator circuit with differential stage, common-source output amplifier and inverter stage.

2 Conventional harvesting interfaces

Common generator interfacing configurations are comparable to the configuration shown in Fig. 1a. Together with this simple structure also conventional circuit principles are usually applied. In case of batteries as power sources, this approach can be very useful and efficient. However, in case of energy harvesting with micro-generators this approach may exhibit crucial drawbacks.

2.1 Conventional rectifier circuits

Conventional full-wave rectifiers suffer a relatively high and unappreciated voltages loss of about 1.4 V. In case of Schottky diodes, that voltage loss could be reduced to around 0.6 V, however, those diodes have intrinsically higher reverse current, which is also not acceptable. Concerning integrated rectifiers, the diodes are realized with MOSFETs, because pn junctions are often not specified in forward direction in CMOS processes. The bulks of PMOS transistors must always be connected to the highest potential and of NMOS to the lowest to avoid leakage currents and latch-up. A dynamic bulk regulation (Fig. 2b), consisting of two transistors, avoids this and must be added to every transistor. Due to this, the number of transistors and leakage increases.

Similarly, conventional full wave (F_W) rectifiers use four MOSFETs connected as diodes. Every half wave is used, but the voltage drop is twice V_t and strongly reduces the available output voltage. A smart improvement is presented in (Ghovanloo and Najafi, 2004), which replaces two diode-connected transistors with two cross-coupled transistors. Thus, only one V_t voltage drop is lost.

2.2 Common conversion concepts

The common approach, as depicted in Fig. 1, is reasonable in case of some energy storage elements, like a battery, as a power source. However, if micro-generators are considered, power can be gathered only when the generator is exited. That means, in order to be able to supply an application independent of the generator excitation, two concerns are necessary. First, if the generator is excited most possible energy has to be harvested. Second, most of that maximum possible energy need to be stored at some energy storage element, like a buffer capacity. Additionally, in order to harvest continuously the maximum possible energy, permanent optimal generator loading for maximum generator output power as well as efficient rectification, and voltage conversion are required (Shengwen et al., 2005).

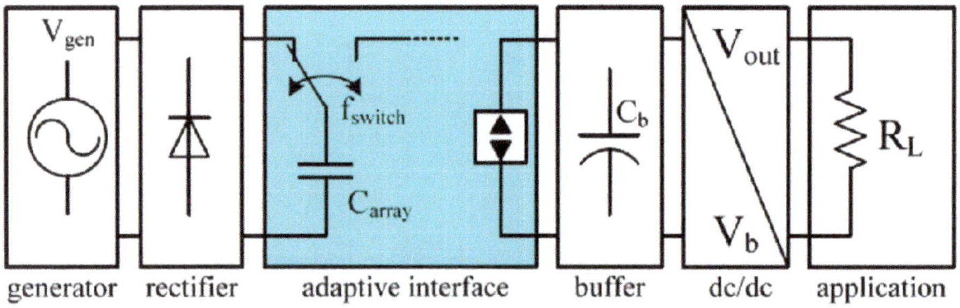

Fig. 4. Building blocks of the two stage approach, which is includes an additional load-adaptive generator interface into the power conversion chain.

3 Proposed harvesting circuitry

The generator interface was especially designed to incorporate with inductive micro-generators (Spreemann, 2006; Kulkarni et al., 2007). The internal source resistance of these inductive generators is between $R_i = 1 - 5\,k\Omega$. The generated open-circuit output voltage has 2.8 V_{pp} at most and frequencies are not higher then $f_{\text{gen}} = 500$ Hz.

3.1 Novel active rectifier

The main goals of this novel rectifier are the reduction of the voltage drop over the MOSFETs and achievement of a high efficiency. This rectifier can be separated into two stages: (i) the negative voltage converter and (ii) the active diode part.

3.1.1 First stage: negative voltage converter (CON)

The first stage of the proposed rectifier circuit is used to convert the negative half waves of the input sinusoidal wave into positive ones. This conversion is done with only four standard CMOS transistors (see Fig. 2a). No special technology, e.g. isolated transistors, is needed to realize this circuit. The left side is realized with PMOS transistors and always delivers the highest potential at V_{in} to node A. The right side consists of NMOS transistors and the low voltage potential is at point B. No additional dynamic bulk regulation transistors are necessary, because this is inherently given in this circuit. Thus, the bulk of the PMOS transistors can be directly connected to point A and the NMOS to point B. This reduces the number of required transistors.

The ability to convert nearly the entire voltage applied at the input to the output is the main advantage of this circuit. The voltage drop with this rectifier is less than 10 mV (Peters et al., 2007) and the negative half wave is now converted into a positive one. But this circuit can not be used to charge a storage capacitor, because the current direction is not controlled and current back flow occurs. Due to this a second stage is necessary to realize a useful rectifier.

3.1.2 Second stage: active diode current barrier (AD)

The main function of the second rectifier stage is to control the current direction. This could be in the simplest case a conventional diode-connected MOSFET. Using this, a full wave rectifier with only one voltage drop V_t over the transistor results (CON_D). This circuit is comparable to the full wave rectifier with switches (FW_S), but it is less prone to latch-ups and uses only 7 compared to 12 transistors.

In order to achieve very high output voltages and high efficiencies the diode should be replaced by another element, the active diode. This active diode works nearly as an ideal diode, with current flowing in only one direction and nearly no voltage drop. The main difference to an ideal diode is the current consumption of the device. A complex realization of an active diode is presented in Lehmann and Moghe (2005). A control circuit is used to determine the conducting and non conducting time of a MOSFET used as a switch. Lam et al. (2005) needs two active diodes to realize a full wave rectifier. The active diode presented in this paper is simpler and based on a fast and low power comparator circuit (see Fig. 2c). MP is the switch which is controlled by the comparator. If the voltage at the anode is higher compared to the cathode the output of the comparator is 0 V and MP is on. If the cathode is higher the output is high and MP is off. The supply voltage of the comparator is taken from the storage capacitor. Using a PMOS transistor as switch no additional start-up circuit is necessary for this active diode.

The most important part of the active diode is the comparator. A fast comparator with a low power consumption is needed. If the comparator is too slow not all available energy can be transferred to the capacitor and additionally current back flow occurs after the input voltage has passed the maximum value. The implemented comparator (see Fig. 3) is simply based on a differential stage, a common-source output amplifier and an inverter. The input $v_{\text{in,DC}}$ is connected to the cathode (C) (Fig. 2c) and is also the power supply of the comparator. $v_{in,AC}$ is connected to the alternating sinusoidal input wave (anode (A)). The overall current consumption is less than 1 μA for an input voltage of ± 2.75 V. The operation

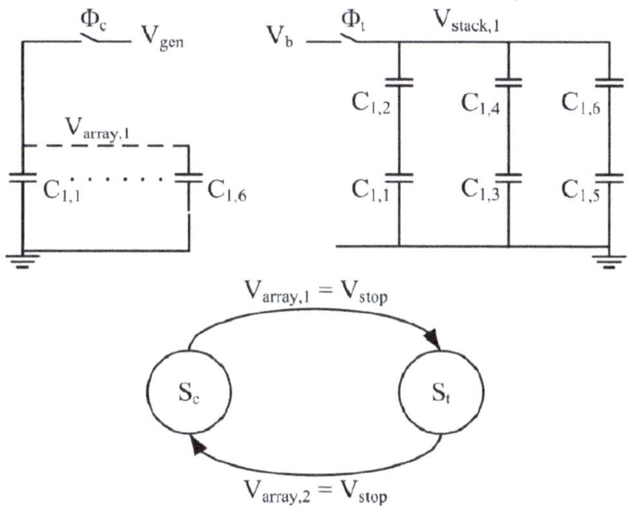

Fig. 5. Schematic and state diagram of a conversion cycle cycle.

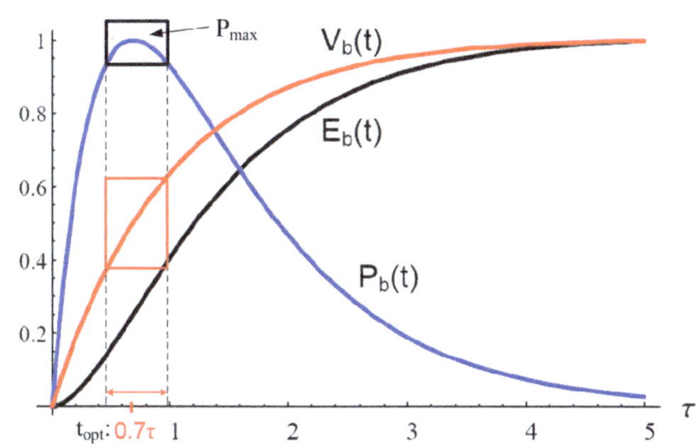

Fig. 6. This plot shows the transient power transfer while an empty capacitor is charged.

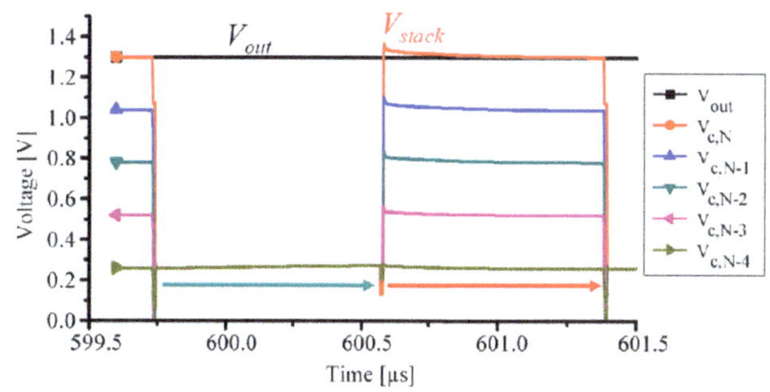

Fig. 7. This graph shows one conversion cycle, whereas V_{array} is the highest voltage within the array.

voltage range is between 1.6 V and 3.75 V. Combination of the negative voltage converter (CON) (Fig. 2a) with the active diode (Fig. 2c) overcomes the mentioned drawback.

3.2 Capacitive generator interface and converter

In order to harvest continuously the highest possible energy, the power electronic is again separated in two stages (see Fig. 4). The first stage is a generator interface and has to be optimized for adaptive impedance matching so as to load the generator as ideal as possible. In addition to that, this first stage also has to transfer the harvested energy most efficiently into a buffer. Thereby, the output voltage of the first stage is neither fixed nor regulated by the converter. For efficient energy transfer, the output voltage is rather adapted to the buffer voltage $V_b(t)$. With appropriate dimensioning of the components, the buffer voltage will vary only little and slowly compared to the generator ac voltage. Thus, the second stage is realized as a dc/dc converter.

The here presented generator-interface works basically as a voltage converter with adaptive input impedance adaptation, efficient energy transfer and implicit voltage up-conversion. By applying power saving circuit techniques, the power consumption of the control circuitry and switch driving should be less then $P_{total} \leq 50\,\mu W$.

3.2.1 Adaptive conversion principle

The proposed adaptive switching converter is based on two capacitor arrays, which toggle complementary-phased with frequency f_{conv} between a charging and a transfer state, S_c and S_t, respectively (see Fig. 5). Whereas, a conversion-cycle is the single succession of S_c and S_t. In the considered converter both arrays consist of k_{max}=6 capacitors, whereas k_{max} directly corresponds to the maximum required voltage conversion ratio. However, the number of array capacitors n_i, which are applied within a single conversion-cycle, is dynamically adapted according to the current required conversion ratio $k_i = V_b / V_{gen}(t)$. Hence, while one array is in charging state S_c, n_i array capacitors out of the k_{max} available array capacitors are connected to the generator in parallel. At this state, the capacitors are charged until the capacitor voltages V_{array} reach a certain voltage limit V_{stop}, as shown in Fig. 7. Afterwards, the array transitions to transfer state S_t. Therefore, the array capacitors have to be re-configured so as to obtain a stacked voltage V_{stack}, which is higher than the buffer voltage V_b of an external buffer. In order to obtain the dynamic adaptation of the optimal conversion ratio k_i, the converter works in an oversampling manner. Thus, if $f_{gen} \ll f_{conv}$. is given, the conversion ratio can be adapted quite continuously within a generator voltage period $1/f_{gen}$.

3.2.2 Adaptive impedance matching

High internal generator resistance R_i requires a particular converter design since an exceeding current load would cause a dramatic voltage breakdown, which would cause reduced

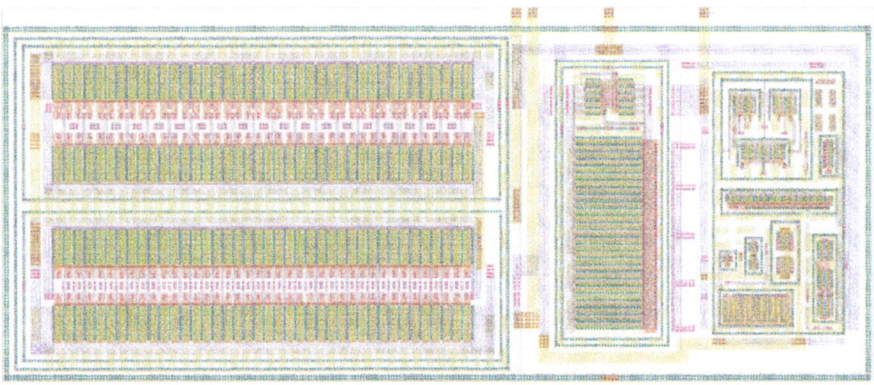

Fig. 8. The layout of the active diode enhanced rectifier, whereas the dimensions are $75\times40\,\mu$m.

generator output power. Due to the low output currents of energy harvesting generators (50–500 μA) capacitive voltage conversion is preferred rather than inductive conversion. However, charging a capacitor allows generally only at $t_{opt}\approx\ln(2)$ maximum power transfer from the generator into the capacitor (see Fig. 6). The generator-interface matching situation at t_{opt} is equivalent to the impedance matching situation where $P_{gen}(R_i{=}R_{load}){=}P_{max}$ is achieved. Therefore, for optimal operation the capacitors charging level should stay around a optimal voltage of $V_{ch,opt}\approx V_{gen}(t)$, which corresponds to t_{opt}. This means, within S_c the generator charges the capacitors up to $V_{stop}{=}V_{ch,opt}{+}\Delta V_{ch}$. Then the capacitors are toggled to S_t and discharged onto an output buffer C_b until the capacitor voltage reaches , ΔV_{ch} and ΔV_{disch}. The consequence of keeping the caps only in a small charging range of 0.5 $V_{gen}(t)\pm\Delta V$ requires an increased number of capacitors. Thus, twice as many capacitors are implemented to boost the generator voltage up to a certain voltage.

3.2.3 Voltage conversion and array configuration

Due to the fact, that charge transfer from one capacitor to another one causes intrinsic losses due to series resistances (Gobbi et al., 2006). But by minimizing the initial voltage gap between V_{stack} and V_b the charge transfer losses are minimized. Hence, the optimal conversion ratio k_i is obtained by

$$k_i = \frac{V_b}{V_{ch,opt} + \Delta V_{ch}} = \frac{V_b}{V_{stop}} \qquad (1)$$

The strategy for stacking is that a maximum quantity $n_i{\to}k_{max}$ of capacitors is used, whereas within each stack-stage n_i/k_i capacitors are connected in parallel (see Fig. 5). Thus, the maximum capacitance for a certain conversion ratio is used, which reduces the conversion frequency f_{conv}.

4 Results

Generally, by application of the proposed energy harvesting circuit concepts more energy can be extracted out of the same micro-generator. The power and voltage losses are reduced and the load matching condition is improved.

4.1 Rectification efficiencies

The main aspects of rectifiers are the maximum output voltage which can be reached and the efficiency η_{rect} , which is simply the quotient of the output and the input energy over one period. The maximum achievable output voltage is especially important for energy harvesting with inductive micro-generators.

The circuits have been simulated and implemented with Cadence using a 0.35 μm HV CMOS C35 process from austriamicrosystems (see Fig. 8). The voltage and efficiency values strongly depend on the used frequency, storage capacitor and ohmic load. Figure 9 shows the transient behavior of the discussed rectifiers. For these simulations the frequency is 125 kHz, a capacitance of 500 pF and a load of 20 kΩ are used. For useful applications a larger capacitance value should be used to reduce the voltage ripple. But this value is more suitable for showing the excellent performance of the proposed rectifiers.

4.1.1 Rectifier output voltage comparison

Figure 9 shows the input and the output wave form of the discussed rectifiers. In this case the open circuit input voltage was ±2.75 V. The negative voltage converter in combination with a PMOS diode (CON_D) has nearly the same voltage characteristic as the FW_S. Both rectifiers with the active diode (HW_AD and CON_AD) reach nearly the input voltage. The minimum voltage of the CON_AD rectifier is also very high, because every half wave is used and the capacitor is charged up twice at each period.

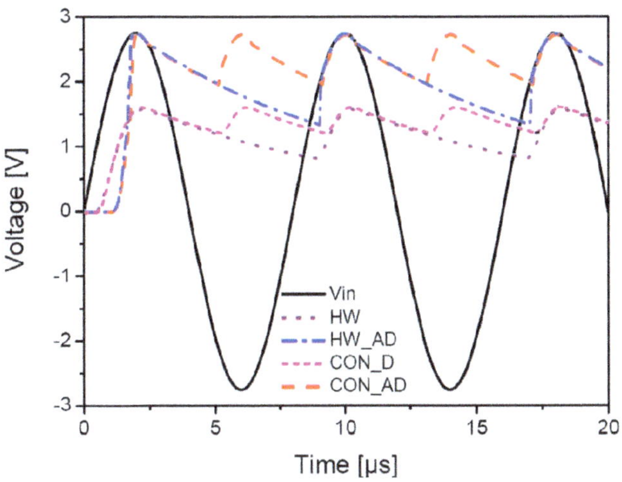

Fig. 9. Simulation results of the output signal of the compared rectifiers. The output voltage range of CON_D is in average less than the output voltage range of CON_AD.

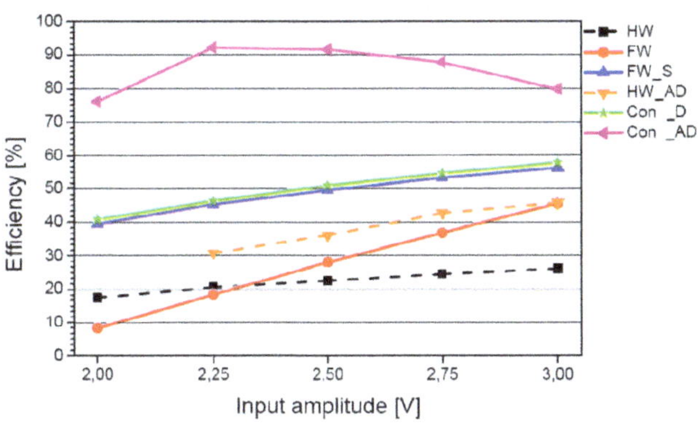

Fig. 10. Efficiencies of the compared rectifiers plotted versus the input voltage.

4.1.2 Enhanced rectification efficiency

Beside the attainable output voltage the efficiency is also of importance. The efficiency of the different rectifiers versus the input amplitude is shown in Fig. 10. The use of an active diode in a half wave rectifier shows a strong increase of the efficiency (HW vs. HW_AD). The highest efficiencies could be reached using the negative voltage converter in combination with the active diode (CON_AD). The efficiency in this example is always larger than 75%.

4.2 Interface matching and conversion results

Within all simulations and calculations, the generator was included as a sine wave source with an ohmic serial resistance. Because energy harvesting applications are considered, there

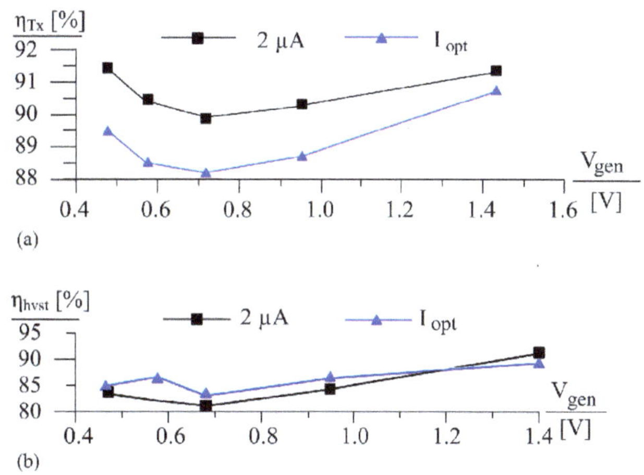

Fig. 11. The simulated efficiencies in graph **(a)** show the ratio of harvested power compared to maximum possible power, and **(b)** shows the overall conversion efficiency.

are two important efficiencies:

$$\eta_{Tx} = \frac{1}{(t_2 - t_1)} \frac{\int_{t_1}^{t_2} V_b I_{ch}\, dt}{\frac{\hat{V}_{gen}}{4R_i}} \tag{2}$$

$$\eta_{hvst} = \eta_{Tx} \times \eta_{conv} = \eta_{Tx} \frac{\int_{t_1}^{t_2} V_{load} I_{load}\, dt}{\int_{t_1}^{t_2} V_{gen} I_{gen}\, dt} \tag{3}$$

Where Eq. (2) defines the input efficiency η_{Tx}, which compares the theoretical maximum generator power to the actual converter input power. The higher η_{Tx} is, the better achieved is impedance matching. In addition to this, the efficiency η_{hvst} in Eq. (3) gives a measure of the complete harvesting efficiency. This efficiency is mainly affected by internal converter losses and capacitor charge transfer losses. As Fig. 11a illustrates, the interface input power almost equals the maximum possible output power.

Both graphs prove that this novel interface approach and conversion method operates as required. Generally, in order to characterize the efficiency by interfacing a generator, not only η_{conv} can be considered, but also η_{Tx} is of importance. Therefore, Fig. 11b shows the overall harvesting efficiency η_{hvst} over the generator voltage range.

5 Conclusions

A novel rectifier concept consisting of a voltage converter and an active diode was presented. No additional bulk regulation transistors are needed and nearly the entire generator voltage appears at the rectifier output. In combination with a MOSFET diode the efficiency is about 50%. In combination with an active diode rectifier efficiencies over 90% can be reached. The maximum output voltage is only a few mV below the input voltage.

It was also shown, that high efficient power extraction and voltage up-conversion is feasible in energy harvesting environments. Especially the realization of adaptive impedance matching and almost continuous maximum generator power enhances the usage of micro-generators further. These advantages give the opportunity of using smaller generators to supply autonomous applications. One drawback might be the relatively large quantity of capacitances, which need to be integrated. Thus, a lot of chip area is necessary and parasitic effects could decrease the overall efficiency.

Acknowledgements. This work is supported by the German Research Foundation (Deutsche Forschungsgemeinschaft – DFG) under Grant Number GR1322 and MA 2193/7-1.

References

Ghovanloo, M. and Najafi, K.: Fully integrated wideband high-current rectifier for inductively powered devices, IEEE J. Solid State Circuits, 39, 1976–1984, 2004.

Gobbi, L., Cabrini, A., and Torelli, G.: Impact of Parasitic Elements on CMOS Charge Pumps: a Numerical Analysis, IEEE International Symposium on Circuits and Systems (ISCAS), May 2006.

Kulkarni, S., Koukharenko, E., Tudor, J., Beeby, S., O'Donnell, T., and Roy, S.: Fabrication and Test of Integrated Micro-Scale Vibration Based Electromagnetic Generator, International Solid-State Sensors Actuators and Microsystems Conference, 2007.

Lam, Y.-H., Ki, W.-H., and Tsui, C.-Y.: Integrated Low-Loss CMOS Active Rectifier for Wirelessly Powered Devices, IEEE Trans. Circuits and Systems, 53(12), 1378–1382, 2006.

Lehmann, T. and Moghe, Y.: On-chip active power rectifiers for biomedical applications, IEEE International Symposium on Circuits and Systems, 1, 732–735, 2005.

Peters, C., Kessling, O., Henrici, F., Ortmanns, M., and Manoli, Y.: CMOS Integrated Highly Efficient Full Wave Rectifier, IEEE IEEE International Symposium on Circuits and Systems (ISCAS), 2415–2418, New Orleans, 2007.

Shengwen, X., Ngo, K. D. T., Nishida, T., Gyo-Bum, Ch., and Sharma, A.: Converter and controller for micro-power energy harvesting, 20th Annual Applied Power Electronics Conference and Exposition, 1, 226–230, March 2005.

Spreemann, D., Folkmer, B., Maurath, D., and Manoli, Y.: Tuneable transducer for low frequency vibrational energy harvesting, EuroSensors 2006, September 2006.

36

Analog circuits using FinFETs: benefits in speed-accuracy-power trade-off and simulation of parasitic effects

M. Fulde[1], J. P. Engelstädter[2,3], G. Knoblinger[2], and D. Schmitt-Landsiedel[1]

[1]Institute of Technical Electronics, Technical University Munich, Germany
[2]Infineon Technologies AG, Villach, Austria
[3]Department of Electrical Engineering, RWTH Aachen University, Aachen, Germany

Abstract. Multi-gate FET, e.g. FinFET devices are the most promising contenders to replace bulk FETs in sub-45 nm CMOS technologies due to their improved sub threshold and short channel behavior, associated with low leakage currents. The introduction of novel gate stack materials (e.g. metal gate, high-k dielectric) and modified device architectures (e.g. fully depleted, undoped fins) affect the analog device properties significantly. First measurements indicate enhanced intrinsic gain (g_m/g_{DS}) and promising matching behavior of FinFETs. The resulting benefits regarding the speed-accuracy-power trade-off in analog circuit design will be shown in this work. Additionally novel device specific effects will be discussed. The hysteresis effect caused by charge trapping in high-k dielectrics or self-heating due to the high thermal resistor of the BOX isolation are possible challenges for analog design in these emerging technologies. To gain an early assessment of the impact of such parasitic effects SPICE based models are derived and applied in analog building blocks.

1 Introduction

Scaling limitations of standard planar bulk CMOS technologies such as short channel effects or leakage currents enforce the introduction of new device concepts like silicon-on-insulator (SOI) and/or multi-gate transistors.

Especially 3-dimensional FinFET devices as shown in Fig. 1 offer a large variety of interesting features. Here the transistor consists of vertical (fully-depleted) Si fins on a buried oxide (BOX), surrounded by the gate stack. The channel width per fin of this tri-gate device results in $W=2H_{\text{fin}}+W_{\text{fin}}$. FinFET devices combine the advantages of fully depleted (FD) SOI devices with the features of a multi-gate device. So they show low short channel effects (SCE), however the requirements on the Si layer thickness (i.e. the fin height) on top of the BOX are relaxed compared to planar (single-gate) FD devices. On the other hand the fin width has to be chosen as small as possible ($W_{\text{fin}} \leq 2/3L_{\text{min}}$) to keep the short channel effects under control, (Colinge, 2004). Using small fin pitches (i.e. $\leq 2H_{\text{fin}}+W_{\text{fin}}$) enables comparable or even higher area efficiency in terms of the current per device width ratio. Table 2 shows an example for a prototype FinFET technology, targeting a 45 nm LP process, (Schulz et al., 2005).

Of course also novel process features can be implemented in a FinFET technology. Metal gates eliminate poly depletion effects and enable undoped Si fins. HALO implants are no longer necessary due to the low SCE. High-k dielectrics can be used to reduce gate tunneling currents while bringing up other new issues such as increased flicker noise or reduced mobility. The introduction of these new materials and device concepts implicates changes in analog transistor characteristics as output resistance or matching behavior. Sect. 2 covers the basic analog transistor properties of FinFETs compared to standard bulk devices. Also their specific impact on circuit design is discussed. New materials and device architectures cause also new effects, such as transient variations in transistor current due to self heating or charge trapping. In Sect. 3 these effects are shown, concerning their impact on transistor behavior, modeling aspects and their influence on analog circuits.

2 Analog device properties and impact on circuit design

The feasibility of digital and analog multi-gate circuits has already been proven, (Knoblinger et al., 2005; Pacha et al., 2006). But these circuits do not take advantage of special multi-gate device features, as discussed in this section. The FinFET devices presented here are based on the technology

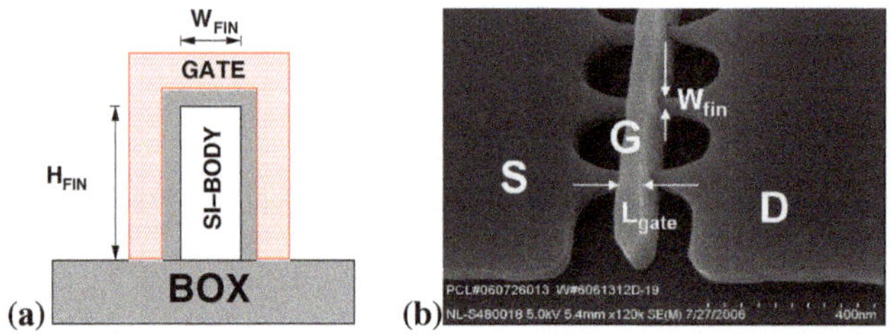

Fig. 1. Schematic X-section **(a)** and TEM image **(b)** of FinFET.

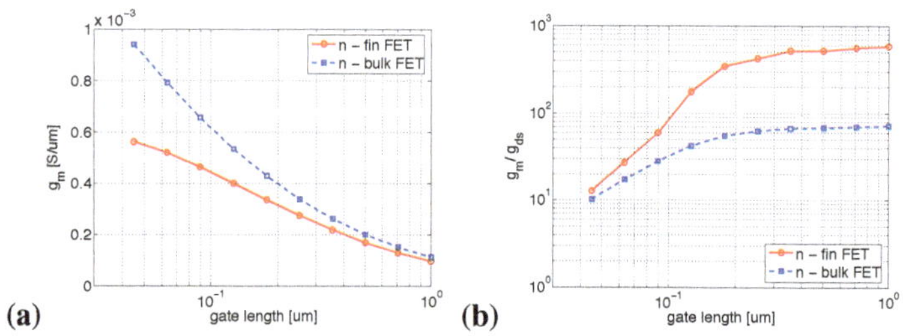

Fig. 2. g_m and g_m/g_{ds} of bulk- and FinFET.

Table 1. Example for FinFET technology.

Technology features		
W_{fin} [nm]	30	
H_{fin} [nm]	60	
$L_{\min}$ [nm]	45	
Gate electrode	TiSiN	
Dielectric	HfSiON	
Fin doping [cm^{-3}]	10^{15}	
Electrical features	NMOS	PMOS
L_{gate} [nm]	60	60
V_{DD} [V]	1.0	1.0
I_{ON} [μA/μm]	560	330
I_{OFF} [nA/μm]	1.4	0.027
S [mV/dec]	69	72
DIBL [mV]	46	44

shown in Table 2, the bulk devices are based on a comparable 45 nm process.

2.1 Small signal parameters

The transconductance g_m and the output conductance g_{DS} are the basic DC transistor properties that determine the performance of analog circuits. Figure 2.1 compares g_m of FinFET and bulk nFET at $V_{GS}=V_T+200\,\text{mV}$ and $V_{DS}=V_{DD}$. The g_m of the FinFET is reduced for short channel lengths up to 30%. This is attributed to the lowered mobility on the rough fin sidewall and the high resistance of the connection from the source-drain landing pads to the active channel region under the gate. Lowered g_m values implicate lowered transit frequency values and decreased RF performance. A possible solution for this issue is a selective epitaxial growth (SEG) process to fill the gaps between gate and source-drain pads and to reduce the high source-drain resistance.

The output conductance of FinFETs is expected to be very low due to the missing HALO implants. HALO implants introduce an additional barrier at the drain side of the channel region that can be modulated by the drain-source voltage. This effect degrades significantly the output resistance of bulk devices in modern CMOS technologies. For a fair comparison we use the intrinsic transistor gain defined as g_m/g_{DS} here. Figure 2.1 shows that the gain of the FinFET is improved, despite the lower g_m. For typical analog device dimensions the gain is about 2–10 times higher. Taking g_m and g_{DS} into account we conclude that FinFETs are best suited for low power analog circuits with frequencies below 10 GHz.

2.2 Benefits in speed-power trade-off

To investigate the benefits of the improved intrinsic gain on circuit level a standard two stage Miller compensated OTA is simulated. The following specifications are used: DC voltage gain $A_{V0}\geq 50\,\text{dB}$, gain bandwidth product $GBW\geq 10\,\text{MHz}$ for a 5pF capacitive load and a phase margin of 60°. To compare FinFET and bulk implementation a figure-of-merit is defined as gain bandwidth to power ratio: FOM$=GBW/P$. The starting point of the comparison is a bulk design using only $3L_{\min}$ devices that fulfills the given specifications. This design is compared to a FinFET implementation using devices with the same channel length of $3L_{\min}$. Due to the high intrinsic transistor gain, the FinFET version shows an improved voltage gain (about 30 dB higher) and reaches the same gain bandwidth product, see the Bode diagram in Fig. 3.

As the specifications do not require such a high gain, the FinFET implementation can use shorter channel lengths to reach higher g_m values. The improved g_m can then be used to enhance the gain bandwidth product or to decrease the power consumption. Figure 3 shows two FinFET implementations using devices with $1.4L_{\min}$ channel length. Both OTAs have a voltage gain of around 50dB, while one is optimized for high gain bandwidth and the other for low power. Table 2 shows the quantitative results of the comparison. Using the improved output conductance the FOM of the FinFET implementation can be improved by about 30%, although the g_m is reduced.

2.3 Matching behavior

Random variations such as fluctuations of channel/gate dopants, oxide charge or surface roughness lead to current mismatch of nominal identical devices. In today's CMOS technologies the mismatch of the threshold voltage V_T is the dominant effect. The variation of V_T can be approximated

Table 2. Characteristic values of different implementations.

Version	3 L_{min} bulk	3 L_{min} FinFET	1.4 L_{min} FinFET	1.4 L_{min} FinFET
GBW [MHz]	10.8	10.6	14.7	10.4
A_{V0} [dB]	48.4	81.3	47.1	47.6
P [W]	53.2	55.6	56.8	41.9
FOM	0.203	0.191	0.259	0.248
Improvement	–	–6%	+28%	+22%

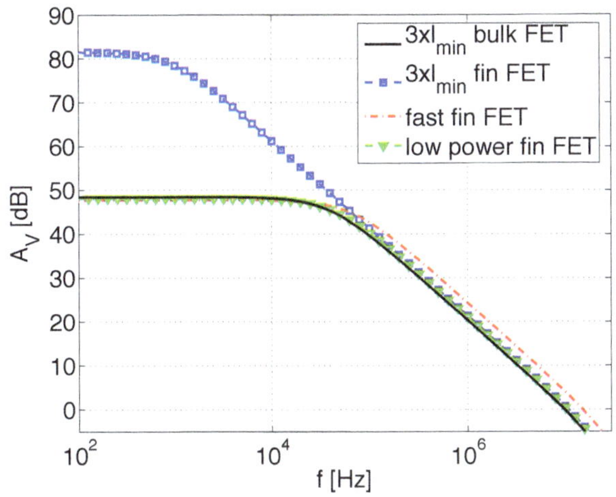

Fig. 3. Bode diagram of different implementations.

as $\sigma_{V_T}=\frac{A_{V_T}}{\sqrt{WL}}$, (Pelgrom et al., 1989). A_{V_T} is a technology constant proportional to the oxide thickness and the channel doping: $A_{V_T} \propto t_{ox}\sqrt[4]{N_D}$. Thus metal gate FinFETs with undoped body are expected to show good matching behavior. This assumption can be proven by measurements as shown in Fig. 4. Compared to bulk devices with poly gate, the matching constant of FinFETs with metal gate amounts roughly to the half. It is important to note that the matching behavior degrades for very narrow fins in the current state of technology. Here other effects are dominant, e.g. the variations of the source-drain resistance, gate misalignment or line edge and sidewall roughness. Technology optimization is still necessary. E.g. SEG could improve the matching behavior by decreasing the source-drain resistance.

2.4 Benefits in speed-accuracy-power trade-off

Of course matching is one of the most important device properties for analog circuit design. It can be shown that in many cases the speed-accuracy-power trade-off is limited in general by the matching behavior (Kinget and Steyaert, 1996):

$$\frac{\text{Speed} \cdot \text{Accuracy}}{\text{Power}} \propto \frac{1}{C_{ox}A_{V_T}^2} \tag{1}$$

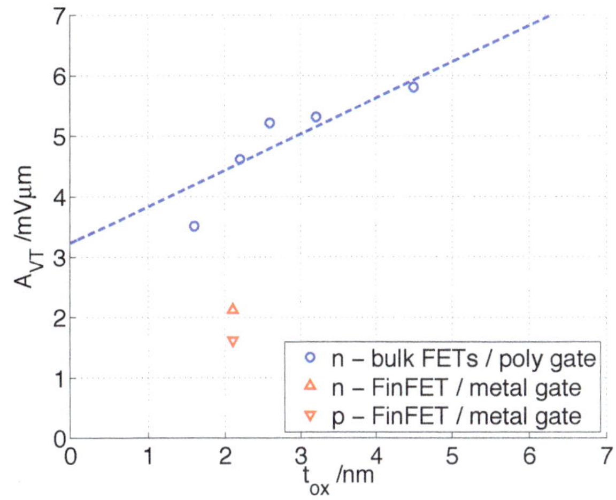

Fig. 4. Matching constant in different technology nodes (Gustin et al., 2006; Parvais et al., 2006; Decoutere, 2006).

Obviously an improved matching constant A_{V_T} can be used to reduce the area consumption or enhance the resolution without the drawback of increased power consumption. As concrete example for this relationship we use a current source in a current steering DA converter. The DAC specifications for resolution and yield determine the tolerable current variation $\frac{\sigma(I)}{I}$ of the current sources (van den Bosch et al., 2001). These matching requirements can be converted into area requirements:

$$WL_{\text{current source}} = \frac{1}{2\left(\frac{\sigma(I)}{I}\right)^2}\left(A_\beta^2 + \frac{4A_{V_T}^2}{(V_{GS}-V_T)^2}\right) \tag{2}$$

Assuming that the mobility mismatch given by A_β^2 is negligible compared to the V_T mismatch, a reduction of A_{V_T} by a factor of two enables a reduction of the current source area by a factor of four.

3 Simulation of parasitic effects

Besides the beneficial analog properties presented above, FinFETs show some new effects that have to be considered in the design of analog circuits. As example we discuss self heating and charge trapping. Both effects lead to transient variations of the transistor current and influence the behavior of analog circuits.

3.1 Self heating – modeling and scaling behavior

The local temperature increase of active devices due to their dissipated power is called self heating (SH). Important transistor parameters, e.g. V_T or the mobility μ depend on the temperature and thus on the dissipated power. For high values of V_{DS} and V_{GS} the reduction of the mobility is the dominant effect, so the transistor current is decreased by SH. Self

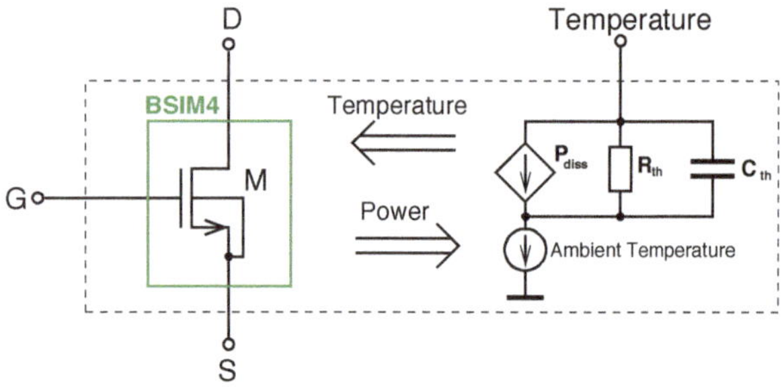

Fig. 5. Self heating equivalent circuit.

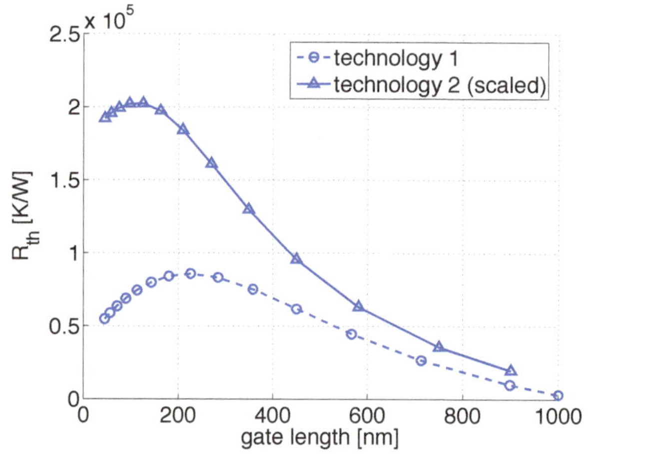

Fig. 6. Impact of technology scaling on R_{th}.

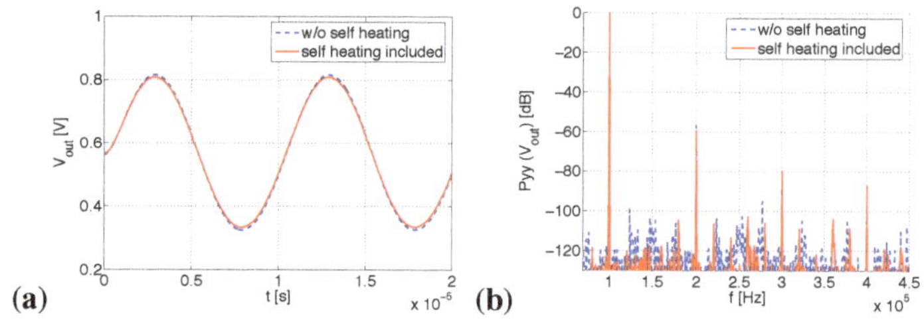

Fig. 7. OTA output voltage with and without SH in time **(a)** and spectral domain **(b)**.

heating is more severe in SOI technologies due to the low thermal conductivity of the surrounding oxide that acts as thermal insulator. To quantify SH usually the thermal resistance R_{th} is used. The time dependence of SH can be described with the thermal capacitance C_{th} as shown in the equivalent circuit model for a transistor with self heating in Fig. 5. The equivalent circuit is based on electro-thermal coupling: the dissipated power is sensed and applied to the thermal RC network which determines the actual device temperature which then again influences the transistor parameters. To avoid complex measurements or device simulations for the extraction of the thermal resistance, in (Bertolissi, 2004) an alternative approach is shown. Here a thermal network of complex 3d FinFET structures is constructed based on small building blocks where a analytic calculation of R_{th} is possible. This model is used here to investigate the impact of technology scaling on the thermal resistance. A scenario with relaxed design rules is compared to a scaled technology as shown in Fig. 6. Fin dimensions and pitch, metal/contact dimensions and pitch as well as the BOX thickness are reduced. The thermal resistance of a test device with 2 fins is increased up to a factor of two in case on the scaled technology, although the density of contacts with high thermal conductivity increases. So SH will become even more severe in future SOI technologies.

3.2 Impact of self heating on analog circuit performance

Using the equivalent circuit model, the impact of SH on FinFET circuits is investigated. R_{th} is obtained from simulations as described above, C_{th} from measurement and device simulations (Molzer et al., 2006) revealing a time constant of about 100ns and a maximum current reduction of about 10%. First a single-ended two stage OTA as presented above is analyzed in terms of it's transient response. A sinusoidal differential input signal is applied with a frequency of $f_{\text{sig}} \ll 1/\tau_{th}$. Figure 3.2 shows the impact of self heating on the output signal. The signal amplitude is decreased by about 2% through SH whereas the waveform is not affected. This is confirmed by the spectral analysis of this signal in Fig. 3.2. The second and third harmonic of the signal are even a touch lower in case of self heating, which can be explained by the decreased amplitude of the output signal. Consequently the impact of SH on the gain and the linearity of the amplifier is negligible here, although the current of a single transistor can be decreased by nearly 10%. This is attributed to the bias points typically chosen for amplifiers with moderate bandwidths. To reach a high g_m/I_D ratio V_{GS} is typically set to values in the range of 50 mV–300 mV above V_T leading to quite low current densities.

Next the impact of SH on comparator circuits is shown. A simple comparator structure with differential input stage, MOS diode loads and common source output is used as example. The adjacent latch is not considered here. In the simulation the differential input voltage is pulsed from a large (positive) value to a small (positive) residual voltage V_{res}, see Fig. 3.3. Without self heating, the comparator output voltage follows the input as intended. If self heating is included, the differential output voltage is negative for a certain period of time. This can be explained with the history of the comparator. Before the step of the input voltage, the devices providing the positive part of the differential output signal suffer much more from self heating than the devices providing the negative part. This situation remains for a short time after the step to small input values near common mode, yielding to strong thermal mismatch and a wrong comparator decision if the latch is activated at this point of time, see Fig. 3.3. The

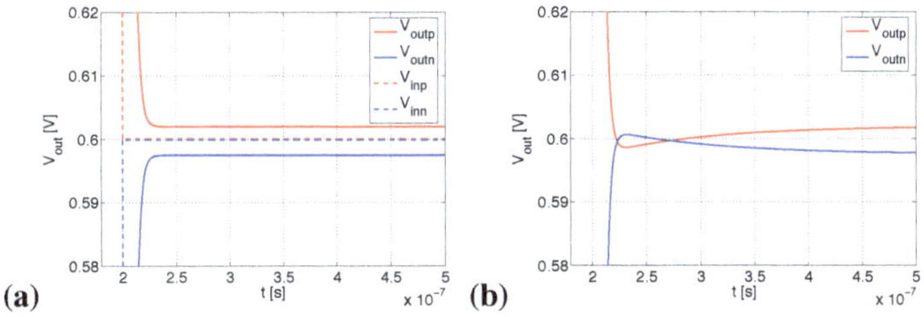

Fig. 8. Comparator input and output voltage without SH **(a)** and with SH **(b)** included.

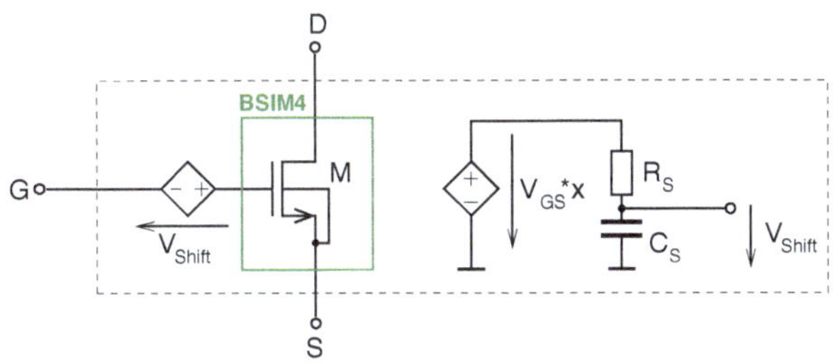

Fig. 9. Charge trapping equivalent circuit.

maximum residual voltage which causes wrong decisions is depending on the input signal, the comparator gain and the sampling time. In this scenario a V_{res} of 0.16 mV leads to a wrong decision for about 50ns. Assuming a reference voltage of 1V, self heating limits the performance of this comparator to a resolution of 12bit or sampling frequencies below 20 MHz. Thus SH is a potential issue for high speed, high resolution AD converters.

3.3 Charge trapping – modeling and impact on analog circuits

Another -not FinFET specific- effect is charge trapping in high-k materials. Under strong positive (in case of nFET) gate bias charges tunnel through the dielectric barrier and occupy free states in the dielectric material. Thereby the threshold voltage of the device is increased. Detrapping of charges is possible leading to a hysteresis effect as shown in Fig. 10. Material properties such as the dielectric itself (e.g. HfO_2) or the interface layer, temperature and bias conditions determine the amount of trapped charges (Ribes et al., 2005). Measured V_T shifts reach values in the range of 100 mV, while values around 10 mV are supposed to be tolerable for digital circuits. Time constants are reported from μs up to ms. For circuit simulations, an equivalent circuit model can be used (Fulde et al., 2006; Tewksbury and Lee, 1994) as shown in Fig. 9. The gate-source voltage is sensed and used to calculate the steady-state V_T shift via a voltage controlled voltage source. This voltage is applied to a RC network that creates an exponential time dependence and then subtracted from the gate-source voltage. Similar to the self heating effect, charge trapping causes transient variations of

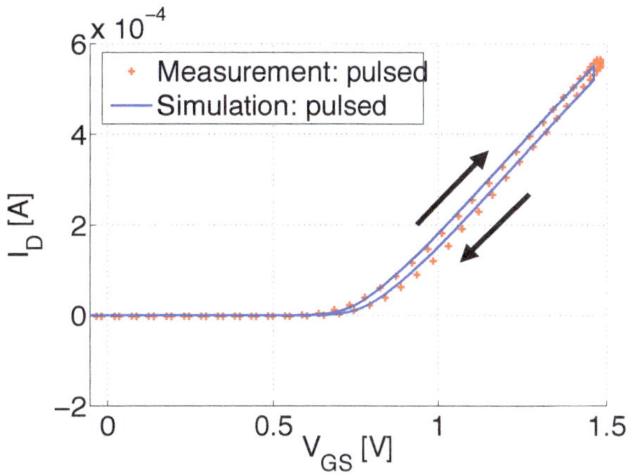

Fig. 10. Measurement and simulation of charge trapping effect.

the transistor current. Therefore the impact of charge trapping on circuit behavior is comparable. The linearity and gain of operational amplifiers is nearly not affected because the input devices see only a small signal oscillation of the gate-source voltage around the bias point. So the amount of charge trapping is almost constant over time. However charge trapping can cause erroneous comparator decisions as discussed above. The maximum residual voltage V_{res} that is necessary to modify the comparator decision is in the range of the maximum steady-state V_T shift. That means V_{res} can be much higher than in case of self-heating, e.g. in the mV regime. Fulde et al. (2006) shows how a dynamic V_T shift of a few mV can degrade the resolution of a 12bit SAR converter.

4 Conclusions

FinFETs offer a lot of interesting device features and show beneficial analog device properties. Primarily the low output conductance and the good matching behavior can be used to improve the figure-of-merit of typical analog applications as operational amplifiers or AD/DA converters. Nevertheless technology optimization, e.g. SEG, is still needed to eliminate parasitic effects like the high source-drain resistance or the matching issues for very narrow fins. Additionally new effects have to be considered in the design of analog circuits. Self heating and charge trapping influence transistor behavior significantly and introduce transient variations. Simple approaches based on equivalent circuit models enable an early assessment of the impact of such effects on circuit performance, revealing possible limitations for high speed, high resolution converters.

Acknowledgements. The authors would like to thank A. Kerber, F. Bauer, M. Kropfitsch, P. Bogner, C. Kropf and K. von Arnim for their valuable support.

References

Bertolissi, L.: Modeling and Simulation of Self-Heating Effects in Deep Sub-Micron Silicon on Insulator (SOI) Technologies for Analog Circuits, Ph.D. thesis, Universita Degli Studi di Udine, 2004.

Colinge, J.-P.: Novel Gate Concepts for MOS devices, in: IEEE Proceedings of the 34th European Solid-State Device Research conference, ESSDERC, 45–49, 2004.

Decoutere, S.: Perspectives of Downscaled CMOS for RFIC and mm-wave Applications, in: IEEE MTT-S International Microwave Symposium, 2006.

Fulde, M., Schmitt-Landsiedel, D., and Knoblinger, G.: Transient Variations in Emerging SOI Technologies: Modeling and Impact on Analog/Mixed-Signal Circuits, in: Proceedings of the IEEE International Symposium on Circuits and Systems, ISCAS, 2007.

Gustin, C., Mercha, A., Loo, J., Subramanian, V., Parvais, B., Dehan, M., and Decoutere, S.: Stochastic Matching Properties of FinFETs, IEEE Electron Device Letters, 27, 846–848, 2006.

Kinget, P. and Steyaert, M.: Impact of transistor mismatch on the speed-accuracy-power trade-off of analog CMOS circuits, in: Proceedings of the IEEE Custom Integrated Circuits Conference, CICC 1996, 333–336, 1996.

Knoblinger, G., Kuttner, F., Marshall, A., Russ, C., Haibach, P., Patruno, P., Schulz, T., Xiong, W., Gostkowski, M., Schruefer, K., and Cleavelin, C.: Design and evaluation of basic analog circuits in an emerging MuGFET technology, in: Proceedings of the IEEE International SOI Conference, 39–40, 2005.

Molzer et al., W.: Self Heating Simulation of Multi-Gate FETs, in: IEEE Proceedings of 36th European Solid-State Device Research conference, ESSDERC, 311–314, 2006.

Pacha, C., von Arnim, K., Schulz, T., Xiong, W., Gostkowski, M., Knoblinger, G., Marshall, A., Nirschl, T., Berthold, J., Russ, C., Gossner, H., Duvvury, C., Patruno, P., Cleavelin, R., and Schruefer, K.: Circuit design issues in multi-gate FET CMOS technologies, in: IEEE International Solid-State Circuits Conference, Digest of Technical Papers, 1656–1665, 2006.

Parvais, B., Gustin, C., de Heyn, V., Loo, J., Dehan, M., Subramanian, V., Mercha, A., Collaert, N., Rooyackers, R., Jurczak, M., Wambacq, P., and Decoutere, S.: Suitability of FinFET technology for low-power mixed-signal applications, in: IEEE International Conference on Integrated Circuit Design and Technology, ICICDT '06, 1–4, 2006.

Pelgrom, M., Duinmaijer, A., and Welbers, A.: Matching properties of MOS transistors, IEEE Journal of Solid-State Circuits, 24, 1433–1439, 1989.

Ribes, G., Mitard, J., Denais, M., Bruyere, S., Monsieur, F., Parthasarathy, C., Vincent, E., and Ghibaudo, G.: Review on high-k dielectrics reliability issues, Device and Materials Reliability, IEEE Transactions on, 5, 5–19, 2005.

Schulz, T., Xiong, W., Cleavelin, C., Schruefer, K., Gostkowski, M., Matthews, K., Gebara, G., Zaman, R., Patruno, P., Chaudhry, A., Woo, A., and Colinge, J.: Fin thickness asymmetry effects in multiple-gate SOI FETs (MuGFETs), in: Proceedings of the IEEE International SOI Conference, 154–156, 2005.

Tewksbury, T.L., I. and Lee, H.-S.: Characterization, modeling, and minimization of transient thresholdvoltage shifts in MOSFETs, IEEE Journal of Solid-State Circuits, 29, 239–252, 1994.

van den Bosch, A., Borremans, M., Steyaert, M., and Sansen, W.: A 10-bit 1-GSample/s Nyquist current-steering CMOS D/A converter, IEEE Journal of Solid-State Circuits, 36, 315–324, 2001.

37

A sensing circuit for single-ended read-ports of SRAM cells with bit-line power reduction and access-time enhancement

T. Heselhaus and T. G. Noll

Chair of Electrical Engineering and Computer Systems, RWTH Aachen University, 52062 Aachen, Germany

Abstract. The conventional sensing scheme of single-ended read-only-ports as integrated in 8T-SRAM cells suffers from low performance compared to double-ended complementary sensing schemes. In the proposed sensing scheme the pre-charge voltage of the single-ended read-bit-line is set to a level above the threshold voltage of the sensing device with an adjustable margin. This margin is minimized to speed up the read access on the one hand and kept large enough to provide a sufficient bit-line noise margin on the other hand. The pre-charge voltage level of the proposed sensing circuit tracks the threshold voltage of the sensing device under process variations in order to maintain a minimum required bit-line noise margin. To avoid unnecessary bit-line discharging, the proposed sensing scheme employs a modified 8T-SRAM cell. Compared to the conventional 8T-SRAM cell, the read port of the proposed cell provides a virtual ground line running in parallel to the bit-lines. An internal driver of the sensing circuit releases the virtual ground line during the evaluation period to prevent the charge dissipation resulting in a raised voltage level. The reduced pre-charge level and the increased virtual ground lead to a reduced bit-line voltage swing and thus a bit-line power reduction. Access time, energy dissipation, and noise margin of the proposed sensing circuit are compared with conventional sensing circuits from the literature for different numbers of memory cells connected to the bit-line. It is shown, that for a specific number of memory cells per bit-line the proposed circuit achieves fastest access time at low power operation.

1 Introduction

Single-ended read sensing schemes will become more important in memory architectures of SRAM cells with additional read-only-ports to overcome the reliability problems related to standard 6T-SRAM cells. Though such single-ended ports require only one bit-line, the main power dissipation originates from the full-swing bit-line sensing schemes.

The intention of this work is to set the initial bit-line voltage close to the threshold level of the individual local sensing device and stop the discharging, once the local sensing has detected the data bit, such that no more than necessary of the bit-line charge is dissipated. The expected side effect is a performance upgrade, since the initial bit-line voltage resides close to the sensing threshold level, such that this level can be reached in a shorter evaluation time.

Section 2 presents a modified memory cell, which is required to enable the bit-line swing reduction. In Sect. 3 the proposed sensing circuit is described. In Sect. 4 the proposed sensing scheme is verified with simulations and compared with other sensing circuits from the literature.

2 The proposed 8T-SRAM cell

Figure 1 shows the conventional (Chang et al., 2005) and the proposed 8T-SRAM cell. Both cell variants consist of a conventional cross-coupled inverter-pair connected via two access transistors to a complementary bit-line pair p and $\overline{p}$, which represents the write-port for storing data into the cell by activating the storage-word-line s. The read-only-port consists of two series connected transistors for single-ended read-out with a separate read-word-line r.

Area and most of the wiring of the proposed cell is identical to the conventional cell, however the proposed cell connects the source of the read-only-port to an additional virtual ground line v instead of V_{SS}. This additional virtual ground

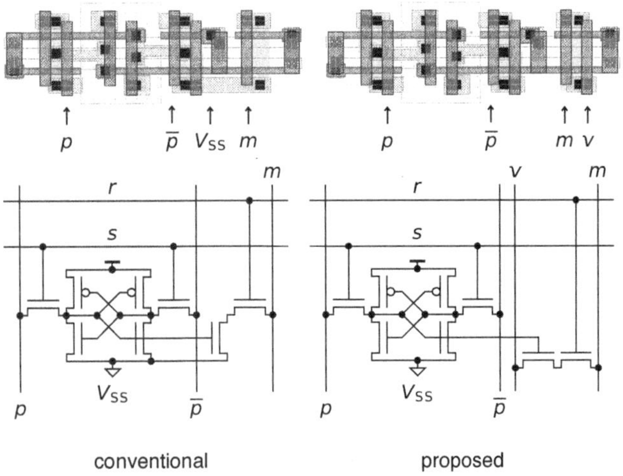

Fig. 1. Conventional and proposed 8T-SRAM cell as schematic and layout view. The additional virtual ground line is labeled as v.

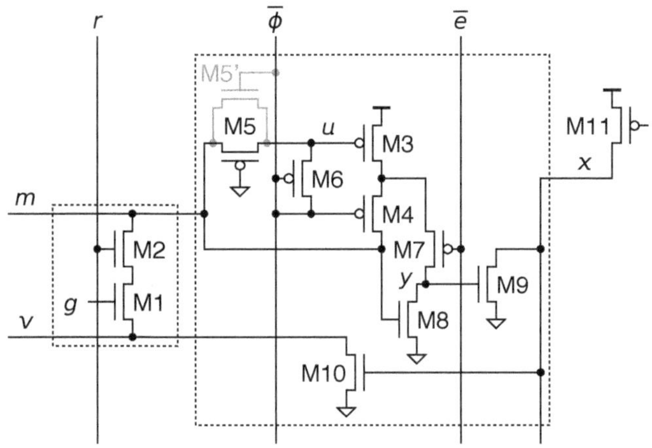

Fig. 2. Schematic of the proposed single-ended read sensing and pre-charge circuit for one memory column and domino connected digit line multiplex.

line runs in parallel to the read-bit-line m and can be easily placed in the thin layout cell without any area penalty. The intention of this additional virtual ground line is to reduce the bit-line swing and power dissipation. The functionality of this virtual ground line will be described in the next section.

3 Pre-charge and sensing circuit

Figure 2 shows the pre-charge and sensing circuit of a single-ended bit-line for one column of memory cells. Only the read-only-port of one memory cell is shown, however, there are N memory cells connected to the bit-lines m and v. The basic idea of the pre-charge circuit is to pre-charge the bit-line m to a defined voltage level above the threshold-level of the sensing transistor M3. Because the drain current of M3 is used to pre-charge the bit-line, M3 should itself turn off as the pre-charge voltage is reached. The pre-charge process is described as follows.

During the pre-charge-phase ($\overline{\phi} = 0$) the word-line is deactivated ($r = 0$) and the sensing is disabled ($\overline{e} = 1$). For an initially discharged bit-line m the gate potentials of M3 and M4 are both zero and the bit-line is charged by the series connected transistors M3 and M4. As the gates of M5 and M6 are at zero, these transistors establish a voltage divider of the actual bit-line voltage. The gate of M3 is connected to the divided voltage u, which is a fraction of the bit-line voltage m. While the bit-line voltage rises, the tapped gate potential u rises as well. The pre-charge process of the bit-line stops, if u rises above $V_{DD} - |V_{th}|$, such that M3 enters the off-region. Hence the final pre-charge voltage $V_{m,max}$ of the bit-line has risen in excess of the sensing threshold level $V_{DD} - |V_{th}|$ of M3. This overshoot can be adjusted by choosing the appropriate dimensioning of M5 and M6.

The intention of the overshoot is to achieve the required noise margin on the bit-line, which must be maintained even under process variations. With the feed-back loop for pre-charging the bit-line via the transistors M3, M4, M5 and M6, the voltage overshoot of $V_{m,max}$ tracks the sensing threshold level of M3. This tracking effect is verified by simulations in Sect. 4.

In fact, in the proposed sensing scheme the signal is detected with two successive sensing stages, realized by M3 and M9. It behaves similar to an inverter sense circuit followed by a domino NMOS pull-down device. However, the nodes x, y and v in the sensing circuit have to be pre-charged or clamped before entering the evaluation phase. During the pre-charge-phase the bit-line m rises to $V_{m,max}$ and shortly M8 begins to discharge the internal node y such that the main output stage M9 turns off. The drain of M9 is connected to a common node x, which could be a global bit-line in a subdivided bit-line memory architecture or a common digit line in some other hierarchical memory architecture (e.g. a column or data word multiplex output). Here, this output x is assumed to be domino-connected, thus x requires an additional pre-charge device M11, which is not part of the sensing circuit. However, M11 pre-charges x to V_{DD} such that M10 turns on and clamps the virtual ground line v to V_{SS}. The circuit is now prepared for sensing.

The pre-charge phase is completed by reverting $\overline{\phi}$ to V_{DD} which turns off M4. Since M6 is connected as a diode, M6 turns off as well, the voltage divider built by M5 and M6 is now no longer active and u follows the bit-line voltage via M5. The transistor M5' may be added to afford a close coupling between m and u over a wide range of bit-line voltages. Actually both M5' and M5 together build up a transmission gate to connect m to the gate of M3. In some cases where M5 is strong enough for a close coupling between m and u, M5' might be omitted. However, in this work M5' is included

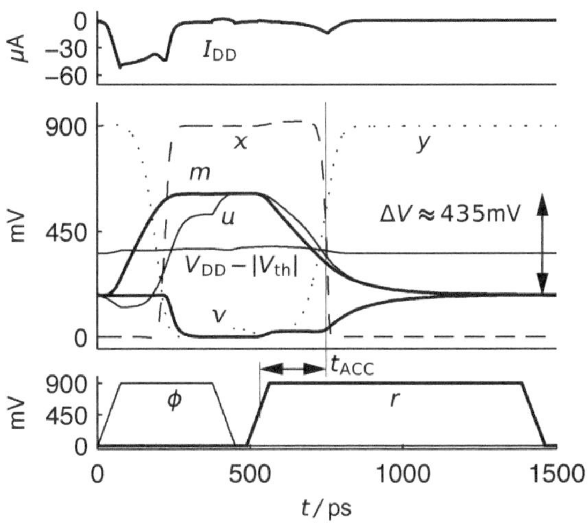

Fig. 3. Simulated waveform for a read-cycle of the proposed sensing circuit (slow corner, 125°C, $V_{DD} = 900$mV).

in each power and performance simulation and comparison. The transistor M5' is minimally dimensioned.

Additionally, $\bar{e}$ is now tied to zero, and M7 connects the sensing transistor M3 to the internal node y. Any leakage of M3 due to the near threshold-level pre-charged bit-line m and node u is compensated now by the weakly dimensioned transistor M8, such that y is kept to V_{SS}. The evaluation-phase starts by activating the word-line r. Depending on the stored data g, the bit-line remains pre-charged or is discharged if $g = V_{DD}$ via M2 and M1 of the memory cell and M10 of the sensing circuit. The gate potential u of M3 falls and M3 now directly turns on, which leads to a steep rising edge on the internal node y. The rising edge on y activates M9, which now discharges the common data output node x.

The power reduction feature is induced by the reduced pre-charge level and the discharge inhibition with the additional virtual ground line v of the proposed 8T-SRAM cell and transistor M10 of the sensing circuit. Since M10 is connected to the data output x, which gets zero if the selected memory cell discharges the bit-line m, M10 turns off and inhibits further charge dissipation to V_{SS} as soon the data bit is detected. Because the bit-lines m and v are designed similar in the layout view, the parasitic capacitances of both bit-lines are similar. Thus, once M10 turns off, the remaining charge on the bit-line m will be balanced between m and v. The final bit-line potential of m would therefore settle to nearly half of the bit-line potential, when the data bit was detected. The final potential depends on the ratio of the parasitic capacitances of m and v.

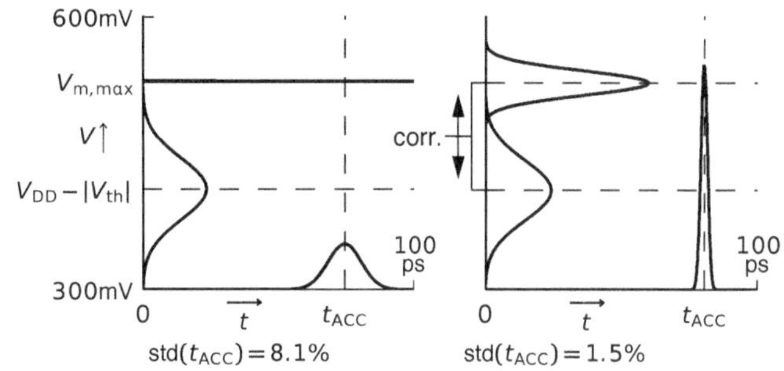

Fig. 4. Verification of the sensing threshold level tracking. The left diagram shows the resulting access time profile with a fixed bit-line pre-charge voltage, while the right diagram shows the results when the sensing circuit includes the voltage divider M5/M6 for the tracking (Monte Carlo simulations, 25°C, $V_{DD} = 900$mV).

4 Simulation results

The sensing circuit was simulated for a 40nm general purpose bulk CMOS technology with a nominal supply voltage of 900mV. All circuit simulations are based on transistor schematics. Extracted parasitics from a layout view are added to the bit-lines and to the virtual ground line corresponding to the connected number of cells $N \in \{8, 16, 32, 64, 128\}$. Internal nodes of the sensing circuit are loaded with estimated parasitic capacitances rated between 0.4fF and 0.7fF depending on the assumed wiring complexity. The transistor dimensioning of the proposed sensing circuit was basically designed with the minimal allowed gate width of $w = 120$nm. Only the driving/sensing transistors M3, M4, M9, and M10 have a gate width of $w = 480$nm.

Simulation using slow corner device models at 125°C and a minimal supply voltage of 900mV are used to verify the function of the proposed sensing circuit under worst case timing conditions. Figure 3 shows the simulated pre-charge and evaluation-phase of a read cycle with a stored "1" on the selected memory cell ($g = V_{DD}$). The achieved bit-line swing ΔV is approximately 435mV, a little bit less than $V_{DD}/2$ such that bit-line energy savings of 50% compared to a full swing bit-line sensing scheme can be expected.

Tracking the pre-charge voltage according to the sensing threshold level of M3 under process variations was functionally verified by Monte Carlo simulations. For this case process variations apply to M3 only, while all other transistors operate in the typical corner. Figure 4 shows the probability density functions (PDF) of sensing threshold level variations and the resulting access time profile without tracking on the left side and with tracking on the right side. On the ordinate the mean sensing threshold level V_{th}, the mean pre-charge level $V_{m,max}$, and their PDFs are plotted. The abscissas show the mean access times t_{ACC} and their PDFs for both cases. The standard deviation of the access time PDF was reduced from 8.1% to 1.5%.

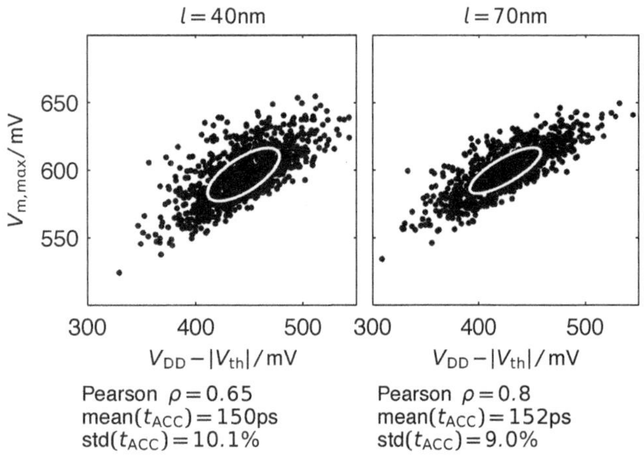

Fig. 5. Simulated pre-charge level correlation to M3-threshold-level variations of 1000 Monte Carlo runs for two different gate lengths of the voltage divider M5/M6 (25°C, $V_{DD} = 900$mV).

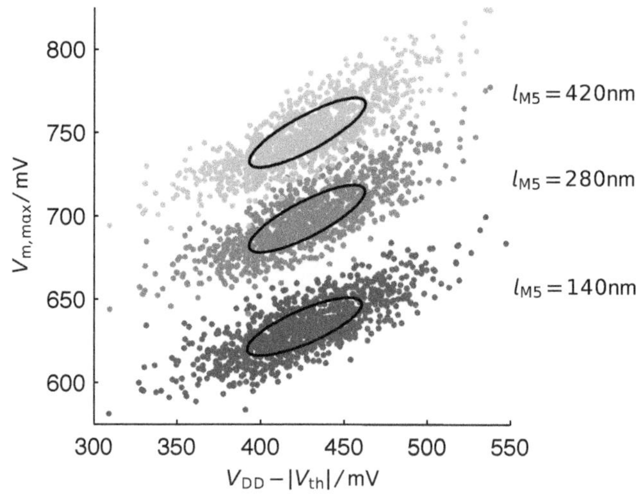

Fig. 6. Monte Carlo simulated pre-charge level for different l_{M5} in correlation to the M3-threshold-level ($l_{M6} = 70$nm, 1000 iterations, 25°C, $V_{DD} = 900$mV).

When considering process variations of all transistors of the sensing circuit this also affects the fraction of the voltage divider M5/M6. Figure 5 shows the simulated correlation of the sensing threshold level variation to the pre-charge level variation. The left side of Fig. 5 shows the voltage level distributions, where all transistors of the sensing circuit in Fig. 2 are designed with a gate lengths of $l_{M5} = l_{M6} = 40$nm. Since device mismatches of the voltage divider transistors M5 and M6 disturb the correlation of the sensing threshold level to the pre-charge level, a reduction of the variability of M5 and M6 can improve the correlation. Choosing a gate length of $l_{M5} = l_{M6} = 70$nm for the voltage divider M5/M6 improves the correlation coefficient to $\rho = 0.8$ compared to $\rho = 0.65$ (for $l_{M5} = l_{M6} = 40$nm). Reducing the variability of the voltage divider reduces the standard deviation of the access time from 10.1% to 9%, while the mean access time remains nearly constant.

Increasing the bit-line noise margin was simulated by enlarging the gate length of M5 up to $l_{M5} = 420$nm for a series of Monte Carlo simulations, while keeping the gate length of M6 fixed to $l_{M6} = 70$nm. The results shown in Fig. 6 exhibit a similar distribution of the pre-charge level in correlation to the sensing threshold voltage like depicted on the right diagram in Fig. 5. However, the dot clouds are shifted to higher pre-charge levels with increasing gate length l_{M5}. The ellipses inside the dot clouds denote the standard error ellipse of the two dimensional density function. The resulting improvement of the mean bit-line noise margin with increasing the gate length $l_{M5} = 70\text{nm} \rightarrow \{140\text{nm}, 280\text{nm}, 420\text{nm}\}$ are respectively $\Delta V_{BL,margin} = \{35\text{mV}, 100\text{mV}, 150\text{mV}\}$.

The proposed sensing circuit was compared to the following sensing circuits (refer to Table 1). As explained for the proposed sensing circuit, the sensing circuits for comparison are simulated using transistor schematics with estimated parasitics for their internal nodes (0.4...0.7fF). The transistors are minimally dimensioned ($l = 40$nm, $w = 120$nm), except for pre-charge and other driving devices ($w = 480$nm). For any number of cells $N \in \{8, 16, 32, 64, 128\}$ the extracted parasitics from a column of conventional 8T-SRAM cells as shown in Fig. 1 are attached to the bit-line m of each reference circuit schematic. The circuit simulations of the reference circuits are applied for the same simulation corners, temperature and supply voltage as for the proposed sensing circuit. The cycle period was 2500ps. Figure 7 shows the reference circuits B and C as implemented for comparison. For the reference circuits domino-read and standard inverter the gate widths of the sensing transistor M3 and the

Table 1. Proposed and reference sensing circuits for comparison.

Designation	
A	The proposed sense circuit as shown in Fig. 2
B	A domino style sensing of the local bit-line m with a PMOS transistor, which implicitly implements the digit line multiplex (see Fig. 7) (Takeda et al., 2004)
C	Sensing with an inverter followed by an NMOS transistor for digit line multiplex (Cosemans et al., 2007). A similar sensing circuit implies a static CMOS NAND gate instead of an inverter for early digit line multiplex (Chang et al., 2007).
D	An AC coupled sense amplifier as PMOS replacement in a domino read path (Qazi et al., 2010). For the reduced bit-line swing due to word-line pulsing 400mV is assumed. To compare this circuit, the driving part for the global bit-line is included.
E	An AC coupled sense amplifier (Verma and Chandrakasan, 2009). For the reduced bit-line swing due to word-line pulsing 200mV is assumed.

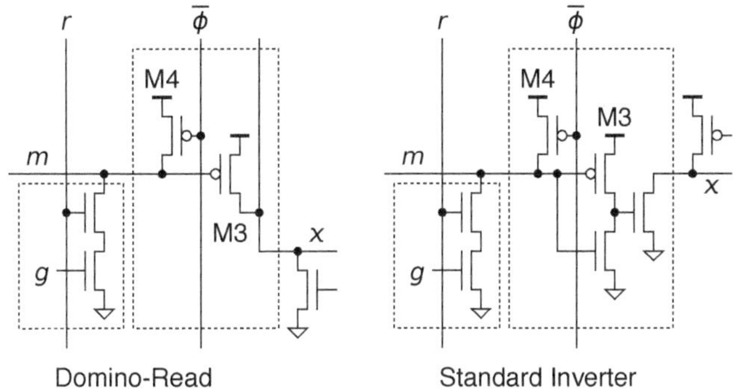

Fig. 7. Schematics of the domino-read (B) and inverter-read (C) sensing circuits for comparison.

Table 2. Simulated results for $N = 128$ cells (slow corner, 125°C, $V_{DD} = 900\text{mV}$).

	A	B	C	D	E
t_{ACC} / ps	300	408	512	401	325
$t_{ACC,SA}$ / ps	82	45	53	94	126
E_{Sense} / fJ	5.08	1.12	2.21	5.58	9.95
E_{BL} / fJ	5.85	13.40	13.42	6.19	2.94
E_{tot} / fJ	10.93	14.53	15.62	11.77	12.89
$V_{BL,margin}$ / mV	120	181	379	96	53
Transistor count[a]	10	3	5	13	14

[a] without transistors of memory cells.

pre-charge transistor M4 were set to $w = 480\text{nm}$, all others were designed with minimal gate width of $w = 120\text{nm}$. Table 2 shows the simulated results of the proposed circuit A versus the reference circuits B...E for $N = 128$ cells. The access time t_{ACC} of the proposed circuit A is faster than B, C, D and E for $N = 128$ cells. The access time t_{ACC} consists of a bit-line dependent contribution $t_{ACC,BL}$ and an access time offset $t_{ACC,SA}$, which depends on the internal complexity of the sensing circuit. Apart from circuit B and C, the proposed circuit shows a smaller intrinsic access time $t_{ACC,SA}$ compared to circuit D and E.

The sense error immunity to bit-line noise was simulated with current pulses to inject charge to the bit-line during an evaluation period. Read evaluation faults are detected within an evaluation period of 2500ps. The tolerable bit-line noise $V_{BL,margin}$ is shown in Table 2 for each circuit.

Table 2 shows that the bit-line energy per cycle of circuit A is similar to circuit D. This results from nearly the same bit-line swing achieved in circuit A and D. The energy dissipation E_{Sense} of circuit A is less than that of circuit D and E. Since circuit B and C operate with full swing bit-lines, but dissipate the least energy E_{Sense}, there must be a crossover for a certain number of cells. Figure 8a shows that this crossover point can be identified at $N \geq 64$ cells.

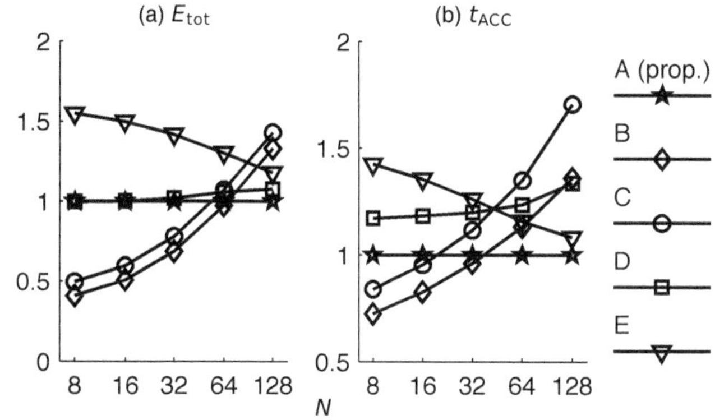

Fig. 8. Simulated t_{ACC} and E_{tot} for different number of connected cells N (normalized to the proposed circuit A, slow corner, 125°C, $V_{DD} = 900\text{mV}$).

Table 3. Comparison of the average access time and relative standard deviation of the proposed sensing circuit (with $l_{M5} = l_{M6} = 70\text{nm}$) for $N = 64$ using 1000 Monte Carlo simulations.

	A	B	C	D	E
mean $(t_{ACC})^a$	152 ps	167 ps	205 ps	169 ps	157 ps
std $(t_{ACC})^a$	9.0%	11.1%	9.5%	10.1%	10.6%
mean $(t_{ACC})^b$	153 ps	164 ps	204 ps	170 ps	158 ps
std $(t_{ACC})^b$	9.5%	14.7%	10.6%	10.6%	11.3%

[a] constant temperature 25°C and supply voltage $V_{DD} = 900\text{mV}$.
[b] normally distributed temperature and supply voltage variations, temperature $\in [-55°\text{C}...125°\text{C}]$, $V_{DD} \in [810\text{mV}...990\text{mV}]$.

A similar crossover point for the access time is located at $N \approx 40$ cells (see Fig. 8b). Simulations indicate smaller access times of circuit E for $N > 256$ cells. However, the proposed sensing circuit dissipates the least amount energy and shows the fastest access time compared to all reference circuits for $N \in [64...128]$.

Table 3 shows the mean access time over 1000 Monte Carlo iterations for each circuit and its standard deviation relative to the mean access time. The proposed sensing circuit provides the fastest mean access time and lowest access time variation compared to the reference circuits.

To implement the proposed sensing circuit in a hierarchical memory architecture, the sensing circuit on all but the last hierarchy level needs a small modification to provide the signaling scheme consisting of a digit line and a virtual ground line. The modification affects the output stage of the proposed sensing circuit and requires an additional internal inverter for driving the virtual ground switch. In Fig. 9 such a modified sensing circuit and its application in a hierarchical memory architecture is shown. The source terminal of the output transistor is now connected to a virtual ground line xv which enables the same signaling scheme as used for the

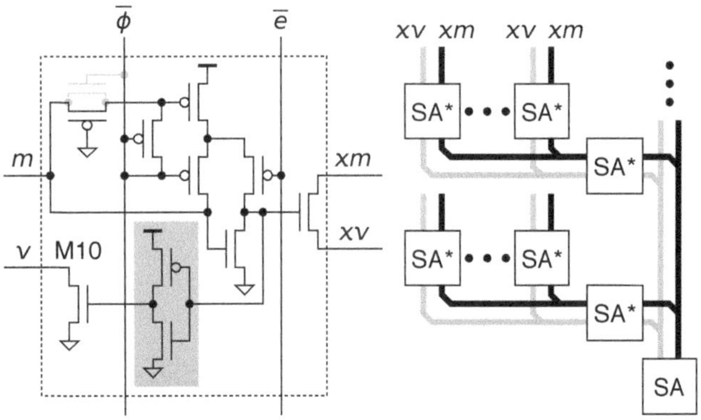

Fig. 9. The modified sensing circuit with a virtual ground on the output side and its implementation in a hierarchical memory architecture. SA* denotes the modified sensing circuit.

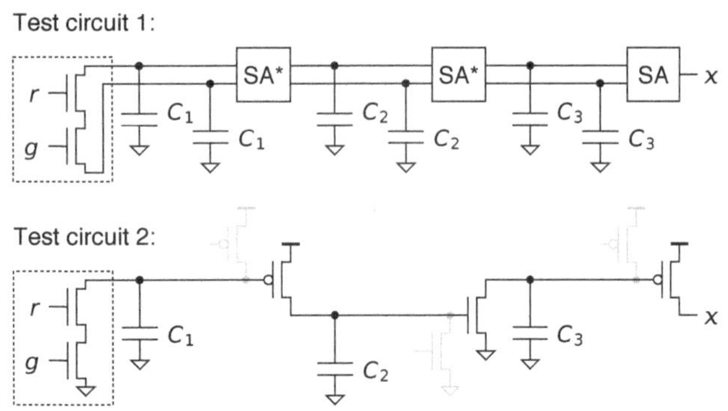

Fig. 10. Test circuits for sensing scheme comparison. Test circuit 1 shows the proposed sensing scheme with virtual grounds. Test circuit 2 shows a domino-style sensing scheme. The gray shaded transistors are used to pre-charge the digit-lines and are not relevant for the access time comparison. The dotted box encloses the read-only port of the (modified) 8T-SRAM cell.

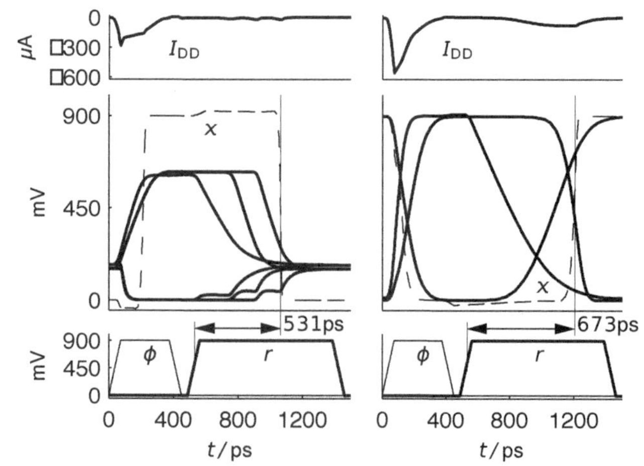

Fig. 11. Comparison of simulated waveforms for a read-cycle of the proposed sensing circuit on the left and a domino-style full-swing signaling circuit on the right (slow corner, 125°C, $V_{DD} = 900$mV).

modified 8T-SRAM cell. Since now the output signal xm will not fall below the threshold voltage of the virtual ground switch M10, the output signal xm cannot be used to control the gate of M10 anymore. Thus, it is required to generate the control signal for the virtual ground switch inside the sensing circuit with an additional inverter (highlighted in Fig. 9 by a gray shaded box).

To show the performance improvement of the proposed signaling scheme, two test circuits representing the data signal path of a hierarchical memory architecture were designed and compared by simulations. The first test circuit employs the proposed sensing circuits and pairs of digit line and virtual ground line on each level of hierarchy. The second test circuit employs a domino-style sensing circuit similar to the left circuit of Fig. 7. However, in domino-style sensing circuits successive stages alternatively employ NMOS and PMOS sensing devices. The test circuits are depicted in Fig. 10. For fair comparison, each corresponding bit-line, digit-line, and virtual ground of the test circuits are equally loaded with parasitic capacitances on each hierarchy level. Figure 11 shows the simulated waveform for read-cycles of both sensing circuits. The left read-cycle was simulated for the proposed signaling scheme with included virtual ground concept. This waveform shows the same properties as the waveform in Fig. 3: the virtual grounds are successively released as soon as the data bit is detected. The reduced voltage swing of approximately 475mV leads to digit-line energy savings of approximately 50%. The right waveform shows the full-swing signals of the domino-style sensing circuits. The access time of the proposed sensing circuit ($t_{ACC} = 531$ ps) is smaller than the access time of the domino-style sensing scheme ($t_{ACC} = 673$ps). This leads to an access time improvement of approximately 21%.

5 Conclusions

A new sensing circuit for a single-ended read-only-port of SRAM cells is introduced. Instead of pre-charging the bit-line to V_{DD}, the proposed circuit sets the pre-charge level close to the threshold level $V_{DD} - |V_{th}|$ of the individual sensing device, while ensuring a process variation tolerant bit-line noise margin. With a small modification of the 8T-SRAM cell to provide an additional bit-line v, the charge dissipation of the bit-line m during evaluation can automatically be stopped by the proposed circuit, once the data bit is detected by the sensing circuit. Both effects together reduce the bit-line swing and the bit-line power dissipation. For $N = 64\ldots128$ cells, the proposed circuit achieves fewest energy compared to the reference circuits along with good performance improvements. The proposed sensing circuit is also suited for a hierarchical memory architecture utilizing the optimized pre-charge level and virtual ground concepts.

References

Chang, L., Fried, D. M., Hergenrother, J., Sleight, J. W., Dennard, R. H., Montoye, R. K., Sekaric, L., McNab, S. J., Topol, A. W., Adams, C. D., Guarini, K. W., and Haensch, W.: Stable SRAM Cell Design for the 32 nm Node and Beyond, in: Symposium on VLSI Technology, 128–129, 2005.

Chang, L., Nakamura, Y., Montoye, R. K., Sawada, J., Martin, A. K., Kinoshita, K., Gebara, F. H., Agarwal, K. B., Acharyya, D. J., Haensch, W., Hosokawa, K., and Jamsek, D.: A 5.3 GHz 8T-SRAM with Operation Down to 0.41 V in 65 nm CMOS, in: Symposium on VLSI Circuits, 252–253, 2007.

Cosemans, S., Dehaene, W., and Catthoor, F.: A Low-Power Embedded SRAM for Wireless Applications, IEEE Journal of Solid-State Circuits, 42, 1607–1617, 2007.

Qazi, M., Stawiasz, K., Chang, L., and Chandrakasan, A.: A 512 kb 8T SRAM Macro Operating Down to 0.57V with An AC-Coupled Sense Amplifier and Embedded Data-Retention-Voltage Sensor in 45 nm SOI CMOS, in: IEEE International Solid-State Circuits Conference, 350–351, 2010.

Takeda, K., Hagihara, Y., Aimoto, Y., Nomura, M., Uchida, R., Nakazawa, Y., Hirota, Y., Yoshida, S., and Saito, T.: Per-Bit Sense Amplifier Scheme for 1GHz SRAM Macro in Sub-100 nm CMOS Technology, in: IEEE International Solid-State Circuits Conference, 1, 502–542, 2004.

Verma, N. and Chandrakasan, A. P.: A High-Density 45 nm SRAM Using Small-Signal Non-Strobed Regenerative Sensing, IEEE Journal of Solid-State Circuits, 44, 163–173, 2009.

38

An X-Band low-power and low-phase-noise VCO using bondwire inductor

K. Hu, F. Herzel, and J. C. Scheytt

IHP GmbH, Im Technologiepark 25, 15236 Frankfurt (Oder), Germany

Abstract. In this paper a low-power low-phase-noise voltage-controlled-oscillator (VCO) has been designed and, fabricated in 0.25 μm SiGe BiCMOS process. The resonator of the VCO is implemented with on-chip MIM capacitors and a single aluminum bondwire. A tail current filter is realized to suppress flicker noise up-conversion. The measured phase noise is -126.6 dBc/Hz at 1 MHz offset from a 7.8 GHz carrier. The figure of merit (FOM) of the VCO is -192.5 dBc/Hz and the VCO core consumes 4 mA from a 3.3 V power supply. To the best of our knowledge, this is the best FOM and the lowest phase noise for bondwire VCOs in the X-band. This VCO will be used for satellite communications.

1 Introduction

For satellite communications, such as HDTV, internet-via-satellite and digital video broadcasting service (DVB-RCS), a low-phase-noise VCO is a prime requirement for the frequency synthesizer. Following the specifications given in (Follmann et al., 2008), phase noise must be lower than -110 dBc/Hz at 1 MHz offset from the carrier frequency. With the integrated resonator, this specification is hard to meet, particularly if the performance must be guaranteed over a wide tuning range. A VCO phase noise of -120 dBc/Hz at 1 MHz offset is, therefore, highly desirable.

With respect to cost, silicon compares favourably to III-V technologies. However, it is difficult to reach the phase noise specification with a fully integrated VCO for silicon technology due to the poor quality factor of the on-chip inductor. Many studies have been presented on minimizing VCO phase noise, e.g. Hegazi (2001); Ferndahl (2005). However, the overall performance of these VCOs, e.g. power consumption, tuning range, phase noise performance, can still not satisfy the specifications. In order to reduce phase noise and maintain relatively low power consumption, bondwire inductors have been used (Craninckx et al., 1995; Kim et al., 2008). By doing so, a figure of merit (FOM) of -190 dBc/Hz has been reported so far.

A differential CMOS VCO implemented with nMOSFETs and pMOSFETs was presented in (Craninckx et al., 1997). Unlike in VCOs using one transistor type only, in this current re-using topology no mid-point in the inductor is required for biasing purpose. We adopt this topology by using a bondwire inductor instead of an integrated coil. This approach minimizes mismatch in the VCO as only one bondwire is used rather than two in other topologies. Moreover, a tail current filter is introduced to reduce the flicker noise up-conversion from the current source. This approach significantly improves the close-in frequency phase noise of the VCO.

2 VCO design

The VCO core employs two nMOSFETs and two pMOSFETs. The combination of NMOS and PMOS transistors gives a negative resistance from both transistor pairs. Consequently, to provide the same negative resistance, the combined NMOS and PMOS structure can efficiently halve the power consumption, which greatly fits the purpose of low power design. Furthermore, by controlling the supply voltage, the signal swing in the VCO core can be well restricted below the transistor breakdown voltage. It ensures a more reliable operation, which is a very important requirement for satellite communications (Tiebout, 2006).

Another convenient feature of this VCO topology is that only one inductor is required for the resonator. This eliminates the inductance mismatch which appears in topologies using two bondwire inductors. The mismatch of resonator inductors in these topologies will result in asymmetric signal

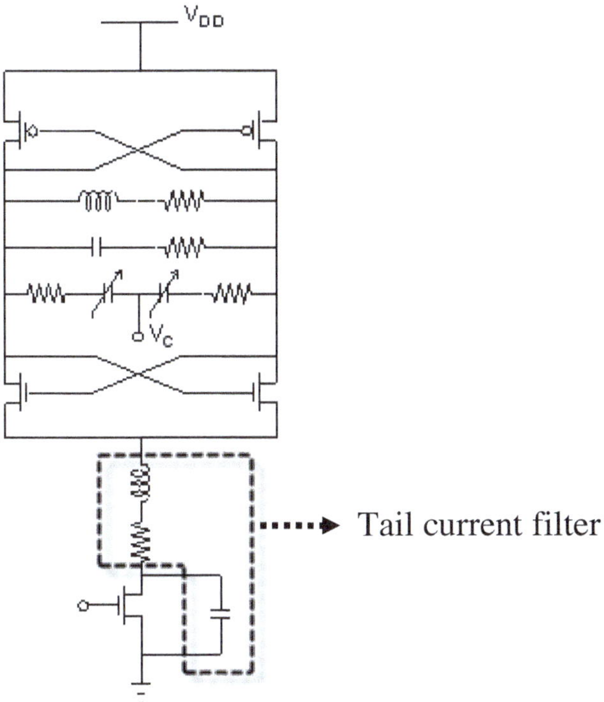

Fig. 1. Schematic of the X-Band VCO.

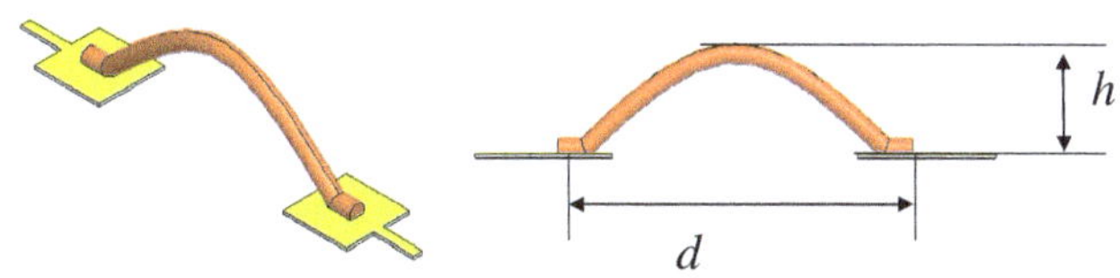

Fig. 2. Bondwire model. h is wire loop height, d is pad distance.

Table 1. EM simulation results for bondwire.

Pad Distance	Wire Radius 20 μm
250 μm	325 pH
300 μm	386 pH
400 μm	502 pH
500 μm	635 pH
600 μm	751 pH

swing across the tank and, consequently, degenerate the oscillator phase noise.

Additionally, the phase noise close to the carrier frequency is mainly dominated by the flicker noise up-conversion, i.e. flicker noise from current source and VCO core transistors. As a result of the odd mode virtual ground at the differential common mode node, the flicker noise from current source is up-converted to $2 \cdot f_0$ and mixed down to f_0 (oscillation frequency), because of the differential pairs mixing action. The up-converted flicker noise will be delivered to the resonator as AM noise (Rael et al., 2000), and the AM noise will be converted to FM noise in the varactor (Muer et al., 2000). Thus, for the suppression of flicker noise from current source, tail current filtering technique (Hegazi et al., 2001) is introduced to short the 2nd harmonic at common mode node. As a result, the close-in frequency phase noise can be largely improved.

3 Bondwire modelling

As mentioned above, there are many studies about improving the phase noise performance for VCO designs, but most of them cannot meet both the low phase noise and the low power consumption requirements. The modified Leeson's Formula (Masini et al., 2001):

$$L(\Delta\omega) = 10 \cdot \log \left\{ \frac{2 \cdot F \cdot k \cdot T}{P_s} \cdot \left[1 + (\frac{1}{2Q} \cdot \frac{\omega_c}{\Delta\omega})^2 \right] \cdot \left(1 + \frac{\Delta\omega_{1/f^3}}{|\Delta\omega|} \right) + \frac{\pi^2}{2} \cdot \left(\frac{K_v \cdot V_m}{\Delta\omega} \right) \right\}, \quad (1)$$

gives a guideline for VCO optimization. Here F is the noise figure of the transistor, $\Delta\omega$ is the offset frequency, ω_c is the center frequency, P_s is the power across the tank, Q is the loaded Q of the tank, K_v is the VCO gain and V_m is the total low frequency noise e.g. noises from DC sources or tuning lines. It is evident from Eq. (1) that the most efficient way of reducing VCO phase noise is to increase the loaded Q of the resonator tank. For the LC resonator VCO, the loaded Q of the tank can be written as:

$$Q = \frac{Q_L \cdot Q_C}{Q_L + Q_C}. \quad (2)$$

In Eq. (2), Q_L is the quality factor of the tank inductor, and Q_C is the quality factor of the tank capacitor. As a result of using the MIM capacitors, Q_C is generally much higher than Q_L. Thus, the loaded Q of the tank is mainly determined by the inductor quality factor. However, due to the lossy silicon substrate, on-chip inductor in the X-Band can only have a quality factor around 10. A different choice for on-chip inductors takes advantage of the parasitic inductance, which is usually associated with bondwire in IC packaging. The series resistance of the bondwire is very low, which leads to a high quality factor. Furthermore, the parasitic capacitance value is given by the bond pads. Thus, by using optimized bondpad size the bondwire inductor can have a very high self-resonance frequency. Figure 2 shows the structure of the bondwire inductor. The radius of the bondwire is 12.5 μm. The pad distance and wire loop height are defined as design parameters to simulate bondwire inductance and quality factor. In order to obtain accurate prediction results, full wave

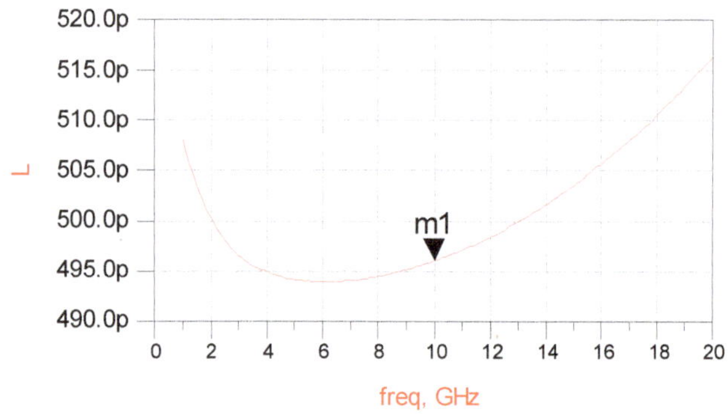

Fig. 3. Inductance for 480 μm length bondwire (EM simulation).

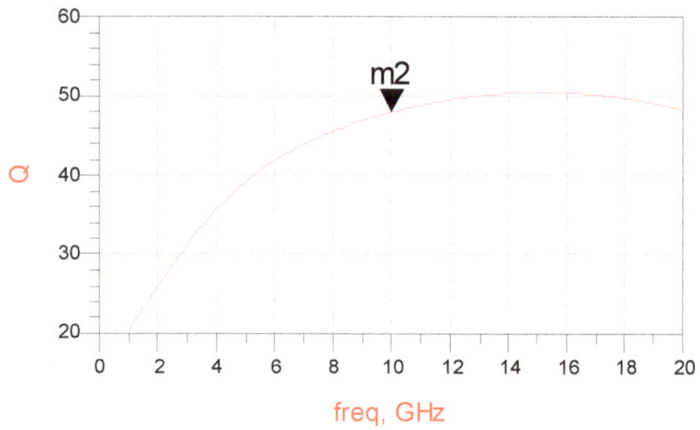

Fig. 4. Quality factor for 480 μm length bondwire (EM simulation).

EM simulator (Ansoft HFSS v11) is used for the bondwire simulations.

Table 1 shows the simulated inductance for different bondwire lengths. The size of the bondpad affects the inductor self-resonance frequency significantly, which is caused by the parasitic capacitance between pad and substrate.

For the requirement of the X-band VCO design, a 500 pH inductor corresponds to an aluminum bondwire with the length of 480 μm. With optimized bondpad size, the simulated quality factor can reach 48 at 10 GHz. Figures. 3 and 4 show the simulated inductance and quality factor for the bondwire inductor used in the X-band VCO design.

4 Simulation results

As shown in Fig. 5, the simulated result demonstrates 8 dB phase noise improvement at 100 kHz offset and 4 dB phase noise improvement at 1 MHz offset with the tail current filtering technique. The simulated phase noise for the final VCO design is −128.3 dBc/Hz at 1 MHz offset and −105 dBc/Hz at 100 kHz offset from a 10.2 GHz carrier. Obviously, the up-conversion of flicker noise has been greatly reduced by introducing the tail current filtering structure.

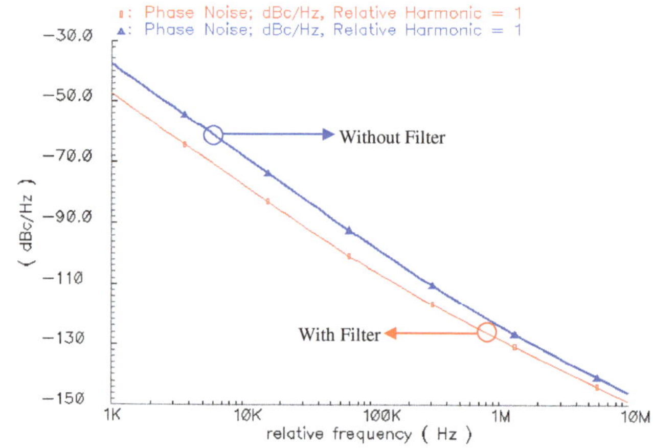

Fig. 5. Phase noise with and without tail current filter.

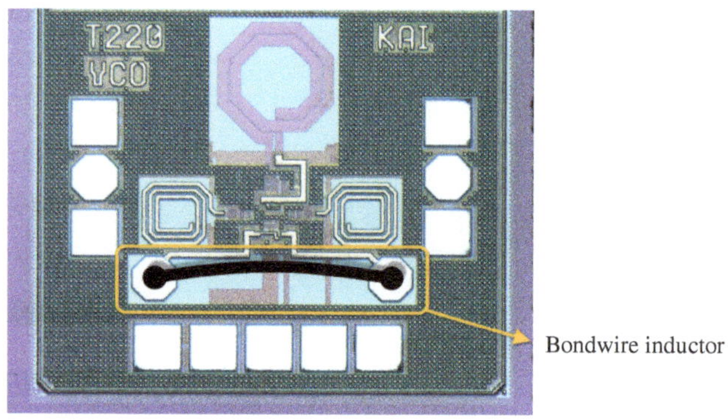

Fig. 6. Chip photograph.

5 Measurement results

A photograph of the VCO chip is shown in Fig. 6. The VCO operates from a 3.3V supply and consumes 4 mA current. With the 50 Ohm output buffer, the VCO delivers an output power from −1 to 2 dBm over the whole tuning range. The measured tuning frequency and output power is shown in Fig. 7.

The VCO phase noise is measured by Aeroflex phase noise meter with the delay line method. As shown in Fig. 8, the phase noise at 1 MHz offset is −126.6 dBc/Hz and shows a very good matching with the simulation results. At 100 kHz offset from the 7.8 GHz carrier, the measured phase noise is −100 dBc/Hz. This relative high phase noise at 100 kHz is due to the flicker noise of the MOSFETs.

The figure of merit for a VCO is defined by Eq. 3 (Tiebout, 2006), where $\omega_0$is the angular oscillation frequency, $L(\Delta\omega)$ is the phase noise at offset $\Delta\omega$ and $P_{\text{diss}}$is the dc power consumption (mW) of the VCO. The measured X-band VCO has a FOM of −192.5 dBc/Hz. To the best of the authors' knowledge, this is the best FOM among published X-band VCOs based on silicon technology.

Table 2. Performance of reported VCO.

	F_C (GHz)	P_{out} (dBm)	L@1 MHz (dBc/Hz)	Offset (MHz)	FOM (dB)	Technology	Topology
H. Jacobsson/RFICS	11.8	−7	−123	1	−183.9	0.5 μm SiGe	4× coulpled VCO array
H. Jacobsson/RFICS	5.9	−9	−126	1	−185.5	0.25 μm SiGe	Coulpled Colpitts
Orsatti/CICC 1999	0.9	NA	−112.7	0.1	−185.3	NA	External LC resonator
Svelto/CICC 2000	1.9	Na	−148	3	−187	NA	Bondwire VCO
K. J. Kim/ISED 2008	1.7	NA	−135.3	1	−190	0.18 μm CMOS	Bondwire VCO
This work	8.8	1.92	−125.3	1	−192.5	0.25 μm SiGe	CC/Bondwire

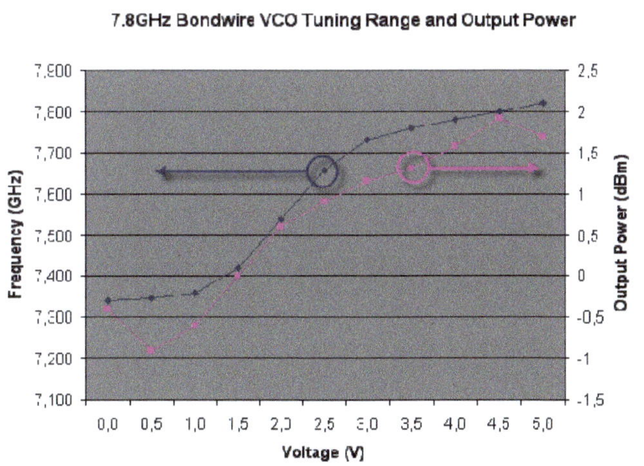

Fig. 7. Measured output power and tuning curve.

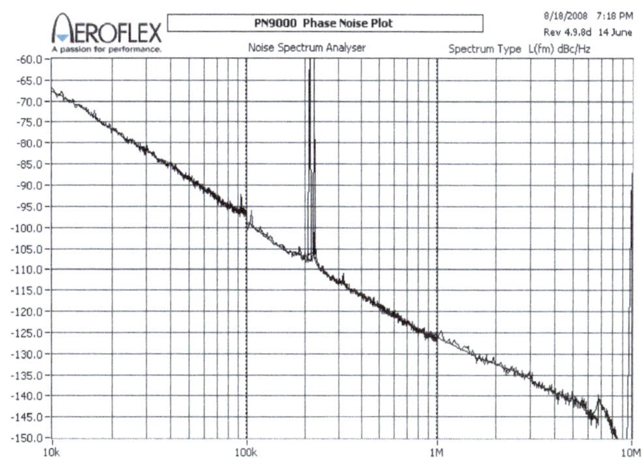

Fig. 8. Measured phase noise for 7.8 GHz VCO.

$$\text{FOM} = L(\Delta\omega) - 20 \cdot \log(\frac{\omega_0}{\Delta\omega}) + 10 \cdot \log(\frac{P_{\text{diss}}}{1\,\text{mW}}) \quad (3)$$

Table 2 lists the performance of state-of-art VCOs based on silicon technology.

6 Conclusions

We have presented an X-band bondwire VCO in 0.25 μm SiGe BiCMOS technology. The bondwire was optimized by using full-wave EM simulation. The tail current noise is compensated by an integrated low pass filter. Due to the use of the complementary MOSFETs, only one bondwire is required. This improves the symmetry of the differential circuit. The presented VCO achieved a phase noise of less than −125 dBc/Hz at 1 MHz offset. The tuning range is from 7.4 to 7.8 GHz. The figure of merit is −192.5 dBc/Hz. To the authors' knowledge, this is the best value reported so far for silicon-based X-band VCOs. In a future design, the complementary MOSFETs will be replaced with NPN and PNP hetero-bipolar transistors (HBTs) in a complementary SiGe BiCMOS technology (Heinemann et al., 2003). This will reduce flicker noise (Niu, 2005) and improve radiation hardness (Cressler et al., 1998).

Acknowledgements. This work was supported by the European Space Agency (ESA) and the German DLR (Deutsches Zentrum für Luft- und Raumfahrt). The authors thank the IHP technology team for the fabrication of the test chip.

References

Follmann, R., Köther, D., Kohl, T., Engels, M., Podrebersek, T., Heyer, V., Schmalz, K., Herzel, F., Winkler, W., Osmany, S., and Jagdhold, U.: A single SiGe chip fractional-N 275 MHz…20 GHz PLL with integrated 20 GHz VCO, IEEE MTT-S International Microwave Symposium, Atlanta, 355–358, June, 2008.

Hegazi, E., Sjoland, H., and Abidi, A.: A Filtering Technique to Lower Oscillator Phase Noise, IEEE International Solid-State Circuits Conference, 364–365, 2001.

Jacobsson, H., Hansson, B., Berg, H., and Gevorgian, S.: Very Low Phase-Noise Fully-Integrated Coupled VCOs, Radio Frequency Integrated Circuits Symposium, 467–470, 2002.

Bao, M., Li, Y., and Jacobsson, H.: A 21.5/43 GHz Dual-Frequency Balanced Colpitts VCO in SiGe Technology, IEEE Journal of Solid-State Circuits, 1352–1355, 2004.

Zirath, H., Jacobsson, H., Bao, M., Ferndahl, M., and Kozhuharov, R.: MMIC-Oscillator Designs for Ultra Low Phase Noise, IEEE

Compound Semiconductor Integrated Circuit Symposium, 204–207, 2005.

Ferndahl, M. and Zirath, H.: Broadband 7GHz VCO in mHEMT Technology, in: Proc. Asia-Pacific Microwave Conference, 4, December, 2005.

Craninckx, J. and Steyaert, M.: A 1.8-GHz CMOS low-phase-noise voltage-controlled oscillator with prescaler, IEEE Journal of Solid-State Circuits, 1474–1482, December, 1995.

Kim, K., Ahn, K. H., and Lim, T. H.: Low Phase Noise Bond Wire VCO for DVB-H, 4th IEEE International Symposium on Electronic Design, Test and Applications, 103–106, 2008.

Craninckx, J., Steyaert, M., and Miyakawa, H.: A Fully Integrated Spiral-LC CMOS VCO Set with Prescaler for GSM and DCS-1800 Systems, in: Proc. IEEE Custom Integrated Circuits Conference, 403–406, 1997.

Tiebout, M.: Low Power VCO Design in CMOS, Springer Berlin Heidelberg, 11 and 74, 2006.

Rael, J. J. and Abidi, A. A.: Physical Process of Phase Noise in Differential LC Oscillators, in: Proc. Custom Integrated Circuit Conference, 569–572, May, 2000.

Muer, B. D., Borremans, M., Steyaert, M., and Puma, G. L.: A 2-GHz Low-Phase Noise Integrated LC-VCO Set With Flicker-Noise Upconversion Mechanism, IEEE j. Solid-State Circuits, 1034–1038, July, 2000.

Hegazi, E., Sjoland, H., and Abidi, A: A Filtering Technique to Lower Qscillator Phase Noise, ISSCC, Session 24, 23.4, 2001.

Masini, L., Pozzoni, M., Caliumi, A., Tomasini, L., Morigi, D., and Lemaire, F.: A Fully Integrated Silicon-Germanium X-Band VCO, online available: http://amsacta.cib.unibo.it/archive/00000048/01/G_13_2.pdf, 2001.

Heinemann, B., Barth, R., Bolze, D., Drews, J., and Formanek, P.: A complementary BiCMOS technology with high speed npn and pnp SiGe:C HBTs, IEEE International Electron Devices Meeting, 5.2.1–5.2.4, 2003.

Niu, G.: Noise in SiGe HBT RF Technology: Physics, Modeling, and Circuit Implementations, in: Proc. IEEE, 93, 1583–1597, 2005.

Cressler, J. D.: SiGe HBT technology: a New Contender for Si based RF and Microwave Circuit Applications, IEEE Transactions on Microwave Theory and Techniques, 46, 572–589, 1998.

Fast evaluation of nonlinear functions using FPGAs

S. Lachowicz[1] **and H.-J. Pfleiderer**[2]

[1]Edith Cowan University, Perth, WA, Australia
[2]Ulm University, Ulm, Germany

Abstract. The paper presents a novel method of evaluating the square root function in FPGA. The method uses a linear approximation subsystem with a reduced size of a look-up table. The reduction in the size of the lookup table is twofold. Firstly, a simple linear approximation subsystem uses the lookup table only for the node points. Secondly, a concept of a variable step look-up table is introduced, where the node points are not uniformly spaced, but the spacing is determined by how close to the linear function the approximated function is. The proposed method of evaluating nonlinear function and specifically the square root function is practical for word lengths of up to 24 bits. The evaluation is performed in one clock cycle.

1 Introduction

Field Programmable Gate Arrays (FPGA) are becoming widely used in Digital Signal Processing (DSP) systems. Modern FPGAs exhibit very high performance, very large number of IOs and high capacity memory on the chip. The development is rapid and risks are low since the IP can be easily modified. For the low to medium production volume it is not economical to develop an ASIC and, if needed, there are companies like Altera who offer a convenient solution of migration from FPGA to ASIC.

The evolution of FPGAs has enabled to implement more and more sophisticated DSP subsystems in the form of dedicated hardware on the device. In late 1990s FPGAs usually had just a few hardware multipliers per chip. In early 2000s, hardwired multipliers in large numbers (up to ~500) with clock speeds of up to 100 MHz became available. Mid 2000s saw the DSP algorithms like full FIR filters implemented, and currently, powerful arithmetic, such as fast square root and divide, floating point operations and many more are being implemented.

In DSP systems nonlinear functions, such as trigonometric, square root or Bessel functions are often encountered. There are several solutions to this problem, for instance, using a CORDIC core, or look-up tables. A lot of effort is made to reduce the size of the look-up tables because it can become prohibitive for high word lengths. For some functions an existing approach is to create hierarchical lookup tables. In this work we present a different approach to reducing the look-up table size. It is proposed to introduce a variable step of the look-up table, so that the number of entries is higher for more nonlinear segments of the curve and smaller for the parts closer to a linear function. The step length is always a power of 2, hence, 8, 16, 32, etc.

The remainder of the paper is organised as follows. Section 2 presents several methods of computation of nonlinear functions. In Sect. 3 the method of linear approximation is described in more detail. Section 4 deals with the non-uniform step lookup table generation and the associated hardware, and presents the experimental results of square root function implementation. It is followed by conclusions.

2 Nonlinear function computation

2.1 Look-up tables

In general the look-up tables are used for single precision (Hassler, 1995; Sarma, 1995; Wong, 1995; Ercegovac, 2000). For longer word lengths the size of the table becomes prohibitively large. Sometimes the look-up tables are used with conjunction of linear approximation system which reduces the size of the table considerably. When used, the look-up table method is the fastest, and if the higher precision is required it is used as an initial approximation for iterative methods.

2.2 CORDIC core

COordinate Rotation DIgital Computer (CORDIC) is a very popular IP core used for various computations (Volder, 1959). It is based on the iterative rotation of a vector in Cartesian coordinate system (Givens rotations):

$$x^{(2)} = \cos\theta \left(x^{(1)} - y^{(1)} \tan\theta\right), \tag{1a}$$

Table 1. CORDIC functions (after DSPedia).

	Rotation Mode: d_i=sign($z^{(i)}$); $z^{(i)} \rightarrow 0$	Vectoring Mode: d_i=-sign($x^{(i)}y^{(i)}$); $y^{(i)} \rightarrow 0$
Circular $\mu = 1$ $e^{(i)} = \tan^{-1}2^{-i}$	x, y, z → CORDIC → K(x.cos z - y.sin z), K(y.cos z + x.sin z), 0 For cos z & sin z, set x = 1/K, y = 0	x, y, z → CORDIC → $K(x^2 + y^2)^{1/2}$, 0, $z + \tan^{-1}(y/x)$ For $\tan^{-1}$ y, set x = 1, z = 0
Linear $\mu = 0$ $e^{(i)} = 2^{-i}$	x, y, z → CORDIC → x, y + (x.z), 0 For multiplication, set y = 0	x, y, z → CORDIC → x, 0, z + (y/x) For division, set z = 0
Hyperbolic $\mu = -1$ $e^{(i)} = \tanh^{-1}2^{-i}$	x, y, z → CORDIC → K^*(x.cosh z - y.sinh z), K^*(y.cosh z + x.sinh z), 0 For cosh z & sinh z, set x = 1/K^*, y = 0	x, y, z → CORDIC → $K^*(x^2 - y^2)^{1/2}$, 0, $z + \tanh^{-1}(y/x)$ For $\tanh^{-1}$y, set x = 1, z = 0

Table 2. CORDIC accuracy (after DSPedia).

	Predicted			Simulated		
n/b	8	9	10	8	9	10
3	5.09	5.31	5.43	5.32	5.71	5.80
4	6.03	6.59	6.98	6.22	6.88	7.34
5	6.28	7.13	7.88	6.52	7.56	8.27
6	6.21	7.17	8.10	6.55	7.11	8.12
7	6.06	7.05	8.04	6.42	7.29	8.29
8	5.92	6.91	7.91	6.33	7.29	8.31
9	5.78	6.78	7.78	6.00	7.00	8.00

$$y^{(2)} = \cos\theta \left(y^{(1)} - x^{(1)} \tan\theta \right). \tag{1b}$$

CORDIC core is widely offered as an IP by the FPGA vendors such as XILINX. Table 1 summarises the functions which can be computed by CORDIC.

As the CORDIC is an iterative method, it requires many clock cycles to achieve the required accuracy. Table 2 presents the accuracy depending on n – number of iterations and b – the number of bits used in the computations.

Also, in terms of used resources the number of slices used by the CORDIC in an FPGA chip is quite considerable.

2.3 Series expansion

Series expansion is an attractive option of calculating a nonlinear function (Ercegovac, 2000). Some of the more useful functions are shown below:

$$\frac{1}{1-x} = 1 + x + x^2 + x^3 + \dots, \tag{2a}$$

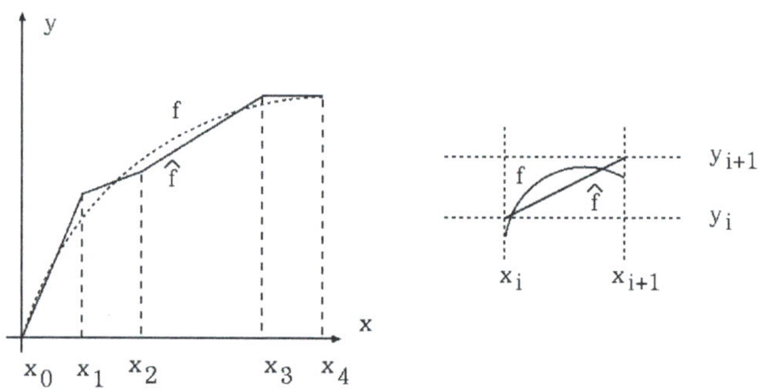

Fig. 1. Piecewise linear approximation of a nonlinear function.

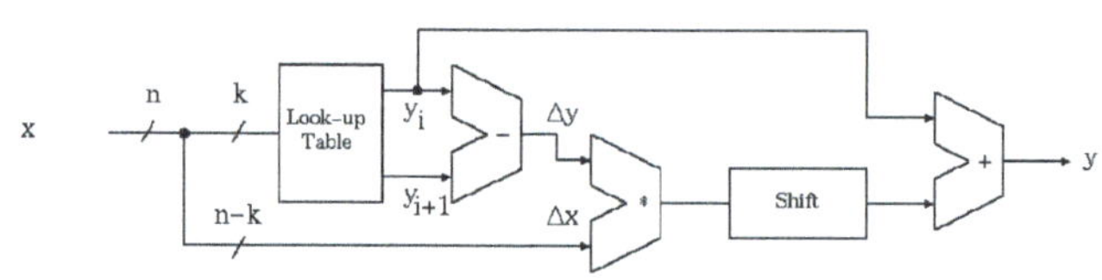

Fig. 2. Hardware for linear function approximation with a fixed step of a look-up table.

$$\cos x = 1 - \frac{1}{2}x^2 + \frac{1}{24}x^4 - \frac{1}{720}x^6 + \dots, \tag{2b}$$

$$\sqrt{x+1} = 1 + \frac{1}{2}x - \frac{1}{8}x^2 + \frac{1}{16}x^3 - \dots. \tag{2c}$$

The method is not iterative, therefore, in principle, the implementation does not have to be pipelined. However, usually the number of required multipliers is high, especially for higher accuracy. Also, preprocessing is generally required to make the argument close to 1. It is worth to be noted that

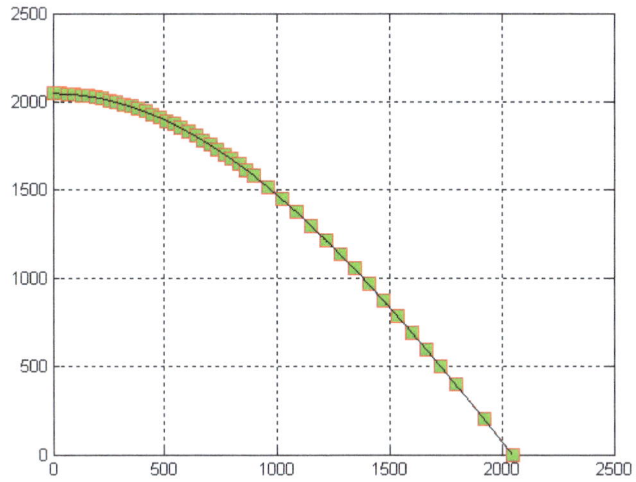

Fig. 3. Node points of the look-up table for the cosine function.

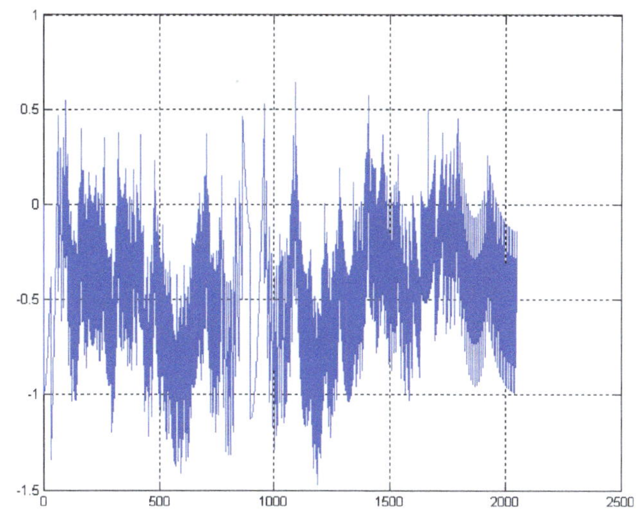

Fig. 4. Approximation error for the cosine function.

many multiplications do not need to be "large", i.e. 53×53 for double precision. Rectangular multipliers can be used where appropriate. After some preprocessing most of the multiplications are $n/2 \times n/4$, where n is the total number of bits.

2.4 Newton-Raphson iteration

Newton-Raphson iteration is a rapid convergence method often used for functions such as $1/x$ and $1/\mathrm{sqrt}(x)$. The general form is:

$$q_{n+1} = q_n - \frac{f(q_n)}{f'(q_n)}. \tag{3a}$$

The following formulae are applicable to $1/x$ and $1/\mathrm{sqrt}(x)$, respectively:

$$q_{n+1} = q_n(2 - xq_n), \tag{3b}$$

$$q_{n+1} = \frac{1}{2} q_n(3 - xq_n^2). \tag{3c}$$

The method is characterised by quadratic convergence and the number of significant bits roughly doubles with each iteration.

3 Linear approximation with look-up tables

One method of reducing the size of the look-up table is to use a linear approximation circuit. Every function can be approximated by a piecewise linear function as presented in Fig. 1.

Between the node points the function can be expressed using the line equation:

$$\widehat{f}(x) = y_i + \frac{y_{i+1} - y_i}{x_{i+1} - x_i}(x - x_i). \tag{4}$$

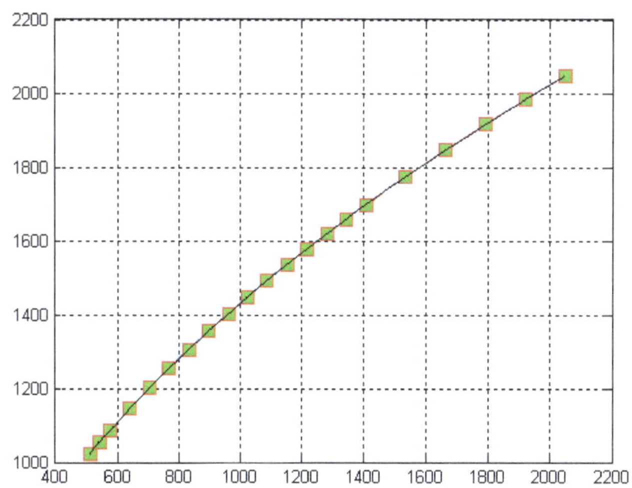

Fig. 5. Node points of an 11-bit look-up table for the sqrt function.

The hardware mapping is presented in Fig. 2. Here the steps of the look-up table are evenly distributed, so the input word x is split into two parts: the more significant part is an address for the look-up table and the less significant part is the increment of x within the current step of the look-up table. The node values y_i and y_{i+1} are read from the look-up table, the increment Δy is calculated and is subsequently multiplied by the increment Δx. The shift replaces the division as the difference $x_{i+1} - x_i$ is always constant and equal to the number being a power of two.

4 Minimum size look-up table generation

Nonlinear functions usually have a different degree of nonlinearity throughout the range of interest. Therefore, to achieve a particular accuracy while using the linear

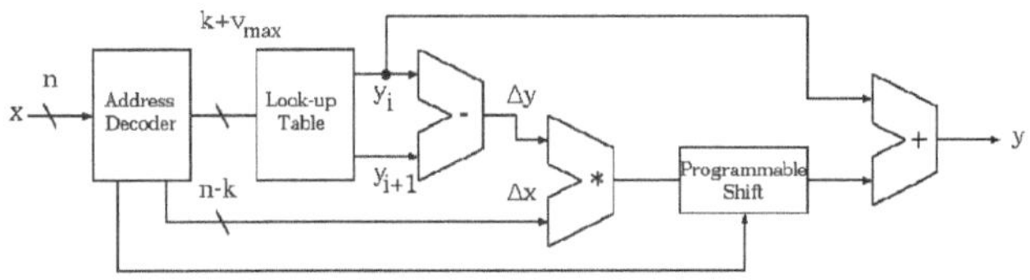

Fig. 6. Hardware for a linear function approximation with a variable step look-up table.

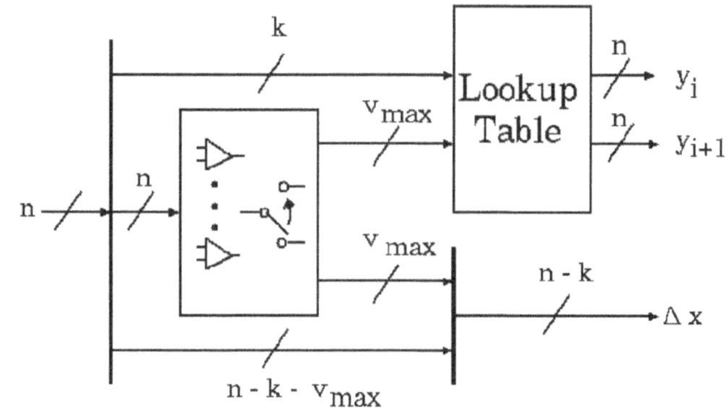

Fig. 7. The address decoder for the circuit from Fig. 6.

approximation, the distance between the node points in the look-up table can be varied. Where the function is more "linear", the distance between the node points can be increased. In the case of the fixed step table, the step corresponding to the most "non-linear" part of the function has to be maintained throughout the whole look-up table.

Figure 3 presents an example of a look-up table generated for the cosine function, in 11-bit resolution and with the accuracy of ± 1 bit. It can be observed that where the function exhibits higher degree of nonlinearity, the step of the look-up table is reduced.

The look-up table is generated as follows:

1. Each step in the look-up table must be of the length equal to the power of 2.

2. The predetermined accuracy of the linear approximation within one step range is verified for the initial value (large) of the step.

3. If the accuracy is not met then the step is reduced by half and the verification is repeated.

4. If the accuracy criterion is met, the algorithm moves to the next step.

5. The next step can be the same length, shorter or longer, however the longer step can only be made if the remaining part of the range is divisible by the new value of the step length. This ensures that the x value can be split into the look-up table address part and the increment part.

As an example of accuracy of the proposed method of a look-up table generation, let us consider a cosine function from Fig. 3. The number of entries in the look-up table is 45. The accuracy of the approximation has been compared with the full look-up table of 2048 entries. The RMS error in respect to the "accurate" (floating point) curve is 0.35 LSB compared with the RMS error of the full look-up table of 0.28 LSB. Figure 4 shows the actual approximation error throughout the input range.

Square root is an important function often encountered in DSP. Figure 5 presents the look-up table for 11-bits resolution with ± 1 bit accuracy. The look-up table has 21 entries, and for 10-bits the linear approximation circuit generates an exact result. Similar table for a 16-bit accuracy has 262 entries. The full look-up tables would require 1024 and 65 536 entries respectively.

The hardware implementing the variable step look-up table linear approximation differs from the fixed-step as it has to split the input word x into variable length address part and the increment Δx and the fixed shift is replaced by programmable shift dependent on the step of the look-up table. The hardware is shown in Fig. 6.

Figure 7 presents the details of the address decoder in the circuit from Fig. 6. An array of comparators determines the current step of the look-up table and splits the input word x into the address part and the increment part. It also controls the number of positions for the shift. The number of comparators in the array is equal to the number of step changes in the look-up table.

Before the input word is presented to the linear approximation circuit the number has to be normalised. For the square root the input number should be within the range $\{1\ldots4\}$ for the output within the range $\{1\ldots2\}$. This operation could be referred to as an internally floating-point operation.

As an example of application of the linear approximation circuit with variable step look-up table, a 16-bit square root $\sqrt{x^2+y^2+z^2}$has been implemented in Spartan 3/1000. The resources used are 242 slices (3%), 1 block RAM (4%) and 4 hardware multipliers (16%). Three of the multipliers are used to calculate the squares and one in the linear approximation circuit. The on-chip hardware multipliers are 18×18, and for the linear approximation circuit a 9×9 multiplier would be sufficient. Therefore, if the variable word length multiplier was available (Pfaender, 2007), the resources could be further reduced. The same circuit was also implemented in Virtex 2 FPGA and the operation was verified using ChipScope. The circuit exhibits latency of approximately 11 ns.

5 Conclusions

A novel method of nonlinear function computation in FPGA has been developed. The standard linear approximation circuit has been modified in order to use a look-up table with variable step. This allows to minimise the size of the look-up table and therefore the resources needed for the implementation. The method can be used directly for word lengths up to about 24 bits, and can also be used as an initial value generator for Newton-Raphson iterations if a higher accuracy is required.

Acknowledgements. The support of the German Academic Exchange Office (DAAD) and Ulm University is gratefully acknowledged.

References

Ercegovac, M. D., Lang, T., Muller, J.-M., and Tisserand, A.: Reciprocation, Square Root, Inverse Square Root, and Some Elementary Functions Using Small Multipliers, IEEE Trans. Computers, 49(7), 628–637, 2000.

Hassler, H. and Takagi, N.: Function Evaluation by Table Look-Up and Addition, in: Proc. 12th IEEE Symp. Computer Arithmetic, edited by: Knowles, S. and McAllister, W., 10–16, July 1995.

Pfaender, O. A., Nopper, R., Pfleiderer, H.-J., Zhou, S., and Bermak, A.: Comparison of reconfigurable structures for flexible word-length multiplication, Adv. Radio Sci., 6, 2008.

Sarma, D. D. and Matula, D. W.: Faithful Bipartite ROM Reciprocal Tables, in: Proc. 12th IEEE Symp. Computer Arithmetic, edited by: Knowles, S. and McAllister, W., 17–28, July 1995.

Volder, J.: The cordic trigonometric computing technique, IRE Trans. Electron. Comput., 3, 330–334, 1959.

Wong, W. F. and Goto, E.: Fast Evaluation of the Elementary Functions in Single Precision, IEEE Trans. Computers, 44(3), 453–457, March 1995.

40

Impact of on-chip inductance on power supply integrity

M. Eireiner[1]**, S. Henzler**[2]**, X. Zhang**[1]**, J. Berthold**[2]**, and D. Schmitt-Landsiedel**[1]

Institute for Technical Electronics, Technical University Munich, Germany
[2]Infineon Technologies AG, Munich, Germany

Abstract. Based on product related scenarios, the impact of on-chip inductance on power supply integrity is analyzed. The impact of varying current profiles is shown to be minimal. In a regular power grid with regular bump connections, the impact of on-chip inductance on the cycle average of the supply voltage can be neglected, even for a worst case estimation of on-chip inductance. Whereas, the maximum transient power supply drop can be significantly underestimated by neglecting on-chip inductance. The impact of on-chip inductance in a System-on-Chip (SoC) environment also can be neglected if the on-chip inductance is conservativly estimated.

1 Introduction

With ongoing technology scaling, power supply integrity has become a major concern even in low power designs. Lowered supply voltages, increasing current densities, increasing operating frequencies, and increasing sheet resistances have decreased the noise margins for every process technology. This trend is expected to continue with technology scaling (Nassif and Fakhouri, 2002; Larsson, 1999; Mezhiba and Friedman, 2004). Power supply noise can be divided into two effects. The first is called IR-Drop and describes the effect of a voltage drop over the parasitic resistance of the power grid, caused by the current flowing through the grid. The second effect is called di/dt noise and describes voltage bounces caused by transient current surges over the parasitic inductance of the power grid. The parasitic inductance of the power grid can be divided into on-chip inductance, coming from all the wires on the chip, and off-chip inductance, coming from all package and board power connections. The paper will address the question, whether on-chip inductance has a significant impact on the total power supply noise bugdet. The rest of the paper is organized as follows. In Sect. 2 and Sect. 3 the movitation and the simulation setup are outlined.

Fig. 1. Schematic of two wires with different effective inductance L_{eff}, having the same geometric dimensions

Simulation results for a worst case estimate of on-chip inductance in a regular power gird are presented in Sect. 4. An improved estimation of the effective on-chip inductance and the resulting simulation results in a SoC environment are presented in Sect. 5. Finally, the paper ends with a conclusion.

2 Motivation

As stated before, power supply noise is either caused by the parasitic resistance or the parasitic inductance of the power supply grid. The extraction of the parasitic resistance of a wire connection, or the supply grid as in our case, is straight forward. The wire is divided into several filaments, and the resistance of all the filaments is added up. The resistance extraction is state of the art and does not pose a significant challenge. However, the resulting inductance of a current loop consisting of two wire segments is given by its self and mutual inductance, as described in Eq. (1).

$$L_{\text{eff}} = L_{11} + L_{22} - 2 \cdot L_{12} \tag{1}$$

Where L_{11} and L_{22} describe the self, and L_{12} the mutual inductance of the current paths, respectively. While L_{11} and L_{22} are set by the geometric dimensions, such as width, height, and length, of the wire, L_{12} is given by the coupling and therefore by the acutal layout of the current loop. This leads to the fact, that two wires having the same geometric dimensions, width, height, and length, can have different effective inductances, as depicted in Fig. 1. Knowing this, it is obvious that the proper extraction of on-chip inductance poses a significant challenge at the design flow. Additionally, by

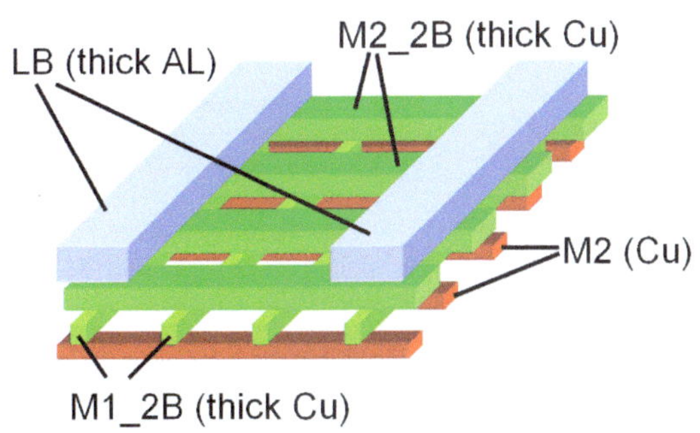

Fig. 2. Schematic of the used 4 layers power grid.

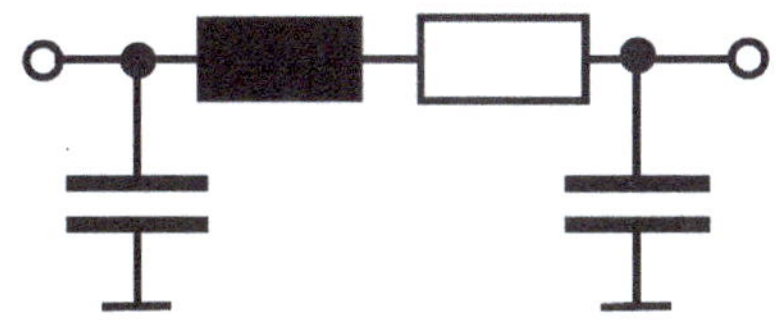

Fig. 3. Schematic of a single R-L-C pi-model.

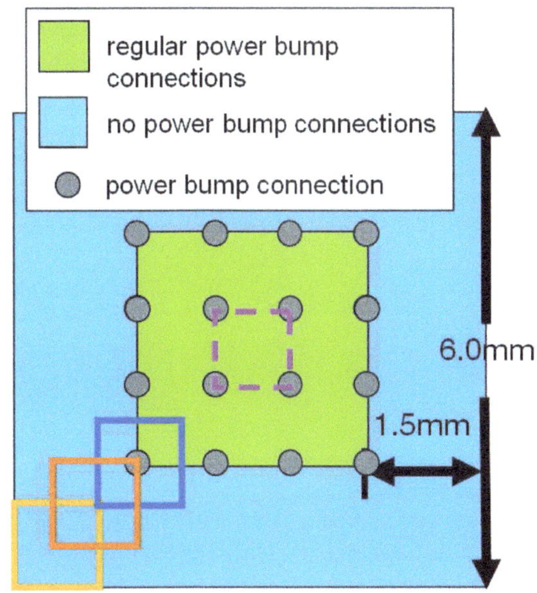

Fig. 4. Schematic of a SoC power grid, with regular power bump connections in the center and no package connections for power delivery in the outer (blue) region.

implementing distributed inductances into to the simulation setup the complexity increases significantly and increases the simulation time. Previous publications claimed that on-chip inductance has a significant impact on the design of power distribution grids (N. Srivastava, 2005). In (Mezhiba and Friedman, 2004) it was shown that the proportion of on-chip inductive noise increases with technology scaling. This paper addresses the question whether on-chip inductance has to be taken into account when analysing delay degradation due to power supply noise.

3 Simulation setup

The simulation setup considered in this analysis is similar to the setup in (Eireiner et al., 2006). As example we take a low power SoC in a 90 nm technology with an ARM926 core (Lueftner et al., 2006). The area of the entire SoC is roughly 6 mm×6 mm, the area of the ARM core is 1 mm×1 mm. The ARM core is operated at a maximum frequency f_{fast}=254 MHz and a slower frequency of f_{slow}=130 MHz. The frequency change from f_{slow} to f_{fast} is taken as exemplary di/dt noise event. During regular operation of the ARM core a peak current of $0.52A$ is assumed. The power grid consists of four metal layers, as depicted in Fig. 2, each orthogonal to its vertical neighbors, with contacts at the overlapping regions. A paired power grid structure (Mezhiba and Friedman, 2002) is used for the metal layers LB, M12B, and M22B to minimized inductive effects. The grid is modeled as a distributed R-L-C circuit built from a pi-model, as depicted in Fig. 3. The values for pi-model were choosen such that every instance of the pi-model represents a filament of the power grid 5 μm long.

As packaging technology a flipchip package is assumed. The bump connections are modeled as ideal inductances with values of $0.5nH$, $0.75nH$, or $1.0nH$ each. In Fig. 4 a schematic of a typical SoC power distribution grid is depicted. In the green region in the center of the SoC regular bump connections, depicted as grey circles, are available for power delivery. In the blue outer region no bump connections are available for power delivery due to I/O and ESD restrictions. In our analysis two different grid scenarios, which can occur in a SoC design, are analysed. The first is called "regular grid". It describes the case in which the on-chip power grid is connected to the package with regular bump connections, as depicted in Fig. 5. As second case, a so called "SoC grid" is analysed. In this scenario cores are placed at least partly in the outer region, where no bump connections can be used for power delivery, see Fig. 4. The coloured rectangles in Fig. 4 represent three exemplary positions of the ARM core, "lower left", "middle", and "center" for the yellow, orange, and blue rectangle, respectively. The dashed purple rectangle in Fig. 4 and Fig. 5 indicate the same area. For inductance extraction the tool FastHenry was used (Kamon et al., 1994). With FastHenry the self inductance and the mutual inductances of all parts of a current loop and the resulting effective inductance are determined. As shown in (Saint-Laurent and Swaminathan, 2004) and verified in (Eireiner et al., 2006), the cycle average of the supply voltage determines the delay of logic circuits. Therefore, the cycle average of the supply voltage, V_{supply}, is taken as key metric for delay degradations, where V_{supply} is defined as the difference between V_{DD} and V_{SS}, $V_{\text{supply}}=V_{DD}-V_{SS}$. As mentioned

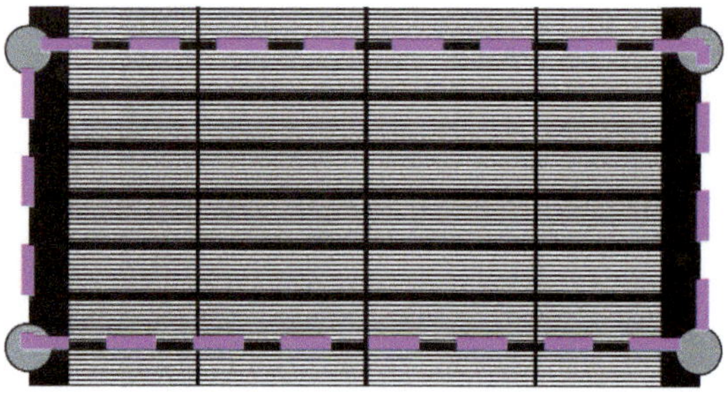

Fig. 5. Schematic of a on-chip grid with regular bump connections, top view of "regular grid".

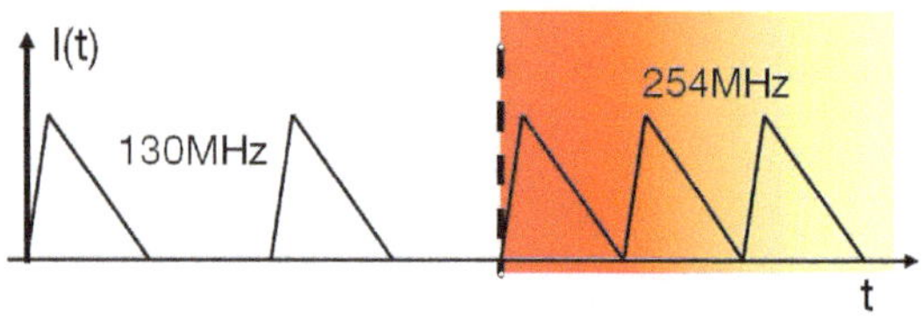

Fig. 6. Schematic of current profiles, emulating a frequency change from 130 MHz to 254 MHz, the colored area indicates the cycles of highest di/dt noise

before, a change of frequency from 130 MHz to 254 MHz was taken as di/dt noise event. The frequency change was emulated by applying current profiles as shown in Fig. 6 to the power grid. The colored area indicates the cycle in which the highest di/dt noise occurs. For simplicity an even current distribution over the entire core is assumed. In a post processing step the cycle and the position within the grid with the highest supply noise was determined.

4 Regular grid simulations

In the following section the simulation results for the "regular grid" will be discussed. To make a worst case estimation of the on-chip inductance, mutual inductances were omitted. This gives an upper bound for the inductance, since it assumes an infinitely wide current loop. In Fig. 7 the maximum transient drop of V_{supply} over the non switching cap, for a bump inductance $L_{\text{bump}}=0.5nH$, is displayed. The non switching capacitance is given by all gate and parasitic capacitances of, as well as the wiring connected to the non switching gates. The solid black lines are the simulation results for the two cases without on-chip inductance (L_0) and with the above mentioned worst case estimate of the on-chip inductance (L_{WC}). Their graphs are associated to the left y axis. On the right y axis the difference between the two black lines is plotted as dashed red line. Depending on the non switching capacitance, the difference is roughly between +100 mV and −10 mV, and therefore in the range of 10% of the absolute values. Therefore, on-chip inductance can have significant impact on the maximal voltage drop. This

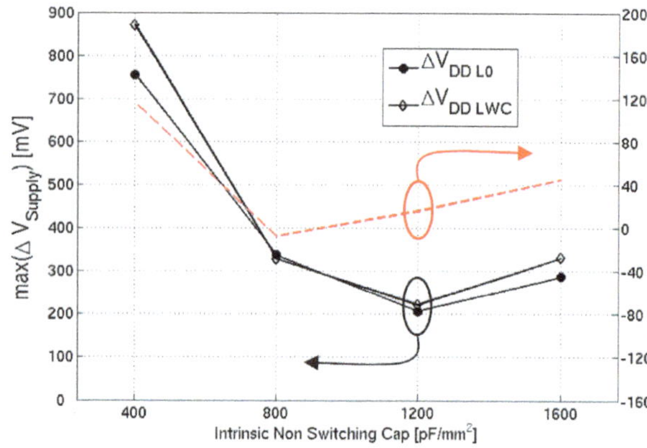

Fig. 7. Simulation results of maximum supply noise for varying intrinsic capacitance and resulting difference between simulations with worst case and zero on-chip inductance.

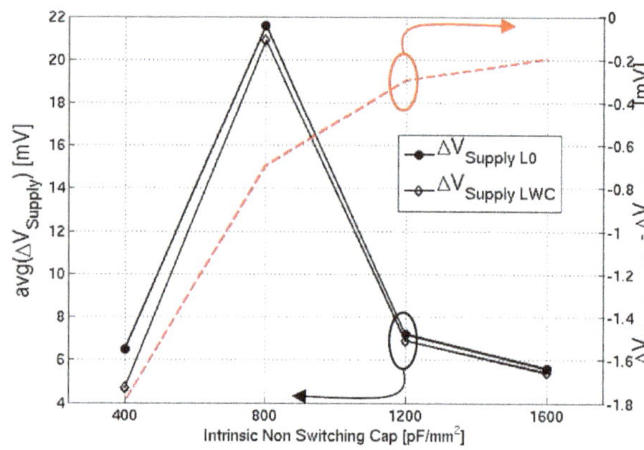

Fig. 8. Simulation results of cycle average of supply noise for varying intrinsic capacitance and resulting difference between simulations with worst case and zero on-chip inductance.

is in line with previous publications (Mezhiba and Friedman, 2004). Figure 8 displays, similar to Fig. 7, the maximum deviation of the cycle average of V_{supply}. Again, on the left y axis the simulation results for the two cases L_0 and L_{WC} are plotted. On the right axis the difference between the two cases is displayed. Comparing Fig. 8 and Fig. 7 yields two major differences. The first is that the absolute values are an order of magnitude smaller for the cycle average as they are for the maximum transient drop. Second, the difference between the two cases, indicated by the red dashed line, is much smaller for the cycle average not only in terms of absolute value but also for the relative deviation.

Therefore, even for a worst case estimation of on-chip inductance it has only a neglible effect on the cycle average of V_{supply} and therefore can be neglected in timing considerations. Figure 8 and Fig. 7 show the simulation results for $L_{\text{bump}}=0.5$ nH. For bump inductances of $L_{\text{bump}}=0.75$ nH and $L_{\text{bump}}=1.0$ nH similar results are obtained and are omit-

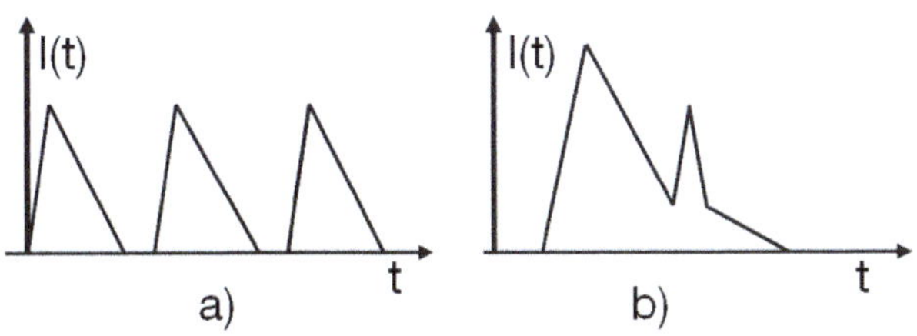

Fig. 9. More realistic current profiles **(a)** the current goes to zero by 90% of the cycle time **(b)** the current profile has a second peak caused by the falling clock edge.

ted here. As a next step, the impact of the current waveform was examined. Therefore, more realistic current profiles, as depicted in Fig. 9 were applied to the simulation setup. The modified current profiles resulted only into minimal changes of the simulation results.

5 SoC grid simulations

In this section a methodology for estimating the average effective inductance of the used ARM core is presented. Following this, simulation results of the SoC for the different inductance cases are presented and discussed.

5.1 Estimation of effective inductance

Initial simulations of the SoC grid showed a considerable deviation for the L_0 and the L_{WC} case even for the cycle average. Therefore, a more realistic estimate of the effective inductance was conducted. For this reason the wiring capacitance distribution of the ARM core, as depicted in Fig. 10, was used to estimate the wire length distribution of the core. The capacitance distribution in Fig. 10 is roughly a straight line on a double logarithmic scale. Hence, the distribution can be described as exponential function and therefore follows Rents Rule (Bakoglu, 1990), which makes it representative for most integrated circuit designs. This resulting wire length distribution was used to make conservative estimation for the effective inductance. To make the estimate as conservative as possible, these assumptions and simplifications were made: All wires with capacitances smaller than 4fF are omitted. With an estimated capacitance per micrometer of $C{=}150\frac{aF}{\mu m}$ this is equal to omitting all wires shorter than $26\mu m$. The chosen capacitance per micrometer is the smallest of all metal layers and leads to an overestimate of the wire length distribution, which in turn leads to an overestimate of the effective inductance. Additionally, for all nets a fanout of one, prohibiting branching and thus forcing maximum wire lengths, was assumed. Another simplification is that the load is assumed to be concentrated at the end of the wire, forcing maximal long current loops. Also a worst case orientation of all wires was assumed, again maximizing the possible current loop. The current distribution was again assumed to be

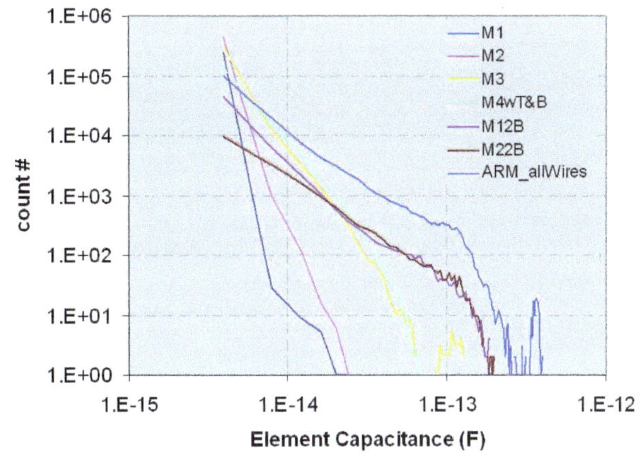

Fig. 10. Wiring capacitance distribution of the investigated ARM core. Capacitances smaller 4fF were omitted.

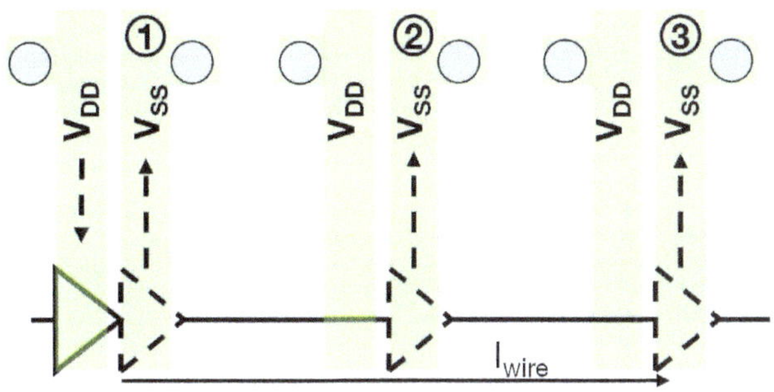

Fig. 11. Schematic for varying current return path with increasing wire length.

even, this might not be a worst case estimate, but was neccesarry for the feasibility of the study. With the resulting wire length distribution the probability for the different current return paths was calculated. In Fig. 11 a schematic is depicted, which shows a driving buffer, solid black triangular, and different load buffers, dashed black triangulars. Depending on the length of the connection, l_{wire}, from the driving to the load buffer the current return path can change from ① to ②, or ③. An inductance extraction for varying current return paths was conducted for all four metal layers used in the power grid wiring. The results are depicted in Fig. 12. The numbering for the current return paths was introduced since the pitches of the different metal layers vary over two orders of magnitude. The numbering was done according to Fig. 11. Figure 12 shows a steep increase of the inductance from the first to the second return path, followed by a saturation region where the inductance is almost independent on the current return path. This is due to the reason that only for the first current return path a strong inductive coupling between the V_{DD} and V_{SS} line exists, which decreases the effective inductance. Going to the second current return path increases the distance between the V_{DD} and V_{SS} line from

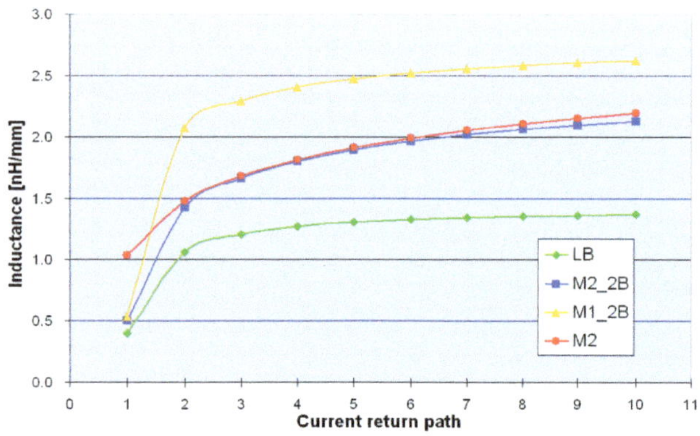

Fig. 12. Effective inductance for varying current return path. The x axis is the counted current return path starting at 1 for the direct return path, see Fig. 11.

Table 1. Probability of different current return paths and resulting effective inductance

		current return path				
		1	2	≥ 3	L_{eff}	L_{WC}
LB	pdf	99.2%	0.7%	0.1%		
	L $[\frac{nH}{mm}]$	0.4	1.2	1.4	0.41	1.91
M22B	pdf	82.6%	13.2%	4.2%		
	L $[\frac{nH}{mm}]$	0.5	1.4	2.1	0.68	2.1

a few hundreds of nanometers, spacing between neighboring lines, to tens or hundreds of micrometers, pitch of two pairs. The only metal layer where this effect does not exist is the lowest metal layer M2. This metal layer is used for the power wiring of the standard cells. Therefore, the minimal spacing of a V_{DD} and V_{SS} line is given by the standard cell height, which is in the range of a couple of micrometers. This leads to an increased minimum inductance. The results from Fig. 12 and the wire length distribution, derived from Fig. 10, are combined to a weighted mean which is then taken as effective inductance L_{eff}. In Table 1 the probability (pdf) of each current return path, the respective inductance of that return path, and the resulting effective inductance, as well as the worst case estimate (L_{WC}), which was used initially, are displayed for the metal layers LB and M22B. For the metal layer LB the effective inductance is $L_{\text{eff}}=0.41\frac{nH}{mm}$ which is almost identical to the inductance of the first current return path. For the M22B layer the effective inductance increases from the inductance of the first return path by 36% to $L_{\text{eff}}=0.68\frac{nH}{mm}$. Compared to the worst case estimate, L_{WC}, used earlier, the increase is still moderate. These inductance values are now used for the simulations in the SoC grid, the results are presented in the next section.

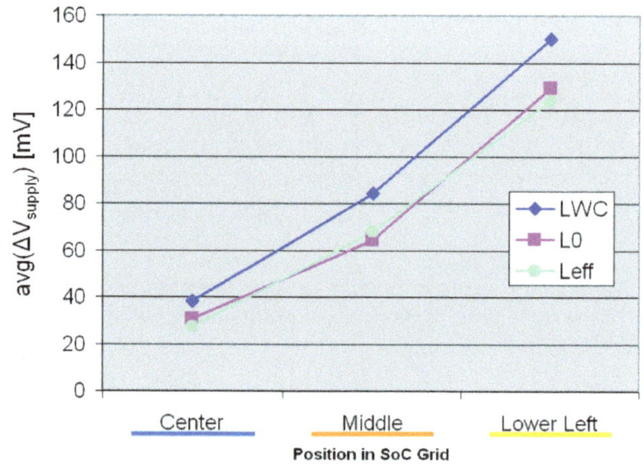

Fig. 13. Simulation results of the cycle average of supply voltage deviation, for varying core positions within the SoC grid and varying on-chip inductance values.

5.2 Simulation results

With the inductance values from Sect. 5.1 simulations in the SoC grid were performed. The simulation results are displayed in Fig. 13. On the x axis the possition within the SoC grid is plotted. The colored underlines relate to the colors used in Fig. 4. On the y axis the cycle average of the maximum supply voltage deviation is displayed for the three cases of no on-chip inductance (L_0), worst case estimate of on-chip inductance (L_{WC}), and effective inductance estimate of on-chip inductance (L_{eff}). It can be seen from Fig. 13 that for all three cases supply noise increase. This is due to the increased IR-Drop coming from the increased distance between core and the nearest power bump. The simulation results also show that an increase of up to 20 mV in cycle average power noise can be observed for L_{WC} compared to L_0. However, for the conservative estimate L_{eff} the cycle average of power noise is very close to L_0. The deviation is smaller than $\pm 5mV$ in absolute values and well below $\pm 5\%$ normalized to the total noise budget. Since the on-chip inductance also in the SoC grid has a small impact on the cycle average of the power supply noise, it can also be neglected during timing considerations in SoC enviroments.

6 Conclusions

In this paper the effect of on-chip inductance on power supply integrity has been examined. Its effect on the maximum transient voltage drop as well as the impact on the cycle average of the supply voltage was presented. In accordance with previous publications, the impact of on-chip inductance on the maximum transient voltage drop was shown to be considerable. Therefore, on-chip inductance has to be taken into account during reliability considerations and on-chip signaling, for which the maximum value is of concern. However, on-chip inductance was shown to have neglible effect on the cycle average of the supply voltage. This holds true for a grid with regular bump connections as well as for a grid scenario where there are no regular power bump connections. Therefore, on-chip inductance can be neglected during timing considerations of logic circuits for which the cycle average of the supply voltage and not the maximum voltage drop is the key metric.

References

Bakoglu, H. B.: Circuits, Interconnections, and Packaging for VLSI, Addison-Wesley Publication Company, 416–419, 1990.

Eireiner, M., Henzler, S., Missal, T., Berthold, J., and Schmitt-Landsiedel, D.: Power Supply Network Design: A Case Study Driven Approach, Adv. Radio Sci., URSI Kleinheubacher Berichte, 2006.

Kamon, M., Tsuk, M., and White, J.: FASTHENRY: a multipole-accelerated 3-D inductance extraction program, IEEE Transactions on Microwave Theory and Techniques, 1750–1758, 1994.

Larsson, P.: Power Supply Noise in Future IC's: A Crystal Ball Reading, IEEE Custom Integrated Circuits Conference, 467–474, 1999.

Lueftner, T., Berthold, J., Pacha, C., Georgakos, G., Sauzon, G., Hoemke, O., Beshenar, J., Mahrla, P., Just, K., Hober, P., Henzler, S., Schmitt-Landsiedel, D., Yakovleff, A., Klein, A., Knight, R., Acharya, P., Mabrouki, H., Juhoor, G., and Sauer, M.: A 90 nm CMOS Low-Power GSM/EDGE Multimedia-Enhanced Baseband Processor with 380 MHz ARM9 and Mixed-Signal Extensions, 2006 ISSCC Digest of Digital Papers, 252–253, 2006.

Mezhiba, A. and Friedman, E.: Inductive Properties of High-Performance Power Distribution Grids, IEEE Transactions on Very Large Scale Integraton (VLSI) Systems, 2002.

Mezhiba, A. V. and Friedman, E. G.: Scaling Trends of On-Chip Power Distribution Noise, IEEE Transactions on Very Large Scale Integration (VLSI) Systems, 12, 386–394, 2004.

N. Srivastava, X. Qi, K. B.: Impact of On-Chip Inductance on Power Distribution Network Design for Nanometer Scale Integrated Circuits, Proceedings of the Sixth International Symposium on Quality Electronic Design, 2005.

Nassif, S. R. and Fakhouri, O.: Technology Trends in Power-Grid-Induced Noise, in: Proceedings of the 2002 international workshop on System-level interconnect prediction, 55–59, ACM Press, 2002.

Saint-Laurent, M. and Swaminathan, M.: Impact of Power-Supply Noise on Timing in High-Frequency Microprocessors, IEEE Transactions on Advanced Packaging, 27, 135–144, 2004.

Countermeasures against NBTI degradation on 6T-SRAM cells

E. Glocker, D. Schmitt-Landsiedel, and S. Drapatz

Lehrstuhl für Technische Elektronik, Technische Universität München, Theresienstraße 90, 80333 München, Germany

Abstract. In current process technologies, NBTI (negative bias temperature instability) has the most severe aging effect on static random access memory (SRAM) cells. This degradation effect causes loss of stability. In this paper countermeasures against this hazard are presented and quantified via simulations in 90 nm process technologies by the established metrics SNM_{read}, SNM_{hold}, I_{read} and Write Level. With regard to simulation results and practicability best candidates are chosen and, dependent on individual preferences at memory cell design, the best countermeasure in each case is recommended.

1 Motivation

The ongoing miniaturization in modern CMOS technologies leads to a more and more challenging SRAM design: during read operation the cell must not flip to prevent data loss. During write operation the memory cell must flip to write new data in the cell. Thus there is only a certain area, where both a reliable read and write operation is possible. On top of variations, degradations are making this area smaller.

To illustrate the effect of variations and degradations on SRAM cells, a Monte Carlo analysis was performed (Fig. 1).

In this analysis 10^5 SNM_{read} simulations (Seevinck et al., 1987) are performed for $0.4\,V < V_{DD} < 1.2\,V$: the transistors are exposed to variations (based on measured σ values), then $-50\,mV$ NBTI degradation is added to one pullup transistor. This represents a realistic worst case, as the expected end-of-life shift after 10 years at 1.32 V and 125 °C is -31 mV. Counting the number of not functional SRAM cells ($SNM_{read} < 0$) the yield (fraction of functional memory cells) is determined. For the 256 Bit SRAM array it is assumed that the failure of one cell leads to the failure of the whole array. When the cells are only influenced by variations, the yield of the 256 Bit SRAM array decreases rapidly below $V_{D,min} = 0.75\,V$. To design functional SRAM arrays despite variations, different assist techniques have been developed. In contrast to variations, degradations occur with increasing magnitude during operating time. The $V_{DD,min}$ value, where the yield of the 256 Bit array starts to decrease, rises from 0.75 V to $V_{D,min} = 0.8\,V$ when the SRAM array is exposed to variations plus $-50\,mV$ NBTI degradation. So $V_{D,min}$ is increased by approx. ΔV_{th}. $V_{D,min}$ would rise more if additional other degradations and/or a higher NBTI degradation were taken into account. As a result, countermeasures against NBTI are necessary to guarantee long-time operational SRAM cells. One first countermeasure is e.g. the Guard Band (Sect. 3.4), where V_{DD} is limited to a value above $V_{D,min}$ to assure that the cell is not operated with a V_{DD} with a high failure probability after degradation.

After describing the consequences of the NBTI degradation on the SRAM cell in Sect. 2, countermeasures against NBTI degradation are presented (Sect. 3) and the best candidates in terms of simulation results and practicability are chosen and compared to each other (Sect. 4). Depending on the individual preferences at memory cell design, the best countermeasure in each case is recommended.

2 Consequences of NBTI degradation on 6T-SRAM memory cells

BTI (bias temperature instability) degradation distinguishes between PBTI (positive BTI) and NBTI (negative BTI) degradation. NBTI affects pMOS transistors with negativ potential on the gate referred to the potential on source and drain (Fig. 2). NBTI weakens pMOS transistor: Positive charges arise in the gate oxide and the absolute threshold voltage $|V_{th}|$ rises. It is harder to turn the transistor on. NBTI can be modeled with a rise of $|V_{th}|$ by ΔV_{th}. PBTI influences nMOS transistors with high-k gate oxide, thus in process technologies under 65 nm. So NBTI has the most damaging effect for current technologies (Drapatz et al., 2009a).

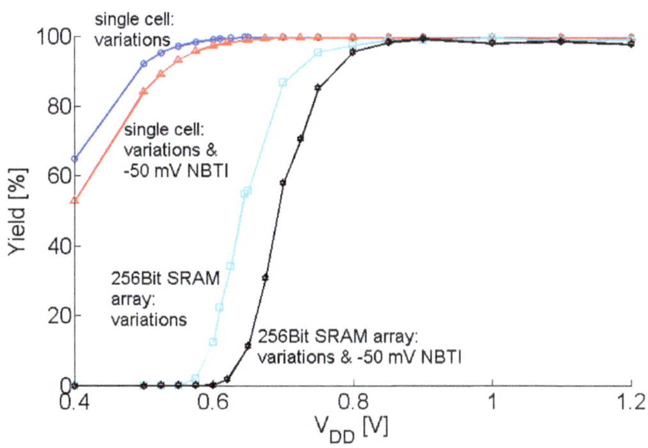

Fig. 1. Yield considering that the SRAM memory cell is exposed to variations and variations plus NBTI degradation of −50 mV, respectively for a single cell and a 256 Bit SRAM array. (based on 10^5 simulations). NBTI degradation increases $V_{D,min}$ (lowest voltage before the yield decreases rapidly) by approx. 50 mV.

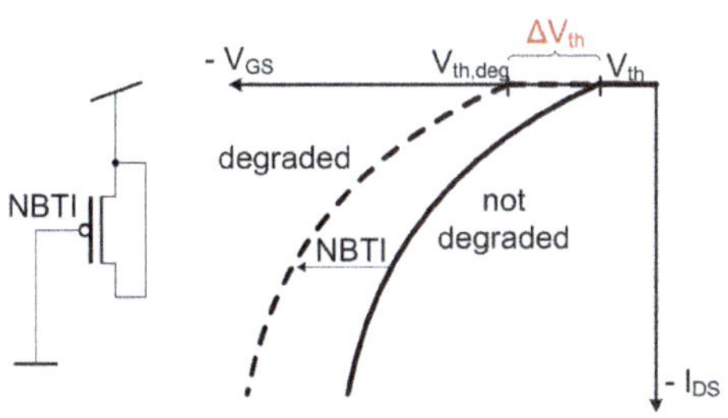

Fig. 2. Left: NBTI stress condition for single pMOS transistor; right: influence of NBTI: higher absolute threshold voltage. The degraded transistor is weaker.

By means of the NBTI-Calculator Formula, adapted to single transistor degradation measurements, the actual existing ΔV_{th} can be determined with a model based on Yan et al. (2009), Huard et al. (2010) with appropiate parameters for the used process technology.

$$|\Delta V_{th}| = A \cdot t^n \cdot V_{DD}^{V_0} \cdot \exp(-\frac{E_a}{kT}) \tag{1}$$

ΔV_{th} rises exponentially with increasing T, exponentially with increasing V_{DD} and with a logarithmic dependence on stress time t. For the used 90 nm pMOS transistor model the ΔV_{th} for the nominal point ($V_{DD} = V_{WL} = V_{BL(B)} = V_{DD,core} = 1.2$ V and $T = 25$ °C) is $\Delta V_{th} \approx -4$ mV after five years and $\Delta V_{th} \approx -5$ mV after ten years. For the realisic worst case point ($V_{DD} = V_{WL} = V_{BL(B)} = V_{DD,core} = 1.2\,\text{V} + 10\% = 1.32$ V and $T = 125$ °C) $\Delta V_{th} \approx -26$ mV after five years and $\Delta V_{th} \approx -31$ mV after ten years. ΔV_{th} will increase for future process technologies. $\Delta V_{th} = -100$ mV is chosen as a worst case value for this technology and the SPICE simulations are performed for ΔV_{th} between −100 mV and 0 mV.

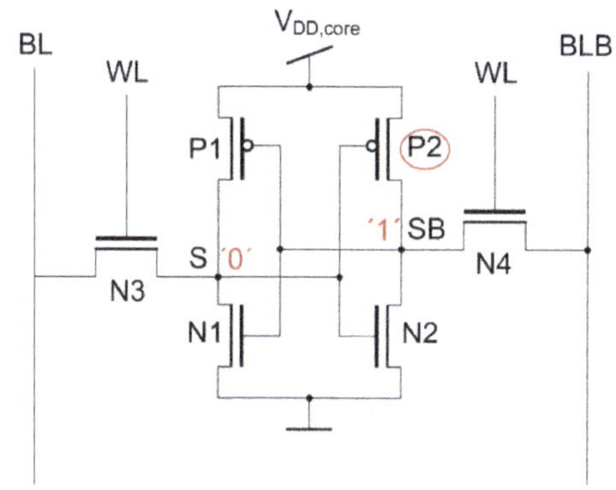

Fig. 3. NBTI stress condition in hold and read operation (the cell is in the "0" state (S: "0", SB: "1")) is only fulfilled for pMOS P2. In write operation the position of "1" and "0" switch, so P1 fullfills the stress condition.

Table 1. Metrics in the nominal point ($V_{DD} = 1.2$ V, $T = 25$ °C) for the non-degraded and the degraded cell (worst case: $\Delta V_{th} = -100$ mV): The most impact is on read stability.

	non-degraded cell $\Delta V_{th} = 0$ mV	degraded cell $\Delta V_{th} = -100$ mV
SNM_{read}	0.117 V	−0.016 V (−13.7%)
SNM_{hold}	0.388 V	−0.026 V (−6.7%)
I_{read}	6.598×10^{-5} A	-7×10^{-9} A (−0.01%)
Write Level	0.656 V	+0.039 V (+5.9%)

In this paper the established metrics SNM_{read}, SNM_{hold}, I_{read} and Write Level are used: SNM describes how much additional noise voltage is necessary to flip the cell. I_{read} is a measure for the speed of the read operation. Write Level describes the voltage necessary to flip the cell (Seevinck et al., 1987; Drapatz et al., 2009b).

In the hold and read operation, the SRAM cell is assumed to be in the "0" state (S: "0", SB: "1") (Fig. 3). So the NBTI stress condition is fulfilled for the pMOS transistor P2. In the write operation the positions of "1" and "0" switch and the stress condition is fulfilled for P1 after that.

Table 1 shows the metric values in the nominal point. The influence of NBTI on the writability (Write Level increases) and the read speed (decrease of I_{read} is small) is unproblematic. So the presented simulation results concentrate on the reduction of the stability.

3 NBTI countermeasures

3.1 Limited temperature

A decreasing temperature leads to improvement of the hold and read stability (0 °C, without considering ΔV_{th}: +1.8%, respectively +6.8%) and to lower ΔV_{th}. For $T = -100$ °C the

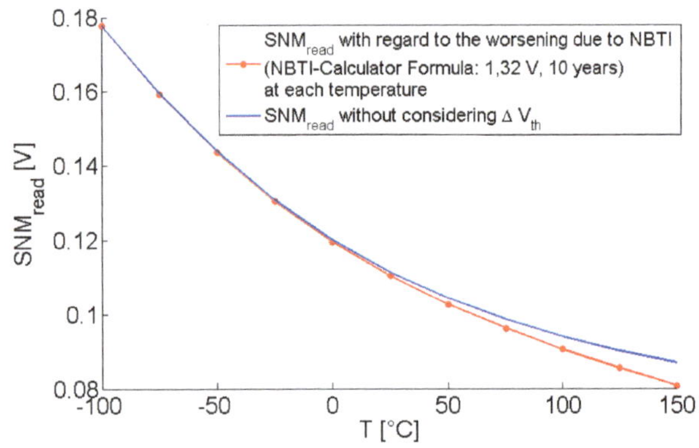

Fig. 4. Read stability over temperature considering both the simulation results at a constant ΔV_{th} and the potential rise of ΔV_{th} with V_{DD} (NBTI Calculator Formula, worst-case point). The lower the temperature, the higher the stability. So no optimal temperature limit can be determined.

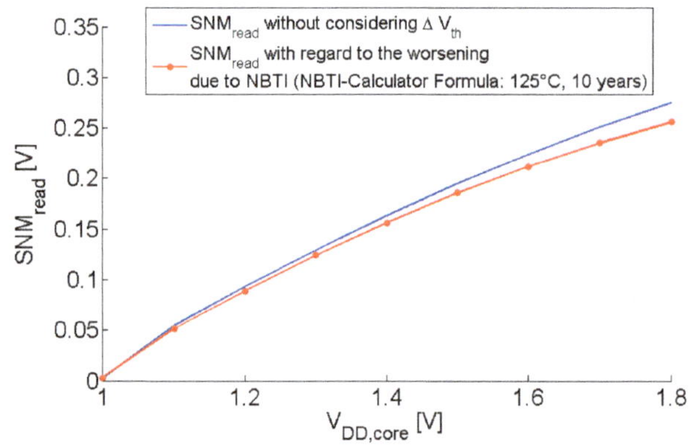

Fig. 5. Read stability over core voltage considering both the simulation results at a constant ΔV_{th} (= 0 mV) and the exponentially rise of ΔV_{th} with V_{DD} (NBTI Calculator Formula, worst-case point). The higher the core voltage, the higher the stability. So no optimal core voltage can be determined.

Table 2. Metrics for various temperatures for the nominal point ($V_{DD} = 1.2$ V, 1 year). Percent values in comparison to the stability at 0 °C ($SNM_{read,0°C}$=0.125 V, $SNM_{hold,0°C} = 0.394$ V). The most impact is on the read stability.

	50 °C	100 °	125 °C
SNM_{read}	0.109 V (−13%)	0.097 V (−23%)	0.091 V (−27%)
SNM_{hold}	0.381 V (−3%)	0.369 V (−7%)	0.363 V (−8%)

degradation formula would yield $\Delta V_{th} \simeq 0$ mV, so the pMOS transistors would not be degraded by NBTI. This is however far away from the operating conditions, where the measurements for fitting of the formula were performed, and an operating temperature $T = -100$ °C is not practicable anyhow. But in general it has to be considered that at higher temperatures the cells are less stable. So one can limit the temperature, although this narrows down the SRAM operating range (compare Table 2 and Fig. 4). Additionally, the NBTI degradation is getting worse with higher temperature. This can be seen in Fig. 4, when a second plot was added to the nominal plot. It considers an NBTI-related ΔV_{th} (worst case, calculated for 10 years at 1.32 V) that occurs due to the increased temperature. For this voltage and time the decrease in stability by raised T is so large that the NBTI degradation, also increased with T, does not affect the result much. It is not possible to determine an optimal temperature limit (Fig. 4): The lower T, the higher is the stability. So one must choose a suitable temperature limit for each particular case (e.g. Table 2, values given for the nominal point).

Table 3. Metrics for various core voltages for the nominal point (25 °C, 1 year). Percent values in comparison to $V_{DD,core} = 1.2$ V. The most improvement is on read stability.

	1.3 V	1.4 V	1.5 V
SNM_{read}	0.15 V (+25%)	0.18 V (+50%)	0.21 V (+75%)
SNM_{hold}	0.41 V (+5%)	0.42 (+8%)	0.44 (+13%)

3.2 Core boosting

The stability is improving with increasing core voltage $V_{DD,core}$: For e.g. 1.5 V instead of 1.2 V core voltage, the read stability can be improved by 75% (compare Table 3 and Fig. 5). On the other hand, it is important to note that the NBTI degradation is getting worse with higher $V_{DD,core}$. This is shown in Fig. 5, where again a second plot was added to the nominal plot. It considers an NBTI-related ΔV_{th} (worst case, calculated for 10 years at $T = 125$ °C) that occurs due to the increased $V_{DD,core}$. For this temperature and time the increased stability by raised $V_{DD,core}$ is so large that the also increased NBTI degradation does not affect the result much. It is not possible to determine an optimal $V_{DD,core}$ (Fig. 5): The higher $V_{DD,core}$, the higher the stability. But higher $V_{DD,core}$ is also associated with higher power dissipation and greater leakage current. So one must choose a suitable $V_{DD,core}$ for each particular case (e.g. from Table 3, values given for the nominal point).

To implement the Core Boosting, an additional voltage is necessary to supply the SRAM memory cell.

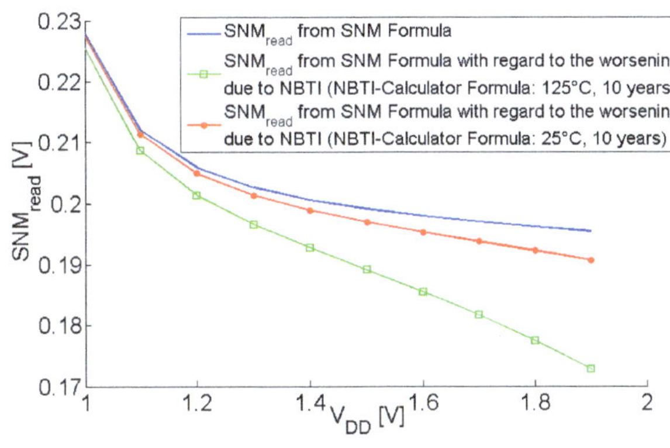

Fig. 6. Read stability over V_{DD} considering both the simulation results at a constant ΔV_{th} and the potential rise of ΔV_{th} with V_{DD} (NBTI Calculator Formula). The larger V_{DD}, the smaller the read stability. So no optimal V_{DD} can be determined.

3.3 Burn-In

Circuits that fail after short operating time are typically affected and weakened by variations. To prevent failures of those circuits in the field, Burn-In is typically used to make those circuits fail before they go to the customer. This is achieved through applying higher supply voltage and temperature for a defined period. Burn-In can be adopted as a NBTI countermeasure in the sense of pre-aging: If a specific ΔV_{th} can be achieved directly after production, ΔV_{th} increases only a little bit over the operating time, because ΔV_{th} rises with a logarithmic dependence over time. ΔV_{th} is approx. 5 mV after operating the cell 10 years at $V_{DD} = 1.2$ V and $T = 25$ °C. To achieve $\Delta V_{th} \approx -5$ mV during assembly, a Burn-In step at e.g. 2 V and 175 °C for 5 s would be necessary. The SRAM cell is centered without the Burn-In, i.e. in matters of stability and writability the cell has the best possible performance. After the Burn-In, the cell is not centered anymore. To achieve both, best possible performance and the use of Burn-In, the cell has to be adapted. For the example above, the width of the pMOS transistors need to be increased from 120 nm to 125 nm (required area rises by approx. 1%). Now the enhanced SRAM cell has the same performance at $\Delta V_{th} \approx -5$ mV (because of Burn-In) as the non-enhanced cell at $\Delta V_{th} = 0$ mV. The exact Burn-In and Enhancement parameters must be chosen for each particual case in subject to the desired accuracy.

3.4 Guard band

Below $V_{D,min}$ there is a drastic yield drop, i.e. safe operation is not possible (Fig. 1). So the minimal V_{DD} is limited to a value above $V_{D,min}$. $V_{D,min}$ depends on the existing degradation and variation and on the size of the SRAM array. As a result the limitation of V_{DD} narrows down the operating range. This is detrimental, especially for the hold operation, because there V_{DD} is typically lowered to reduce leakage current and power consumption. Nevertheless, the Guard Band is the up-to-date countermeasure in industry for all kinds of degradation, because it is easy to implement. An optimal V_{DD} limit must maximize the operating range and ensure good performance. To measure the performance the results of the SNM Formula (Seevinck et al., 1987) at constant ΔV_{th} and the exponentially rise of ΔV_{th} with V_{DD} were taken into account (Fig. 6): The greater V_{DD}, the smaller the read stability, where the slope increases below $V_{DD} = 1.2$ V. When V_{DD} is lowered from 1.2 V to 1 V, the read stability improves by approx. 11%), but read speed, hold stability and writability weaken. In summary an optimal V_{DD} border must be determined for each particular case.

Table 4. Metrics for various wordline voltages for the nominal point ($T = 25$ °C). Percent values in comparison to $V_{WL} = 1.2$ V. The read stability is improved, but read speed and writability decrease.

	0.7 V	0.9 V	1 V
SNM$_{read}$	0.286 V (+144%)	0.218 V (+86%)	0.184 V (+57%)
I_{read}	2.4×10^{-5} A (−64%)	4.1×10^{-5} A (−38%)	4.97×10^{-5} A (−25%)
Write Level	0.148 V (−77%)	0.35 V (−47%)	0.447 V (−32%)

3.5 WL Boosting

WL "Boosting" actually means a lower WL voltage to improve read stability. For e.g. $V_{WL} = 0.7$ V instead of 1.2 V the read stability can be improved by 144% (the hold stability remains unchanged). On the other hand, read speed and writability worsen by 64% and 77%. (Table 4). It is not possible to determine an optimal V_{WL}, because there is an approx. linear relation between each metric and V_{WL}. So one must choose a suitable V_{WL} for each particular case (e.g. from Table 4, values given for the nominal point). It is also possible to implement different WL voltages during read and write, but at the cost of increased complexity of the peripheral circuits. In any case, an additional voltage is necessary to supply the WL driver circuits.

3.6 Symmetric degradation

The NBTI stress condition is fulfilled only for the pMOS of the "1" side of the cell. Long hold of this "1" lets the cell become assymmetric: P2 is degraded, while P1 is new. The intension of the symmetric degradation is to avoid this asymmetry by forcing a symmetric degradation of the cell. For this, the save state is re-programmed after a certain time to ensure that both pMOSFETs experience the same ΔV_{th}. This countermeasure is not expedient, because due to the logarithmic dependence on stress time the difference between symmetric and asymmetric cells is small. But considering a NBTI degradation with recovery characteristic (Drapatz et al., 2010), that is not considered in this paper, this countermeasure potentially may lead to a higher improvement.

3.7 Lower precharge level

In read operation the node of the "0" side rises. As a consequence, it is easier to flip the cell. As a countermeasure one can lower the precharge level of the "0" side bitline. But in practice it is not possible to just lower the precharge level of the "0" side, because the position of "0" is unknown before the read operation. The only possible implementation is the lowering of the overall precharge level. However, a lower precharge level of the "1" side leads to an easier flipping of the SRAM cell during read. If the overall precharge level is lowered, the weakening of the "1" side node voltage is greater than the improvement of the "0" side node voltage: At $\Delta V_{th} = -100\,\mathrm{mV}$ and $V_{BL,precharge} = 0.7\,\mathrm{V}$ the node voltage of the "0" side increases from 0 V to 0.202 V and at $V_{BL,precharge} = 1.2\,\mathrm{V}$ the voltage increases by additional 0.004 V to 0.206 V. At $\Delta V_{th} = -100\,\mathrm{mV}$ and $V_{BL,precharge} = 0.7\,\mathrm{V}$ the node voltage of the "1" side decreases from 1.2 V to 1.02 V, whereas at $V_{BL,precharge} = 1.2\,\mathrm{V}$ the voltage decrease is only 1.198 V, i.e. 0.178 V less than at $V_{BL,precharge} = 0.7\,\mathrm{V}$. The simulations of the read stability confirm this result: SNM_{read} decreases by approx. 25% at $V_{BL,precharge} = 0.7\,\mathrm{V}$. The read stability is not improved, so lowering the precharge level is not expedient.

3.8 Alternative SRAM design: 8T, 6T with 8T size ratio

In read operation the 6T-SRAM cell experiences a read stability problem, because the node of the "0" side sees strong disturb: If the node voltage of the "0" side reaches V_{th} of the pulldown transistor of the "1" side, the node voltage of the "1" side is lowered. As a consequence the "0" of the "0" side can be overwritten by a "1".

The 8T-SRAM cell (Fig. 7) ensures a read-disturb-free operation, because this read stability problem does not exist. Data output and data retention are separated from each other via separate read and write signal lines. So the 8T cell is as stable in the read operation as the 6T cell in the hold operation. The required area for the 8T-SRAM cell is approx. 30% bigger than the area for the 6T design due to 2 additional transistors (Zhang, unpublished data; Bauer, 2009). During write operation, the 8T-SRAM array is disturbed by a parasitic read operation. The voltages of the wordline and bitlines of the 6 core transistors of the 8T cell for the half-selected columns in write operation are identical to the voltages of the 6T cell in read operation (Fig. 7). As a result the half-selected columns experience the same loss of stability in the write operation as the 6T cell in read operation. To prevent this loss of stability and the potential data loss, an array architecture without a bitline multiplexer is needed, where all cells connected to the

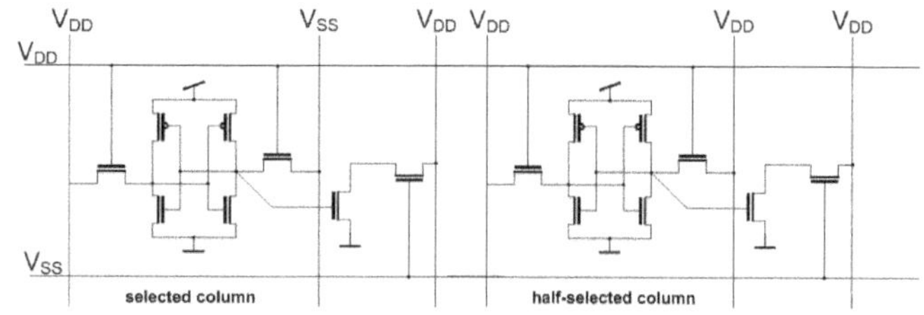

Fig. 7. 8T-SRAM memory cell write operation for the selected (left) and the half-selected (right) column. The half-selected column experiences the same loss of stability as the 6T cell in read operation, due to a parasitic read operation.

same wordline are written at the same time. This mux-free array architecture leads to further increase of the required area (Bauer, 2009).

Another countermeasure is an enhanced 6T-SRAM cell. If e.g. a 6T cell with the same area requirements as the 8T cell is chosen, the read and hold stability increase by approx. 15% and 3%. The increase of the stability is smaller for the enhanced 6T cell than for the 8T cell. So in comparison to the 8T-SRAM design the enhanced 6T cell is not expedient.

3.9 Body biasing

The absolute threshold voltage $|V_{th}|$ rises, because of the NBTI degradation. With decreasing bulk-substrate voltage $V_{BS} < 0$ the pMOS threshold voltage V_{th} decreases (von Arnim et al., 2005). So with $V_{BS} < 0$ the rise of $|V_{th}|$ can be reversed: For $V_{BS} = 0.7\,\mathrm{V}$ and $\Delta V_{th} = -50\,\mathrm{mV}$ the SRAM memory cell has the same metric values as for $V_{BS} = 1.2\,\mathrm{V}$ and $\Delta V_{th} = 0\,\mathrm{mV}$. For greater ΔV_{th} the stability and read speed is increased and the writability is decreased. The choice of a particular body biasing voltage must be fitted to the occuring ΔV_{th}. Besides it is possible to rise V_{BS} only in read and/or hold operation. Although this leads to a varrying V_{BS}. So the great capacitance of the n-well must be transhipped depending on the operating state.

4 The best NBTI countermeasures

With regard to our simulation results and practicability, Core and WL Boosting, Burn-In, 8T-SRAM design and Guard Band are chosen as the best NBTI countermeasures. They are compiled in Table 5 and compared to each other in the following. The 8T-SRAM design is the most useful NBTI countermeasure: The stability problem of the 6T-SRAM cell does not occur anymore, but 2 additional transistors (required area is approx. 30% greater) are needed. The Burn-In ensures that ΔV_{th} only rises a little during operating time by an additional assembly step at higher T and V_{DD}. So the stability is approximately constant and the SRAM memory cell stays functional over operating time. The required area rises by approx. 1%, because the pMOS transistors must be

Table 5. Summary of the best NBTI countermeasures, numerical results for a 90 nm technology.

Countermeasure	Result	Positive aspects	Negative aspects
Core Boosting ($V_{DD,core} = 1.5$ V, nominal point)	SNM$_{read}$=0.21 V (+75%) SNM$_{hold}$=0.43 V (+13%) $I_{read} = 6.99 \times 10^{-5}$ A (+6%) Write Level = 0.55 V (+17%) determine optimal $V_{DD,core}$ for each particular case,	increase of stability, easy to implement	decrease of writability, additional voltage to supply cell, greater $V_{DD,core}$ ⇒ increase of power consumption/leakage current
WL Boosting ($V_{WL} = 0.7$ V, nominal point)	SNM$_{read}$=0.29 V (+144%) SNM$_{hold}$ unchanged $I_{read} = 2.4 \times 10^{-5}$ A (−64%) Write Level = 0.15 V (−77%) determine optimal V_{WL} for each particular case,	increase of read stability, easy to implement, decrease of V_{WL} ⇒ decrease of power consumption/ leakage current	decrease of writability, decrease of read speed, additional voltage to supply cell
Burn-In	pMOS widened by 5 nm ⇒ size ratio rises by approx. 1%, choose Burn-In for each particular case e.g. for operation by $V_{DD} = 1.2$ V $T = 25$ °C: 5 s at 2 V and 175 °C	Over operating time: ΔV_{th} decreases only little ⇒ stability remains more or less constant	enhance of the cell ⇒ size ratio rises by approx. 1%, additional manufacturing step
Guard Band	determine optimal V_{DD} for each particular case e.g. $\Delta V_{th} = -50$ mV, 256 Bit SRAM array: $V_{D,min} \approx 0.8$ V	easy to implement (just border V_{DD}), up-to-date countermeasure	greater V_{DD} ⇒ increase of power, consumption/leakage current, narrows down operating range
8T SRAM Design	size ratio rises by approx. 30%, compared to 6T, read-disturb-free operation	separate read- & write signal lines ⇒Separation of data output and -retention	mux-free array architecture, increase of size-ratio by approx. 30%

widened. For implementation of the Core and WL Boosting an additional voltage (plus additional periphery and wire connection) to supply the cell is necessary in each case. Both countermeasures increase the read stability. The Core Boosting also increases the hold stability. The WL Boosting deteriorates the writability and read speed, while the Core Boosting leads to higher power consumption and greater leakage current. An optimal $V_{DD,core}$ or V_{WL} respectively, can not be chosen because of the approximately linear dependence of the metrics. So a suitable voltage must be chosen for each particular case. It is not recommended to implement the Core and the WL Boosting at the same time, because three voltages to supply the array would be needed. The Guard Band is the easiest countermeasure in terms of implementation: Only the miminmal V_{DD} must be guarded to be above $V_{DD,min}$. This narrows down the operating range of the SRAM and increases power consumption and leakage current. A suitable minimal V_{DD} must be chosen for each particular case.

5 Summary and conclusions

In this paper countermeasures against NBTI degradation that mostly impacts the stability of the cell were presented. With regard to simulations results and practicability the best candidates were chosen and compared to each other. Because it is not possible to define an optimal countermeasure, the best countermeasure, depending on the individual preferences in memory design, is recommended as follows:

1. *Area is not the first priority:* 8T SRAM Design
2. *Additional expenses during assembly is okay:* Burn-In
3. *Additional expenses for additional periphery and wire connection is acceptable:* WL or Core Boosting
4. *No design-change of the cell is preferred:* Guard Band

References

Bauer, F.: A Design Space Comparison of 6T and 8T SRAM Core-Cells, in: 18th International Workshop PATMOS 2008, 10–12 September, Revised Selected Papers, Lecture Notes in Computer Science, pp. 116–125, 2009.

Drapatz, S., Fischer, T., Hofmann, K., Amirante, E., Huber, P., Ostermayr, M., Georgakos, G., and Schmitt-Landsiedel, D.: Fast stability analysis of large-scale SRAM arrays and the impact of NBTI degradation, Solid-State Circuits Conference, ESSCIRC 2009, 35th European, pp. 92–95, 2009a.

Drapatz, S., Georgakos, G., and Schmitt-Landsiedel, D.: Impact of negative and positive bias temperature stress on 6T-SRAM cells, Adv. Radio Sci., 7, 191–196, doi:10.5194/ars-7-191-2009, 2009b.

Drapatz, S., Hofmann, K., Georgakos, G., and Schmitt-Landsiedel, D.: Impact of fast-recovering NBTI degradation on stability of large-scale SRAM arrays, Solid-State Device Research Conference, ESSDERC 2010. 40th European, 2010.

Huard, V., Chevallier, R., Parthasarathy, C., Mishra, A., Ruiz-Amador, N., Persin, F., Robert, V., Chimeno, A., Pion, E., Planes, N., Ney, D., Cacho, F., Kapoor, N., Kulshrestha, V., Chopra, S., and Vialle, N.: Managing SRAM reliability from bitcell to library level, in IRPS, IEEE International, 2010.

Seevinck, E., List, F. J., and Lohstroh, J.: Static Noise Margin Analysis of MOS SRAM Cells, IEEE Journal of solid-state circuits, SC-22(5), 748–754, 1987.

von Arnim, K., Borinski, E., Seegebrecht, P., Fiedler, H., Brederlow, R., Thewes, R., Berthold, J., and Pacha, C.: Efficiency of Body Biasing in 90-nm CMOS for Low-Power Digital Circuits, IEEE Journal of Solid-State Circuits, 40(7), 1549–1556 , 2005.

Baoguang Yan, Qingguo Fan, Bernstein, J. B., Jin Qin, and Jun Dai: Reliability simulation and circuit-failure analy- sis in analog and mixed-signal applications, Dev. and Mat. Rel., IEEE Transactions on, 9(3), 339–347 , 2009.

Zhang, M.: Performance Comparison of SRAM Cells Implemented in 6, 7 and 8-Transistor Cell Topologies, Electrical and Computer Engineering, University of California, Davis, unpublished data.

A low power clock generator with adaptive inter-phase charge balancing for variability compensation in 40-nm CMOS

U. Schulze, M. Broich, O. Weiss, and T. G. Noll

Chair of Electrical Engineering and Computer Systems, RWTH Aachen University, Germany

Abstract. Power dissipation besides chip area is still one main optimization issue in high performance CMOS design. Regarding high throughput building blocks for digital signal processing architectures which are optimized down to the physical level a complementary two-phase clocking scheme (CTPC) is often advantageous concerning ATE-efficiency. The clock system dissipates a significant part of overall power up to more than 50% in some applications.

One efficient power saving strategy for CTPC signal generation is the charge balancing technique. To achieve high efficiency with this approach a careful optimization of timing relations within the control is inevitable.

However, as in modern CMOS processes device variations increase, timing relations between sensitive control signals can be affected seriously. In order to compensate for the influence of global and local variations in this work, an adaptive control system for charge balancing in a CTPC generator is presented. An adjustment for the degree of charge recycling is performed in each clock cycle. In the case of insufficient recycling the delay elements which define duration and timing position of the recycling pulse are corrected by switchable timing units.

In a benchmark with the conventional clock generation system, a power reduction gain of up to 24.7% could be achieved. This means saving in power of more than 12% for a complete number-crunching building block.

1 Introduction

Designing high performance building blocks for digital signal processing systems is facing increased challenges in modern CMOS generations especially concerning leakage and process parameter variation (Pang et al., 2009; Das et al., 2009; Nassif et al., 2007; Jaffari and Anis, 2006). These challenges of ultra deep submicron technologies have a great impact on well established design strategies. Nevertheless the demand of efficient implementations concerning area, throughput and power dissipation has to be satisfied.

For highly pipelined architectures in many cases a complementary two-phase clocking scheme (CTPC) is the most efficient choice for realizing the synchronization. In such building blocks power dissipation of the logic can often be reduced to 60%, in some applications even down to 50% of overall power (see Fig. 1, left diagram) This makes the clock system to be the main energy consumer.

In order to reduce clock power an approach for signal generation which performs charge balancing between complementary phases is very promising (Noll and de Man, 1992; Kyriakis-Bitzaros and Nikolaidis, 1997). With this strategy the power efficiency in the clock system can be be enhanced by one third (Fig. 1, right diagram). For a proper functionality of the charge balancing technique, the control signals for buffers and balancing transistors have to be in well adjusted timing relations to each other. Even small shifts between the signals can reduce the amount of recycled charge significantly or cause shorts towards the power supply. As in modern CMOS technologies process variations increase and will in future generations increase even more, power saving clock signal generation will become more complicated.

To overcome the difficulty of timing shifts between sensitive control signals caused by variability we propose in this work an adaptive approach. In a first step, the quality of the charge balancing event is evaluated. If it is not working well timing relations between the control signals are corrected by inserting or skipping little delays resulting in an optimized adjustment of the signals and thus in optimum recycling performance.

In Sect. 2 of this paper the principle of charge balancing is explained. Section 3 describes two main imperfections which can occur within the balancing process. Then in Sect. 4 a new gate for equilibrium detection is presented. Section 5 gives an overview of the complete clock generator implementaion. Finally, in Sect. 6 a benchmarking is made towards two conventional clock systems without parameter adaption.

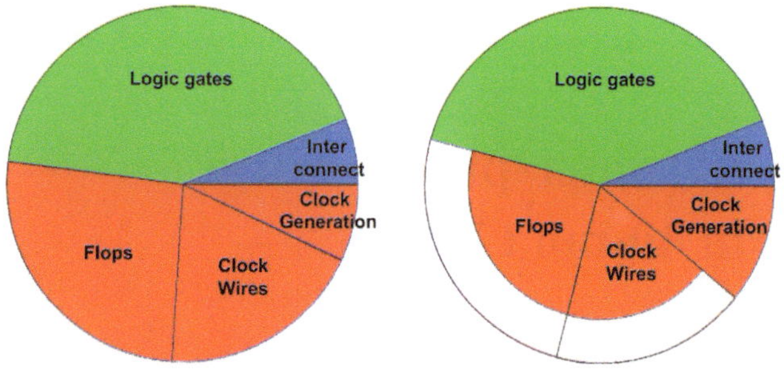

Fig. 1. Power breakdown of FIR-filter with/-out charge balancing.

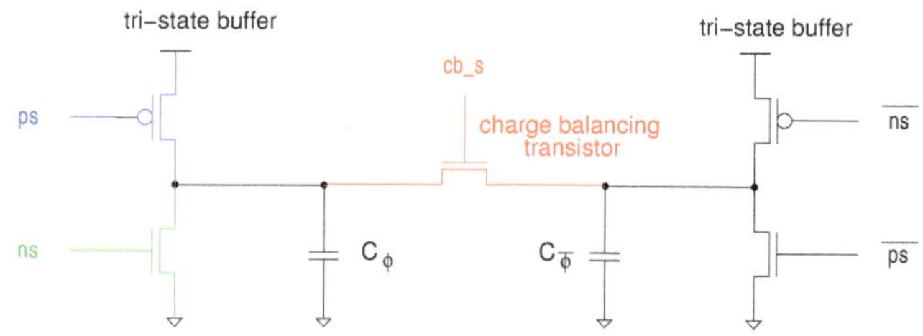

Fig. 2. Charge balancing circuit for complementary clock phases.

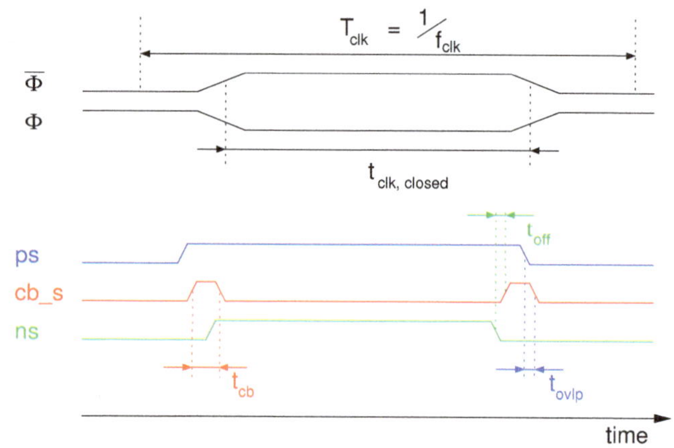

Fig. 3. Timing of control pulses and resulting clock signal phases.

2 Inter-phase charge balancing: basic principle

The mechanism of charge balancing between two complementary phases of one clock signal has been introduced in Noll and de Man (1992). Figure 2 shows the basic circuit including two capacitors which represent the capacitive loads of both complementary signals Φ and $\overline{\Phi}$. The parasitic elements include capacitances of global and local wiring and the clock inputs of all pipeline elements.

Shorting both phases at the moment of attended switching will equalize the charges on both parasitic capacitors. The voltage equilibrium can thus be reached without taking any current from the supply nodes in a best case. If the hardware overhead for the charge balancing circuits is neglected about 50% of power saving within the clock network is possible.

In Fig. 3 both complementary phases Φ and $\overline{\Phi}$ of one clock signal within a CTPC system and the according control signals are illustrated. The balancing mechanism works as follows: to initiate a clock signal transition at first, both tri-state buffers have to be closed making the clock trees floating for a short amount of time t_{off} (see Fig. 3, control signals *ps* and *ns*).

After that, the charge balancing transistor is opened with a balancing pulse of length t_{cb} (see Fig. 3, control signal *cb_s*) until the equilibrium is almost reached.

Finally, with a little, well adjusted overlapping time t_{ovlp} (see Fig. 3, control signals *ns* and *ps*) two of the four tri-state transistors get an impulse to conduct for the rest of the half-cycle. This event will complete the transition.

The optimal choice of t_{ovlp} results in a fast clock signal transition without any short-circuit currents. During this time the charge balancing transistor begins to close while the tri-state buffers already start conducting increasing the risk of a *VDD*-to-*GND* short. In modern CMOS technologies this overlap time has to be reduced or eliminated in order to get good robustness against process variations.

3 Two kinds of signal degradations and the reasons

In the case of well adjusted charge balancing the *cb_s* pulse in Fig. 3 is as long as necessary to recycle about half of the charge which is stored in the clock capacitances. This leads to the best power-delay efficiency, because the clock transition time is as short as possible at the maximum recycled charge.

When the balancing time is shortened, for example due to a slower path for the event that activates the balancing transistor, the full power saving potential cannot be exploited. A clock signal waveform as illustrated in Fig. 4 will be the result. In this case the power gain towards a conventional clock system decreases. But the timing constraints for the clock transitions are fulfilled and the functionality of the logical paths in the building block is guaranteed.

An opposite situation will occur when the balancing impulse gets too long due to signal run time degradations reverse to the case explained above. The resulting waveforms are shown in Fig. 5. In this case the power gain is marginally increased. However, the functionality of the system cannot be guaranteed any longer, because clock transitions get very long which may cause timing violations for critical paths in the logic.

Timing shifts in the control signal paths are mainly caused by different propagation delay variations in the involved logic gates. The main sources of varying delays are on-chip variation (OCV) of transistors, temperature drift, supply noise and process corner variations. If e.g. the system is optimized for the slow-corner, the balancing impulses in the fast-corner may degrade and in disadvantageous cases become too short to get the full voltage swing.

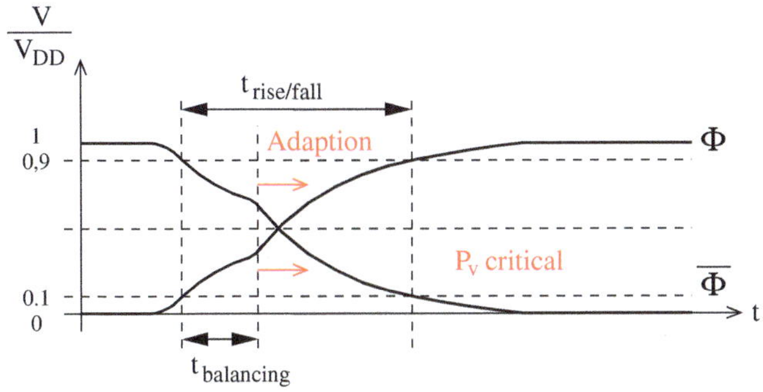

Fig. 4. Charge balancing time is too short.

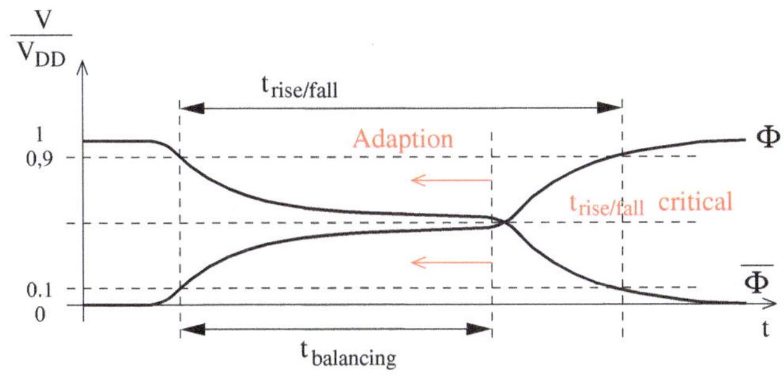

Fig. 5. Charge balancing time is too long.

4 Adapting the charge balancing time

To overcome the above described drawbacks we propose in this work a clock generator which is able to adapt the process of charge balancing individually for every produced chip over the relevant temperature range and in a wide interval of supply voltages. For the adaption functionality it is inevitable to fulfill some kind of measurement towards the recycling event.

As quantification of the recycled charge is complicated to realize we developed a circuit which allows an observation of the charge balancing process.

4.1 Equilibrium detection gate

With a simple binary information, if a voltage level near the equilibrium has been reached before the tri-state buffers are activated, an adaption decision can be made.

Figure 6 shows the new gate which is able to make a full-swing transition while both inputs, which are the complementary clock phases Φ and $\overline{\Phi}$, reach only a level of half the supply voltage – the equilibrium.

Suppose a rising transition of clock signal $\overline{\Phi}$. At the beginning input $\Phi = 1$ and input $\overline{\Phi} = 0$. From this it follows that both n-transistors on the left side conduct, because their gates are connected to Φ. Both left outputs are on logical "0". Accordingly the outputs on the right side are both on logical "1".

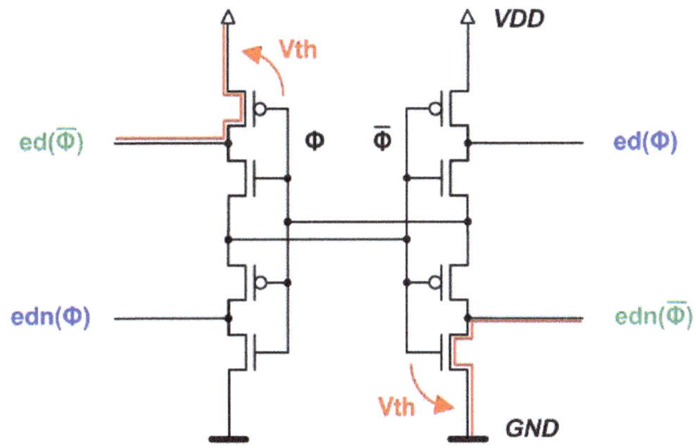

Fig. 6. Equilibrium detection gate.

When Φ is falling until $U_\Phi - U_{\overline{\Phi}} \approx V_{th}$ all four transistors in the middle of the gate, the devices which have no connection to *VDD* or *GND* get closed. When Φ is reaching a voltage level of $U_\Phi = VDD - V_{th}$ the upper left p-transistor begins to conduct pulling the green equilibrium detection signal $ed(\overline{\Phi})$ to *VDD*. Analoguous the other green output $edn(\overline{\Phi})$ on the lower right position of the gate is pulled down to *GND*. These synchronous switching events mark the detection point for the equilibrium.

Later on, when $U_\Phi - U_{\overline{\Phi}} \approx -V_{th}$ the blue output signals $ed(\Phi)$ and $edn(\Phi)$ begin to switch. These events are not relevant for the detection of a rising edge of $\overline{\Phi}$ as explained above. $ed(\Phi)$ and $edn(\Phi)$ can in the following half-cycle detect the equilibrium of a rising edge of Φ.

4.2 Timing relation for the equilibrium detection

A detection of the equilibrium is not enough to make a decision if the balancing process is well adjusted or not. To get an information about the quality of the charge recycling it is important at what time the equilibrium is reached.

For this purpose a capture flip-flop is required which has to be clocked by the control signal of the according tri-state buffer. When the tri-state buffer is activated the balancing event ends and from then on charge is taken from the power supply. At this moment the voltage equilibrium should be reached if the balancing process is properly adjusted.

Therefore the capture flip-flop is closed by the tri-state activation signal (see Fig. 7). If the equilibrium detection signal comes too late it cannot enter into the capture flip-flop. The output of the flip-flop remains on logical "0" and gives the message: "balancing time is too short".

In this case the balancing impulse will be made a little longer by inserting more delay elements Δt into the timing path for the charge balancing impulse and for the tri-state enable signal.

A too long balancing time is difficult to detect, so every 16 cycles the balancing impulse is shortened by one step. If this

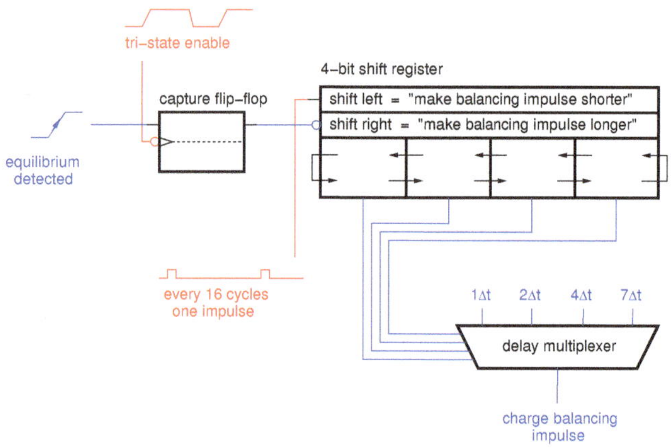

Fig. 7. Charge balancing adaption logic.

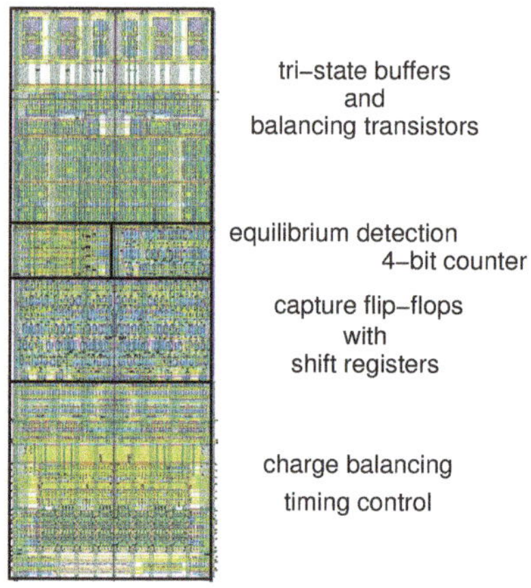

Fig. 8. Layout of the clock generator in 40-nm CMOS.

shortening was not necessary it is corrected in the following cycle by the adaption mechanism explained above.

5 VLSI implementation of the clock generator with parameter adaption

With the above described logic path the clock generator with parameter adaption has been implemented. For the realization in a 40-nm CMOS technology there are some further details to regard.

In order to obtain an efficient implementation a trade-off between hardware complexity and timing shift granularity has to be found. Therefore, we decided not to add the same delay increment Δt with every step of the delay multiplexer. For a small balancing time the increment towards the next longer one is chosen small, while the increment is gradually increased with longer balancing times. With this non-equidistant delay steps an overall number of four different delays is sufficient (see Fig. 7).

For a robust functionality of the system steep transitions of the equilibrium detection signals are important. On the one hand, this can be enhanced by high threshold voltage devices for the four middle transistors in Fig. 6 which are not connected to *VDD* or *GND*. The higher their threshold voltage is the earlier they are switched off when a clock transition occurs and the outputs of the detection gate are earlier prepared for switching. On the other hand, the four transistors connected to *VDD* or *GND* are low threshold voltage devices for fast switching in the moment the equilibrium is reached.

Figure 8 shows the layout of the 40-nm CMOS implementation of the clock generator. The adaption circuitry, positioned in the middle, accounts for about 30% of the complete area. As the clock generator usually consumes much less than 5% of the chip area of a signal processing building block, the overhead for the parameter adaption feature is almost negligible.

Table 1. Power dissipation @400 MHz; 40 °C; 650 mV.

Corner case	fixed charge balancing	adaptive charge balancing	gain
Slow	558 μW	584 μW	−4.5%
Typical	774 μW	627 μW	19%
Fast	948 μW	714 μW	24.7%

6 Benchmarking

In this section we will compare a conventional clock generator and one with charge balancing but without parameter adaption to the proposed new clock generator with parameter adaption for the charge balancing feature.

In Table 1 the power dissipation of both charge balancing implementations is summarized for the classical process corners at a total clock capacitance of $4 \times C_\Phi$ with $C_\Phi = 1$ pF.

One drawback of the system without parameter adaption is, that the fixed timing for the balancing pulses has to be a compromise over all corner cases. The adaptive system switches into another delay multiplexer position for each of the three corner cases. In the slow-corner, for which the frequency constraint of the application has to be met, the conventional system is optimized and has a small advantage over the adaptive system. But in the typical- and fast-corner the power efficiency of the adaptive system is by about 20% superior.

In Fig. 9 the power dissipation of both clock systems is illustrated over the relevant temperature interval for three different capacitive loads of a clock tree at slow/650 mV.

The system with fixed charge balancing, which is well adjusted for the timing worst case at a temperature of 0 °C, gets out of the recycling optimum with rising temperature. The adaptive system switches into a longer balancing time, when the circuit becomes faster with increasing temperature, which means a higher recycling gain.

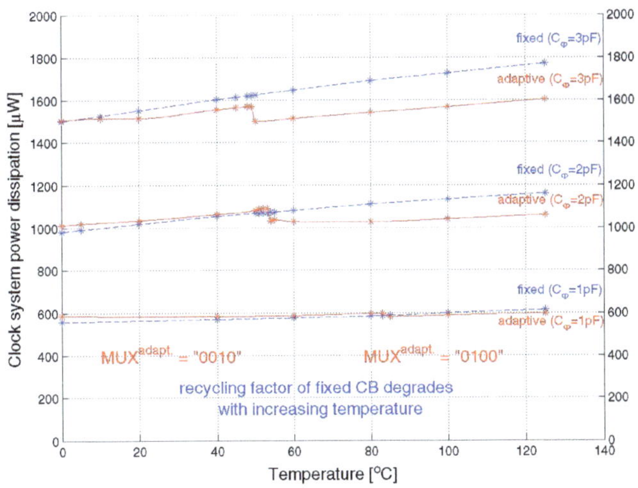

Fig. 9. Power dissipation over the relevant temperature interval.

Table 2. Clock power: Slow-corner vs. Monte Carlo simulation.

Corner case	no charge bal.	fixed charge bal.	adaptive charge bal.
Slow	719 μW	558 μW	584 μW
gain	–	23%	19%
Monte Carlo Global: slow	746 μW	705 μW	659 μW
gain	–	5%	11%

Finally, a monte-carlo benchmark is presented in Table 1 over all three clock systems mentioned above. The clock system with fixed charge balancing suffers much from device variations. A worst scenario taken from monte-carlo simulations reduces the gain towards the conventional clock system by more than three-quarters. To some extent the adaption logic is able to compensate these degradations. The resulting gain of the adaptive approach is more than twice as high as without parameter adaption.

7 Conclusion

In a clock system with complementary phases the power dissipation can be reduced by about 30% with the implementation of a charge balancing approach. One drawback of this approch, the compromise which has to be made for the balancing time over all process corners can be very good compensated by an adaption logic. The adaptive clock generator proposed in this work has an about 20% lower power dissipation in the typical- and fast-corner towards the same clock generator without adaption feature. This means a total power reduction within a high performance number-crunching unit of more than 10%.

Verifying the clock system with monte carlo-simulation reveals, that a charge balancing approach suffers much from process variation. The gain towards a conventional clock system can be more than doubled by applying a parameter adaption to the charge balancing mechanism.

For future research an alternative adaption logic may be possible by using an equilibrium detection gate based on a short circuit measurement.

References

Das, B. P., Amrutur, B., Jamadagni, H. S., Arvind, N. V., and Visvanathan, V.: Within-Die Gate Delay Variability Measurement Using Reconfigurable Ring Oscillators, IEEE Transactions on Semiconductor Manufacturing, 22, 256–267, 2009.

Kyriakis-Bitzaros, E. D. and Nikolaidis, S.: Design of Low Power CMOS Drivers Based on Charge Recycling Technique, IEEE Symposium on Circuits and Systems, pp. 1924–1927, 1997.

Jaffari, J. and Anis, M.: Variability-Aware Device Optimization under I_{on} and Leakage Current Constraints, International Symposium on Low Power Electronics and Design, 2006.

Pang, L.-T., Qian, K., Spanos, C. J., and Nikolic, B.: Measurement and Analysis of Variability in 45 nm Strained-Si CMOS Technology, IEEE Journal of Solid-State Circuits, 44, 2233–2243, 2009.

Nassif, S., Bernstein, K., Frank, D. J., Gattiker, A., Haensch, W., Ji, B. L., Nowak, E., Pearson, D., and Rohrer, N. J.: High Performance CMOS Variability in the 65 nm Regime and Beyond, IEEE International Electron Devices Meeting, pp. 569–571, 2007.

Noll, T. G. and de Man, E.: Pushing the Performance Limits due to Power Dissipation of Future ULSI Chips, IEEE International Symposium on Circuits and Systems, 4, 1652–1655, 1992.

43

A low-noise current preamplifier in 120 nm CMOS technology

H. Uhrmann, W. Gaberl, and H. Zimmermann

Institute of Electrical Measurements and Circuit Design, Vienna University of Technology, Austria

Abstract. In this paper we examine the impact of deep submicron CMOS technology on analog circuit design with a special focus on the noise performance and the ability to design low-noise preamplifiers. To point out, why CMOS technology can grow to a key technology in low-noise and high-speed applications, various amplifier stages, applied in literature, are compared. One, that fits as a current preamplifier for low-noise applications, is the current mirror. Starting from the basic current mirror, an enhanced current preamplifier is developed, that offers low-noise and high-speed operation. The suggested chip is realized in 0.12 μm CMOS technology and needs a chip area of 100 μm$\times$280μm. It consumes about 15 mW at a supply voltage of 1.5 V. The presented current preamplifier has a bandwidth of 750 MHz and a gain of 36 dB. The fields of application for current preamplifiers are, for instance, charge amplifiers, amplifiers for low-voltage differential signaling (LVDS) based point-to-point data links or preamplifiers for photodetectors.

1 Introduction

The heading towards high-speed circuits and low power dissipation combined with a cost-effective design in digital CMOS technology leads to lower structure sizes. As a result of mixed-signal design in System-on-Chips (SoCs), where analog and digital circuits are assembled on one chip, analog designers have to concern about this technology. New CMOS technologies open up some advantages but entail severe drawbacks as well. Short gate lengths, which simultaneously means small parasitic capacitances, boost the transit frequency of the CMOS transistors to a high level. Furthermore, circuit designs with low power dissipation and cheap fabrication costs are possible. The disadvantages have to be considered as well. A severe problem is the low supply voltage, that is defined by the break-through voltage of the transistors. The low supply voltage affects the signal headroom too, which is decreasing drastically. Further issues are the increase of 1/f-noise with smaller gate lengths and the dramatic decrease of the Early voltage (Chang, 1991). Another challenge is, that gate leakage and its mismatch are a serious problem in new technologies like 65 nm CMOS (Annema et al., 2005). The improvement of matching by increasing the gate area, results in an enhancement of the parasitic capacitances and thus in a limitation of the speed. The above mentioned characteristics of deep sub-micron CMOS technologies force a careful circuit design for a proper system performance.

Sometimes current-mode circuits can have some advantages compared to voltage-mode circuits in new CMOS technologies. Current-mode circuits provide the opportunity for low supply voltage, have a tunable input impedance and are less susceptible to power and ground fluctuations. Furthermore current-mode signal processing is a very attractive approach due to the simplicity in implementing operations, such as addition, subtraction and multiplication by a constant (Yuan, 2006).

Current amplification, as an essential part in analog circuit design is necessary for various devices and sensor interfaces, measurement equipment or signal-processing circuits. A reliable detection of currents in the range of nanoampere to micro-ampere is a complex problem depending on the requirements of the target system. Small currents appear in various tasks, which may have many different specifications. Those specifications concern bandwidth, power consumption, supply voltage, linearity, noise, et cetera. Feasible applications for current amplifiers are preamplifiers for measurement systems, charge amplifiers, current-mode filters or optical sensors and amplifiers for low-voltage differential signaling (LVDS) based point-to-point data links or high-speed bus systems.

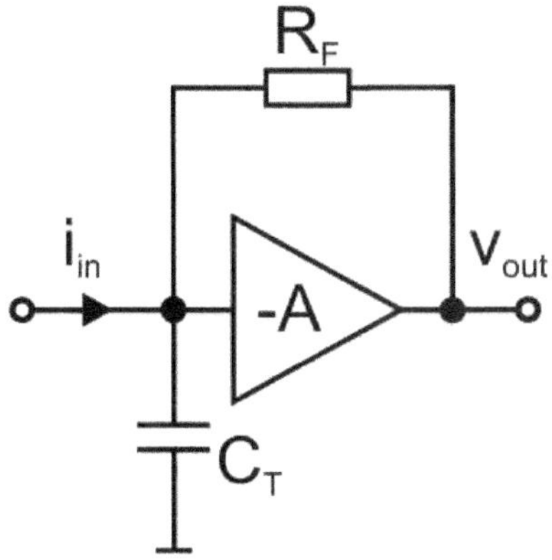

Fig. 1. General schematic of a transimpedance amplifier.

The possibilities that allow the transformation of low input-signal currents are presented in Sect. 2, the proposed circuit is described in Sect. 3 and measurement results are given in Sect. 4.

2 Current amplifiers

In literature, two common approaches of current amplification are mentioned. Both, the transimpedance amplifier and the current mirror, are compared to each other.

2.1 The common approach: transimpedance amplifier

A common approach to convert the current-mode input signal to a proportional voltage signal is the use of a transimpedance amplifier (TIA). A conventional TIA, as shown in Fig. 1, has the simplified transfer function

$$\frac{v_{\text{out}}}{i_{\text{in}}} = G(s) = -\left(\frac{A}{1+A}\right)\frac{R_F}{1+sC_T\left(\frac{R_F}{A+1}\right)}, \tag{1}$$

where A denotes the open-loop voltage gain, R_F and C_T give the feedback resistance and the total input capacitance, respectively. It easily can be seen, that the cut-off frequency can be increased by lowering the feedback resistance R_F. Unfortunately, R_F has an undesirable direct impact on the overall gain, that is also lowered.

An estimation of the equivalent input noise current is presented in Park and Papavassiliou (1999). The examination of the noise sources in the general transimpedance amplifier shows the influence of the circuit parameters on the noise performance (Eq. 2). The equivalent input noise current is given by

$$\left\langle i_{eq}^2 \right\rangle = \frac{(2\pi)^2 C_T^2 \overline{v_n^2} B_{eq}^3}{3} + \frac{(\overline{v_n^2} + \overline{v_f^2}) B_{eq}}{R_F^2}, \tag{2}$$

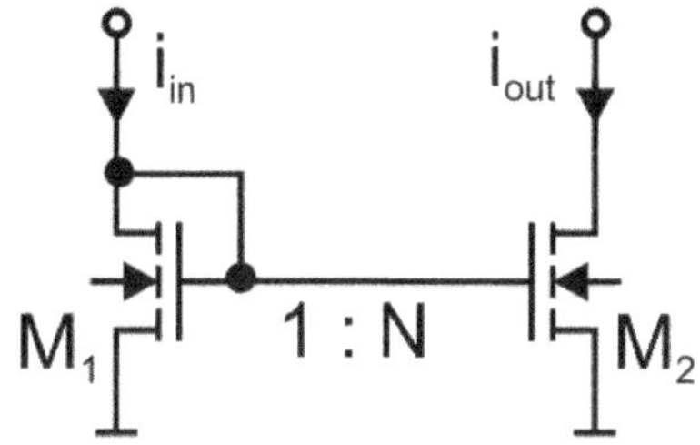

Fig. 2. Basic current mirror.

where B_{eq} is the noise bandwidth and $\overline{v_n^2}$ and $\overline{v_f^2}$ are denoted by (Γ=2/3 in the saturation region)

$$\overline{v_n^2} = \frac{4kT\Gamma}{g_m} \quad \text{and} \quad \overline{v_f^2} = 4kTR_F\ . \tag{3}$$

The first term of Eq. (2) is dominant for high frequencies, hence, C_T has to be as low as possible. The second term of Eq. (2) shows that a reduction of the feedback resistor R_F leads to higher noise. The above considerations show, that the value of R_F has to fit for the trade-off between bandwidth, gain and noise. The use of a current preamplifier offers a way out of these limitations (Yuan, 2006).

2.2 Current mirror

Besides various current-mode amplifier structures, the current mirror is a basic cell, that is commonly used for current amplification. A simple current mirror, as shown in Fig. 2, consists of two transistors. The ratio of their currents is given by

$$\frac{i_{\text{out}}}{i_{\text{in}}} = N = \frac{\frac{W_2}{L_2}}{\frac{W_1}{L_1}}, \tag{4}$$

assuming, that v_{GS} is equal for both transistors. To obtain good matching, the channel length L is usually kept equal for both transistors. Other important properties with a positive impact on current preamplifiers when using current mirrors are the low input resistance and the high output resistance (Laker and Sansen, 1994).

Current amplifiers are already topic in recent publications. The potential advantage of a current preamplifier circuit combined with a transimpedance amplifier compared to a classical transimpedance amplifier is shown in Sturm et al. (2005). An optoelectronic receiver IC for optical storage applications is presented. They suggest a two-stage design, consisting of a current preamplifier and a following TIA, which is called ITIA. The current preamplifier is a current mirror, that is made of two bipolar NPN transistors. A noise performance calculation, that compares a classical transimpedance amplifier with the proposed ITIA shows that the input referred current noise can be lower for the ITIA structure. For high-frequency operation the classical

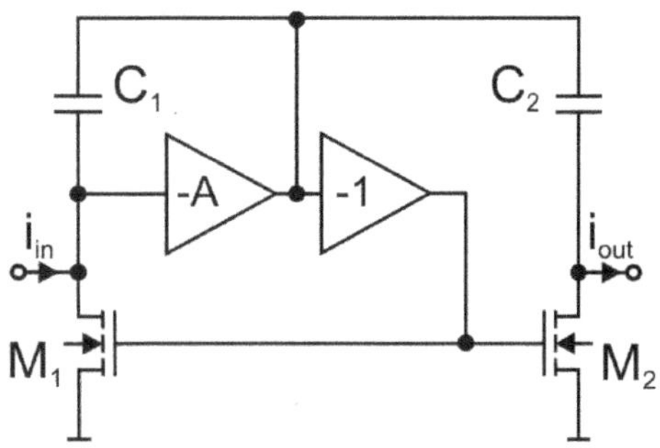

Fig. 3. A low-noise current preamplifier in CMOS technology.

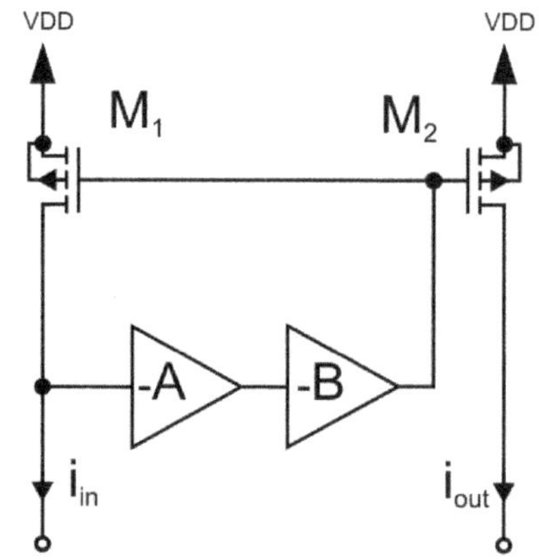

Fig. 4. Block diagram of the low-noise current preamplifier.

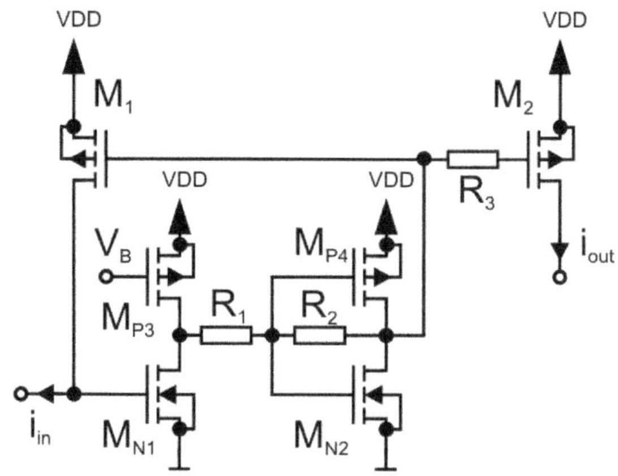

Fig. 5. Schematic of the current preamplifier.

TIA structure suffers from some poor properties concerning noise, as already mentioned. A high bandwidth needs a low feedback resistor, which increases the noise contribution and lowers the transimpedance gain. Especially for low input currents, that need a high gain for further signal processing and a high signal-to-noise ratio the ITIA shows a much better noise performance than the classical TIA. This comparison demonstrates, that current preamplifiers lead to a better noise performance than the classic TIA input stage, if low input-signal currents and high frequencies are demanded.

A low-noise amplifier (LNA) topology for Optoelectronic Integrated Circuits (OEICs) is shown in Fig. 3 using 0.18 μm CMOS technology by Giardina et al. (2006). The proposed LNA contains a current mirror, an amplifier stage and an inverting stage. The capacitors C_1 and C_2 are used to enhance the bandwidth of the amplifier. The inverting stage adds a phase shift of π to drive the gates of M_1 and M_2 in the correct phase. Beside the presentation of the OEIC preamplifier the benefits of decreasing gate lengths for OEICs were emphasized. They concluded, that decreasing gate lengths improve the noise performance of CMOS circuits at a given input capacitance, this means that the lowest noise level is limited by the minimal gate length.

3 Circuit design

In this paper, we propose a current preamplifier designed in 0.12 μm CMOS technology. The presented current-mode preamplifier is based on a current mirror as well. As shown in Fig. 4, the current mirror consists of transistors M_1 and M_2, whose width ratio leads to a current amplification. The injected input current i_{in} is mirrored to i_{out} as a result of the common gate-source voltage. For the enhancement of the speed of the current mirror preamplifier, additional voltage-mode amplifiers are used to regulate the gates of the current mirror transistors. For this operation the change of the potential at the input node, caused by the fluctuating input currents, is used. An increasing current in transistor M_1 causes a slight voltage reduction on the input node and vice versa. This current dependent voltage alteration is amplified and used in the feed-forward loop, consisting of stages A and B, to drive the gates. This technique rises the bandwidth and allows high-speed operation.

To keep an eye on the noise performance of the current preamplifier, another important item, that concerns the design of voltage amplifiers A and B, has to be observed. To obtain the highest signal-to-noise ratio, the first amplifier (amplifier A) should have maximum gain. The second amplifier (B), as depicted in Fig. 4, is used to change the phase of the voltage signal and to adjust the gain of the feed-forward circuit.

In Giardina et al. (2006) the circuit details for the blocks in Fig. 3 were not described. Here, we introduce a solution. In addition we exploit PMOS transistors (M_1, M_2) in the current amplifier (Figs. 4 and 5). The PMOS transistors allow the reverse current direction, that is needed in several sensor applications. Moreover a PMOS transistor, in the technology used, adds less noise to the preamplifier as well and affects the overall noise performance in a positive manner.

In Fig. 5 the detailed final circuit of the current preamplifier, except biasing, is shown. Above considerations are

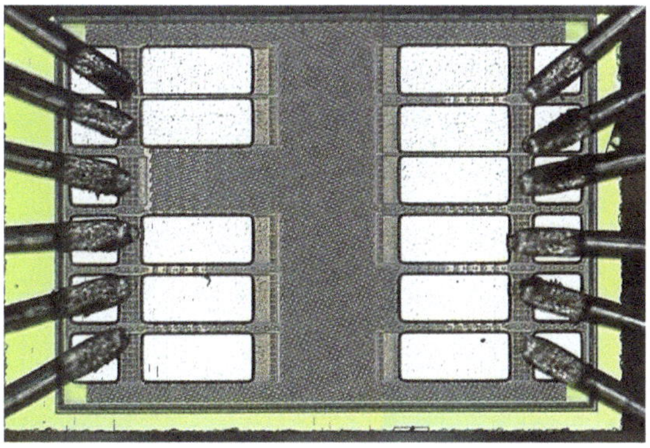

Fig. 6. Chip photo of the current preamplifier.

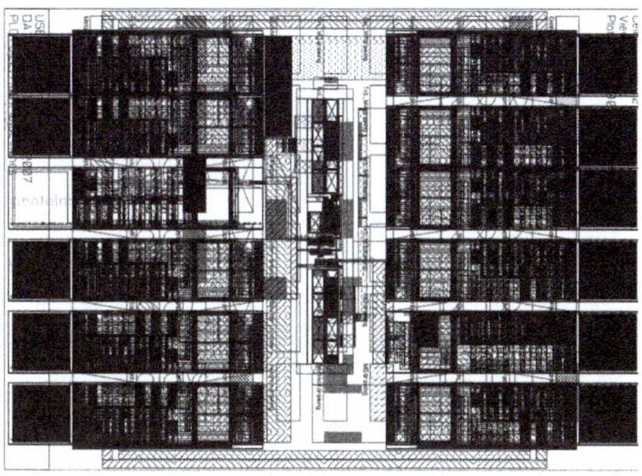

Fig. 7. Layout of the current preamplifier.

directly observed in the circuit design. To achieve high gain and to maximize the signal-to-noise ratio in the feed-forward path, a common-source amplifier is used for the first stage. The following amplifier has the structure of an inverter stage in feedback configuration with R_2. The ratio $\frac{R_2}{R_1}$ adjusts the overall gain of both voltage amplifiers, that has an impact on bandwidth and operation point of the current-mode preamplifier.

The cut-off frequency of the entire preamplifier is set by resistor R_3, as shown in Fig. 5. The resistor and the parasitic gate-source capacitance of the transistor M_2 build a first-order low-pass, whose bandwidth defines the cut-off frequency of the preamplifier. The circuit performance and measurement results are presented in the following section.

4 Measurement results

The above described current-mode preamplifier was simulated and fabricated in 0.12 μm CMOS technology. Since the chip photograph in Fig. 6 does not show any details due to the planarization and passivation layers, the layout of the chip is shown in Fig. 7. The measured power consumption

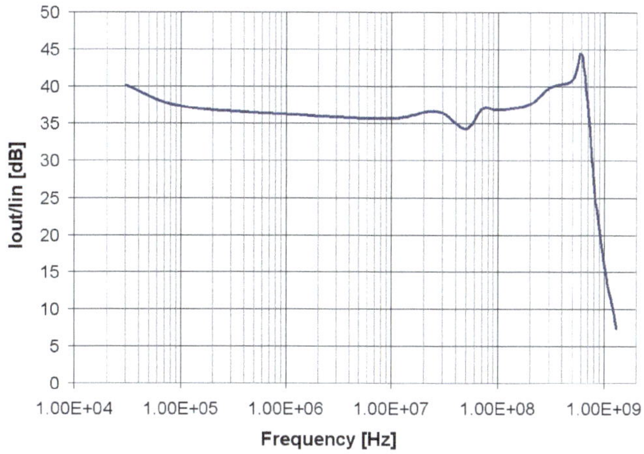

Fig. 8. Measured frequency response.

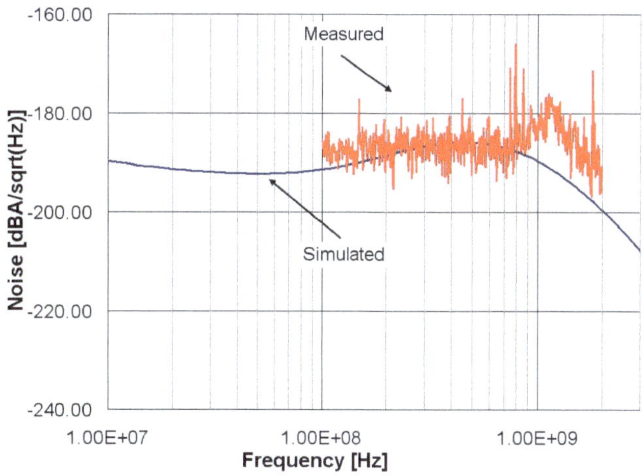

Fig. 9. Measured versus simulated spectral current noise density.

of the chip was 15 mW at a supply voltage of 1.5 V. The active area is 100 μm$\times$280 μm. The achieved gain is approximately 36 dB and the bandwidth is about 750 MHz. Figure 8 shows the measured frequency characteristic.

Figure 9 shows the simulated spectral noise density versus the measured spectral current noise density at the output. The measured output spectral current noise density matches to the simulated values very well, and results in 380 pA/$\sqrt{\text{Hz}}$. The output noise integrated over the bandwidth of 750 MHz is equal to 10 μA. The equivalent input referred current noise is 165 nA.

Table 1 shows a comparison of the presented current preamplifier to recent publications. Although the gain is not the highest, which is difficult to compare because of multistage amplifier design of Sturm et al. (2005) and de Jong et al. (2002), we achieved the highest bandwidth. We fabricated in the newest technology, which has only 1.5 V supply

Table 1. Comparison to other publications.

Publication	Sturm et al. (2005)[1]	Giardina et al. (2006)	de Jong et al. (2002)[2]	This Work
Gain	108 dB	48 dB	max. 66 dB	36 dB
Cut-Off Frequency	260 MHz	600 MHz	250 MHz	750 MHz
Technology	0.5 μm BiCMOS	0.18 μm CMOS	0.6 μm CBiMOS3	0.12 μm CMOS
Supply Voltage	5 V	1.5 V	5 V	1.5 V
Power Consumption	300 mW	12.5 mW	170 mW	15 mW
Active Area	n.s.[3]	360×260 μm^2	n.s.[3]	100×280 μm^2

[1] two stages
[2] four stages
[3] not specified

Table 2. Performance summary of the proposed current-mode preamplifier.

Gain	36 dB
Cut-Off frequency	750 MHz
Technology	0.12 μm CMOS
Supply voltage	1.5 V
Power consumption	15 mW
Active area	100×280 μm^2
Integrated In-Band Output Current Noise	10 μA
Equivalent Integrated In-Band Input Noise Current	165 nA

voltage. Furthermore we designed the chip with the smallest active area. Unfortunately Sturm et al. (2005) and de Jong et al. (2002) did not mention their active area, but it can be estimated that a multistage amplifier design occupies a larger area. In Table 2 the performance of the proposed current preamplifier is summarized.

5 Conclusions

In this work we present a low-noise current preamplifier based on a current mirror in 0.12 μm CMOS technology. It can be used for current-mode circuits, measurement systems, charge amplifiers, optical sensors, et cetera. Furthermore the classical transimpedance amplifier is compared to this new approach. Motivations to realize a low-noise current-mode amplifier in deep sub-micron CMOS processes are highlighted. The obtained observations were used to realize a low-noise preamplifier employing a feed-forward path to broaden the bandwidth. The insertion of PMOS transistors as current mirror has a positive impact on the noise performance. The designed preamplifier stage has a gain of 36 dB and a bandwidth of 750 MHz. The amplifier with the size of 100 μm×280 μm consumes about 15 mW at 1.5 V power supply.

Acknowledgements. The authors thank A. Bertl and L. Dörrer from Infineon Technology Austria AG in Villach for the initiation of this work and for chip processing. Partial financial funding from Infineon Technologies Austria AG and from the Austrian Federal Ministry for Transport, Innovation, and Technology in the project SOFT-RoC via FFG is gratefully acknowledged.

References

Annema, A.-J., Nauta, B., van Langevelde, R., and Tuinhout, H.: Analog Circuits in Ultra-Deep-Submicron CMOS, IEEE J. Solid-State Circuits, 40(1), 132–142, 2005.

Chang, Z. Y.: Low-Noise Wide-Band Amplifiers in Bipolar and CMOS Technologies, Kluwer Academic Publishers, ISBN: 0-7923-9096-2, 1991.

de Jong, G. W., Bergervoet, J. R. M., Brekelmans, J. H. A., and van Mil, J. F. P.: A DC-to-250 MHz Current Pre-Amplifier with Integrated Photo-Diodes in Standard BiCMOS, for Optical-Storage Systems, IEEE International Solid-State Circuits Conference, 1, 362, 2002.

Giardina, M., Stek, A., de Jong, G. W., and Bergervoet, J. R. M.: Design of Low-Noise CMOS OEIC for Blu-ray Disc Optical Storage Systems, European Solid-State Circuits Conference, 9, 271–274, 2006.

Laker, K. R. and Sansen, W. M. C.: Design of Analog Integrated Circuits and Systems, McGraw-Hill Companies, ISBN: 0-07-113458-1, 1994.

Park, S. M. and Papavassiliou, C.: On the Design of Low-Noise Giga-Hertz Bandwidth Preamplifiers for Optical Receiver Applications, IEEE Conference on Electronics, Circuits and Systems, 2, 785–788, 1999.

Sturm, J., Leifhelm, M., Schatzmayr, H., Groiß, S., and Zimmermann, H.: Optical Receiver IC for CD/DVD/Blue-Laser Application, IEEE J. Solid-State Circuits, 40(7), 1406–1412, 2005.

Yuan, F.: Low-Voltage CMOS Current-Mode Preamplifier: Analysis and Design, IEEE Trans. Circuits and Systems I, 53, 26–39, 2006.

44

Temperature modeling and emulation of an ASIC temperature monitor system for Tightly-Coupled Processor Arrays (TCPAs)

E. Glocker[1], S. Boppu[2], Q. Chen[3], U. Schlichtmann[3], J. Teich[2], and D. Schmitt-Landsiedel[1]

[1]Lehrstuhl für Technische Elektronik, Technische Universität München (TUM), Theresienstr. 90, 80333 München, Germany
[2]Lehrstuhl für Informatik 12 (Hardware-Software-Co-Design), Friedrich-Alexander-Universität Erlangen-Nürnberg (FAU), Cauerstr. 11, 91058 Erlangen, Germany
[3]Lehrstuhl für Entwurfsautomatisierung, Technische Universität München (TUM), Arcisstr. 21, 80333 München, Germany

Correspondence to: E. Glocker (elisabeth.glocker@tum.de)

Abstract. This contribution provides an approach for emulating the behaviour of an ASIC temperature monitoring system (TMon) during run-time for a tightly-coupled processor array (TCPA) of a heterogeneous invasive multi-tile architecture to be used for FPGA prototyping. It is based on a thermal RC modeling approach. Also different usage scenarios of TCPA are analyzed and compared.

1 Introduction

The demand for increasing performance and at the same time decreasing feature size in modern CMOS technologies lead to higher power densities influencing chip temperature. The chip temperature also dependens on thermal conductivity of chip and package. High temperature reduces circuit speed and causes transistor degradation. Normally the workload is distributed in a non-uniform way in different circuit blocks leading to local temperature hot spots. Altogether, this can greatly affect the lifetime and the reliability of a chip (Brooks et al., 2007; Semenov et al., 2006). This problem becomes even more critical in multi-core scenarios: Core temperature increases, because it depends not only on its own power density, but also on the activity and the power density of neighbour cores. Monitoring of operating temperature during run-time makes it possible to control and limit it to increase lifetime and reliability and to find the best suitable application-core pairs during resource allocation.

This paper presents an approach for emulating an ASIC temperature monitoring system (TMon) for Tightly-Coupled Processor Array (TCPA) tiles of a heterogeneous invasive multi-tile architecture to be used for FPGA prototyping. Before the emulating principle is shown, TCPA usage scenarios are simulated and analysed in Sect. 3 with detailed temperature simulations based on thermal RC modeling approach for an ASIC layout. Section 4 shows then the emulation principle and first results for using this method.

1.1 Invasive computing, temperature emulation

Static and central management of computing systems may not scale, since SoC architectures with thousand or even more processors on a single chip can be foreseen in the near future. Also, imperfections like parameter variations and ageing lead to increasingly unreliable components. Invasive computing (Teich et al., 2011) is a resource-aware parallel computing paradigm that provides a self-organizing task management to deal with this: Applications get the ability to explore the system and dynamically spread onto processors in the "invasion" phase and also give back unused resources depending on the current needs. The system is a heterogeneous invasive multi-tile architecture that includes computation, I/O and memory tiles. Computation tiles are either TCPA or RISC tiles that may have different number and (for RISC tiles) types of processors. The individual tiles are connected via a network-on-chip (NoC). Resource allocation considers application constraints such as required number and type of resources including physical hardware properties like temperature. The run-time support system and an underlying hardware management unit handle the resource allocation in close-collaboration. A closed-loop control system is formed, including monitors, run-time support system, hardware management unit and applications.

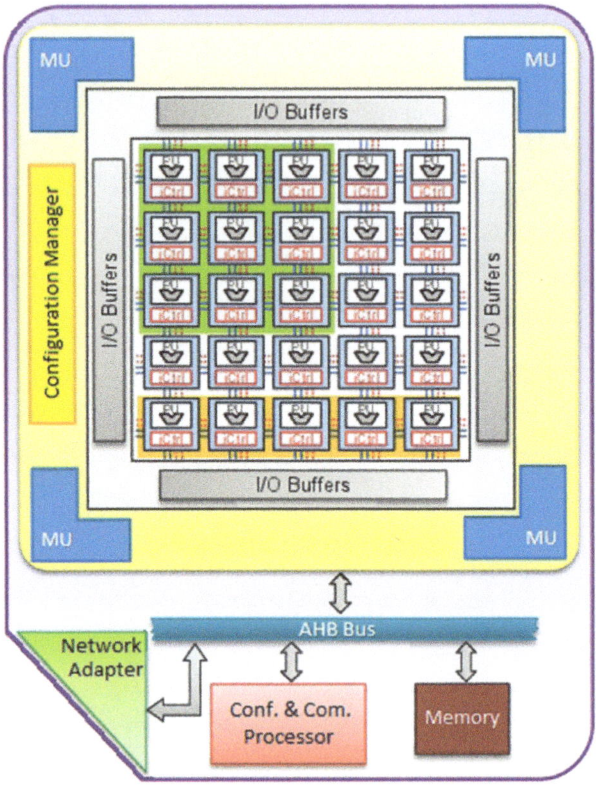

Figure 1. TCPA tile with a 5 × 5 processing elements (PEs) array. Initiation of an invasion may be done from the corner PEs (master PEs).

In preparation of an ASIC implementation, FPGA prototyping is required for tuning design parameters and design verification. However, monitor modules with analog-circuit components, like temperature monitors, cannot be directly implemented on prefabricated FPGA structures. Also, since power dissipation and the evolution of temperature on FPGA differ from that in an ASIC implementation, the FPGA will show very different behaviour compared to the envisioned ASIC. Nevertheless, the interaction of hardware, run-time support system and application during resource allocation, taking monitoring data into account, has to be verified and tested. Thus, an emulation of the behaviour of an ASIC TMon on the FPGA platform is required.

1.2 Related work

Exploration of thermal behaviour is done offline either with detailed simulators like in Skadron et al. (2004) or virtual platforms (e.g. Bartolini et al., 2013). Online thermal exploration during runtime for FPGA prototyping relies on existing temperature sensors as done in Shen and Qiu (2011) that do not emulate the ASIC temperature. Also coupled hardware/software solutions with e.g. a host-computer running a thermal modeling tool like in Atienza et al. (2006) are available that could be much slower than the prototyped system. Therefore, a solution as fast as real temperature monitors is required. Nevertheless related work dealing with offline detailed ASIC temperature simulations (e.g. Skadron et al., 2004) can be adapted for temperature monitor emulation.

2 Tightly-Coupled Processor Arrays (TCPAs)

A TCPA is a massively parallel architecture containing tightly-coupled light-weight processing elements (PEs) (Kissler et al., 2006). An example TCPA tile (array size of 5 × 5 PEs) is shown in Fig. 1. The invasion of new applications always starts at the corner PEs (master PEs). Each PE has an invasion controller (iCtrl), a dedicated hardware component for invasion. Each TCPA tile contains a "Conf. & Com. Processor" managing the communication and configuration of regions of claimed PEs with proper programs. Reconfigurable I/O Buffers, tile local Memory, a Configuration Manager and management units (MU) are also present. In the TCPA tile one MU is present at each corner of the array to supervise invasion requests and application execution on the master PEs. Communication outside of the TCPA tile is handled with a Network Adapter (Hannig et al., 2013, 2014).

The amount of instruction memory and control overhead for individual PEs is kept as small as possible. Thus a tight coupling, cycle-based communcation by point-to-point connections among PEs (from neighbour to neighbour) is realized. Individual PEs do not have direct global memory access (Lari et al., 2012): Data transfer is realized to and from the array through the border PEs that have a connection to the banks of the surrounding memory buffers. The number of invaded PEs can be dynamically reduced and increased depending on the requirements of the corresponding application with different invasion strategies (Muddasani et al., 2012). TCPAS are typically used as hardware accelerators in an MPSoC for execution of computationally intensive loop programs like streaming applications (video/image processing) (Muddasani et al., 2012).

3 ASIC temperature simulation for TCPAs

From a temperature point of view, it is feasible to carefully analyse and optimise the placements of active PEs. By using temperature monitor data, it is possible to control the temperature by load-balancing (e.g. proper resource allocation) under different control targets: One control target might be to keep the temperature as low as possible. Obviously the temperature should be limited to avoid overheating. An even lower temperature is desirable to reduce the effect of PE ageing over time and thus increase PEs lifetime and reliability. Another control target might be to keep the temperature difference inside the array low to ensure similarly low PE ageing. Proper resource allocation can be used to outbalance high temperatures in the past with low ones in the present.

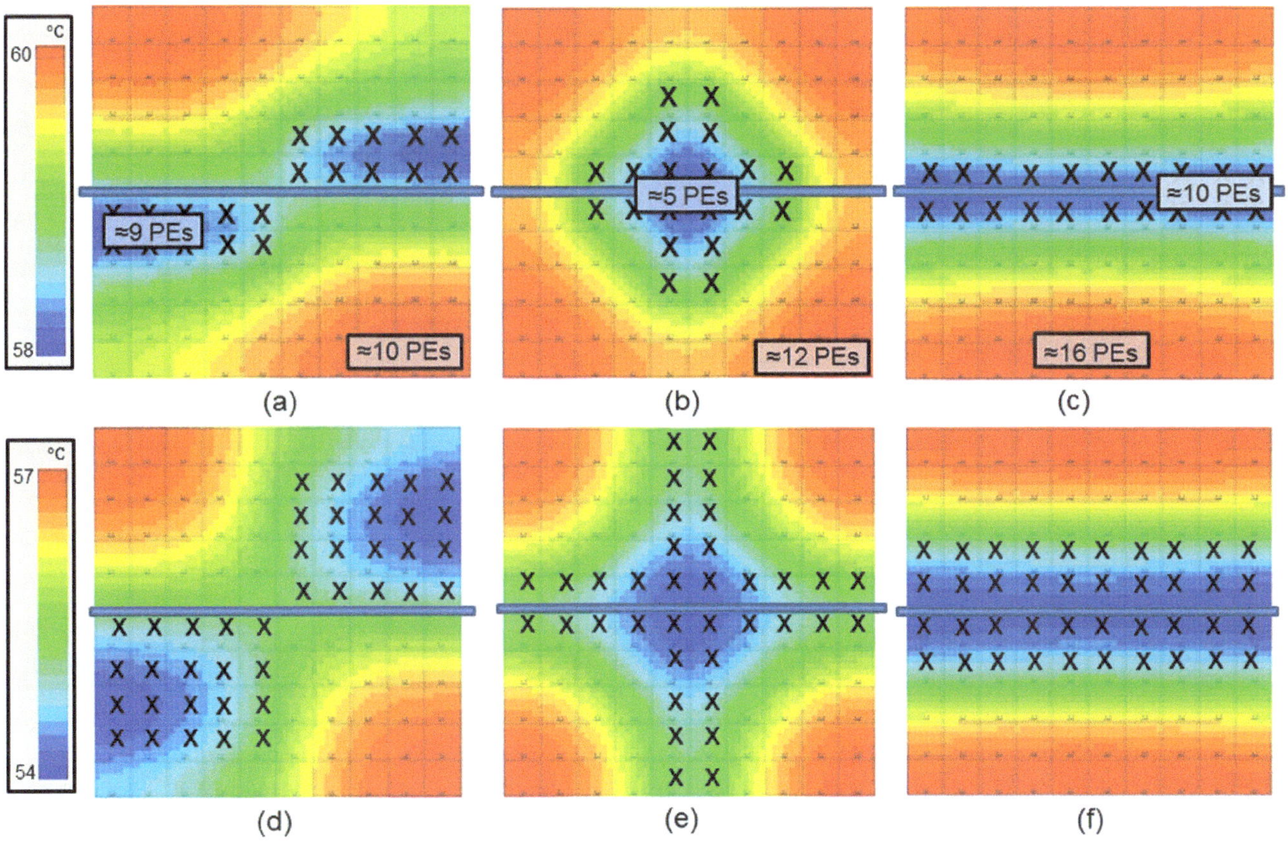

Figure 2. TCPA usage scenario of 80 % in **(a)–(c)** and 60 % usage in **(d)–(f)**: Two applications are running on the TCPA (border between the applications: straight line). Each of them has 40 invaded PEs for 80 % and 30 for 60 % usage (unused PEs are marked with an 'X'). Different positions of the unused PEs lead to different temperature evolutions. For (a)–(c) the size of the area affected with hottest and coolest temperatures is given (unit is PEs).

3.1 Simulation framework

Temperature simulations, based on a thermal RC model, are performed for steady-state temperatures with HotSpot (Huang et al., 2004). More details on the thermal RC model are given in Glocker and Schmitt-Landsiedel (2013). Steady-state temperature means that (with a constant load) equilibrium temperature distribution is reached. That means the temperature does not depend on time anymore. The steady-state temperature is a good estimation, if the task execution time is large enough (Brooks et al., 2007). TCPAs are typically used for intensive loop programs, so the execution time for applications is typically very long. The accuracy of HotSpot has been intensively studied: For considered floorplans, it showed good agreement with finite-element simulator and test-chip results where the deviations for steady-state temperatures was found to be always less than 5.8 % (Huang et al., 2008; Skadron et al., 2004).

In this section simulations are done for an ASIC TCPA layout with an array size of 10×10 PEs and they are done for a size of 5×5 PEs in Sect. 4. Each PE contains a power consuming region (about 83 % of total PE size) and a region that does not consume power (wiring, space in between PEs) of about 17 % of total PE size. Those individual PE blocks are placed next to each other to form the desired PE array. Possible invasion strategies are linear, meander walk, rectangular and random fashion as explained in Lari et al. (2012). It is assumed that each PE is either used (constant power consumption) or not used (just very small leakage power consumption). The TCPA tile is designed in TSMC 65 nm. Ambient temperature is 45 °C and chip thickness is 500 μm. Convection capacitance is 140.4 $\mathrm{JK^{-1}}$, convection resistance is 0.1 $\mathrm{KW^{-1}}$, silicon thermal conductivity is 100 $\mathrm{Wm^{-1}\,K^{-1}}$ and silicon specific heat is 1.75 $\mathrm{MJm^{-3}\,K^{-1}}$ (Skadron et al., 2004).

3.2 Evaluation of TCPA usage scenarios

If the complete 10×10 TCPA array is used (full-use scenario), the resulting temperature is approximately constant at 63 °C (highest temperature in tile middle 62.8 °C, lowest temperature at tile corners 62.6 °C). This corresponds to the temperature worst-case scenario.

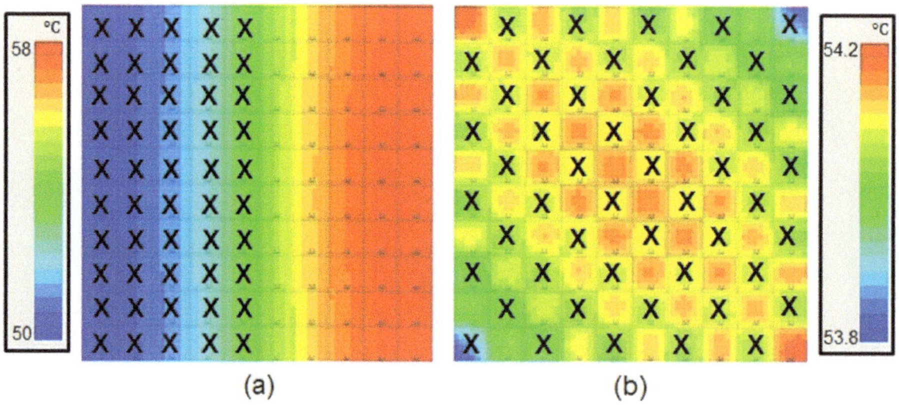

Figure 3. A 50 % usage scenario for one application is possible by using half of the array (PEs marked with an 'X' are unused) or by using the complete array (PEs marked with an 'X' are unused by the application itself, but still need to transfer data from one active PE to the next).

Figure 4. The emulated TMon gives out maximum temperatures for four 9-PE region temperatures (the dashed lines show two of those 9-PE regions. The other two are found accordingly). For finding all four 9-PE region temperature, the TCPA tile has to be partioned further into four 4-PE regions at the corners (light red background colour), a 1-PE region in the middle (light blue background colour) and four 2-PE regions in between (grey background colour).

A typical usage scenario (standard scenario) is 80 % usage as shown in Fig. 2a–c for two applications with 40 invaded PEs each. For Fig. 2(a) the improvement compared to full-use scenario is 4 % for hottest and 7 % for coolest temperature. Different positions of the unused PEs lead to different temperature evolutions (Fig. 2b and 2c). The temperature difference compared to Fig. 2a is small: For hottest temperature it is 0.3 °C for both. For coolest temperature it is 0.7 °C for Fig. 2b and no improvement for Fig. 2c. But the area size affected by hottest and coolest temperatures change: In Fig. 2a the area affected by hottest temperature has the size of about 10 PEs. For Fig. 2b it has the size of about 12 PEs and in Fig. 2c of about 16 PEs (worst case, incease of 6 PEs compared to Fig. 2a). The area affected by coolest temperature has a size of about 9 PEs in Fig. 2a, about 10 PEs in Fig. 2c and about 5 PEs in Fig. 2b (best case, decrease of 4 PEs compared to Fig. 2a).

If the usage decreases to 60 % as shown in Fig. 2d–f, the changing size of the area with hottest respectively coolest temperature is still visible for different scenarios. For Fig. 2d the temperature improvement compared to full-use scenario is 8 % for hottest and 14 % for coolest temperature. Compared to Fig. 2d, the temperature improvement for hottest temperature is 2 °C for both Fig. 2e and f. The coolest temperature is worsen by −1 °C for both.

A 50 % usage scenario for one application is shown in Fig. 3. Either one half of the TCPA array can be used (Fig. 3a). If it is known, that the other half of the array is not needed for some time, it may be possible to use the complete array for the application (Fig. 3b). The latter scenario is better for two reasons: The highest temperature is lowered by 6 % and the temperature distribution is more uniform. Note, that those PEs not directly used by the application, still need to transfer data from one active PE to the next and would consume an insignificant amount of power.

4 Emulation of TCPA temperature monitor system

The basic idea behind the emulation of the TMon is based on the fact that a PE's temperature depends on the PEs own activity and the activities of neighbour PEs. If we can determine the temperature based on PE's own activity and the influence of neighbouring PEs on this temperature, then it is possible to find the resulting PE temperature.

4.1 Emulation principle

For emulation, a TCPA array size of 5 × 5 was chosen. Because of the high number of PEs in a TCPA tile, the TMon does not give out each PE's temperature, but the temperature for regions of 9 PEs (9-PE region temperature) as shown in Fig. 4 for two out of four possible 9-PE regions with

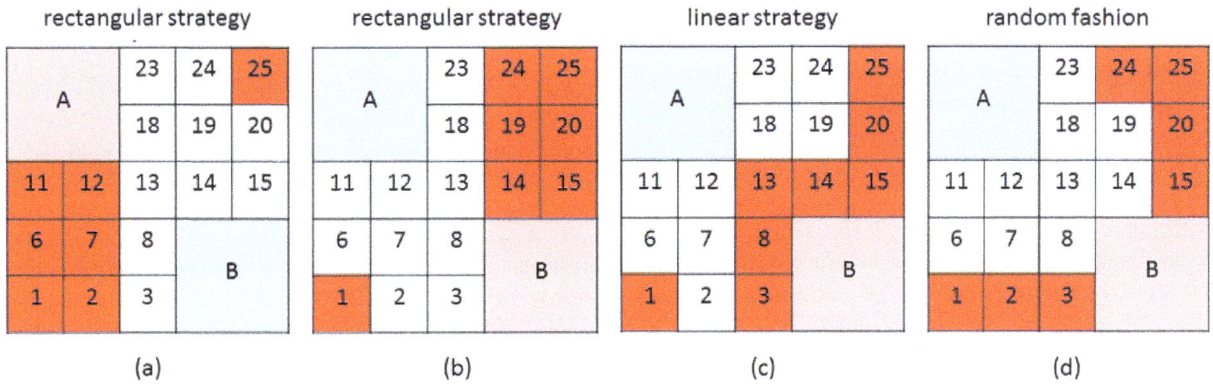

Figure 5. By using the emulated TMon, best suitable PEs for new applications can be found. The dark red colored PEs are currently used. A new application must be started in either region A or B. The light red and blue coloring of region A resp. B show the hotter or cooler region because of the activity of the curently used dark red colored PEs. The new application should always start in light blue colored region.

dashed lines. The region size of 9 PEs is feasible since invasion can start only at one of the four master PEs in the corners. By using the four 9-PE region temperatures the best suitable master PE (from temperature point of view) can be determined. A standard usage scenario would use 80 % of the PEs. For a 5 × 5 array with 4 applications running, 80 % means 5 PEs for each application. So information about the 9-PE region of each corner is normally needed to find a big enough region with adequate PEs for an application and adequate PEs around those application PEs. The output of the emulated monitor system are maximum temperatures. Besides resource allocation of new applications, the temperature monitor data can be used during application runtime: If an application wants to invade additional PEs, the 9-PE region temperatures can be used to find a suitable direction (in terms of temperature evolution) for the invasion.

To find the 9-PE region temperatures, the TCPA tile is partioned further as shown in Fig. 4: The resulting regions are four 4-PE regions at the corners (light red background colour), a 1-PE region in the middle (light blue background colour) and four 2-PE regions in between (grey background colour). The temperature for a 4-PE region is found without considering the influence of PEs outside the 4-PE borders (single 4-PE region temperature). In a second step, the single 4-PE region temperature is used to find the corresponding 9-PE region temperature without considering influence of PEs outside the 9-PE borders (single 9-PE region temperature). Next, the influence of PEs outside this 9-PE border on the 9-PE region temperature is found (neighbour region influence). In a last step, the single 9-PE region temperature and neighbour region influence are added up to find the resulting 9-PE region temperature (final 9-PE region temperature). This has to be done for all four 9-PE regions.

For finding those region temperatures and the neighbour region influence, results of detailed simulations of steady-state temperatures (for all possible positions for active PEs) are obtained and mapped to Look-up-table (LUT) entries. The output of the emulated monitor system are the four final 9-PE region temperatures. For finding a suitable direction for invasion during application runtime, the four single 4-PE region temperatures could be used in addition to the final 9-PE region temperatures. This would lead to a more detailed view on the temperature evolution inside the array.

4.2 Results

The maximum temperatures for PE regions obtained from detailed simulations of ASIC temperatures and results from TMon emulation (that emulates the behaviour of an ASIC temperature monitor) differ by maximally 0.5 °C. A more accurate implementation is possible at the expense of higher memory usage and may lead to increased signal lengths and slower reaction to the current system state. Likewise, a less accurate implementation is possible. A balance between accuracy, implementation speed and memory usage has to be determined with regard to an optimised closed-loop control system. This is subject of future work.

4.3 Usage example

In the following, the usage of our method to determine the best suitable master PE is demonstrated for the four examples in Fig. 5. The control target is to keep the temperature for newly invaded PEs as low as possible. In all examples, a new application wants to invade four PEs with rectangular strategy. The invasion must start with a master PE and the used PEs can be those in either region A or B. The light red and blue coloring of region A resp. B shows the results determined from the emulated ASIC TMon.

For Fig. 5a the dark red colored PEs are used by the already running applications (rectangular strategy). The temperature in region A is higher than that in region B. According to the emulated ASIC TMon region B should be taken for the new application. For detailed simulations, the results are similar. The difference from emulation to detailed simulation results are only −0.18 °C for region A and +0.09 °C for region B. Likewise for Fig. 5b, the temperature in region B is higher. Emulated and detailed simulation results lead to the decision for region A with a difference in emulation results of +0.09 °C for region A and −0.18 °C for region B. The dark red colored PEs are used with linear strategy in Fig. 5c and with random fashion distribution in Fig. 5d. The emulated and detailed simulation results lead both to the decision for the cooler region A. The difference in emulation results for Fig. 5c are +0.23 °C for region A and −0.19 °C for region B. For Fig. 5d the difference in emulation results are +0.25 °C for region A and +0.01 °C for region B.

In return the decisions during resource allocation (like the choice of a master PE in this example) influence the temperature evolution of the TCPA tile, since newly invaded PEs become active in addition to the ones already in use. Also different active PE placements result in different temperature evolutions. The complete closed-loop control system is formed containing monitors, run-time support system and hardware management unit (performing resource allocation in close-collaboration) and applications. If more than one TCPA tile is considered, using monitor data becomes even more important: Temperature monitor data can be used for selecting the best suitable TCPA tile first. Afterwards a suitable master PE can be selected.

5 Conclusions

This paper presented an emulation approach for an ASIC TCPA (of an invasive multi-tile architecture) TMon for FPGA prototyping. The TMon models the behaviour of an ASIC temperature sensor. It was shown that such a TMon can be used for resource allocation. From temperature point of view, it is feasible to carefully analyse and optimise the current active PE placements.

Subject of future work includes power emulation and implementing the emulated ASIC TMon on the invasive multi-tile architecture FPGA prototyping platform for forming the closed-loop control system. The interaction with other system components involved in the closed-loop control system has to be studied closely to find a good balance between accuracy, speed of implementation and memory usage.

Acknowledgements. This work was supported by the German Research Foundation (DFG) as part of the Transregional Collaborative Research Centre "Invasive Computing" (SFB/TR 89).

Edited by: D. Killat

References

Atienza, D., Del Valle, P. G., Paci, G., Poletti, F., Benini, L., De Micheli, G., and Mendias, J. M.: A Fast HW/SW FPGA-Based Thermal Emulation Framework for Multi-Processor System-on-Chip, Design Automation Conference (DAC'06), pp. 618–623, 2006.

Bartolini, A., Cacciari, M., Tilli, A., and Benini, L.: Thermal and Energy Management of High-Performance Multicores: Distributed and Self-Calibrating Model-Predictive Controller, IEEE T. Parall. Distr., 24, 170–183, 2013.

Brooks, D., Dick, R. P., Joseph, R., and Li Shang: Power, Thermal, and Reliability Modeling in Nanometer-Scale Microprocessors, IEEE Micro, 27, 49–62, 2007.

Glocker, E. and Schmitt-Landsiedel, D.: Modeling of temperature scenarios in a multicore processor system, Adv. Radio Sci., 11, 219–225, doi:10.5194/ars-11-219-2013, 2013.

Hannig, F., Schmid, M., Lari, V., Boppu, S., and Teich, J.: System Integration of Tightly-Coupled Processor Arrays using Reconfigurable Buffer Structures, Proceedings of the ACM International Conference on Computing Frontiers 2013 (CF'13), ACM, 2:1–2:4, 2013.

Lari, V., Boppu, S., Tanase, A., and Reiche, O.: Invasive Tightly-Coupled Processor Arrays: A Domain-Specific Architecture/Compiler Co-Design Approach, ACM Transactions on Embedded Computing Systems (TECS), forthcoming, 2014.

Huang, W., Stan, M. R., Skadron, K., Sankaranarayanan, K., Ghosh, S., and Velusamy, S.: Compact Thermal Modeling for Temperature-Aware Design, Design Automation Conference (DAC'04), pp. 878–883, 2004.

Huang, W., Sankaranarayanan, K., Skadron, K., Ribando, R. J., and Stan, M. R.: Accurate Pre-RTL Temperature-Aware Design Using a Parameterized, Geometric Thermal Model, IEEE T. Comput., 57, 1277–1288, 2008.

Kissler, D., Hannig, F., Kupriyanov, A., and Teich, J.: A highly parameterizable parallel processor array architecture, Proceedings of the IEEE International Conference on Field Programmable Technology 2006 (FPT'06), pp. 105–112, 2006.

Lari, V., Narovlyanskyy, A., Hannig, F., and Teich, J.: Decentralized dynamic resource management support for massively parallel processor arrays, 2011 IEEE International Conference on Application-Specific Systems, Architectures and Processors (ASAP'11), pp. 87–94, 2011.

Lari, V., Muddasani, S., Boppu, S., Hannig, F., and Teich, J.: Design of Low Power On-chip Processor Arrays, 2012 IEEE 23rd International Conference on Application-Specific Systems, Architectures and Processors (ASAP'12), pp. 165–168, 2012.

Muddasani, S., Boppu, S., Hannig, F., Kuzmin, B., Lari, V., Teich, J.: A Prototype of an Invasive Tightly-Coupled Processor Array, 2012 Conference on Design and Architectures for Signal and Image Processing (DASIP'12), pp. 1–2, 2012.

Semenov, O., Vassighi, A., and Sachdev, M.: Impact of Self-Heating Effect on Long-Term Reliability and Performance Degradation in CMOS Circuits, IEEE T. Device Mat. Re., 6, pp. 17–27, 2006.

Shen, H. and Qiu, Q.: An FPGA based Distributed Computing System with Power and Thermal Management Capabilities, Proceedings of the 20th International Conference on Computer Communications and Networks (ICCCN'11), pp. 1–6, 2011.

Skadron, K., Stan, M. R., Sankaranarayanan, K., Huang, W., Velusamy, S., and Tarjan, D.: Temperature-Aware Microarchitecture: Modeling and Implementation, ACM Transactions on Architecture and Code Optimization, 1, pp. 94–125, 2004.

Teich, J., Henkel, J., Herkersdorf, A., Schmitt-Landsiedel, D., Schröder-Preikschat, W., and Snelting, G.: Invasive Computing: An Overview, edited by: Hübner, M. and Becker, J., Multiprocessor System-on-Chip – Hardware Design and Tool Integration, Berlin, Heidelberg, Germany, Springer, pp. 241–268, 2011.

Implementation of a digital trim scheme for SAR ADCs

J. Bialek[1], **A. Wickmann**[2], **F. Ohnhaeuser**[2], **G. Fischer**[1], **R. Weigel**[1], **and T. Ussmueller**[1]

[1]University of Erlangen-Nuremberg, Departement of Electrical Engineering, Erlangen, Germany
[2]Texas Instruments Deutschland GmbH, Germany

Correspondence to: J. Bialek (jan@bialek-net.de)

Abstract. Successive approximation register (SAR) analog-to-digital Converters (ADC) are based on a capacitive digital-to-analog converter (CDAC) (McCreary and Gray, 1975). The capacitor mismatch in the capacitor array of the CDAC impacts the differential non-linearity (DNL) of the ADC directly. In order to achieve a transfer function without missing codes, trimming of the capacitor array becomes necessary for SAR ADCs with a resolution of more than 12 bit. This article introduces a novel digital approach for trimming. DNL measurements of an 18 bit SAR ADC show that digital trimming allows the same performance as analog trimming. Digital trimming however reduces the power consumption of the ADC, the die size and the required time for the production test.

1 Introduction

SAR ADCs are widely used in electronic circuits because they cover the important application area for mid resolution and mid speed. In general, SAR ADCs are available with resolutions of 10 bit to 18 bit. The advantage of the SAR architecture is the possibility to capture a snapshot value of an analog signal. This behavior cannot be achieved with $\Delta\Sigma$-Converters since they convert the input voltage over a period of time instead of the input voltage of a single point in time. Whereas Pipeline ADCs are not manufactured with high resolutions because of rising demands of amplifier linearity with rising resolution.

Fig. 1 shows the basic principle of a SAR ADC. When the conversion cycle of the ADC is started, the sample and hold element (S&H) stores the input voltage V_{in}. The input voltage is compared to the output voltage of the DAC, which is usually implemented as a CDAC. The control logic adjusts the input word of the DAC according to the comparator decision. This cycle is repeated for each bit of resolution of the ADC, which means that an 18 bit ADC needs 18 conversion cycles to output the result. At the end of the conversion cycle the difference between the input voltage V_{in} and the DAC output voltage V_{DAC} is less than one Least Significant Bit (LSB).

The basic schematic of a 3 bit SAR ADC is shown in Fig. 2. The capacitors are used for sampling the input voltage as well as for the CDAC. The position of the switches in Fig. 2 represents the sampling mode. When switch S_C is opened, the trapped charge on the capacitors is proportional to the input voltage V_{in}. During the conversion cycle, the switches S_1 through S_3 are switched between the reference voltage V_{ref} and ground until the voltage V_A at node A is equal to the voltage at the non-inverting comparator input, which is in this case ground potential. At the end of the conversion the position of the switches S_1 through S_3 resemble the digital output word corresponding to the input voltage V_{in}. The voltage V_A can the be written as

$$V_A = -V_{\text{in}} + \frac{k_1 V_{\text{ref}} C}{2C} + \frac{k_2 V_{\text{ref}} \frac{C}{2}}{2C} + \frac{k_3 V_{\text{ref}} \frac{C}{4}}{2C} \quad (1)$$

where k_1 through k_3 are either 0 or 1 according to the position of the switches S_1 through S_3. Switch S_4 is not used for the conversion of the input voltage, as it is switched to ground potential at the beginning of the conversion. Its only purpose is to raise the total value of the capacitor array to $2C$. Equation! 1 shows the importance of capacitor matching. The voltage V_A is not a function of a capacitor value, but is a function of the ratio of the capacitor values and the value of the capacitor array.

The capacitors in the capacitor array are binary weighted. If their ratios are not binary, the conversion result will be wrong. Due to variances in fabrication processes, the capacitor ratios are never perfectly binary weighted. These errors result in deviation of the DNL. Examples of a perfect and erroneous DNL are shown in Fig. 3. The DNL can be measured using a staircase measurement. The input voltage of the ADC is set to a certain value where an output code transition

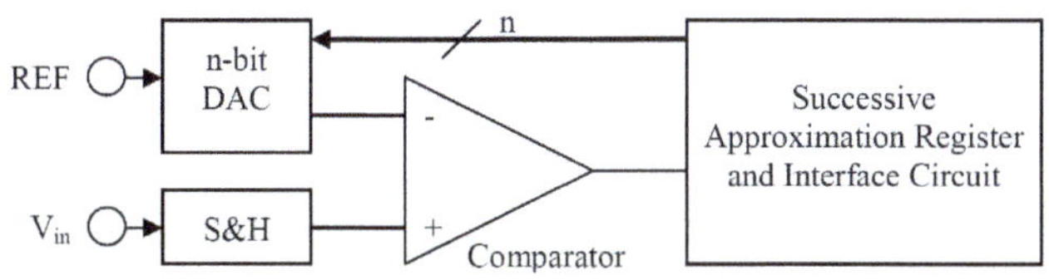

Fig. 1. Basic principle of a SAR ADC

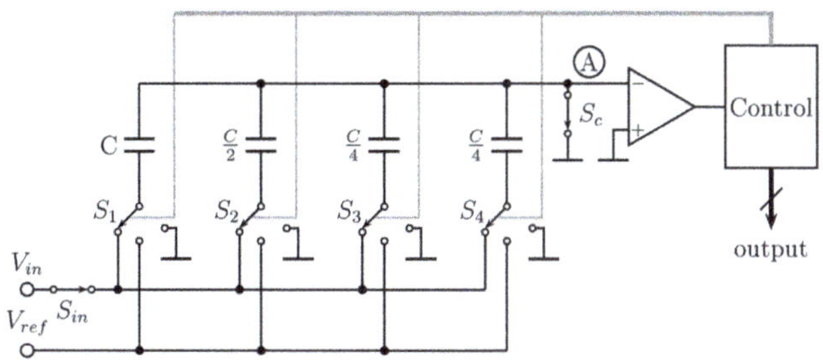

Fig. 2. Schematic of a 3 bit single ended SAR ADC with CDAC

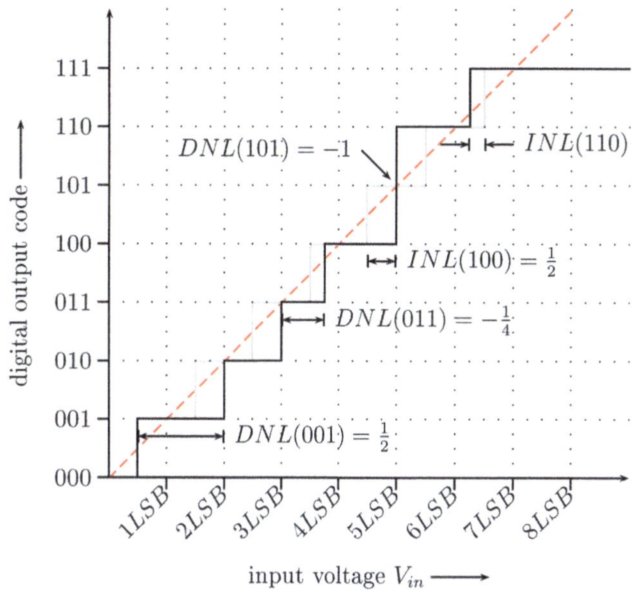

Fig. 3. Ideal (light) and erroneous (dark) output characteristic of an ADC

occurs. Now the input voltage will be raised until the next output code transition occurs. The difference between these two input voltage values shall be measured in multiples of the LSB. The perfect DNL has a code transition every LSB, which means that every output code has the length of one LSB. The code length varies in the non-perfect DNL. If a code is completely missing, the DNL of this code equals −1. The integral non-linearity (INL) of a given output code is the sum of all the preceding DNL errors. These errors caused by the mismatch of capacitor ratios can be corrected by trimming. If all DNL errors can be corrected to equal zero, the INL errors are also zero.

2 Analog trimming

Trimming must be done in-package because parasitics between the die and the package influence the performance of the ADC. It is not possible to estimate these parasitics of ADCs with a resolution of 16 bit and more because air inclusions within the package make a large contribution to the parasitics. Therefore, analog in-package trimming was introduced (Ohnhaeuser et al. , 2010). A second capacitor array is added to the circuitry which is used for trimming the first capacitor array. This circuit can be seen in Fig. 4. During the production test of the assembled die, the DNL is measured and the trim capacitors are switched to fit the DNL. This procedure has to be repeated at least once more to get adequate results due to inaccuracies of the trim capacitors. Because the trim capacitors are not used during sampling, this procedure also produces a gain error, which has to be corrected afterwards. The trim capacitor array occupies space on the die and also adds to the production costs because of the increased test time.

3 Digital trimming

The capacitors of the CDAC represent the major code transitions. If the error for each capacitor is known, it is possible to implement an arithmetic unit within the ADC that corrects the error (Reinhold et al. , 2011). This makes the trim array obsolete. This method only works, if the output codes are unique, i.e. no codes are missing. Therefore, a dynamic error correction is introduced, which adds a stage of redundancy. Several different forms of error correction units have been described in previous publications (Ogawa et al. , 2008; Bacrania , 1986; Cho et al. , 2009). The schematic of the improved capacitor array can be seen in Fig. 5. The dynamic error correction consists of two capacitors C_{DUP} and C_{DDWN} (Ohnhaeuser and Huemer , 2008; Srinivasan and Godbole , 2002). In this example, the dynamic error correction is placed between the second and the third bit decision. In dependence on the second bit decision the capacitor C_{DUP} is switched to the reference voltage or the capacitor C_{DDWN} is switched to ground potential to increase or decrease the voltage at the comparator input. The rest of the conversion executes as before. After the conversion, the value of the dynamic error correction is added to the SAR output. In order to guarantee converging voltages at the comparator input, the capacitances of C_{DUP} and C_{DDWN} are chosen to be a little smaller than C_2.

At the end of the conversion, the errors of each switched capacitor and the dynamic error correction are added to the output word.

Figure 6 shows an example of a possible implementation. In this case, the complete error is calculated at once which requires a large arithmetic unit. Also it will take some time until the corrected result will be available. Therefore, a finite

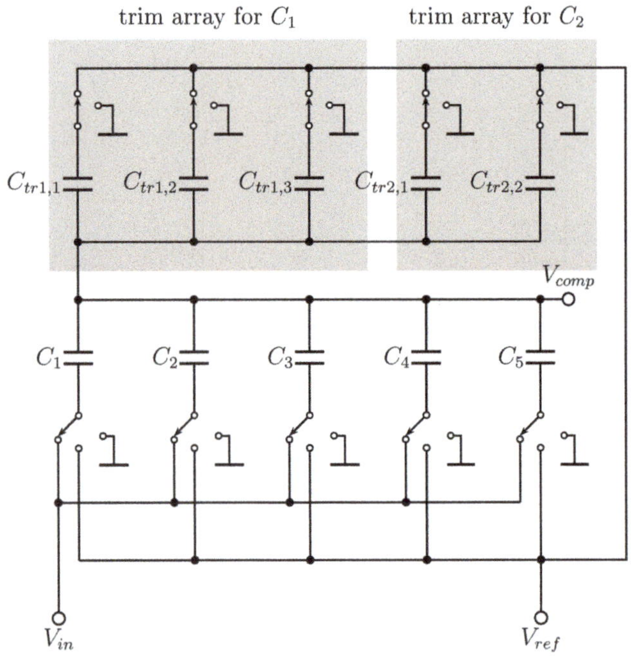

Fig. 4. CDAC with capacitor array for trimming C_1 and C_2

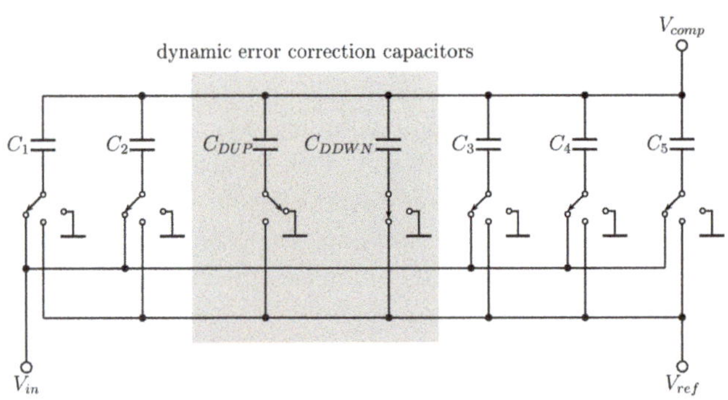

Fig. 5. CDAC with dynamic error correction

state machine (FSM) approach to digital trimming will be faster and the arithmetic unit will be smaller (Bialek , 2009). After each bit decision, the error of the current bit decision is added to the present complete error if the bit decision is logic "1". Otherwise the complete error remains unaltered. The same is done for the dynamic error correction stage. Since the lower bits are usually not trimmed the complete error will be readily calculated several clock cycles before the end of the conversion. This has two advantages. First, the final addition of the total error and the SAR output can be done right after the last bit decision. The data can be processed faster than calculating everything at the end. Therefore, the sampling rate does not change. Second, it is possible to introduce additional trim capacitors that have a weight of a half and a quarter LSB into the CDAC. These capacitors will be switched according to the total error before the last conver-

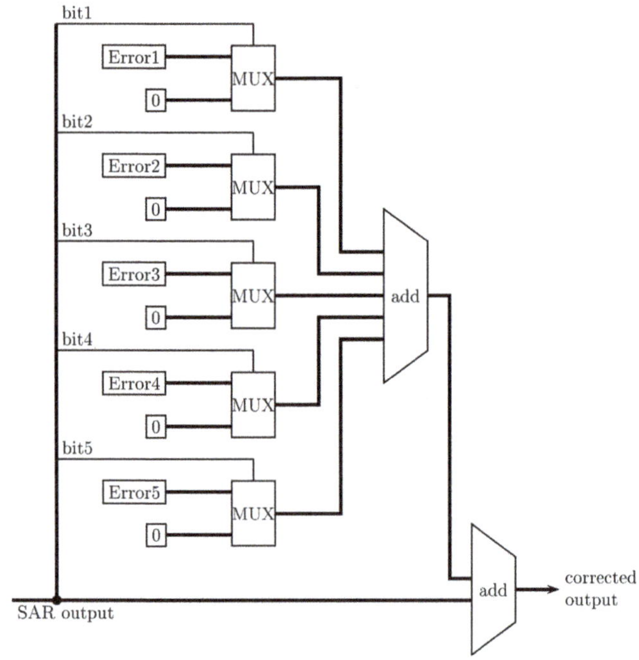

Fig. 6. Principal of digital trimming

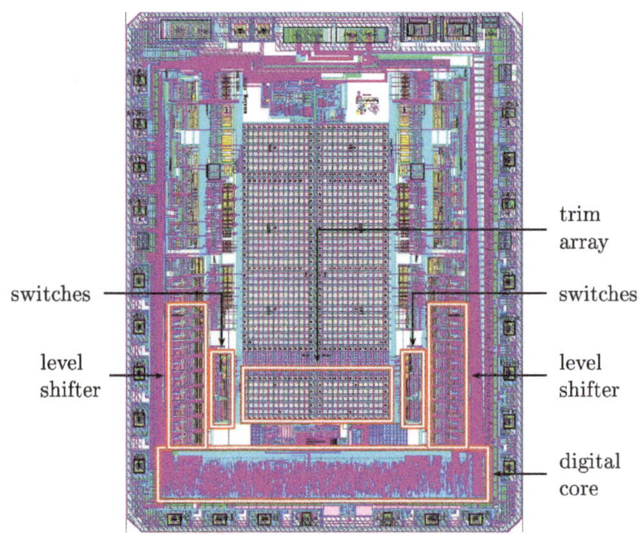

Fig. 7. Layout of 18 bit SAR ADC with trim related circuitry

sions according to the fractional sum. This increases the trim resolution to $\frac{1}{8}$ LSB.

4 Measurements

Digital trimming was implemented in an 18 bit SAR ADC. The top level layout of the die is shown in Fig 7. The digital circuitry is located close to the bottom of the layout. Digital trimming occupies approximately one third of the digital core. The die area can be shrinked about 10 % because

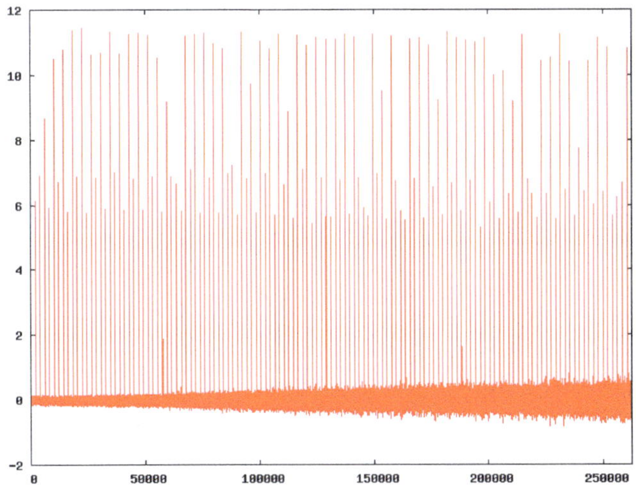

Fig. 8. DNL of an untrimmed 18 bit SAR ADC

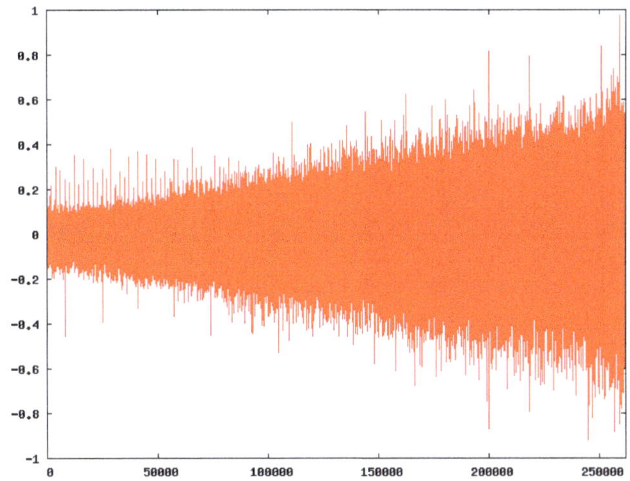

Fig. 9. Digitally trimmed DNL of the same SAR ADC as shown in Fig. 8

trim capacitors, level shifters and switches as shown in Fig. 7 are no longer necessary. Figure 8 shows the untrimmed and Fig. 9 shows the trimmed DNL of the same ADC in 18 bit mode. The untrimmed DNL shows mostly positive values because of the dynamic error correction shown in Fig. 5. The capacitance of the dynamic error correction capacitors is designed to ensure that no output codes are missing. Thus, resulting in large positive DNL values at major code transitions that happen prior to the dynamic error correction. After trimming, the DNL is always in between +1 LSB and −1 LSB. This means that no codes are missing and digital trimming achieves the desired result. The DNL in Figs. 8 and 9 becomes larger with codes coming closer to the full scale. This behavior is not caused by digital trimming, since the trimmed and untrimmed DNL show this characteristic. Possible causes for this might be noise within the ADC or noise in the reference path.

5 Conclusions

In this paper, a new digital trim method for SAR ADCs was introduced. The digital trim was implemented into an ADC as a design investigation and DNL measurements were performed. The DNL of the trimmed SAR ADC shows a similar characteristic of a state-of-the art SAR ADC and it was shown, that it is possible to decrease the die size. Digital trimming does not add parasitics to the CDAC and does not alter the mismatch of the capacitor array. The power consumption can be reduced because constant recharging of a trim capacitor array is no longer necessary. Also, it is not necessary to re-measure the DNL after the trim values are stored in the part. Approximately 50 % of the production costs result from the production test, of which another 50 % are caused by trimming. Reducing the time for production test will significantly cut the production costs of SAR ADCs.

Acknowledgements. The authors would like to thank the Bavarian Research foundation (BFS) for funding this research project.

References

Bacrania, K.: A 12-bit Successive-Approximation-Type ADC with Digital Error Correction, IEEE J. Solid-St. Circ., 21, 1016–1025, 1986

Bialek, J.: Implementierung eines digitalen Trimverfahrens für SAR ADCs, Diploma Thesis at the University of Erlangen-Nürnberg, Erlangen, 2009

Cho, S.-H., Lee, C.-K., Sung, B.-R.-S., and Ryu, S.-T.: Digital error correction technique for binary decision successive approximation ADCs, Electron. Lett., 45, 395–397, 2009

McCreary, J. L. and Gray, P. R.: All-MOS Charge Redistribution Analog-to-Digital Conversion Techniques – Part 1, IEEE J. Solid-St. Circ., 10, 371–379, 1975

Ogawa, T., Kobayashi, H., Hotta, M., Takahashi, Y., San, H., and Takai, N.: SAR ADC Algorithm with Redundancy, IEEE Asia Pacific Conference on Circuits and Systems 2008 (APCCAS2008), 268–271, 2008

Ohnhaeuser, F. and Huemer, M.: Methods to Eliminate Dynamic Errors in High-Performance SAR A/D Converter, In the CD-ROM Proceedings of the International Symposium on Circuits and Systems (ISCAS2008), Seattle, USA, 2398–2401, 2008.

Ohnhaeuser, F., Allinger, M., and Huemer, M.: Trim Techniques for DC Specifications for A/D converters based on successive approximation, AEU-Int. J. Electron. C., 790–793, 2010

Reinhold M., Ohnhaeuser F., and Badenius, M.: Digital trimming of SAR ADCs, US Patent 8 049 654, Texas Instruments, 2011.

Srinivasan, C. and Godbole, K.: Error correction architecture to increase speed and relax current drive requirements of SAR ADC, US Patent 6 747 589, Texas Instruments, 2002.

Permissions

All chapters in this book were first published in ARS, by Copernicus Publications; hereby published with permission under the Creative Commons Attribution License or equivalent. Every chapter published in this book has been scrutinized by our experts. Their significance has been extensively debated. The topics covered herein carry significant findings which will fuel the growth of the discipline. They may even be implemented as practical applications or may be referred to as a beginning point for another development.

The contributors of this book come from diverse backgrounds, making this book a truly international effort. This book will bring forth new frontiers with its revolutionizing research information and detailed analysis of the nascent developments around the world.

We would like to thank all the contributing authors for lending their expertise to make the book truly unique. They have played a crucial role in the development of this book. Without their invaluable contributions this book wouldn't have been possible. They have made vital efforts to compile up to date information on the varied aspects of this subject to make this book a valuable addition to the collection of many professionals and students.

This book was conceptualized with the vision of imparting up-to-date information and advanced data in this field. To ensure the same, a matchless editorial board was set up. Every individual on the board went through rigorous rounds of assessment to prove their worth. After which they invested a large part of their time researching and compiling the most relevant data for our readers.

The editorial board has been involved in producing this book since its inception. They have spent rigorous hours researching and exploring the diverse topics which have resulted in the successful publishing of this book. They have passed on their knowledge of decades through this book. To expedite this challenging task, the publisher supported the team at every step. A small team of assistant editors was also appointed to further simplify the editing procedure and attain best results for the readers.

Apart from the editorial board, the designing team has also invested a significant amount of their time in understanding the subject and creating the most relevant covers. They scrutinized every image to scout for the most suitable representation of the subject and create an appropriate cover for the book.

The publishing team has been an ardent support to the editorial, designing and production team. Their endless efforts to recruit the best for this project, has resulted in the accomplishment of this book. They are a veteran in the field of academics and their pool of knowledge is as vast as their experience in printing. Their expertise and guidance has proved useful at every step. Their uncompromising quality standards have made this book an exceptional effort. Their encouragement from time to time has been an inspiration for everyone.

The publisher and the editorial board hope that this book will prove to be a valuable piece of knowledge for researchers, students, practitioners and scholars across the globe.

List of Contributors

O. A. Pfänder
Microelectronics Department, University of Ulm, Germany

H.-J. Pfleiderer
Microelectronics Department, University of Ulm, Germany

S. W. Lachowicz
School of Engineering and Mathematics, Edith Cowan University, Perth, Western Australia

S. Jayapal
Chair of Microelectronics, Department of Microsystems Engineering (IMTEK), University of Freiburg, Georges-Koehler-Allee 102, 79110 Freiburg, Germany

Y. Manoli
Chair of Microelectronics, Department of Microsystems Engineering (IMTEK), University of Freiburg, Georges-Koehler-Allee 102, 79110 Freiburg, Germany

M. Anis
Institute for Microelectronics, Technical University of Kaiserslautern, Germany

R. Tielert
Institute for Microelectronics, Technical University of Kaiserslautern, Germany

S. K. Lakshmanan
Institute of Integrated Sensor Systems, TU-Kaiserslautern, Germany

A. Koenig
Institute of Integrated Sensor Systems, TU-Kaiserslautern, Germany

C. Puttmann
Heinz Nixdorf Institute, University of Paderborn, Germany

J. Shokrollahi
Chair for System Security, Ruhr-University of Bochum, Germany

M. Porrmann
Heinz Nixdorf Institute, University of Paderborn, Germany

U. Rückert
Heinz Nixdorf Institute, University of Paderborn, Germany

T. Jungeblut
System and Circuit Technology, Heinz Nixdorf Institute, University of Paderborn, Germany

M. Grünewald
System and Circuit Technology, Heinz Nixdorf Institute, University of Paderborn, Germany

M. Porrmann
System and Circuit Technology, Heinz Nixdorf Institute, University of Paderborn, Germany

U. Rückert
System and Circuit Technology, Heinz Nixdorf Institute, University of Paderborn, Germany

Th. Fischer
Lehrstuhl für Technische Elektronik, Technische Universität München, München, Germany

A. Olbrich
Infineon Technologies AG, München, Germany

G. Georgakos
Infineon Technologies AG, München, Germany

B. Lemaitre
Infineon Technologies AG, München, Germany

D. Schmitt-Landsiedel
Lehrstuhl für Technische Elektronik, Technische Universität München, München, Germany

B. Heyne
University of Dortmund, Information Processing Lab, Otto-Hahn-Str. 4, 44227 Dortmund, Germany

J. Götze
University of Dortmund, Information Processing Lab, Otto-Hahn-Str. 4, 44227 Dortmund, Germany

C. Boit
TUB Berlin University of Technology, Berlin, Germany

R. Schlangen
TUB Berlin University of Technology, Berlin, Germany

A. Glowacki
TUB Berlin University of Technology, Berlin, Germany

U. Kindereit
TUB Berlin University of Technology, Berlin, Germany

T. Kiyan
TUB Berlin University of Technology, Berlin, Germany

U. Kerst
TUB Berlin University of Technology, Berlin, Germany

T. Lundquist
DCG Systems, Fremont, CA, USA

S. Kasapi
NVIDIA, Santa Clara, CA, USA

H. Suzuki
Hamamatsu Photonics KK, Japan

E. Franell
Institute for Technical Electronics, Technical University Munich, Germany

S. Drueen
Infineon Technologies AG, Munich, Germany

H. Gossner
Infineon Technologies AG, Munich, Germany

D. Schmitt-Landsiedel
Institute for Technical Electronics, Technical University Munich, Germany

Ph. Teichmann
Institute for Technical Electronics, Technical University Munich, Munich, Germany

J. Fischer
Institute for Technical Electronics, Technical University Munich, Munich, Germany

F. Chouard
Institute for Technical Electronics, Technical University Munich, Munich, Germany

D. Schmitt-Landsiedel
Institute for Technical Electronics, Technical University Munich, Munich, Germany

U. Vishnoi
Chair of Electrical Engineering and Computer Systems, RWTH Aachen University, Aachen, Germany

T. G. Noll
Chair of Electrical Engineering and Computer Systems, RWTH Aachen University, Aachen, Germany

A. Mucha
Corporate Research and Technologies, Siemens AG, Munich, Germany
Lehrstuhl für Technische Elektronik, Technische Universität München, Munich, Germany

M. Schienle
Corporate Research and Technologies, Siemens AG, Munich, Germany

D. Schmitt-Landsiedel
Lehrstuhl für Technische Elektronik, Technische Universität München, Munich, Germany

M. Lüders
Munich University of Technology, Munich, Germany

B. Eversmann
Texas Instruments Deutschland, Freising, Germany

D. Schmitt-Landsiedel
Munich University of Technology, Munich, Germany

R. Brederlow
Texas Instruments Deutschland, Freising, Germany

G. Bertotti
Chair of Sensor and Actuator Systems, TU Berlin, Berlin, Germany
ARCES, University of Bologna, Bologna, Italy

A. Laifi
Chair of Sensor and Actuator Systems, TU Berlin, Berlin, Germany

E. Di Gioia
Chair of Sensor and Actuator Systems, TU Berlin, Berlin, Germany

M. Masoumi
Chair of Sensor and Actuator Systems, TU Berlin, Berlin, Germany

N. Dodel
Chair of Sensor and Actuator Systems, TU Berlin, Berlin, Germany

E. F. Scarselli
ARCES, University of Bologna, Bologna, Italy

R. Thewes
Chair of Sensor and Actuator Systems, TU Berlin, Berlin, Germany

B. Neumann
Chair of Electrical Engineering and Computer Systems, RWTH Aachen University, 52062 Aachen, Germany

T. von Sydow
Chair of Electrical Engineering and Computer Systems, RWTH Aachen University, 52062 Aachen, Germany

H. Blume
Chair of Electrical Engineering and Computer Systems, RWTH Aachen University, 52062 Aachen, Germany

T. G. Noll
Chair of Electrical Engineering and Computer Systems, RWTH Aachen University, 52062 Aachen, Germany

Th. Fischer
Lehrstuhl für Technische Elektronik, Technische Universität München, München, Germany

T. Nirschl
Infineon Technologies AG, München, Germany

B. Lemaitre
Infineon Technologies AG, München, Germany

D. Schmitt-Landsiedel
Lehrstuhl für Technische Elektronik, Technische Universität München, München, Germany

D. Schmidt
Microelectronic System Design Research Group, University of Kaiserslautern, Erwin-Schrödinger-Straße, 67663 Kaiserslautern, Germany

T. Brack
Microelectronic System Design Research Group, University of Kaiserslautern, Erwin-Schrödinger-Straße, 67663 Kaiserslautern, Germany

U. Wasenmüller
Microelectronic System Design Research Group, University of Kaiserslautern, Erwin-Schrödinger-Straße, 67663 Kaiserslautern, Germany

N.Wehn
Microelectronic System Design Research Group, University of Kaiserslautern, Erwin-Schrödinger-Straße, 67663 Kaiserslautern, Germany

M. Keller
Chair of Microelectronics, Department of Microsystems Engineering (IMTEK), University of Freiburg, Georges-Koehler-Allee 102, D-79110 Freiburg, Germany

A. Buhmann
Chair of Microelectronics, Department of Microsystems Engineering (IMTEK), University of Freiburg, Georges-Koehler-Allee 102, D-79110 Freiburg, Germany

M. Kuderer
Chair of Microelectronics, Department of Microsystems Engineering (IMTEK), University of Freiburg, Georges-Koehler-Allee 102, D-79110 Freiburg, Germany

Y. Manoli
Chair of Microelectronics, Department of Microsystems Engineering (IMTEK), University of Freiburg, Georges-Koehler-Allee 102, D-79110 Freiburg, Germany

W. Kraus
Institute for Technical Electronics, Technische Universität München, Germany

D. Schmitt-Landsiedel
Institute for Technical Electronics, Technische Universität München, Germany

S. Drapatz
Lehrstuhl für Technische Elektronik, Technische Universität München, Germany
Infineon Technologies AG, München-Neubiberg, Germany

G. Georgakos
Infineon Technologies AG, München-Neubiberg, Germany

D. Schmitt-Landsiedel
Lehrstuhl für Technische Elektronik, Technische Universität München, Germany

J. Wittmann
Robert Bosch Center for Power Electronics, Reutlingen University, Reutlingen, Germany

A. Seidel
Robert Bosch Center for Power Electronics, Reutlingen University, Reutlingen, Germany

B. Wicht
Robert Bosch Center for Power Electronics, Reutlingen University, Reutlingen, Germany

N. Fichtner
Institute for High Frequency Engineering, Technische Universität München, Arcisstr. 21, 80333 München, Germany

X. Zhou
Institute for High Frequency Engineering, Technische Universität München, Arcisstr. 21, 80333 München, Germany

P. Russer
Institute for High Frequency Engineering, Technische Universität München, Arcisstr. 21, 80333 München, Germany

M. Weis
Technische Universität München, Lehrstuhl für Technische Elektronik

R. Emling
Technische Universität München, Lehrstuhl für Technische Elektronik

D. Schmitt-Landsiedel
Technische Universität M¨unchen, Lehrstuhl für Technische Elektronik

B. Dimov
Institut für Mikroelektronik- und Mechatronik-Systeme gGmbH, Ilmenau, Germany

V. Boos
Institut für Mikroelektronik- und Mechatronik-Systeme gGmbH, Ilmenau, Germany

T. Reich
Institut für Mikroelektronik- und Mechatronik-Systeme gGmbH, Ilmenau, Germany

Ch. Lang
Institut für Mikroelektronik- und Mechatronik-Systeme gGmbH, Ilmenau, Germany

E. Hennig
Institut für Mikroelektronik- und Mechatronik-Systeme gGmbH, Ilmenau, Germany

R. Sommer
Institut für Mikroelektronik- und Mechatronik-Systeme gGmbH, Ilmenau, Germany

D. Schmidt
Microelectronic System Design Research Group, University of Kaiserslautern, Germany

M. Krämer
Networked Systems Group, University of Kaiserslautern, Germany

T. Kuhn
Networked Systems Group, University of Kaiserslautern, Germany

N. Wehn
Microelectronic System Design Research Group, University of Kaiserslautern, Germany

E. Glocker
Lehrstuhl für Technische Elektronik, Technische Universität München, Theresienstraße 90, 80333 München, Germany

D. Schmitt-Landsiedel
Lehrstuhl für Technische Elektronik, Technische Universität München, Theresienstraße 90, 80333 München, Germany

S. A. Figur
EADS Deutschland GmbH, EADS Innovation Works Germany, 81663 München, Germany

F. van Raay
Fraunhofer Institute of Applied Solid-State Physics (IAF), Tullastr. 72, 79108 Freiburg, Germany

R. Quay
Fraunhofer Institute of Applied Solid-State Physics (IAF), Tullastr. 72, 79108 Freiburg, Germany

L. Vietzorreck
TU München, Arcisstr. 21, 80333 München, Germany

V. Ziegler
EADS Deutschland GmbH, EADS Innovation Works Germany, 81663 München, Germany

N. Pour Aryan
Lehrstuhl für Technische Elektronik, Technische Universität München, Germany

L. Heiß
Lehrstuhl für Technische Elektronik, Technische Universität München, Germany

D. Schmitt-Landsiedel
Lehrstuhl für Technische Elektronik, Technische Universität München, Germany

G. Georgakos
Infineon Technologies AG, M¨unchen-Neubiberg, Germany

M. Wirnshofer
Lehrstuhl für Technische Elektronik, Technische Universität München, Germany

A. Bräckle
University of Stuttgart, Institute of Electrical and Optical Communications Engineering, Stuttgart, Germany

L. Rathgeber
University of Stuttgart, Institute of Electrical and Optical Communications Engineering, Stuttgart, Germany

F. Sieger
University of Stuttgart, Institute of Electrical and Optical Communications Engineering, Stuttgart, Germany

S. Heck
University of Stuttgart, Institute of Electrical and Optical Communications Engineering, Stuttgart, Germany

M. Berroth
University of Stuttgart, Institute of Electrical and Optical Communications Engineering, Stuttgart, Germany

S. Tontisirin
Technische Universität Kaiserslautern, Germany

R. Tielert
Technische Universität Kaiserslautern, Germany

M. Eireiner
Institute for Technical Electronics, Technical University Munich, Germany

S. Henzler
Infineon Technologies AG, Munich, Germany

T. Missal
Institute for Technical Electronics, Technical University Munich, Germany

J. Berthold
Infineon Technologies AG, Munich, Germany

D. Schmitt-Landsiedel
Institute for Technical Electronics, Technical University Munich, Germany

T. Föhn
University of Stuttgart, Institute for Electrical and Optical Communications Engineering, Stuttgart, Germany

C. Fischer
University of Stuttgart, Institute for Electrical and Optical Communications Engineering, Stuttgart, Germany

M. Berroth
University of Stuttgart, Institute for Electrical and Optical Communications Engineering, Stuttgart, Germany

D. Maurath
Chair of Microelectronics, Department of Microsystems Engineering (IMTEK), University of Freiburg, Georges-Koehler-Allee 102, 79110 Freiburg, Germany

C. Peters
Chair of Microelectronics, Department of Microsystems Engineering (IMTEK), University of Freiburg, Georges-Koehler-Allee 102, 79110 Freiburg, Germany

T. Hehn
Chair of Microelectronics, Department of Microsystems Engineering (IMTEK), University of Freiburg, Georges-Koehler-Allee 102, 79110 Freiburg, Germany

M. Ortmanns
Chair of Microelectronics, Department of Microsystems Engineering (IMTEK), University of Freiburg, Georges-Koehler-Allee 102, 79110 Freiburg, Germany

Y. Manoli
Chair of Microelectronics, Department of Microsystems Engineering (IMTEK), University of Freiburg, Georges-Koehler-Allee 102, 79110 Freiburg, Germany

M. Fulde
Institute of Technical Electronics, Technical University Munich, Germany

J. P. Engelstädter
Infineon Technologies AG, Villach, Austria
Department of Electrical Engineering, RWTH Aachen University, Aachen, Germany

G. Knoblinger
Infineon Technologies AG, Villach, Austria

D. Schmitt-Landsiedel
Institute of Technical Electronics, Technical University Munich, Germany

T. Heselhaus
Chair of Electrical Engineering and Computer Systems, RWTH Aachen University, 52062 Aachen, Germany

T. G. Noll
Chair of Electrical Engineering and Computer Systems, RWTH Aachen University, 52062 Aachen, Germany

K. Hu
IHP GmbH, Im Technologiepark 25, 15236 Frankfurt (Oder), Germany

F. Herzel
IHP GmbH, Im Technologiepark 25, 15236 Frankfurt (Oder), Germany

J. C. Scheytt
IHP GmbH, Im Technologiepark 25, 15236 Frankfurt (Oder), Germany

S. Lachowicz
Edith Cowan University, Perth, WA, Australia

H.-J. Pfleiderer
Ulm University, Ulm, Germany

M. Eireiner
Institute for Technical Electronics, Technical University Munich, Germany

S. Henzler
Infineon Technologies AG, Munich, Germany

X. Zhang
Institute for Technical Electronics, Technical University Munich, Germany

J. Berthold
Infineon Technologies AG, Munich, Germany

D. Schmitt-Landsiedel
Institute for Technical Electronics, Technical University Munich, Germany

E. Glocker
Lehrstuhl für Technische Elektronik, Technische Universität München, Theresienstraße 90, 80333 München, Germany

D. Schmitt-Landsiedel
Lehrstuhl für Technische Elektronik, Technische Universität München, Theresienstraße 90, 80333 München, Germany

S. Drapatz
Lehrstuhl für Technische Elektronik, Technische Universität München, Theresienstraße 90, 80333 München, Germany

U. Schulze
Chair of Electrical Engineering and Computer Systems, RWTH Aachen University, Germany

M. Broich
Chair of Electrical Engineering and Computer Systems, RWTH Aachen University, Germany

O. Weiss
Chair of Electrical Engineering and Computer Systems, RWTH Aachen University, Germany

T. G. Noll
Chair of Electrical Engineering and Computer Systems, RWTH Aachen University, Germany

H. Uhrmann
Institute of Electrical Measurements and Circuit Design, Vienna University of Technology, Austria

W. Gaberl
Institute of Electrical Measurements and Circuit Design, Vienna University of Technology, Austria

H. Zimmermann
Institute of Electrical Measurements and Circuit Design, Vienna University of Technology, Austria

E. Glocker
Lehrstuhl für Technische Elektronik, Technische Universität München (TUM), Theresienstr. 90, 80333 München, Germany

S. Boppu
Lehrstuhl für Informatik 12 (Hardware-Software-Co-Design), Friedrich-Alexander-Universität Erlangen-Nürnberg (FAU), Cauerstr. 11, 91058 Erlangen, Germany

Q. Chen
Lehrstuhl für Entwurfsautomatisierung, Technische Universität München (TUM), Arcisstr. 21, 80333 München, Germany

U. Schlichtmann
Lehrstuhl für Entwurfsautomatisierung, Technische Universität München (TUM), Arcisstr. 21, 80333 München, Germany

J. Teich
Lehrstuhl für Informatik 12 (Hardware-Software-Co-Design), Friedrich-Alexander-Universität Erlangen-Nürnberg (FAU), Cauerstr. 11, 91058 Erlangen, Germany

J. Bialek
University of Erlangen-Nuremberg, Departement of Electrical Engineering, Erlangen, Germany

A. Wickmann
Texas Instruments Deutschland GmbH, Germany

F. Ohnhaeuser
Texas Instruments Deutschland GmbH, Germany

G. Fischer
University of Erlangen-Nuremberg, Departement of Electrical Engineering, Erlangen, Germany

R. Weigel
University of Erlangen-Nuremberg, Departement of Electrical Engineering, Erlangen, Germany

T. Ussmueller
University of Erlangen-Nuremberg, Departement of Electrical Engineering, Erlangen, Germany

D. Schmitt-Landsiedel
Lehrstuhl für Technische Elektronik, Technische Universität München (TUM), Theresienstr. 90, 80333 München, Germany